Barry University
MAT 091

Charles P. McKeague

*xyz*textbooks

Barry University
MAT 091
Charles P. McKeague

Editorial Director: Patrick McKeague

Sales Manager: Amy Jacobs

Production Manager: Matthew Hoy

Project Manager: Staci Truelson

Developmental Editor: Katherine Heistand Shields

Sales Representative: Bruce Spears

Text Designer: Devin Christ

Cover Image: ©Deejpilot/iStockPhoto.com

Printed in the United States of America

AIE Edition: ISBN-13: 978-1-936368-85-3/ ISBN-10: 1-936368-85-4
Student Edition: ISBN-13: 978-1-936368-65-5/ ISBN-10: 1-936368-65-X

For product information and technology assistance, contact us at
XYZ Textbooks, 1-877-745-3499

For permission to use material from this text or product,
e-mail: **info@mathtv.com**

XYZ Textbooks
1339 Marsh St.
San Luis Obispo, CA 93401
USA

For your course and learning solutions, visit **www.xyztextbooks.com**

Brief Contents

Contents

7 Measurement 321

8 Geometry 371

9 Linear Equations and Inequalities in One Variable 409

Preface

XYZ Textbooks is on a mission to improve the quality and affordability of course materials for mathematics. Our 2013 Concepts with Applications Series brings a new level of technology innovation, which is certain to improve student proficiency.

As you can see, these books are presented in a format in which each section is organized by easy to follow learning objectives. Each objective includes helpful example problems, as well as additional practice problems in the margins, and are referenced in each exercise set. To help students relate what they are studying to the real world, each chapter and section begin with a real-life application of key concepts. You and your students will encounter extensions of these opening applications in the exercise sets.

We have put features in place that help your students stay on the trail to success. With that perspective in mind, we have incorporated this navigational theme into the following new and exciting features.

QR Codes

Unique technology embedded in each exercise set allows students with a smart phone or other internet-connected mobile device to view our video tutorials without being tied to a computer.

10. Noelle recieves $17,000 in two loans. One loan charges 5% interest per year and the other 6.5%. If her total interest after 1 year is $970, how much was each loan?
   $9,000 at 5%, $8,000 at 6.5%

SCAN TO ACCESS

Vocabulary Review

A list of fill-in-the-blank sentences appears at the beginning of each exercise set to help students better comprehend and verbalize concepts.

Vocabulary Review

Choose the correct words to fill in the blanks below.

   prime   composite   remainder   lowest terms

1. A _____ number is any whole number greater than 1 that has exactly two divisors: 1 and the number itself.
2. A number is a divisor of another number if it divides it without a _____.
3. Any whole number greater than 1 that is not a prime number is called a _____ number.
4. A fraction is said to be in _____ if the numerator and the denominator have no factors in common other than the number 1.

Getting Ready For Class

Simple writing exercises for students to answer after reading each section increases their understanding of key concepts.

GETTING READY FOR CLASS

After reading through the preceding section, respond in your own words and in complete sentences.

A. What is a prime number?
B. Why is the number 22 a composite number?
C. Factor 120 into a product of prime factors.
D. How would you reduce a fraction to lowest terms?

Key words

A list of important vocabulary appears at the beginning of each section to help students prepare for new concepts. These words can also be found in blue italics within the sections.

Navigation Skills

A discussion of important study skills appearing in each chapter anticipates students' needs as they progress through the course.

Navigation Skills: Prepare, Study, Achieve

Your instructor is a vital resource for your success in this class. Make note of your instructor's office hours and utilize them regularly. Compile a resource list that you keep with your class materials. This list should contain your instructor's office hours and contact information (e.g., office phone number or e-mail), as well as classmates' contact information that you can utilize outside of class. Communicate often with your classmates about how the course is going for you and any questions you may have. Odds are that someone else has the same question and you may be able to work together to find the answer.

Find the Mistake

Complete sentences that include common mistakes appear in each exercise set to help students identify errors and aid their comprehension of the concepts presented in that section.

Find the Mistake

Each sentence below contains a mistake. Circle the mistake and write the correct word or phrase on the line provided.
1. The set of all inputs for a function is called the range. _____
2. For the relation (1, 2), (3, 4), (5, 6), (7, 8), the domain is {2, 4, 6, 8}. _____
3. Graphing the relation $y = x^2 - 3$ will show that the domain is $\{ x \mid x \geq -3 \}$. _____
4. If a horizontal line crosses the graph of a relation in more than one place, the relation cannot be a function. _____

Landmark Review

A review appears in the middle of each chapter for students to practice key skills, check progress, and address difficulties.

Landmark Review: Checking Your Progress [7.1– 7.3]

Solve the following systems.

1. $y = 2x + 5$
   $y = -3x - 5$
2. $2x - y = -2$
   $2x - 3y$
3. $2x - y = 1$
   $y = -2x + 5$
4. $2x + 5y = 0$
   $x - 5y = -9$
5. $3x - 3y = 4$
   $x = 3y - 1$

6. $-6x + 2y = -9$
   $y = 3x + 4$
7. $x - y = 11$
   $x + y = 1$
8. $-3x + y = -15$
   $-2x + 3y = -17$
9. $2x + 5y = 17$
   $x - 4y = -24$
10. $x + 4y = 5$
    $3x + 8y = 14$

End of Chapter Reviews, Cumulative Reviews, and Tests

Problems at the end of each chapter provide students with a comprehensive review of each chapter's concepts.

CHAPTER 5 Test

Write each ratio as a fraction in lowest terms. [5.1]

1. 48 to 18
2. $\frac{5}{8}$ to $\frac{3}{4}$
3. 6 to $2\frac{4}{7}$
4. 0.14 to 0.4

Find the unknown term in each proportion. [5.3, 5.4]

10. $\frac{4}{7} = \frac{24}{x}$
11. $\frac{2.5}{5} = \frac{1.5}{x}$
12. **Baseball** A baseball player gets 13 hits in his first 20 games of the season. If he continues at the same rate, how many hits will he get in 60 games? [5.5]

Trail Guide Projects

Individual or group projects appear at the end of each chapter for students to apply concepts learned to real life.

CHAPTER 2 Trail Guide Project

Supplies Needed

A Piece of graph paper
B 4 colored pens or pencils

Color by Fractions

This project will provide a visual approach for working with the multiplication and division of fractions. You will need a piece of graph paper and 4 colored pens or pencils.

1. On your graph paper, draw two squares each with a side length of 20 boxes. Fill in the boxes of your first square using your pens or pencils. You may choose any pattern as long as you use all four colors and fill in every box in your square.

XYZ Textbooks is committed to helping students achieve their goals of success. This new series of highly-developed, innovative books will help students prepare for the course ahead and navigate their way to a successful course completion.

Supplements

Instructor

Name: _____

Phone: _____

Email: _____

Office Hours: _____

Teacher Assistant

Name: _____

Phone: _____

Email: _____

Office Hours: _____

On-Campus Tutoring

Location: _____

Hours: _____

Study Group

Name: _____

Phone: _____

Email: _____

Name: _____

Phone: _____

Email: _____

Name: _____

Phone: _____

Email: _____

For the Instructor

Online Homework   XYZ Homework provides powerful online instructional tools for faculty and students. Randomized questions provide unlimited practice and instant feedback with all the benefits of automatic grading. Tools for instructors can be found at *www.xyzhomework.com* and include the following:

- Quick setup of your online class
- More than 3,000 randomized questions, similar to those in the textbook, for use in a variety of assessments, including online homework, quizzes, and tests
- Text and videos designed to supplement your instruction
- Automated grading of online assignments
- Flexible gradebook
- Message boards and other communication tools, enhanced with calculator-style input for proper mathematics notation

MathTV.com   With more than 8,000 videos, MathTV.com provides the instructor with a useful resource to help students learn the material. MathTV.com features videos of most examples in the book, explained by the author and a variety of peer instructors. If a problem can be completed more than one way, the peer instructors often solve it by different methods. Instructors can also use the site's Build a Playlist feature to create a custom list of videos for posting on their class blog or website.

For the Student

Online Homework   XYZ Homework provides powerful online instruction and homework practice for students. Benefits for the student can be found at *www.xyzhomework.com* and include the following:

- Unlimited practice with problems similar to those in the text
- Online quizzes and tests for instant feedback on performance
- Online video examples
- Convenient tracking of class programs

MathTV.com   Students have access to math instruction 24 hours a day, seven days a week, on MathTV.com. Assistance with any problem or subject is never more than a few clicks away.

Online Book   This text is available online for both instructors and students. Tightly integrated with MathTV.com, students can read the book and watch videos of the author and peer instructors explaining most examples. Access to the online book is available free with the purchase of a new book.

Additional Chapter Features

Learning Objectives

We have provided a list of the section's important concepts at the beginning of each section near the section title, noted with an orange capital letter, which is reiterated beside each learning objective header in the section.

Example and Practice Problems

Example problems, with work, explanations, and solutions are provided for every learning objective. Beside each example problem, we have also provided a corresponding practice problem and its answer in the margin.

Margin Notes

In many sections, you will find yellow notes in the margin that correspond to the section's current discussion. These notes contain important reminders, fun facts, and helpful advice for working through problems.

Colored Boxes

Throughout each section, we have highlighted important definitions, properties, rules, how to's, and strategies in colored boxes for easy reference.

Blueprint for Problem Solving

We provide students with a step-by-step guide for solving application problems in a clear and efficient manner.

Facts from Geometry

Throughout the book, we highlight specific geometric topics and show students how they apply to the concepts they are learning in the course.

Using Technology

Throughout the book, we provide students with step-by-step instructions for how to work some problems on a scientific or graphing calculator.

Descriptive Statistics

In some sections, we call out to the use of descriptive statistics as they are applied to the concepts students are learning, and thus, applied to real life.

Vocabulary Review

At the beginning of every exercise set, these fill-in-the-blank sentences help students comprehend and verbalize concepts from the corresponding section.

Problems

Each section contains problems ranging in difficulty that allow students to practice each learning objective presented in the section. The majority of the sections will also include problems in one or more of the following categories:

- Applying the Concepts
- Extending the Concepts or One Step Further
- Improving your Quantitative Literacy
- Getting Ready for the Next Section
- Extending the Concepts

Chapter Summary

At the end of each chapter, we provide a comprehensive summary of the important concepts from the chapter. Beside each concept, we also provide additional example problems for review.

Answers to Odd-Numbered Questions

At the end of the book, we have supplied answers for all odd-numbered problems that appear in the following features:

- Landmark Reviews
- Vocabulary Reviews
- Exercise Sets
- Find the Mistake Exercises

Navigation Skills at a Glance

- Compile a resource list for assistance.
- Set short- and long-term goals for the course.
- Complete homework completely and efficiently.
- Pay attention to instructions, and work ahead.
- Be an active listener while in class.
- Engage your senses when studying and memorizing concepts.
- Study with a partner or group.
- Maintain a positive academic self-image, and avoid burnout.
- Prepare early for the final exam, and reward yourself for a job well done!

Operations with Integers

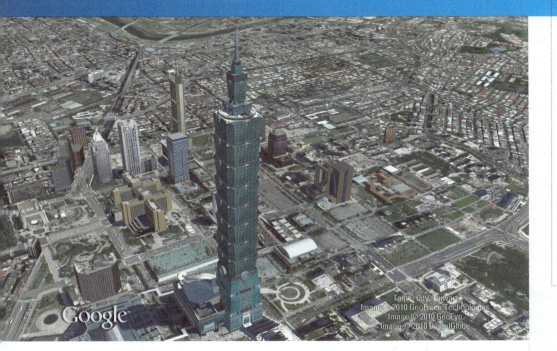

Taipei City, Taiwan
Image © 2010 GeoForce Technologies
Image © 2010 GeoEye
Image © 2010 DigitalGlobe
Google

Imagine riding an elevator up 84 floors in just 37 seconds at a speed of 3,314 feet per minute. If you visit Taipei 101 in Taiwan, you can do just that. Recognized as the fastest high-speed pressurized elevator in the world, the trip from the fifth-floor ticket windows to the eighty-ninth floor observation deck will cost you around $11.00 US. Once there, you will be treated to spectacular views of Taipei and the surrounding rivers and mountains. You can also see the 900-ton tuned mass damper suspended from the ninety-second floor. The damper is a pendulum that serves to keep the skyscraper from excessive movement during typhoons and earthquakes. It serves as a counter to the swaying forces exerted on the tall building by high winds and tremors.

The 106-floor building consists of 101 above-ground floors and 5 floors below ground level. Suppose you were to get in an elevator on the second floor and descend 4 floors. Instinctively, you know you would be on the second floor below ground level. If we assigned the ground floor a value of zero, we could represent your trip with the expression

$$2 + (-4) = -2$$

Now suppose that from there you went up 50 floors and then down 10 floors. That expression looks like

$$(-2) + 50 + (-10) = 38$$

Understanding the mathematics behind these expressions requires that you know how to work with negative numbers, which are one of the topics of this chapter.

OBJECTIVES

A Understand the relationship between positive and negative numbers.

B Find the absolute value and the opposite of a number.

KEY WORDS

negative numbers

origin

inequality

absolute value

opposites

Video Examples
Section 1.1

Note A number, other than 0, with no sign (+ or −) in front of it is assumed to be positive. That is, 5 = +5.

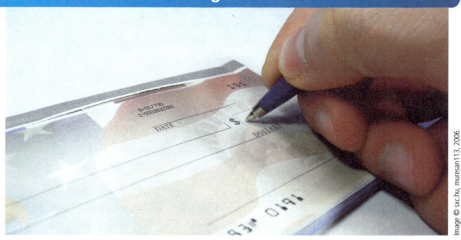

Image © sxc.hu, muresan113, 2006

Suppose you have a balance of $20 in your checkbook and then write a check for $30. You are now overdrawn by $10. How will you write this new balance? One way is with a negative number by writing the balance as −$10.

Check No.	Date	Description of Transaction	Payment/Debit (−)	Deposit/Credit (+)	Balance	
					$20	00
1568	4/15	Bookstore	$30	00		−$10	00

Negative numbers can be used to describe other situations as well—for instance, temperature below zero and distance below sea level.

A Negative Numbers

To see the relationship between negative and positive numbers, we can extend the number line as shown in Figure 1. We first draw a straight line and label a convenient point with 0. This is called the *origin*, and it is usually in the middle of the line. We then label positive numbers to the right (as we have done previously), and negative numbers to the left.

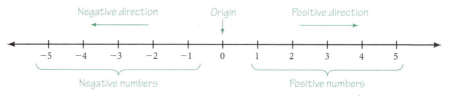

FIGURE 1

The numbers increase going from left to right. If we move to the right, we are moving in the positive direction. If we move to the left, we are moving in the negative direction. Any number to the left of another number is considered to be smaller than the number to its right.

−4 is less than −2 because −4 is to the left of −2 on the number line.

FIGURE 2

We see from the line that every negative number is less than every positive number.
  In algebra, we can use inequality symbols when comparing numbers.

Notation for Inequalities

If a and b are any two numbers on the number line, then

$a < b$ is read "a is less than b."

$a > b$ is read "a is greater than b."

As you can see, the inequality symbols always point to the smaller of the two numbers being compared. Here are some examples that illustrate how we use the inequality symbols.

EXAMPLE 1   Explain the meaning of each expression.

a. $3 < 5$ **b.** $0 > 100$ **c.** $-3 < 5$ **d.** $-5 < -2$

Solution

a. $3 < 5$ is read "3 is less than 5." Note that it would also be correct to write $5 > 3$. Both statements, "3 is less than 5" and "5 is greater than 3," have the same meaning. The inequality symbols always point to the smaller number.

b. $0 > 100$ is a false statement, because 0 is less than 100, not greater than 100. To write a true inequality statement using the numbers 0 and 100, we would have to write either $0 < 100$ or $100 > 0$.

c. $-3 < 5$ is a true statement, because -3 is to the left of 5 on the number line, and, therefore, it must be less than 5. Another statement that means the same thing is $5 > -3$.

d. $-5 < -2$ is a true statement, because -5 is to the left of -2 on the number line, meaning that -5 is less than -2. Both statements $-5 < -2$ and $-2 > -5$ have the same meaning; they both say that -5 is a smaller number than -2.

B Absolute Value and Opposites

It is sometimes convenient to talk about only the numerical part of a number and disregard the sign ($+$ or $-$) in front of it. The following definition gives us a way of doing this.

DEFINITION   absolute value

The *absolute value* of a number is its distance from 0 on the number line. We denote the absolute value of a number with vertical lines. For example, the absolute value of -3 is written $|-3|$.

The absolute value of a number is never negative because it is a distance, and a distance is always measured in positive units (unless it happens to be 0).

EXAMPLE 2   Simplify each expression.

a. $|5|$ **b.** $|-3|$ **c.** $|-7|$

Solution

a. $|5| = 5$ The number 5 is 5 units from 0.

b. $|-3| = 3$ The number -3 is 3 units from 0.

c. $|-7| = 7$ The number -7 is 7 units from 0.

Two numbers that are the same distance from 0 but in opposite directions from 0 are called *opposites*. The notation for the opposite of a is $-a$.

3. Give the opposite of each of the following numbers:

$$3, 1, 8, -4, -6$$

EXAMPLE 3   Give the opposite of each of the following numbers:

$$5, 7, 1, -5, -8$$

Solution

The opposite of 5 is -5.

The opposite of 7 is -7.

The opposite of 1 is -1.

The opposite of -5 is $-(-5)$, or 5.

The opposite of -8 is $-(-8)$, or 8.

We see from this example that the opposite of every positive number is a negative number, and, likewise, the opposite of every negative number is a positive number. The last two parts of Example 3 illustrate the following property:

PROPERTY   Opposite of a Negative Number

If a represents any positive number, then it is always true that

$$-(-a) = a$$

In other words, this property states that the opposite of a negative number is a positive number.

It should be evident now that the symbols $+$ and $-$ can be used to indicate several different ideas in mathematics. In the past, we have used them to indicate addition and subtraction. They can also be used to indicate the direction a number is from 0 on the number line. For instance, the number $+3$ (read "positive 3") is the number that is 3 units from zero in the positive direction. On the other hand, the number -3 (read "negative 3") is the number that is 3 units from 0 in the negative direction. The symbol $-$ can also be used to indicate the opposite of a number, as in $-(-2) = 2$. The interpretation of the symbols $+$ and $-$ depends on the situation in which they are used. For example,

$3 + 5$ The + sign indicates addition.

$7 - 2$ The − sign indicates subtraction.

-7 The − sign is read "negative" 7.

$-(-5)$ The first − sign is read "the opposite of."
              The second − sign is read "negative" 5.

This may seem confusing at first, but as you work through the problems in this chapter you will get used to the different interpretations of the symbols $+$ and $-$.

We should mention here that the set of whole numbers along with their opposites forms the set of *integers*. That is,

$$\text{Integers} = \{\ldots, -3, -2, -1, 0, 1, 2, 3, \ldots\}$$

Answers

3. $-3, -1, -8, 4, 6$

DESCRIPTIVE STATISTICS Displaying Negative Numbers

Below is a table of temperatures in which the temperatures below zero are represented by negative numbers. We will use the information in Table 1 to draw a scatter diagram and a line graph.

Table 1

Record low temperatures for Denver, Colorado

Month	Temperature
January	−21°F
February	−23°F
March	−8°F
April	4°F
May	21°F
June	33°F
July	40°F
August	37°F
September	14°F
October	2°F
November	−5°F
December	−24°F

Notice that the vertical axis in Figure 3 looks like the number line we have been using. To produce the scatter diagram, we place a dot above each month, across from the temperature for that month. For example, the dot above July will be across from 40°. Doing the same for each of the months, we have the scatter diagram shown in Figure 3. To produce the line graph in Figure 4, we simply connect the dots in Figure 3 with line segments.

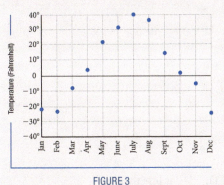

FIGURE 3

FIGURE 4

GETTING READY FOR CLASS

After reading through the preceding section, respond in your own words and in complete sentences.

A. Write the statement "3 is less than 5" in symbols.

B. What is the absolute value of a number?

C. Describe what we mean by numbers that are "opposites" of each other.

D. If you locate two different numbers on the real number line, which one will be the smaller number?

Vocabulary Review

Choose the correct words to fill in the blanks below.

less	absolute value	origin	integers
opposites	more	positive	negative

1. The _____ on a real number line that includes negative numbers is found at zero.

2. A number, other than zero, with no sign in front of it is assumed to be _____.

3. The inequality symbol < means _____ than, whereas the symbol > means _____ than.

4. The _____ of a number is its distance from 0 on the number line.

5. Two numbers that are the same distance from 0 but in opposite directions from 0 are called _____.

6. The opposite of a _____ number is a positive number.

7. The set of whole numbers along with their opposites forms the set of _____.

Problems

A Write each of the following in words.

1. $4 < 7$ **2.** $0 < 10$ **3.** $5 > -2$ **4.** $8 > -8$

5. $-10 < -3.$ **6.** $-20 < -5$ **7.** $0 > -4$ **8.** $0 > -100$

Write each of the following in symbols.

9. 30 is greater than -30.   **10.** -30 is less than 30.   **11.** -10 is less than 0.

12. 0 is greater than -10.   **13.** -3 is greater than -15.   **14.** -15 is less than -3.

Place either < or > between each of the following pairs of numbers so that the resulting statement is true.

15. 3   7

16. 17   0

17. 7   −5

18. 2   −13

19. −6   0

20. −14   0

21. −12   −2

22. −20   −1

23. −9 |9|

24. |12| −7

25. −3 |6|

26. |8| −2

27. 15 |−4|

28. 20 |−6|

29. |−2| |−7|

30. |−3| |−1|

B Find each of the following absolute values.

31. |2|

32. |7|

33. |100|

34. |10,000|

35. |−8|

36. |−9|

37. |−231|

38. |−457|

39. |−42|

40. |−9,500|

41. |−200|

42. |−350|

43. |8|

44. |9|

45. |231|

46. |457|

Give the opposite of each of the following numbers.

47. 3

48. −5

49. −2

50. 15

51. 75

52. −32

53. 0

54. 1

Simplify each of the following.

55. $-(-2)$ **56.** $-(-5)$ **57.** $-(-8)$ **58.** $-(-3)$

59. $-|-2|$ **60.** $-|-5|$ **61.** $-|-8|$ **62.** $-|-3|$

63. What number is its own opposite?

64. Is $|a| = a$ always a true statement?

65. If n is a negative number, is $-n$ positive or negative?

66. If n is a positive number, is $-n$ positive or negative?

Estimating

Work the next problems mentally, without pencil and paper or a calculator.

67. Is -60 closer to 0 or -100?

68. Is -20 closer to 0 or -30?

69. Is -10 closer to -20 or 20?

70. Is -20 closer to -40 or 10?

71. Is -362 closer to -360 or -370?

72. Is -368 closer to -360 or -370?

Applying the Concepts

73. Temperature and Altitude Yamina is flying from Phoenix to San Francisco on a Boeing 737 jet. When the plane reaches an altitude of 33,000 feet, the temperature outside the plane is 61 degrees below zero Fahrenheit. Represent this temperature with a negative number. If the temperature outside the plane gets warmer by 10 degrees, what will the new temperature be?

74. Temperature Change At 11:00 in the morning in Superior, Wisconsin, Jim notices the temperature is 15 degrees below zero Fahrenheit. Write this temperature as a negative number. At noon it has warmed up by 8 degrees. What is the temperature at noon?

75. Temperature Change At 10:00 in the morning in White Bear Lake, Wisconsin, Zach notices the temperature is 5 degrees below zero Fahrenheit. Write this temperature as a negative number. By noon the temperature has dropped another 10 degrees. What is the temperature at noon?

76. Snorkeling Steve is snorkeling in the ocean near his home in Maui. At one point he is 6 feet below the surface. Represent this situation with a negative number. If he descends another 6 feet, what negative number will represent his new position?

Table 2 lists various wind chill temperatures. The top row gives air temperature, while the first column gives wind speed, in miles per hour. The numbers within the table indicate how cold the weather will feel. For example, if the thermometer reads 30°F and the wind is blowing at 15 miles per hour, the wind chill temperature is 9°F.

Table 2

Wind chill temperatures

	Air Temperatures (°F)							
Wind Speed	30°	25°	20°	15°	10°	5°	0°	−5°
10 mph	16°	10°	3°	−3°	−9°	−15°	−22°	−27°
15 mph	9°	2°	−5°	−11°	−18°	−25°	−31°	−38°
20 mph	4°	−3°	−10°	−17°	−24°	−31°	−39°	−46°
25 mph	1°	−7°	−15°	−22°	−29°	−36°	−44°	−51°
30 mph	−2°	−10°	−18°	−25°	−33°	−41°	−49°	−56°

77. Wind Chill Find the wind chill temperature if the thermometer reads 25°F and the wind is blowing at 25 miles per hour.

78. Wind Chill Find the wind chill temperature if the thermometer reads 10°F and the wind is blowing at 25 miles per hour.

79. Wind Chill Which will feel colder: a day with an air temperature of 10°F and a 25-mph wind, or a day with an air temperature of −5°F and a 10-mph wind?

80. Wind Chill Which will feel colder: a day with an air temperature of 15°F and a 20-mph wind, or a day with an air temperature of 5°F and a 10-mph wind?

Table 3 lists the record low temperatures for each month of the year for Lake Placid, New York. Table 4 lists the record high temperatures for the same city.

Table 3
Record low temperatures for Lake Placid, New York

Month	Temperature
January	−36°F
February	−30°F
March	−14°F
April	−2°F
May	19°F
June	22°F
July	35°F
August	30°F
September	19°F
October	15°F
November	−11°F
December	−26°F

Table 4
Record high temperatures for Lake Placid, New York

Month	Temperature
January	54°F
February	59°F
March	69°F
April	82°F
May	90°F
June	93°F
July	97°F
August	93°F
September	90°F
October	87°F
November	67°F
December	60°F

81. Temperature Figure 5 is a bar chart of the information in Table 3. Construct a scatter diagram of the same information. Then connect the dots in the scatter diagram to obtain a line graph of that same information. (Notice that we have used the numbers 1 through 12 to represent the months January through December.)

82. Temperature Figure 6 is a bar chart of the information in Table 4. Construct a scatter diagram of the same information. Then connect the dots in the scatter diagram to obtain a line graph of that same information. (Again, we have used the numbers 1 through 12 to represent the months January through December.)

FIGURE 5

FIGURE 6

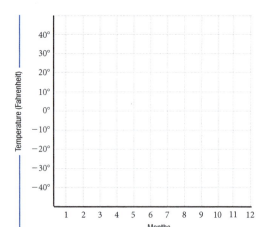

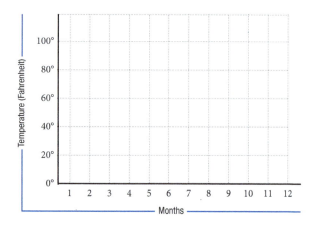

Getting Ready for the Next Section

Add or subtract.

83. $10 + 15$ **84.** $12 + 15$ **85.** $15 - 10$ **86.** $15 - 12$

87. $10 - 5 - 3 + 4$ **88.** $12 - 3 - 7 + 5$ **89.** $[3 + 10] + [8 - 2]$ **90.** $[2 + 12] + [7 - 5]$

Find the Mistake

Each problem below contains a mistake. Circle the mistake and write the correct number(s) or word(s) on the line provided.

1. The expression $-4 < 1$ is read "-4 is more than 1." _____

2. The number -3 appears to the left of the number -24 on the number line. _____

3. The opposite of -18 is 9. _____

4. The absolute value of -36 is -3. _____

Navigation Skills: Prepare, Study, Achieve

At the end of the first exercise set of each chapter, we provide an important discussion of study skills that will help you succeed in this course. Pay special attention to these skills. Ponder each one and apply it to your life. Your success is in your hands.

Studying is the key to success in this course. However, many students have never learned effective skills for studying. Study skills include but are not limited to the following:

- Work done on problems for practice and homework
- Amount of time spent studying
- Time of day and location for studying
- Management of distractions during study sessions
- Material chosen to review
- Order and process of review

Let's begin our discussion with the topic of homework. From the first day of class, we recommend you spend two hours on homework for every hour you are scheduled to attend class. Any less may drastically impact your success in this course. To help visualize this commitment, map out a weekly schedule that includes your classes, work shifts, extracurriculars, and any additional obligations. Fill in the hours you intend to devote to completing assignments and studying for this class. Post this schedule at home and keep a copy with your study materials to remind you of your commitment to success.

1.2 Addition with Negative Numbers

Suppose you are in Las Vegas playing blackjack and you lose $3 on the first hand and then you lose $5 on the next hand. If you represent winning with positive numbers and losing with negative numbers, how will you represent the results from your first two hands? Since you lost $3 and $5 for a total of $8, one way to represent the situation is with addition of negative numbers.

$$(-\$3) + (-\$5) = -\$8$$

From this example we see that the sum of two negative numbers is a negative number. To generalize addition of positive and negative numbers, we can use the number line.

We can think of each number on the number line as having two characteristics: (1) a *distance* from 0 (absolute value) and (2) a *direction* from 0 (positive or negative). The distance from 0 is represented by the numerical part of the number (like the 5 in the number −5), and its direction is represented by the + or − sign in front of the number.

A Addition of Numbers on the Number Line

We can visualize addition of numbers on the number line by thinking in terms of distance and direction from 0. Let's begin with a simple problem we know the answer to. We interpret the sum 3 + 5 on the number line as follows:

1. The first number is 3, which tells us "start at the origin, and move 3 units in the positive direction."

2. The + sign is read "and then move."

3. The 5 means "5 units in the positive direction."

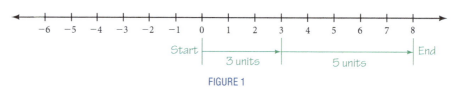

FIGURE 1

Figure 1 shows these steps. To summarize, 3 + 5 means to start at the origin (0), move 3 units in the *positive* direction, and then move 5 units in the *positive* direction. We end up at 8, which is the sum we are looking for: 3 + 5 = 8.

**Video Examples
Section 1.2**

Note This method of adding numbers may seem a little complicated at first, but it will allow us to add numbers we couldn't otherwise add.

EXAMPLE 1   Add $3 + (-5)$ using the number line.

Solution   We start at the origin, move 3 units in the positive direction, and then move 5 units in the negative direction, as shown in Figure 2. The last arrow ends at -2, which must be the sum of 3 and -5. That is,

$$3 + (-5) = -2$$

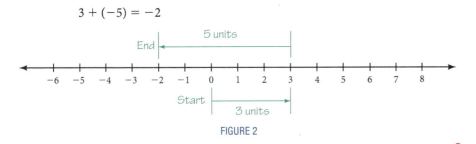

FIGURE 2

EXAMPLE 2   Add $-3 + 5$ using the number line.

Solution   We start at the origin, move 3 units in the negative direction, and then move 5 units in the positive direction, as shown in Figure 3. We end up at 2, which is the sum of -3 and 5. That is,

$$-3 + 5 = 2$$

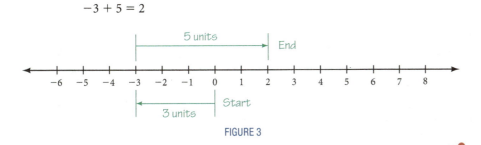

FIGURE 3

EXAMPLE 3   Add $-3 + (-5)$ using the number line.

Solution   We start at the origin, move 3 units in the negative direction, and then move 5 more units in the negative direction. This is shown on the number line in Figure 4. As you can see, the last arrow ends at -8. We must conclude that the sum of -3 and -5 is -8. That is,

$$-3 + (-5) = -8$$

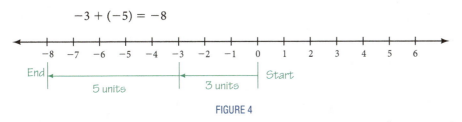

FIGURE 4

Adding numbers on the number line as we have done in these first three examples gives us a way of visualizing addition of positive and negative numbers. We want to be able to write a rule for addition of positive and negative numbers that doesn't involve the number line. The number line is a way of justifying the rule we will write. Here is a summary of the results we have so far:

$$3 + 5 = 8 \qquad\qquad -3 + 5 = 2$$

$$3 + (-5) = -2 \qquad\qquad -3 + (-5) = -8$$

Practice Problems

1. Add $4 + (-6)$ using the number line.

2. Add $-5 + 7$ using the number line.

3. Add $-4 + (-6)$ using the number line.

Answers
1. -2
2. 2
3. -10

Looking over these results, we write the following rule for adding any two numbers:

RULE   Adding Any Two Numbers

1. To add two numbers with the same sign: Add their absolute values, and use the common sign. If both numbers are positive, the answer is positive. If both numbers are negative, the answer is negative.

2. To add two numbers with different signs: Subtract the smaller absolute value from the larger absolute value. The answer will have the sign of the number with the larger absolute value.

The following examples show how the rule is used. You will find that the rule for addition is consistent with all the results obtained using the number line.

4. Add all combinations of positive and negative 8 and 4.

EXAMPLE 4   Add all combinations of positive and negative 10 and 15.

Solution
$$10 + 15 = 25$$
$$10 + (-15) = -5$$
$$-10 + 15 = 5$$
$$-10 + (-15) = -25$$

Notice that when we add two numbers with the same sign, the answer also has that sign. When the signs are not the same, the answer has the sign of the number with the larger absolute value.

Once you have become familiar with the rule for adding positive and negative numbers, you can apply it to more complicated sums.

5. Simplify: $12 + (-4) + (-5) + 6$.

EXAMPLE 5   Simplify: $10 + (-5) + (-3) + 4$.

Solution   Adding left to right, we have
$$10 + (-5) + (-3) + 4 = 5 + (-3) + 4 \qquad 10 + (-5) = 5$$
$$= 2 + 4 \qquad\qquad\quad 5 + (-3) = 2$$
$$= 6$$

6. Simplify:
$[-5 + (-8)] + [7 + (-5)]$.

EXAMPLE 6   Simplify: $[-3 + (-10)] + [8 + (-2)]$.

Solution   We begin by adding the numbers inside the brackets.
$$[-3 + (-10)] + [8 + (-2)] = [-13] + [6] = -7$$

GETTING READY FOR CLASS

After reading through the preceding section, respond in your own words and in complete sentences.

A. Explain how you would use the number line to add 3 and 5.

B. If two numbers are negative, such as -3 and -5, what sign will their sum have?

C. If you add two numbers with different signs, how do you determine the sign of the answer?

D. With respect to addition with positive and negative numbers, does the phrase "two negatives make a positive" make any sense?

Answers
4. $8 + 4 = 12, 8 + (-4) = 4$
   $-8 + 4 = -4, -8 + (-4) = -12$
5. 9
6. -11

Vocabulary Review

Choose the correct words to fill in the blanks in the paragraph below.

   different   smaller   absolute values   sign   same   larger

   To add two numbers with the _____ sign, add their _____, and use the
common sign. If both numbers are positive, the answer is positive. If both numbers are negative, the answer is negative.
To add two numbers with _____ signs, subtract the _____ absolute value
from the _____ absolute value. The answer will have the _____ of the
number with the larger absolute value.

Problems

A Draw a number line from −10 to 10 and use it to add the following numbers.

1. $2 + 3$

2. $2 + (-3)$

3. $-2 + 3$

4. $-2 + (-3)$

5. $5 + (-7)$

6. $-5 + 7$

7. $-4 + (-2)$

8. $-8 + (-2)$

9. $10 + (-6)$

10. $-9 + 3$

11. $7 + (-3)$

12. $-7 + 3$

13. $-4 + (-5)$

14. $-2 + (-7)$

Combine the following by using the rule for addition of positive and negative numbers. (Your goal is to be fast and accurate at addition, with the latter being more important.)

15. $7 + 8$

16. $9 + 12$

17. $5 + (-8)$

18. $4 + (-11)$

19. $-6 + (-5)$

20. $-7 + (-2)$

21. $-10 + 3$

22. $-14 + 7$

23. $-1 + (-2)$

24. $-5 + (-4)$

25. $-11 + (-5)$

26. $-16 + (-10)$

27. $4 + (-12)$

28. $9 + (-1)$

29. $-85 + (-42)$

30. $-96 + (-31)$

31. $-121 + 170$

32. $-130 + 158$

33. $-375 + 409$

34. $-765 + 213$

Complete the following tables.

35.

First Number a	Second Number b	Their Sum a + b
5	−3	
5	−4	
5	−5	
5	−6	
5	−7	

36.

First Number a	Second Number b	Their Sum a + b
−5	3	
−5	4	
−5	5	
−5	6	
−5	7	

37.

First Number x	Second Number y	Their Sum x + y
−5	−3	
−5	−4	
−5	−5	
−5	−6	
−5	−7	

38.

First Number x	Second Number y	Their Sum x + y
30	−20	
−30	20	
−30	−20	
30	20	
−30	0	

Add the following numbers left to right.

39. $10 + (-18) + 4$

40. $-2 + 4 + (-6)$

41. $24 + (-6) + (-8)$

42. $35 + (-5) + (-30)$

43. $-201 + (-143) + (-101)$

44. $-27 + (-56) + (-89)$

45. $-321 + 752 + (-324)$

46. $-571 + 437 + (-502)$

47. $-8 + 3 + (-5) + 9$

48. $-9 + 2 + (-10) + 3$

49. $-2 + (-5) + (-6) + (-7)$

50. $-8 + (-3) + (-4) + (-7)$

51. $15 + (-30) + 18 + (-20)$

52. $20 + (-15) + 30 + (-18)$

53. $-78 + (-42) + 57 + 13$

54. $-89 + (-51) + 65 + 17$

Use the rule for order of operations to simplify each of the following.

55. $(-8 + 5) + (-6 + 2)$

56. $(-3 + 1) + (-9 + 4)$

57. $(-10 + 4) + (-3 + 12)$

58. $(-11 + 5) + (-3 + 2)$

59. $20 + (-30 + 50) + 10$

60. $30 + (-40 + 20) + 50$

61. $108 + (-456 + 275)$

62. $106 + (-512 + 318)$

63. $[5 + (-8)] + [3 + (-11)]$

64. $[8 + (-2)] + [5 + (-7)]$

65. $[57 + (-35)] + [19 + (-24)]$

66. $[63 + (-27)] + [18 + (-24)]$

67. Find the sum of -8, -10, and -3.

68. Find the sum of -4, 17, and -6.

69. What number do you add to 8 to get 3?

70. What number do you add to 10 to get 4?

71. What number do you add to -3 to get -7?

72. What number do you add to -5 to get -8?

73. What number do you add to -4 to get 3?

74. What number do you add to -7 to get 2?

75. If the sum of -3 and 5 is increased by 8, what number results?

76. If the sum of -9 and -2 is increased by 10, what number results?

Estimating

Work Problems 77–84 mentally, without pencil and paper or a calculator.

77. The answer to the problem $251 + 249$ is closest to which of the following numbers?

  a. 500   **b.** 0   **c.** -500

78. The answer to the problem $251 + (-249)$ is closest to which of the following numbers?

  a. 500   **b.** 0   **c.** -500

79. The answer to the problem $-251 + 249$ is closest to which of the following numbers?

  a. 500   **b.** 0   **c.** -500

80. The answer to the problem $-251 + (-249)$ is closest to which of the following numbers?

  a. 500   **b.** 0   **c.** -500

81. The sum of 77 and 22 is closest to which of the following numbers?

  a. -100 **b.** -60 **c.** 60   **d.** 100

82. The sum of -77 and 22 is closest to which of the following numbers?

  a. -100 **b.** -60 **c.** 60   **d.** 100

83. The sum of 77 and -22 is closest to which of the following numbers?

  a. -100 **b.** -60 **c.** 60   **d.** 100

84. The sum of -77 and -22 is closest to which of the following numbers?

  a. -100 **b.** -60 **c.** 60   **d.** 100

Applying the Concepts

85. Checkbook Balance Ethan has a balance of –$40 in his checkbook. If he deposits $100 and then writes a check for $50, what is the new balance in his checkbook?

Check No.	Date	Description of Transaction	Payment/Debit (-)	Deposit/Credit (+)	Balance
					-$40 00
	9/15	Deposit		$100 00	
1568	9/16	CVS Pharmacy	$50 00		

86. Checkbook Balance Justin has a balance of –$20 in his checkbook. If he deposits $70 and then writes a check for $50, what is the new balance in his checkbook?

Check No.	Date	Description of Transaction	Payment/Debit (-)	Deposit/Credit (+)	Balance
					-$20 00
	6/22	Deposit		$70 00	
1775	6/24	Donation	$50 00		

87. **Gambling** While gambling in Las Vegas, a person wins $74 playing blackjack and then loses $141 on roulette. Use positive and negative numbers to write this situation in symbols. Then give the person's net loss or gain.

88. **Gambling** While playing blackjack, a person loses $17 on his first hand, then wins $14, and then loses $21. Write this situation using positive and negative numbers and addition; then simplify.

89. **Stock Gain/Loss** Suppose a certain stock gains 3 points on the stock exchange on Monday and then loses 5 points on Tuesday. Express the situation using positive and negative numbers, and then give the net gain or loss of the stock for this 2-day period.

90. **Stock Gain/Loss** A stock gains 2 points on Wednesday, then loses 1 on Thursday, and gains 3 on Friday. Use positive and negative numbers and addition to write this situation in symbols, and then simplify.

91. **Distance** The distance between two numbers on the number line is 10. If one of the numbers is 3, what are the two possibilities for the other number?

92. **Distance** The distance between two numbers is 8. If one of the numbers is -5, what are the two possibilities for the other number?

Getting Ready for the Next Section

Give the opposite of each number.

93. 2

94. 3

95. -4

96. -5

97. -30

98. -15

99. Subtract 3 from 5.

100. Subtract 2 from 8.

101. Find the difference of 7 and 4.

102. Find the difference of 8 and 6.

Find the Mistake

Each problem below contains a mistake. Circle the mistake and write the correct number(s) or word(s) on the line provided.

1. The problem $6 + (-10) = -4$ is interpreted as, "Start at the origin, move 6 units in the positive direction, and then move 10 units in the positive direction." _____

2. Adding two numbers with different signs will give an answer that has the same sign as the number with the smaller absolute value. _____

3. Adding -8 and -5 gives us 13. _____

4. The sum of -2, 4, -3, and -5 is -2. _____

1.3 Subtraction with Negative Numbers

©iStockphoto.com/serg269

Video Examples
Section 1.3

Note This definition of subtraction may seem a little strange at first. In Example 1, you will notice that using the definition gives us the same results we are used to getting with subtraction. As we progress further into the section, we will use the definition to subtract numbers we haven't been able to subtract before.

Static apnea is a competitive discipline in which a person holds his or her breath underwater for as long as possible, without the assistance of scuba gear. At the time of this book's printing, the current world record is 11 minutes and 35 seconds held by Stéphane Mifsud of France. Static apnea is used in the sport of freediving, where a diver descends to a pre-announced depth without breathing assistance.

Imagine a freediver standing at the edge of a boat that is 5 feet above the surface of the ocean. The freediver jumps into the water and descends to a depth of 15 feet. We can use a number line to show the diver's height. If we consider the diver's height on the boat as 5, then the ocean surface would be 0, and the diver's depth would be -15. What is the difference in the diver's height between the boat and her final depth? Intuitively, we know the difference in the two heights is 20 feet. We also know that the word *difference* indicates subtraction. The difference between 5 and -15 is written

$$5 - (-15)$$

It must be true that $5 - (-15) = 20$. In this section, we will see how our definition for subtraction confirms that this last statement is in fact correct. Let's start with a formal definition for subtraction that uses the rules we have developed for addition to help us solve problems.

A Subtraction of Any Two Numbers

> **DEFINITION** subtraction
>
> If a and b represent any two numbers, then it is always true that
>
> $$a - b = a + (-b)$$
>
> To subtract b add its opposite, $-b$.

In words: Subtracting a number is equivalent to adding its opposite.

Let's see if this definition conflicts with what we already know to be true about subtraction.

From previous experience, we know that

$$5 - 2 = 3$$

We can get the same answer by using the definition we just gave for subtraction. Instead of subtracting 2, we can add its opposite, -2. Here is how it looks:

$$5 - 2 = 5 + (-2)$$ *Change subtraction to addition of the opposite.*

$$= 3$$ *Apply the rule for addition of positive and negative numbers.*

The result is the same whether we use our previous knowledge of subtraction or the new definition. The new definition is essential when the problems begin to get more complicated.

EXAMPLE 1   Subtract: $-7 - 2$.

Solution   We have never subtracted a positive number from a negative number before. We must apply our definition of subtraction.

$$-7 - 2 = -7 + (-2)$$ *Instead of subtracting 2, we add its opposite, -2.*

$$= -9$$ *Apply the rule for addition.*

EXAMPLE 2   Subtract: $12 - (-6)$.

Solution   The first $-$ sign is read "subtract," and the second one is read "negative." The problem in words is "12 subtract negative 6." We can use the definition of subtraction to change this to the addition of positive 6.

$$12 - (-6) = 12 + 6$$ *Subtracting -6 is equivalent to adding 6.*

$$= 18$$ *Add.*

The following table shows the relationship between subtraction and addition:

Subtraction	Addition of the opposite	Answer
$7 - 9$	$7 + (-9)$	-2
$-7 - 9$	$-7 + (-9)$	-16
$7 - (-9)$	$7 + 9$	16
$-7 - (-9)$	$-7 + 9$	2
$15 - 10$	$15 + (-10)$	5
$-15 - 10$	$-15 + (-10)$	-25
$15 - (-10)$	$15 + 10$	25
$-15 - (-10)$	$-15 + 10$	-5

The previous examples illustrate all the possible combinations of subtraction with positive and negative numbers. There are no new rules for subtraction. We apply the definition to change each subtraction problem into an equivalent addition problem. The rule for addition can then be used to obtain the correct answer.

EXAMPLE 3   Combine: $-3 + 6 - 2$.

Solution   The first step is to change subtraction to addition of the opposite. After that has been done, we add left to right.

$$-3 + 6 - 2 = -3 + 6 + (-2)$$ *Subtracting 2 is equivalent to adding -2.*

$$= 3 + (-2)$$ *Add left to right.*

$$= 1$$

Practice Problems

1. Subtract: $-5 - 3$.

Note   A real-life analogy to Example 1 would be: "If the temperature were 7° below 0 and then it dropped another 2°, what would the temperature be then?"

2. Subtract: $10 - (-5)$.

3. Combine: $-5 + 8 - 3$.

Answers
1. -8
2. 15
3. 0

4. Subtract 5 from -8.

EXAMPLE 4   Subtract 3 from -5.

Solution   Subtracting 3 is equivalent to adding -3.

$$-5 - 3 = -5 + (-3) = -8$$

Subtracting 3 from -5 gives us -8.

5. For the airplane described in Example 5, find the difference in temperature inside the plane if it was 70°F at takeoff and drops to -5°F.

EXAMPLE 5   Many of the planes used by the United States during World War II were not pressurized or sealed from outside air. As a result, the temperature inside these planes was the same as the surrounding air temperature outside. Suppose the temperature inside a B-17 Flying Fortress is 50°F at takeoff and then drops to -30°F when the plane reaches its cruising altitude of 28,000 feet. Find the difference in temperature inside this plane at takeoff and at 28,000 feet.

Solution   The temperature at takeoff is 50°F, whereas the temperature at 28,000 feet is -30°F. To find the difference we subtract, with the numbers in the same order as they are given in the problem.

$$50 - (-30) = 50 + 30 = 80$$

The difference in temperature is 80°F.

Subtraction and Taking Away

Some people may believe that the answer to $-5 - 9$ should be -4 or 4, not -14. If this is happening to you, you are probably thinking of subtraction in terms of taking one number away from another. Thinking of subtraction in this way works well with positive numbers if you always subtract the smaller number from the larger. In algebra, however, we encounter many situations other than this. The definition of subtraction, that $a - b = a + (-b)$ clearly indicates the correct way to use subtraction. That is, when working subtraction problems, you should think "addition of the opposite," not "taking one number away from another."

USING TECHNOLOGY   Calculator Note

Here is how we work the subtraction problem shown in Example 1 on a calculator.

Scientific Calculator: 7 $\boxed{+/-}$ $\boxed{-}$ 2 $\boxed{=}$
Graphing Calculator: $\boxed{(-)}$ 7 $\boxed{-}$ 2 $\boxed{\text{ENT}}$

GETTING READY FOR CLASS

After reading through the preceding section, respond in your own words and in complete sentences.

A. Write the subtraction problem $5 - 3$ as an equivalent addition problem.

B. Explain the process you would use to subtract 2 from -7.

C. Write an addition problem that is equivalent to the subtraction problem $-20 - (-30)$.

D. Why is it important to think of subtraction as addition of the opposite?

Answers
4. -13
5. 75

Vocabulary Review

Choose the correct words to fill in the blanks below.

opposite   negative   addition   positive   subtract

1. To _____ b from a, add its opposite, $-b$.

2. Subtracting a number is equivalent to adding its _____.

3. Once you turn a subtraction problem into the addition of the opposite, apply the rule for _____ of positive and negative numbers.

4. When you subtract 10 from -15, you get a _____ answer.

5. When you subtract -10 from 15, you get a _____ answer.

Problems

A Subtract.

1. $7 - 5$

2. $5 - 7$

3. $8 - 6$

4. $6 - 8$

5. $-3 - 5$

6. $-5 - 3$

7. $-4 - 1$

8. $-1 - 4$

9. $5 - (-2)$

10. $2 - (-5)$

11. $3 - (-9)$

12. $9 - (-3)$

13. $-4 - (-7)$

14. $-7 - (-4)$

15. $-10 - (-3)$

16. $-3 - (-10)$

17. $15 - 18$

18. $20 - 32$

19. $100 - 113$

20. $121 - 21$

21. $-30 - 20$

22. $-50 - 60$

23. $-79 - 21$

24. $-86 - 31$

25. $156 - (-243)$

26. $292 - (-841)$

27. $-35 - (-14)$

28. $-29 - (-4)$

Complete the following tables.

29.

First Number x	Second Number y	The Difference of x and y $x - y$
8	6	
8	7	
8	8	
8	9	
8	10	

30.

First Number x	Second Number y	The Difference of x and y $x - y$
10	12	
10	11	
10	10	
10	9	
10	8	

31.

First Number x	Second Number y	The Difference of x and y $x - y$
8	-6	
8	-7	
8	-8	
8	-9	
8	-10	

32.

First Number x	Second Number y	The Difference of x and y $x - y$
-10	-12	
-10	-11	
-10	-10	
-10	-9	
-10	-8	

Simplify as much as possible by first changing all subtractions to addition of the opposite and then adding left to right.

33. $4 - 5 - 6$

34. $7 - 3 - 2$

35. $-8 + 3 - 4$

36. $-10 - 1 + 16$

37. $-8 - 4 - 2$

38. $-7 - 3 - 6$

39. $33 - (-22) - 66$

40. $44 - (-11) + 55$

41. $-900 + 400 - (-100)$

42. $-300 + 600 - (-200)$

43. Subtract -6 from 5.

44. Subtract 8 from -2.

45. Find the difference of -5 and -1.

46. Find the difference of -7 and -3.

47. Subtract -4 from the sum of -8 and 12.

48. Subtract -7 from the sum of 7 and -12.

49. What number do you subtract from -3 to get -9?

50. What number do you subtract from 5 to get 8?

Estimating

Work the next problems mentally, without pencil and paper or a calculator.

51. The answer to the problem $52 - 49$ is closest to which of the following numbers?

   a. 100   **b.** 0   **c.** -100

52. The answer to the problem $-52 - 49$ is closest to which of the following numbers?

   a. 100   **b.** 0   **c.** -100

53. The answer to the problem $52 - (-49)$ is closest to which of the following numbers?

   a. 100   **b.** 0   **c.** -100

54. The answer to the problem $-52 - (-49)$ is closest to which of the following numbers?

   a. 100   **b.** 0   **c.** -100

55. Is the difference $-161 - (-62)$ closer to -200 or -100?

56. Is the difference $-553 - 50$ closer to -600 or -500?

57. The difference of 37 and 61 is closest to which of the following numbers?

   a. -100 **b.** -20 **c.** 20   **d.** 100

58. The difference of 37 and -61 is closest to which of the following numbers?

   a. -100 **b.** -20 **c.** 20   **d.** 100

59. The difference of -37 and 61 is closest to which of the following numbers?

   a. -100 **b.** -20 **c.** 20   **d.** 100

60. The difference of -37 and -61 is closest to which of the following numbers?

   a. -100 **b.** -20 **c.** 20   **d.** 100

Applying the Concepts

61. Temperature On Monday, the temperature reached a high of 28° above 0. That night it dropped to 16° below 0. What is the difference between the high and the low temperatures for Monday?

62. Checkbook Balance Susan has a balance of $572 in her checking account when she writes a check for $435 to pay the rent. Then she writes another check for $172 for textbooks. Write a subtraction problem that gives the new balance in her checking account. What is the new balance in her checking account?

Repeated below is the table of wind chill temperatures that we used previously. Use it for Problems 63–66.

Wind speed	Air Temperature (°F)							
	30°	25°	20°	15°	10°	5°	0°	−5°
10 mph	16°	10°	3°	−3°	−9°	−15°	−22°	−27°
15 mph	9°	2°	−5°	−11°	−18°	−25°	−31°	−38°
20 mph	4°	−3°	−10°	−17°	−24°	−31°	−39°	−46°
25 mph	1°	−7°	−15°	−22°	−29°	−36°	−44°	−51°
30 mph	−2°	−10°	−18°	−25°	−33°	−41°	−49°	−56°

63. Wind Chill If the temperature outside is 15°F, what is the difference in wind chill temperature between a 15-mile-per-hour wind and a 25-mile-per-hour wind?

64. Wind Chill If the temperature outside is 0°F, what is the difference in wind chill temperature between a 15-mile-per-hour wind and a 25-mile-per-hour wind?

65. Wind Chill Find the difference in temperature between a day in which the air temperature is 20°F and the wind is blowing at 10 miles per hour and a day in which the air temperature is 10°F and the wind is blowing at 20 miles per hour.

66. Wind Chill Find the difference in temperature between a day in which the air temperature is 0°F and the wind is blowing at 10 miles per hour and a day in which the air temperature is −5°F and the wind is blowing at 20 miles per hour.

Use the tables below to work Problems 67–70.

Table 1

Record low temperatures for Lake Placid, New York

Month	Temperature
January	−36°F
February	−30°F
March	−14°F
April	−2°F
May	19°F
June	22°F
July	35°F
August	30°F
September	19°F
October	15°F
November	−11°F
December	−26°F

Table 2

Record high temperatures for Lake Placid, New York

Month	Temperature
January	54°F
February	59°F
March	69°F
April	82°F
May	90°F
June	93°F
July	97°F
August	93°F
September	90°F
October	87°F
November	67°F
December	60°F

67. Temperature Difference Find the difference between the record high temperature and the record low temperature for the month of December.

68. Temperature Difference Find the difference between the record high temperature and the record low temperature for the month of March.

69. Temperature Difference Find the difference between the record low temperatures of March and December.

70. Temperature Difference Find the difference between the record high temperatures of March and December.

Getting Ready for the Next Section

Perform the indicated operations.

71. $3(2)(5)$

72. $5(2)(4)$

73. 6^2

74. 8^2

75. 4^3

76. 3^3

77. $6(3 + 5)$

78. $2(5 + 8)$

79. $3(9 - 2) + 4(7 - 2)$

80. $2(5 - 3) - 7(4 - 2)$

81. $(3 + 7)(6 - 2)$

82. $(6 + 1)(9 - 4)$

Find the Mistake

Each problem below contains a mistake. Circle the mistake and write the correct number(s) or word(s) on the line provided.

1. Subtracting 5 from 4 is the same as adding 4 and 5. _____

2. To subtract -1 from 8, we must move 1 unit in the negative direction from 8 on the number line. _____

3. The problem $11 - (-7)$ is read, "11 negative subtract 7." _____

4. To find the difference of -5 and 6, change -5 to 5 and add 6. _____

Landmark Review: Checking Your Progress

Find the absolute value and the opposite of each of the following numbers.

1. 2

2. -11

3. 25

4. -110

Simplify each of the following.

5. $-(-7)$

6. $-(5)$

7. $-|-3|$

8. $-|15|$

Add the following numbers.

9. $5 + 6$

10. $-7 + 4$

11. $7 + (-5)$

12. $-15 + (-5)$

Subtract the following numbers.

13. $10 - 6$

14. $6 - 12$

15. $-5 - 7$

16. $7 - (-8)$

17. $-12 - (-3)$

18. $-27 - (-38)$

1.4 Multiplication with Negative Numbers

OBJECTIVES

A Evaluate exponential expressions with whole numbers.

B Multiply any numbers, including negative numbers.

KEY WORDS

base

exponent

©iStockphoto.com/JayKayl

The cane toad was introduced to Australian agricultural fields in 1935 to help rid sugar cane crops from an invasive beetle. Instead, the toads became the invasive species multiplying from an original number of 102 to a current population of more than one and a half million. The toad preys on and competes with native species, and its skin contains a toxin that is deadly when ingested by native predators and even pets. Scientists are studying how global warming may affect these amphibians. Other cold-blooded animals have a difficult time breathing in warmer temperatures, but the toad's heart and lungs appear to thrive in the heat. On the other hand, once temperatures drop below 15 degrees Celsius (59 degrees Fahrenheit), the toad's system shuts down and he can barely hop. This is good news for much of Southern Australia where weather conditions are too cold for the toad. However, as the Earth warms, the toad population may invade those areas as well.

The following bar chart contains record low temperature readings for various cities in Australia rounded to the nearest integer. Notice that some of these temperatures are represented by negative numbers.

**Video Examples
Section 1.4**

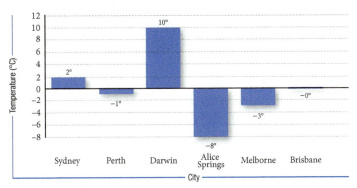

FIGURE 1

Suppose one year Melborne records a low temperature of $-2°F$ and Alice Springs records a low temperature three times lower. How would we calculate the temperature for Alice Springs? The answer is $-6°F$.

$$3(-2) = -6$$

From this we conclude that it is reasonable to say that the product of a positive number and a negative numbers is a negative number.

In order to generalize multiplication with negative numbers, recall that we first defined multiplication by whole numbers to be repeated addition. That is,

$$\underbrace{3 \cdot 2}_{\text{Multiplication}} = \underbrace{2 + 2 + 2}_{\text{Repeated addition}}$$

We begin by looking at exponents, a shorthand notation we will develop.

A   Exponents

Exponents are a shorthand way of writing repeated multiplication. In the expression 2^3, 2 is called the *base* and 3 is called the *exponent*. The expression 2^3 is read "2 to the third power" or "2 cubed." The exponent 3 tells us to use the base 2 as a multiplication factor three times.

$$2^3 = 2 \cdot 2 \cdot 2 \qquad \text{2 is used as a factor three times.}$$

We can simplify the expression by multiplication.

$$
\begin{aligned}
2^3 &= 2 \cdot 2 \cdot 2 \\
&= 4 \cdot 2 \\
&= 8
\end{aligned}
$$

The expression 2^3 is equal to the number 8. We can summarize this discussion with the following definition:

> **DEFINITION**   exponent
>
> An *exponent* is a whole number that indicates how many times the base is to be used as a factor. Exponents indicate repeated multiplication.

For example, in the expression 5^2, 5 is the base and 2 is the exponent. The meaning of the expression is

$$
\begin{aligned}
5^2 &= 5 \cdot 5 \qquad \text{5 is used as a factor two times.} \\
&= 25
\end{aligned}
$$

The expression 5^2 is read "5 to the second power" or "5 squared."

EXAMPLE 1   Name the base and the exponent.
a. 3^2 **b.** 3^3 **c.** 2^4

Solution

a. The base is 3, and the exponent is 2. The expression is read "3 to the second power" or "3 squared."

b. The base is 3, and the exponent is 3. The expression is read "3 to the third power" or "3 cubed."

c. The base is 2, and the exponent is 4. The expression is read "2 to the fourth power."

As you can see from this example, a base raised to the second power is also said to be *squared*, and a base raised to the third power is also said to be *cubed*. These are the only two exponents (^{2} and ^{3}) that have special names. All other exponents are referred to only as "fourth powers," "fifth powers," "sixth powers," and so on.

The next example shows how we can simplify expressions involving exponents by using repeated multiplication.

EXAMPLE 2   Expand and multiply.
a. 3^2 **b.** 4^2 **c.** 3^3 **d.** 2^4

Solution

a. $3^2 = 3 \cdot 3 = 9$

b. $4^2 = 4 \cdot 4 = 16$

c. $3^3 = 3 \cdot 3 \cdot 3 = 9 \cdot 3 = 27$

d. $2^4 = 2 \cdot 2 \cdot 2 \cdot 2 = 4 \cdot 4 = 16$

Finally, we should consider what happens when the numbers 0 and 1 are used as exponents.

> **RULE** Exponent One
>
> Any number raised to the first power is itself. That is, if we let the letter a represent any number, then
> $$a^1 = a$$

> **RULE** Exponent Zero
>
> Any number other than 0 raised to the 0 power is 1. That is, if a represents any nonzero number, then it is always true that
> $$a^0 = 1$$

EXAMPLE 3   Simplify.

a. 5^1 **b.** 9^1 **c.** 4^0 **d.** 8^0

Solution

a. $5^1 = 5$

b. $9^1 = 9$

c. $4^0 = 1$

d. $8^0 = 1$

3. Simplify.
   a. 8^1
   b. 10^0
   c. 16^1
   d. x^0

B   Multiplication of Any Number

The concept of multiplication as repeated addition is very helpful when it comes to developing the rule for multiplication problems that involve negative numbers. For the next example, we look at what happens when we multiply a negative number by a positive number.

EXAMPLE 4   Multiply: $3(-5)$.

Solution   Writing this product as repeated addition, we have

$$3(-5) = (-5) + (-5) + (-5)$$
$$= -10 + (-5)$$
$$= -15$$

The result, -15, is obtained by adding the three negative 5s.

4. Multiply: $5(-2)$.

EXAMPLE 5   Multiply: $-3(5)$.

Solution   In order to write this multiplication problem in terms of repeated addition, we will have to reverse the order of the two numbers. This is easily done, because multiplication is a commutative operation.

$$-3(5) = 5(-3) \qquad \text{Commutative property}$$
$$= (-3) + (-3) + (-3) + (-3) + (-3) \qquad \text{Repeated addition}$$
$$= -15$$

The product of -3 and 5 is -15.

5. Multiply: $-5(2)$.

Answers

3. **a.** 8   **b.** 1   **c.** 16   **d.** 1
4. -10
5. -10

6. Multiply: $-5(-2)$.

Note   We want to be able to justify everything we do in mathematics. The discussion tells *why* $-3(-5) = 15$.

EXAMPLE 6   Multiply: $-3(-5)$.

Solution   It is impossible to write this product in terms of repeated addition. We will find the answer to $-3(-5)$ by solving a different problem. Look at the following problem:

$$-3[5 + (-5)] = -3[0] = 0$$

The result is 0, because multiplying by 0 always produces 0. Now we can work the same problem another way and in the process find the answer to $-3(-5)$. Applying the distributive property to the same expression, we have

$$-3[5 + (-5)] = -3(5) + (-3)(-5) \qquad \textcolor{green}{\text{Distributive property}}$$
$$= -15 + (?) \qquad\qquad \textcolor{green}{-3(5) = -15}$$

The question mark must be 15, because we already know that the answer to the problem is 0, and 15 is the only number we can add to -15 to get 0. So our problem is solved.

$$-3(-5) = 15$$

Table 1 gives a summary of what we have done so far with multiplication.

Table 1

Original numbers have	For example	The answer is
Same signs	$3(5) = \quad 15$	Positive
Different signs	$-3(5) = -15$	Negative
Different signs	$3(-5) = -15$	Negative
Same signs	$-3(-5) = \quad 15$	Positive

From the multiplication examples we have done so far and their summaries in Table 1, we write the following rule for multiplication of positive and negative numbers:

RULE   Multiplication of Any Two Numbers

To multiply any two numbers, we multiply their absolute values.

1. The answer is positive if both the original numbers have the same sign. That is, the product of two numbers with the same sign is positive.

2. The answer is negative if the original two numbers have different signs. The product of two numbers with different signs is negative.

This rule should be memorized. By the time you have finished reading this section and working the problems at the end of the section, you should be fast and accurate at multiplication with positive and negative numbers.

7. Find the following products:
  a. $3(5)$
  b. $-3(-5)$
  c. $3(-5)$
  d. $-3(5)$

EXAMPLE 7   Find the following products:

a. $2(4)$ **b.** $-2(-4)$ **c.** $2(-4)$ **d.** $-2(4)$

Solution

a. $2(4) = 8$ \qquad \textcolor{green}{Like signs; positive answer}

b. $-2(-4) = 8$ \qquad \textcolor{green}{Like signs; positive answer}

c. $2(-4) = -8$ \qquad \textcolor{green}{Unlike signs; negative answer}

d. $-2(4) = -8$ \qquad \textcolor{green}{Unlike signs; negative answer}

Answers

6. 10
7. a. 15 **b.** 15 **c.** -15 **d.** -15

EXAMPLE 8   Simplify: $-3(2)(-5)$.

Solution

$$-3(2)(-5) = -6(-5) \qquad \text{Multiply } -3 \text{ and } 2 \text{ to get } -6$$
$$= 30$$

8. Simplify: $-5(3)(-2)$.

EXAMPLE 9   Use the definition of exponents to expand each expression. Then simplify by multiplying.

a. $(-6)^2$ **b.** -6^2 **c.** $(-4)^3$ **d.** -4^3

Solution

a. $(-6)^2 = (-6)(-6)$ Definition of exponents

       $= 36$ Multiply.

b. $-6^2 = -6 \cdot 6$ Definition of exponents

       $= -36$ Multiply.

c. $(-4)^3 = (-4)(-4)(-4)$ Definition of exponents

       $= -64$ Multiply.

d. $-4^3 = -4 \cdot 4 \cdot 4$ Definition of exponents

       $= -64$ Multiply.

In Example 9, the base is a negative number in Parts a and c, but not in Parts b and d. We know this is true because of the use of parentheses.

9. Use the definition of exponents to expand each expression. Then simplify.
     a. $(-4)^2$
     b. -4^2
     c. $(-3)^2$
     d. -3^2

EXAMPLE 10   Simplify: $-4 + 5(-6 + 2)$.

Solution   Simplifying inside the parentheses first, we have

$$-4 + 5(-6 + 2) = -4 + 5(-4) \qquad \text{Simplify inside parentheses.}$$
$$= -4 + (-20) \qquad \text{Multiply.}$$
$$= -24 \qquad \text{Add.}$$

10. Simplify: $-6 + 2(-4 + 8)$.

EXAMPLE 11   Simplify: $-3(2 - 9) + 4(-7 - 2)$.

Solution   We begin by subtracting inside the parentheses.

$$-3(2 - 9) + 4(-7 - 2) = -3(-7) + 4(-9)$$
$$= 21 + (-36)$$
$$= -15$$

11. Simplify:
     $-2(3 - 5) + 5(-6 - 1)$.

GETTING READY FOR CLASS

After reading through the preceding section, respond in your own words and in complete sentences.

A. Write the multiplication problem $3(-5)$ as an addition problem.

B. How may we apply the distributive property to a multiplication problem?

C. If two numbers have the same sign, then their product will have what sign?

D. If two numbers have different signs, then their product will have what sign?

Answers

8. 30

9. a. 16   **b.** -16 **c.** 9   **d.** -9

10. 2

11. -31

EXERCISE SET 1.4

Problems

A For each of the following expressions, name the base and the exponent.

1. 4^5

2. 5^4

3. 3^6

4. 6^3

5. 8^2

6. 2^8

7. 9^1

8. 1^9

9. 4^0

10. 0^4

Use the definition of exponents as indicating repeated multiplication to simplify each of the following expressions.

11. 6^2

12. 7^2

13. 2^3

14. 2^4

15. 1^4

16. 5^1

17. 9^0

18. 27^0

19. 9^2

20. 8^2

21. 10^1

22. 8^1

23. 12^1

24. 16^0

B Find each of the following products. (Multiply.)

25. $7(-8)$ **26.** $-3(5)$ **27.** $-6(10)$ **28.** $4(-8)$

29. $-7(-8)$ **30.** $-4(-7)$ **31.** $-9(-9)$ **32.** $-6(-3)$

33. $3(-2)(4)$ **34.** $5(-1)(3)$ **35.** $-4(3)(-2)$ **36.** $-4(5)(-6)$

37. $-1(-2)(-3)$ **38.** $-2(-3)(-4)$

Use the definition of exponents to expand each of the following expressions. Then multiply according to the rule for multiplication.

39. a. $(-4)^2$ **40. a.** $(-5)^2$ **41. a.** $(-5)^3$

   b. -4^2 **b.** -5^2 **b.** -5^3

42. a. $(-4)^3$ **43. a.** $(-2)^4$ **44. a.** $(-1)^4$

   b. -4^3 **b.** -2^4 **b.** -1^4

Complete the following tables. Remember, if $x = -5$, then $x^2 = (-5)^2 = 25$.

45.

Number x	Square x^2
-3	
-2	
-1	
0	
1	
2	
3	

46.

Number x	Cube x^3
-3	
-2	
-1	
0	
1	
2	
3	

47.

First Number x	Second Number y	Their Product xy
6	2	
6	1	
6	0	
6	-1	
6	-2	

48.

First Number x	Second Number y	Their Product xy
7	4	
7	2	
7	0	
7	-2	
7	-4	

49.

First Number a	Second Number b	Their Product ab
−5	3	
−5	2	
−5	1	
−5	0	
−5	−1	
−5	−2	
−5	−3	

50.

First Number a	Second Number b	Their Product ab
−9	6	
−9	4	
−9	2	
−9	0	
−9	−2	
−9	−4	
−9	−6	

Use the rule for order of operations along with the rules for addition, subtraction, and multiplication to simplify each of the following expressions.

51. $4(-3 + 2)$

52. $7(-6 + 3)$

53. $-10(-2 - 3)$

54. $-5(-6 - 2)$

55. $-3 + 2(5 - 3)$

56. $-7 + 3(6 - 2)$

57. $-7 + 2[-5 - 9]$

58. $-8 + 3[-4 - 1]$

59. $2(-5) + 3(-4)$

60. $6(-1) + 2(-7)$

61. $3(-2)4 + 3(-2)$

62. $2(-1)(-3) + 4(-6)$

63. $(8 - 3)(2 - 7)$

64. $(9 - 3)(2 - 6)$

65. $(2 - 5)(3 - 6)$

66. $(3 - 7)(2 - 8)$

67. $3(5 - 8) + 4(6 - 7)$

68. $-2(8 - 10) + 3(4 - 9)$

69. $-3(4 - 7) - 2(-3 - 2)$

70. $-5(-2 - 8) - 4(6 - 10)$

71. $3(-2)(6 - 7)$

72. $4(-3)(2 - 5)$

73. Find the product of -3, -2, and -1.

74. Find the product of -7, -1, and 0.

75. What number do you multiply by -3 to get 12?

76. What number do you multiply by -7 to get -21?

77. Subtract -3 from the product of -5 and 4.

78. Subtract 5 from the product of -8 and 1.

Applying the Concepts

79. Day Trading Larry is buying and selling stock from his home computer. He owns 100 shares of Company A and 50 shares of Company B. Suppose that in one day, those stocks had the gain and loss shown in the table below. What was Larry's net gain or loss for the day on those two stocks?

Stock	Number Of Shares	Gain/Loss
Company A	100	-2
Company B	50	$+8$

80. Stock Gain/Loss Amy owns stock that she keeps in her retirement account. She owns 200 shares of Apple Computer and 100 shares of Gap Inc. Suppose that in one month those stocks had the gain and loss shown in the table below. What was Amy's net gain or loss for the month on those two stocks?

Stock	Number Of Shares	Gain/Loss
Apple	200	$+14$
Gap	100	-5

81. Temperature Change A hot-air balloon is rising to its cruising altitude. Suppose the air temperature around the balloon drops 4 degrees each time the balloon rises 1,000 feet. What is the net change in air temperature around the balloon as it rises from 2,000 feet to 6,000 feet?

82. Temperature Change A small airplane is rising to its cruising altitude. Suppose the air temperature around the plane drops 4 degrees each time the plane increases its altitude by 1,000 feet. What is the net change in air temperature around the plane as it rises from 5,000 feet to 12,000 feet?

Baseball Major league baseball has various player awards at the end of the year. One of them is the Rolaids Relief Man of the Year. To compute the Relief Man standings, points are gained or taken away based on the number of wins, losses, saves, and blown saves a relief pitcher has at the end of the year. The pitcher with the most Rolaids Points is the Rolaids Relief Man of the Year. The formula $P = 3s + 2w + t - 2l - 2b$ gives the number of Rolaids points a pitcher earns, where s = saves, w = wins, t = tough saves, l = losses, and b = blown saves. Use this formula to complete the following tables.

83. National League 2010

Pitcher, Team	W	L	Saves	Tough Saves	Blown Saves	Rolaids Points
Heath Bell, San Diego	6	1	47	3	5	
Brian Wilson, San Francisco	3	3	48	5	7	
Carlos Marmol, Chicago	2	3	38	5	5	
Billy Wagner, Atlanta	7	2	37	7	0	
Francisco Cordero, Cincinati	6	5	40	8	0	

84. American League 2010

Pitcher, Team	W	L	Saves	Tough Saves	Blown Saves	Rolaids Points
Rafael Soriano, Tampa Bay	3	2	45	0	3	
Joakim Soria, Kansas City	1	2	43	0	3	
Neftali Feliz, Texas	4	3	40	1	3	
Jonathan Papelbon, Boston	5	7	37	1	8	
Kevin Gregg, Toronto	2	6	37	1	6	

Golf One way to give scores in golf is in relation to par, the number of strokes considered necessary to complete a hole or course at the expert level. Scoring this way, if you complete a hole in one stroke less than par, your score is -1, which is called a *birdie*. If you shoot 2 under par, your score is -2, which is called an *eagle*. Shooting 1 over par is a score of $+1$, which is a *bogie*. A *double bogie* is 2 over par, and results in a score of $+2$.

85. Sergio Garcia's Scorecard The table below shows the scores Sergio Garcia had on the first round of a PGA tournament. Fill in the last column by multiplying each value by the number of times it occurs. Then add the numbers in the last column to find the total. If par for the course was 72, what was Sergio Garcia's score?

	Value	Number	Product
Eagle	-2	0	
Birdie	-1	7	
Par	0	7	
Bogie	$+1$	3	
Double Bogie	$+2$	1	
		Total:	

86. Karrie Webb's Scorecard The table below shows the scores Karrie Webb had on the final round of an LPGA Standard Register Ping Tournament. Fill in the last column by multiplying each value by the number of times it occurs. Then add the numbers in the last column to find the total. If par for the course was 72, what was Karrie Webb's score?

	Value	Number	Product
Eagle	-2	1	
Birdie	-1	5	
Par	0	8	
Bogie	$+1$	3	
Double Bogie	$+2$	1	
		Total:	

Estimating

Work the next problems mentally, without pencil and paper or a calculator.

87. The product $-32(-522)$ is closest to which of the following numbers?

    a. 15,000   **b.** -500 **c.** $-1,500$ **d.** $-15,000$

88. The product $32(-522)$ is closest to which of the following numbers?

    a. 15,000   **b.** -500 **c.** $-1,500$ **d.** $-15,000$

89. The product $-47(470)$ is closest to which of the following numbers?

    a. 25,000   **b.** 420   **c.** $-2,500$ **d.** $-25,000$

90. The product $-47(-470)$ is closest to which of the following numbers?

    a. 25,000   **b.** 420   **c.** $-2,500$ **d.** $-25,000$

91. The product $-222(-987)$ is closest to which of the following numbers?

    a. 200,000   **b.** 800   **c.** -800 **d.** $-1,200$

92. The sum $-222 + (-987)$ is closest to which of the following numbers?

    a. 200,000   **b.** 800   **c.** -800 **d.** $-1,200$

93. The difference $-222 - (-987)$ is closest to which of the following numbers?

    a. 200,000   **b.** 800   **c.** -800 **d.** $-1,200$

94. The difference $-222 - 987$ is closest to which of the following numbers?

    a. 200,000   **b.** 800   **c.** -800 **d.** $-1,200$

Getting Ready for the Next Section

Perform the indicated operations.

95. $35 \div 5$

96. $32 \div 4$

97. $\dfrac{20}{4}$

98. $\dfrac{30}{5}$

99. $12 - 17$

100. $7 - 11$

101. $\dfrac{6(3)}{2}$

102. $\dfrac{8(5)}{4}$

103. $80 \div 10 \div 2$

104. $80 \div 2 \div 10$

105. $\dfrac{15 + 5(4)}{17 - 12}$

106. $\dfrac{20 + 6(2)}{11 - 7}$

107. $4(10^2) + 20 \div 4$

108. $3(4^2) + 10 \div 5$

Find the Mistake

Each problem below contains a mistake. Circle the mistake and write the correct number(s) or word(s) on the line provided.

1. Writing the problem $6(-4)$ as repeated addition gives us $(-4)(-4)(-4)(-4)(-4)(-4)$.

_____

2. Multiplying a negative by a positive and then by another negative will give us a negative answer. _____

3. The problem $(-5)^3$ can also be written as $-3(5)(5)(5)$. _____

4. Work for the problem $-5(3 - 6) - 2(3 + 1)$ looks like the following:

$$-5(3 - 6) - 2(3 + 1) = -5(-3) - 2(4) \quad \text{_____}$$
$$= 15 + 8 \quad \text{_____}$$
$$= 23 \quad \text{_____}$$

1.5 Division with Negative Numbers

©iStockphoto.com/buzbuzzer

Suppose four friends invest equal amounts of money in a moving truck to start a small business. After 2 years the truck has dropped $10,000 in value. If we represent this change with the number $-$10,000, then the loss to each of the four partners can be found with division.

$$(-\$10{,}000) \div 4 = -\$2{,}500$$

From this example, it seems reasonable to assume that a negative number divided by a positive number will give a negative answer. In this section, we will work division problems involving negative numbers.

A Division with Any Number

To cover all the possible situations we can encounter with division of negative numbers, we use the relationship between multiplication and division. If we let n be the answer to the problem $12 \div (-2)$, then we know that

$$12 \div (-2) = n \quad \text{and} \quad -2(n) = 12$$

From our work with multiplication, we know that n must be -6 in the multiplication problem above, because -6 is the only number we can multiply -2 by to get 12. Because of the relationship between the two problems above, it must be true that 12 divided by -2 is -6.

The following pairs of problems show more quotients of positive and negative numbers. In each case, the multiplication problem on the right justifies the answer to the division problem on the left.

$6 \div 3 = 2$	because	$3(2) = 6$
$6 \div (-3) = -2$	because	$-3(-2) = 6$
$-6 \div 3 = -2$	because	$3(-2) = -6$
$-6 \div (-3) = 2$	because	$-3(2) = -6$

These results can be used to write the rule for division with negative numbers.

RULE Division with Negative Numbers

To divide two numbers, we divide their absolute values.

1. The answer is positive if both the original numbers have the same sign. That is, the quotient of two numbers with the same signs is positive.

2. The answer is negative if the original two numbers have different signs. That is, the quotient of two numbers with different signs is negative.

Video Examples
Section 1.5

Practice Problems

1. Divide.
  a. $10 \div 5$
  b. $-10 \div 5$
  c. $10 \div (-5)$
  d. $-10 \div (-5)$

2. Simplify.
  a. $\frac{12}{4}$
  b. $\frac{-12}{4}$
  c. $\frac{12}{-4}$
  d. $\frac{-12}{-4}$

EXAMPLE 1   Divide.
a. $12 \div 4$ **b.** $-12 \div 4$ **c.** $12 \div (-4)$ **d.** $-12 \div (-4)$

Solution

a. $12 \div 4 = 3$ Like signs; positive answer
b. $-12 \div 4 = -3$ Unlike signs; negative answer
c. $12 \div (-4) = -3$ Unlike signs; negative answer
d. $-12 \div (-4) = 3$ Like signs; positive answer

EXAMPLE 2   Simplify.
a. $\frac{20}{5}$ **b.** $\frac{-20}{5}$ **c.** $\frac{20}{-5}$ **d.** $\frac{-20}{-5}$

Solution

a. $\frac{20}{5} = 4$ Like signs; positive answer

b. $\frac{-20}{5} = -4$ Unlike signs; negative answer

c. $\frac{20}{-5} = -4$ Unlike signs; negative answer

d. $\frac{-20}{-5} = 4$ Like signs; positive answer

From the examples we have done so far, we can make the following generalization about quotients that contain negative signs:

If a and b are numbers and b is not equal to 0, then

$$\frac{-a}{b} = \frac{a}{-b} = -\frac{a}{b} \quad \text{and} \quad \frac{-a}{-b} = \frac{a}{b}$$

The last examples in this section involve more than one operation. We use the rules developed previously in this chapter and the rule for order of operations to simplify each.

3. Simplify: $\dfrac{-12 + 2\,(-3)}{10 - 16}$.

EXAMPLE 3   Simplify: $\dfrac{-15 + 5(-4)}{12 - 17}$.

Solution   Simplifying above and below the fraction bar, we have

$$\frac{-15 + 5(-4)}{12 - 17} = \frac{-15 + (-20)}{-5} = \frac{-35}{-5} = 7$$

4. Simplify: $-3(5^2) + 16 \div (-2)$.

EXAMPLE 4   Simplify: $-4(10^2) + 20 \div (-4)$.

Solution   Applying the rule for order of operations, we have

$$-4(10^2) + 20 \div (-4) = -4(100) + 20 \div (-4) \quad \text{Exponents first}$$
$$= -400 + (-5) \quad \text{Multiply and divide.}$$
$$= -405 \quad \text{Add.}$$

GETTING READY FOR CLASS

After reading through the preceding section, respond in your own words and in complete sentences.

A. Write a multiplication problem that is equivalent to the division problem $-12 \div 4 = -3$.

B. Write a multiplication problem that is equivalent to the division problem $-12 \div (-4) = 3$.

C. If two numbers have the same sign, then their quotient will have what sign?

D. Dividing a negative number by 0 always results in what kind of expression?

Answers

1. a. 2   **b.** -2 **c.** -2 **d.** 2
2. a. 3   **b.** -3 **c.** -3 **d.** 3
3. 3
4. -83

Vocabulary Review

Choose the correct words to fill in the blanks below.

same   negative   absolute values   different

1. To divide two numbers, we divide their _____.

2. The quotient of two numbers with the _____ sign is positive.

3. The quotient of two numbers with _____ signs is negative.

4. For a quotient that contains a _____ sign, $\dfrac{-a}{-b} = \dfrac{a}{b}$.

Problems

A Find each of the following quotients. (Divide.)

1. $-15 \div 5$

2. $15 \div (-3)$

3. $20 \div (-4)$

4. $-20 \div 4$

5. $-30 \div (-10)$

6. $-50 \div (-25)$

7. $\dfrac{-14}{-7}$

8. $\dfrac{-18}{-6}$

9. $\dfrac{12}{-3}$

10. $\dfrac{12}{-4}$

11. $-22 \div 11$

12. $-35 \div 7$

13. $\dfrac{0}{-3}$

14. $\dfrac{0}{-5}$

15. $125 \div (-25)$

16. $-144 \div (-9)$

Complete the following tables.

17.

First number a	Second number b	The quotient of a and b $\dfrac{a}{b}$
100	-5	
100	-10	
100	-25	
100	-50	

18.

First number a	Second number b	The quotient of a and b $\dfrac{a}{b}$
24	-4	
24	-3	
24	-2	
24	-1	

19.

First number a	Second number b	The quotient of a and b $\dfrac{a}{b}$
-100	-5	
-100	5	
100	-5	
100	5	

20.

First number a	Second number b	The quotient of a and b $\dfrac{a}{b}$
-24	-2	
-24	-4	
-24	-6	
-24	-8	

Use any of the rules developed in this chapter and the rule for order of operations to simplify each of the following expressions as much as possible.

21. $\dfrac{4(-7)}{-28}$

22. $\dfrac{6(-3)}{-18}$

23. $\dfrac{-3(-10)}{-5}$

24. $\dfrac{-4(-12)}{-6}$

25. $\dfrac{2(-3)}{6-3}$

26. $\dfrac{2(-3)}{3-6}$

27. $\dfrac{4-8}{8-4}$

28. $\dfrac{9-5}{5-9}$

29. $\dfrac{2(-3)+10}{-4}$

30. $\dfrac{7(-2)-6}{-10}$

31. $\dfrac{2+3(-6)}{4-12}$

32. $\dfrac{3+9(-1)}{5-7}$

33. $\dfrac{6(-7)+3(-2)}{20-4}$

34. $\dfrac{9(-8)+5(-1)}{12-1}$

35. $\dfrac{3(-7)(-4)}{6(-2)}$

36. $\dfrac{-2(4)(-8)}{(-2)(-2)}$

37. $(-5)^2 + 20 \div 4$

38. $6^2 + 36 \div 9$

39. $100 \div (-5)^2$

40. $400 \div (-4)^2$

41. $-100 \div 10 \div 2$

42. $-500 \div 50 \div 10$

43. $-100 \div (10 \div 2)$

44. $-500 \div (50 \div 10)$

45. $(-100 \div 10) \div 2$

46. $(-500 \div 50) \div 10$

47. Find the quotient of -25 and 5.

48. Find the quotient of -38 and -19.

49. What number do you divide by -5 to get -7?

50. What number do you divide by 6 to get -7?

51. Subtract -3 from the quotient of 27 and 9.

52. Subtract -7 from the quotient of -72 and -9.

Estimating

Work Problems 53–60 mentally, without pencil and paper or a calculator.

53. Is $397 \div (-401)$ closer to 1 or -1?

54. Is $-751 \div (-749)$ closer to 1 or -1?

55. The quotient $-121 \div 27$ is closest to which of the following numbers?
   a. -150 **b.** -100 **c.** -4 **d.** 6

56. The quotient $1,000 \div (-337)$ is closest to which of the following numbers?
   a. 663   **b.** -3 **c.** -30 **d.** -663

57. Which number is closest to the sum $-151 + (-49)$?
   a. -200 **b.** -100 **c.** 3   **d.** 7,500

58. Which number is closest to the difference $-151 - (-49)$?
   a. -200 **b.** -100 **c.** 3   **d.** 7,500

59. Which number is closest to the product $-151(-49)$?
   a. -200 **b.** -100 **c.** 3   **d.** 7,500

60. Which number is closest to the quotient $-151 \div (-49)$?
   a. -200 **b.** -100 **c.** 3   **d.** 7,500

Applying the Concepts

61. Temperature Line Graph The table below gives the low temperature for each day of one week in White Bear Lake, Minnesota. Draw a line graph of the information in the table.

Low temperatures in White Bear Lake, Minnesota

Day	Temperature
Monday	10°F
Tuesday	8°F
Wednesday	−5°F
Thursday	−3°F
Friday	−8°F
Saturday	5°F
Sunday	7°F

62. Temperature Line Graph The table below gives the low temperature for each day of one week in Fairbanks, Alaska. Draw a line graph of the information in the table.

Low temperatures in Fairbanks, Alaska

Day	Temperature
Monday	−26°F
Tuesday	−5°F
Wednesday	9°F
Thursday	12°F
Friday	3°F
Saturday	−15°F
Sunday	−20°F

Getting Ready for the Next Section

Apply the distributive property to each expression.

63. $5(3 + 7)$

64. $8(4 + 2)$

Simplify.

65. 6^2

66. 12^2

67. $100(75)$

68. $100(53)$

69. $2(100) + 2(75)$

70. $2(100) + 2(53)$

71. a. $4 + 3$

b. $-5 + 7$

c. $8 - 1$

d. $-4 - 2$

e. $3 - 7$

72. a. $5 + 2$

b. $-6 + 7$

c. $9 - 1$

d. $-5 - 3$

e. $2 - 5$

Find the Mistake

Write true or false for each of the following sentences. If false, circle the mistake and write the correct word on the line provided.

1. True or False: Dividing a negative number by a positive number will give a positive number for an answer.

_____

2. True or False: Dividing two numbers with like signs will give a positive answer. _____

The work for the problems below contains mistakes. Circle where the mistake first occurs and write the correct work on the lines provided.

3. Simplify $\dfrac{-6(-6-2)}{-11-1}$.

$$\dfrac{-6(-6-2)}{-11-1} = \dfrac{-6(-4)}{-10} = \dfrac{24}{-10} = \dfrac{-12}{5}$$ _____

4. Simplify $[(-5)(5)-20] \div -3^2$.

$$[(-5)(5)-20] \div -3^2 = -45 \div 9$$ _____

$$= -5$$ _____

A Simplify expressions containing variables.

B Simplify algebraic expressions with similar terms.

C Solve application problems involving algebraic expressions.

KEY WORDS

algebraic expression

similar terms

**Video Examples
Section 1.6**

Suppose you are building a rectangular dog run and want to find the size rectangle that will enclose the largest territory. We will use 24 yards of fencing to enclose the run. The diagram below shows six dog runs, each of which has a perimeter of 24 yards. Notice how the length decreases as the width increases.

Dog runs with a perimeter = 24 yards

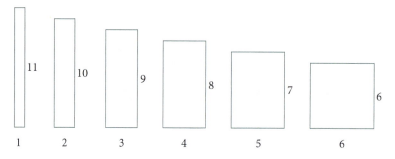

Since area is length times width, we can build a table and a line graph that show how the area changes as we change the width of the dog run.

Area enclosed by rectangle of perimeter 24 yards

Width (yards)	Area (square yards)
1	11
2	20
3	27
4	32
5	35
6	36

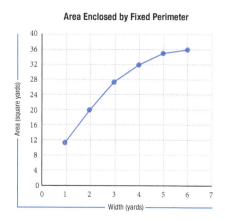

In this section, we want to simplify expressions containing variables—that is, algebraic expressions.

A Expressions with Variables

To begin, let's review how we use the associative properties for addition and multiplication to simplify expressions.

Consider the expression $4(5x)$. We can apply the associative property of multiplication to this expression to change the grouping so that the 4 and the 5 are grouped together, instead of the 5 and the x. Here's how it looks:

$$4(5x) = (4 \cdot 5)x \quad \text{Associative property}$$
$$= 20x \quad \text{Multiply: } 4 \cdot 5 = 20.$$

We have simplified the expression to $20x$, which in most cases in algebra will be easier to work with than the original expression.

Here are some examples:

Practice Problems

EXAMPLE 1 Simplify: $-2(5x)$.

Solution

$$-2(5x) = (-2 \cdot 5)x \quad \text{Associative property}$$
$$= -10x \quad \text{The product of } -2 \text{ and } 5 \text{ is } -10.$$

1. Simplify: $-3(4x)$.

EXAMPLE 2 Simplify: $3(-4y)$.

Solution

$$3(-4y) = [3(-4)]y \quad \text{Associative property}$$
$$= -12y \quad 3 \text{ times } -4 \text{ is } -12.$$

2. Simplify: $2(-5y)$.

We can use the associative property of addition to simplify expressions also.

EXAMPLE 3 Simplify: $(2x + 5) + 10$.

Solution

$$(2x + 5) + 10 = 2x + (5 + 10) \quad \text{Associative property}$$
$$= 2x + 15 \quad \text{Add.}$$

3. Simplify: $(3x + 4) + 12$.

Previously, we introduced the distributive property. In symbols, it looks like this:

$$a(b + c) = ab + ac$$

Because subtraction is defined as addition of the opposite, the distributive property holds for subtraction as well as addition. That is,

$$a(b - c) = ab - ac$$

We say that multiplication distributes over addition and subtraction. Here are some examples that review how the distributive property is applied to expressions that contain variables.

EXAMPLE 4 Simplify: $2(a - 3)$.

Solution

$$2(a - 3) = 2(a) - 2(3) \quad \text{Distributive property}$$
$$= 2a - 6 \quad \text{Multiply.}$$

4. Simplify: $3(n - 4)$.

Here are some examples that use a combination of the associative property and the distributive property.

Answers
1. $-12x$
2. $-10y$
3. $3x + 16$
4. $3n - 12$

5. Simplify: $2(3n + 4)$.

EXAMPLE 5   Simplify: $4(5x + 3)$.
Solution

$$
\begin{aligned}
4(5x + 3) &= 4(5x) + 4(3) && \text{Distributive property} \\
&= (4 \cdot 5)x + 4(3) && \text{Associative property} \\
&= 20x + 12 && \text{Multiply.}
\end{aligned}
$$

6. Simplify: $4(8x + 4y)$.

EXAMPLE 6   Simplify: $5(2x + 3y)$.
Solution

$$
\begin{aligned}
5(2x + 3y) &= 5(2x) + 5(3y) && \text{Distributive property and} \\
& && \text{associative property} \\
&= 10x + 15y && \text{Multiply.}
\end{aligned}
$$

We can also use the distributive property to simplify expressions like $4x + 3x$. Because multiplication is a commutative operation, we can also rewrite the distributive property like this:

$$ b \cdot a + c \cdot a = (b + c)a $$

Applying the distributive property in this form to the expression $4x + 3x$, we have

$$
\begin{aligned}
4x + 3x &= (4 + 3)x && \text{Distributive property} \\
&= 7x && \text{Add.}
\end{aligned}
$$

B Similar Terms

Expressions like $4x$ and $3x$ are called *similar terms* because the variable parts are the same. Some other examples of similar terms are $5y$ and $-6y$ and the terms $7a$, $-13a$, $\frac{3}{4}a$. To simplify an algebraic expression (an expression that involves both numbers and variables), we combine similar terms by applying the distributive property. Table 1 shows several pairs of similar terms and how they can be combined using the distributive property.

Table 1

Original Expression		Apply Distributive Property		Simplified Expression
$4x + 3x$	$=$	$(4 + 3)x$	$=$	$7x$
$7a + a$	$=$	$(7 + 1)a$	$=$	$8a$
$-5x + 7x$	$=$	$(-5 + 7)x$	$=$	$2x$
$8y - y$	$=$	$(8 - 1)y$	$=$	$7y$
$-4a - 2a$	$=$	$(-4 - 2)a$	$=$	$-6a$
$3x - 7x$	$=$	$(3 - 7)x$	$=$	$-4x$

As you can see from the table, the distributive property can be applied to any combination of positive and negative terms so long as they are similar terms.

7. Simplify: $14y + 2y$.

EXAMPLE 7   Simplify: $12x + 3x$.
Solution

$$
\begin{aligned}
12x + 3x &= (12 + 3)x && \text{Distributive property} \\
&= 15x && \text{Add.}
\end{aligned}
$$

Answers
5. $6n + 8$
6. $32x + 16y$
7. $16y$

EXAMPLE 8   Simplify: $4a - 12a$.
Solution

$$4a - 12a = (4 - 12)a \qquad \text{Distributive property}$$
$$= -8a \qquad \text{Subtract.}$$

8. Simplify: $3n - 9n$.

C Algebraic Expressions Representing Area and Perimeter

Below are a square with a side of length s and a rectangle with a length of l and a width of w. The table that follows the figures gives the formulas for the area and perimeter of each.

Square

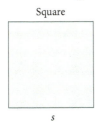

s

Rectangle

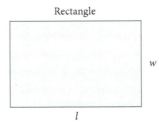

w

l

	Square	Rectangle
Area A	s^2	lw
Perimeter P	$4s$	$2l + 2w$

EXAMPLE 9   Find the area and perimeter of a square with a side 6 inches long.
Solution   Substituting 6 for s in the formulas for area and perimeter of a square, we have

$$\text{Area} = A = s^2 = 6^2 = 36 \text{ square inches}$$
$$\text{Perimeter} = P = 4s = 4(6) = 24 \text{ inches}$$

9. Find the area and perimeter of a square with a side of 8 inches.

EXAMPLE 10   A soccer field is 100 yards long and 75 yards wide. Find the area and perimeter.

75 yd

100 yd

Solution   Substituting 100 for l and 75 for w in the formulas for area and perimeter of a rectangle, we have

$$\text{Area} = A = lw = 100(75) = 7{,}500 \text{ square yards}$$
$$\text{Perimeter} = P = 2l + 2w = 2(100) + 2(75) = 200 + 150 = 350 \text{ yards}$$

10. A NFL football field is 100 yards long and 53.3 yards wide. Find the area and perimeter.

GETTING READY FOR CLASS

After reading through the preceding section, respond in your own words and in complete sentences.

A. Can two rectangles with the same perimeter have different areas?

B. Without actually multiplying, how do you apply the associative property to the expression $4(5x)$?

C. What are similar terms?

D. Explain why $2a - a$ is a, rather than 1.

Answers
8. $-6n$
9. Area $= 64$ in.^{2}, Perimeter $= 32$ in.
10. Area $= 5{,}330$ yd^2, Perimeter $= 306.6$ yd^2

Vocabulary Review

Choose the correct words to fill in the blanks below.

commutative   similar terms   associative   distributive

1. Using the _____ property of multiplication, we can change $3(6x)$ to $18x$.
2. Applying the _____ property to the expression $4(x - 2)$ gives $4x - 8$.
3. Since multiplication is a _____ operation, we can rewrite the distributive property as $a \cdot b + a \cdot c = a(b + c)$.
4. If the variable parts of two expressions are the same, then the expressions are called _____.

Problems

A Apply the associative property to each expression, and then simplify the result.

1. $5(4a)$ **2.** $8(9a)$ **3.** $6(8a)$ **4.** $3(2a)$

5. $-6(3x)$ **6.** $-2(7x)$ **7.** $-3(9x)$ **8.** $-4(6x)$

9. $5(-2y)$ **10.** $3(-8y)$ **11.** $6(-10y)$ **12.** $5(-5y)$

13. $2 + (3 + x)$ **14.** $9 + (6 + x)$ **15.** $5 + (8 + x)$ **16.** $3 + (9 + x)$

17. $4 + (6 + y)$ **18.** $2 + (8 + y)$ **19.** $7 + (1 + y)$ **20.** $4 + (1 + y)$

21. $(5x + 2) + 4$ **22.** $(8x + 3) + 10$ **23.** $(6y + 4) + 3$ **24.** $(3y + 7) + 8$

25. $(12a + 2) + 19$ **26.** $(6a + 3) + 14$ **27.** $(7x + 8) + 20$ **28.** $(14x + 3) + 15$

Apply the distributive property to each expression, and then simplify.

29. $7(x + 5)$

30. $8(x + 3)$

31. $6(a - 7)$

32. $4(a - 9)$

33. $2(x - y)$

34. $5(x - a)$

35. $4(5 + x)$

36. $8(3 + x)$

37. $3(2x + 5)$

38. $8(5x + 4)$

39. $6(3a + 1)$

40. $4(8a + 3)$

41. $2(6x - 3y)$

42. $7(5x - y)$

43. $5(7 - 4y)$

44. $8(6 - 3y)$

B Use the distributive property to combine similar terms.

45. $3x + 5x$

46. $7x + 8x$

47. $3a + a$

48. $8a + a$

49. $-2x + 6x$

50. $-3x + 9x$

51. $6y - y$

52. $3y - y$

53. $-8a - 2a$

54. $-7a - 5a$

55. $4x - 9x$

56. $5x - 11x$

Applying the Concepts

C Area and Perimeter Find the area and perimeter of each square if the length of each side is as given below.

57. $s = 6$ feet

58. $s = 14$ yards

59. $s = 9$ inches

60. $s = 15$ meters

Area and Perimeter Find the area and perimeter for a rectangle if the length and width are as given below.

61. $l = 20$ inches, $w = 10$ inches

62. $l = 40$ yards, $w = 20$ yards

63. $l = 25$ feet, $w = 12$ feet

64. $l = 210$ meters, $w = 120$ meters

Temperature Scales Recall that in the metric system, the scale we use to measure temperature is the Celsius scale. On this scale water boils at 100 degrees and freezes at 0 degrees. When we write 100 degrees measured on the Celsius scale, we use the notation 100°C, which is read "100 degrees Celsius." If we know the temperature in degrees Fahrenheit, we can convert to degrees Celsius by using the formula

$$C = \frac{5(F - 32)}{9}$$

where F is the temperature in degrees Fahrenheit. Use this formula to find the temperature in degrees Celsius for each of the following Fahrenheit temperatures.

65. 68°F   **66.** 59°F   **67.** 41°F   **68.** 23°F   **69.** 14°F   **70.** 32°F

Find the Mistake

Each problem below contains a mistake. Circle the mistake and write the correct number(s) or word(s) on the line provided.

1. Simplifying $-6(3x)$ gives us $18x$. _____

2. To simplify the problem $4(z - 3)$, we use the associative property to get $4z - 12$. _____

3. To simplify $-2x + 5x + 4$, combine the similar terms $-2x$ and $5x$ to get $-10x + 4$. _____

4. A rectangular hockey rink that is $12x$ meters in length and $4x$ meters in width has a perimeter of $48x$ meters.

_____

Exponential Decay

In 2011, Japan was the victim of a devastating earthquake and tsunami. The earthquake caused a multitude of nuclear meltdowns, most notably at the Fukushima I Nuclear Power Plant. Cesium-137 is a radioactive isotope created by nuclear fission. The fallout from the Fukushima I Power Plant contained dangerously high levels of Cesium-137, which caused widespread evacuations of the surrounding areas. This isotope has a half life of approximately 30.17 years. As Cesium-137 undergoes exponential decay, the area surrounding the nuclear disaster will slowly become less radioactive, therefore, less dangerous for humans. Research the nuclear disaster in Japan, as well as the importance of exponents in the half-life of Cesium-137. When will it be safe for residents to return to their homes near the power plant?

Chapter 1 Summary

Absolute Value [1.1]

1. $|3| = 3$ and $|-3| = 3$

The absolute value of a number is its distance from 0 on the number line. It is the numerical part of a number. The absolute value of a number is never negative.

Opposites [1.1]

2. $-(5) = -5$ and $-(-5) = 5$

Two numbers are called opposites if they are the same distance from 0 on the number line but in opposite directions from 0. The opposite of a positive number is a negative number, and the opposite of a negative number is a positive number.

Addition of Positive and Negative Numbers [1.2]

3. $3 + 5 = 8$
$-3 + (-5) = -8$

1. To add two numbers with *the same sign*: Simply add absolute values and use the common sign. If both numbers are positive, the answer is positive. If both numbers are negative, the answer is negative.

2. To add two numbers with *different signs*: Subtract the smaller absolute value from the larger absolute value. The answer has the same sign as the number with the larger absolute value.

Subtraction [1.3]

4. $3 - 5 = 3 + (-5) = -2$
$-3 - 5 = -3 + (-5) = -8$
$3 - (-5) = 3 + 5 = 8$
$-3 - (-5) = -3 + 5 = 2$

Subtracting a number is equivalent to adding its opposite. If a and b represent numbers, then subtraction is defined in terms of addition as follows:

$$a - b = a + (-b)$$

Subtraction   Addition of the opposite

Exponents [1.4]

5. $2^3 = 2 \cdot 2 \cdot 2 = 8$
$5^0 = 1$
$3^1 = 3$

In the expression 2^3, 2 is the *base* and 3 is the *exponent*. An exponent is a shorthand notation for repeated multiplication. The exponent 0 is a special exponent. Any nonzero number to the 0 power is 1.

Multiplication with Positive and Negative Numbers [1.4]

6. $3(5) = 15$
$3(-5) = -15$
$-3(5) = -15$
$-3(-5) = 15$

To multiply two numbers, multiply their absolute values.

1. The answer is *positive* if both numbers have the same sign.

2. The answer is *negative* if the numbers have different signs.

Division with Positive and Negative Numbers [1.5]

The rule for assigning the correct sign to the answer in a division problem is the same as the rule for multiplication. That is, like signs give a positive answer, and unlike signs give a negative answer.

7. $\dfrac{12}{4} = 3$

$\dfrac{-12}{4} = -3$

$\dfrac{12}{-4} = -3$

$\dfrac{-12}{-4} = 3$

Simplifying Expressions [1.6]

We simplify algebraic expressions by applying the commutative, associative, and distributive properties.

8. Simplify.
a. $-2(5x) = (-2 \cdot 5)x = -10x$
b. $4(2a - 8) = 4(2a) - 4(8)$
$= 8a - 32$

Combining Similar Terms [1.6]

We combine similar terms by applying the distributive property.

9. Combine similar terms.
a. $5x + 7x = (5 + 7)x = 12x$
b. $2y - 8y = (2 - 8)y = -6y$

CHAPTER 1 Review

Give the opposite and absolute value of each number. [1.1]

1. $-\dfrac{3}{4}$

2. 16

Place an inequality symbol ($<$ or $>$) between each pair of numbers so that the resulting statement is true. [1.1]

3. $-8 \quad -1$

4. $|-9| \quad |3|$

Simplify each expression. [1.1]

5. $-(-6)$

6. $-|-3|$

Perform the indicated operations. [1.2-1.5]

7. $6 + (-4)$

8. $-76 - (-19)$

9. $(-12)(-7)$

10. $-42 \div 3$

Simplify the following expressions as much as possible. [1.2, 1.3, 1.4, 1.5]

11. $(-5)^2$

12. $(-4)^3$

13. $(-6)(5) - (3)(-7)$

14. $(9 - 2)(2 - 17)$

15. $\dfrac{24 - 3(-5)}{15 - 9}$

16. $\dfrac{-6(3) + 4(-3)}{9 - 6}$

17. $(-2)^4 - [12 + (-5)]$

18. $(-3)(8) - (9)(4)$

19. $9 - 4(9 - 3^3)$

20. $3(15) + 9(-2)$

21. $\dfrac{-9 - 5(3)}{20 - 4(3)}$

22. $5(6 - 10) + 8^2$

23. $(-4)^3 - 5(3 - 7)$

24. $\dfrac{8(5) + 5(6 - 11)}{3(5 + 2)}$

25. Give the sum of -16 and -57.

26. Subtract -9 and -27.

27. What is the product of -11 and 7?

28. Give the quotient of 96 and -8.

29. What is five times the sum of -9 and 13.

30. Three times the difference of 7 and 19 is increased by 9.

31. Gambling A gambler loses $107 Saturday night and wins $357 on Sunday. Give the gambler's net loss or gain as a negative or positive number.

32. Temperature On Friday, the temperature reaches a high of 47° above 0 and a low of 3° below 0. What is the difference between the high and low temperatures for Friday?

The snapshot shows the number of text messages sent and voice minutes used by different age groups in one month. Use the information to answer the following questions.

33. Write an expression to describe the difference between the number of text messages sent by 35-44 year-olds and 18-24 year-olds. Solve the expression.

34. Write an expression to describe the difference between the number of voice minutes used by 35-44 year-olds and 24-34 year-olds. Solve the expression.

Give the opposite and absolute value of each number. [1.1]

1. 27

2. $-\dfrac{1}{4}$

Place an inequality symbol ($<$ or $>$) between each pair of numbers so that the resulting statement is true. [1.1]

3. $-4 \quad 3$

4. $|4| \quad |-6|$

Simplify each expression. [1.1]

5. $-(-4)$

6. $-|-14|$

Perform the indicated operations. [1.2, 1.3]

7. $9 + (-12)$

8. $-84 - (-48)$

9. $(-8)(-9)$

10. $(63) \div (-3)$

Simplify the following expressions as much as possible. [1.2, 1.3, 1.4, 1.5]

11. $(-3)^3$

12. $(-2)^2$

13. $(-5)(2) - (-6)(4)$

14. $(7 - 4)(9 - 15)$

15. $\dfrac{-6 + 2(-4)}{2 - 5}$

16. $\dfrac{-4(5) + 4(7)}{5 - 7}$

17. $(-3)^3 - (9 + 7)$

18. $(-4)(7) - (-5)(-6)$

19. $5 - 3(7 - 2^3)$

20. $2(-8) + 6(-3)$

21. $\dfrac{-8 + 4(2)}{15 - 9(3)}$

22. $4(2 - 5) + 5^2$

23. $(-3)^2 - 4(7 - 4)$

24. $\dfrac{7(-3) - 4(8 - 5)}{3(9 + 2)}$

25. Give the sum of -9 and -32.

26. Subtract -7 and -24.

27. What is the product of -9 and -4?

28. Give the quotient of 56 and -7.

29. What is three times the sum of 12 and -17.

30. Two times the difference of 6 and 11 is increased by 4.

31. **Gambling** A gambler loses $215 Saturday night and wins $156 on Sunday. Give the gambler's net loss or gain as a negative or positive number.

32. **Temperature** On Friday, the temperature reaches a high of 14° above 0 and a low of 3° below 0. What is the difference between the high and low temperatures for Friday?

The illustration shows the number of animals housed at some of the large zoos in the United States. Use the information to answer the following questions.

SEEING SPOTS AND STRIPES
www.buzzle.com

San Diego Zoo, CA	4000
Columbus Zoo, OH	5000
Phoenix Zoo, AZ	1200
Disney Animal Kingdom, FL	1700
Houston Zoo, TX	1000
Maryland Zoo, MD	2000

* Rounded number of animals per zoo

33. Write an expression to describe the difference between the number of animals at the Maryland Zoo and the Columbus Zoo. Solve the expression.

34. Write an expression to describe the difference between the number of animals at the Houston Zoo and the San Diego Zoo. Solve the expression.

Multiplication and Division of Fractions

The Roman Colosseum in Rome, Italy is a testament to the advancements made in engineering and architecture by the ancient Roman Empire. Initiated by the Roman Emperor Vespasianin in the year 70 AD, it was completed in 80 AD after his death. The next Emperor, Titus, opened the Colosseum to the public. It is said the inaugural ceremony lasted for more than 100 days. The Roman Colosseum is an amphitheater that had the capacity to hold 50,000 people, a huge number at that time in history.

For comparison, the table and bar chart below shows the seating capacity for other coliseums.

Seating Capacity	
Los Angeles Coliseum	134,254
Oakland Coliseum	55,528
Roman Colosseum	50,000
Nassau Veteran's Memorial Coliseum	15,000
The Coliseum at Caesars Palace	4,100
The Coliseum, St. Petersburg, FL	2,000

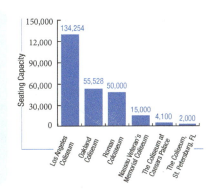

What fraction of the capacity of the Los Angeles Coliseum represents the capacity of the Roman Colosseum? Problems like this one require multiplying fractions, which is one of the topics of this chapter.

Video Examples
Section 2.1

Note   As we mentioned in Chapter 1, when we use a letter to represent a number, or a group of numbers, that letter is called a variable. In the definition for a fraction, we are restricting the numbers that the variable b can represent to numbers other than 0. As you know we want to avoid writing an expression that would imply division by the number 0.

2.1   The Meaning and Properties of Fractions

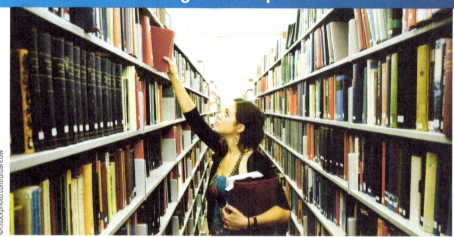

©iStockphoto.com/urbancow

The information in the table below shows enrollment for Cal Poly State University in California. The pie chart was created from the table. Both the table and pie chart use fractions to specify how the students at Cal Poly are distributed among the different schools within the university.

Cal Poly State University Enrollment

School	Fraction Of Students
Agriculture	$\frac{2}{9}$
Architecture and Environmental Design	$\frac{1}{9}$
Business	$\frac{1}{9}$
Engineering	$\frac{5}{18}$
Liberal Arts	$\frac{3}{20}$
Science and Mathematics	$\frac{3}{20}$

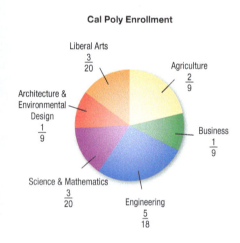

From the table, we see that $\frac{1}{9}$ (one-ninth) of the students are enrolled in the College of Business. This means one out of every nine students at Cal Poly is studying Business. The fraction $\frac{1}{9}$ tells us we have 1 part of 9 equal parts. That is, the students at Cal Poly could be divided into 9 equal groups, so that one of the groups contained all the business students and only business students.

Figure 1 shows a rectangle that has been divided into equal parts, four different ways. The shaded area for each rectangle is $\frac{1}{2}$ the total area.

Now that we have an intuitive idea of the meaning of fractions, here are the more formal definitions and vocabulary associated with fractions.

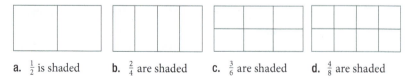

a. $\frac{1}{2}$ is shaded   b. $\frac{2}{4}$ are shaded   c. $\frac{3}{6}$ are shaded   d. $\frac{4}{8}$ are shaded

FIGURE 1

A Identifying Parts of a Fraction

DEFINITION fraction

A *fraction* is any number that can be put in the form $\frac{a}{b}$ (also sometimes written a/b), where a and b are numbers and b is not 0.

Some examples of fractions are

$$\frac{1}{2} \qquad \frac{3}{4} \qquad \frac{7}{8} \qquad \frac{9}{5}$$

One-half   *Three-fourths*   *Seven-eighths*   *Nine-fifths*

DEFINITION terms

For the fraction $\frac{a}{b}$, a and b are called the *terms* of the fraction. More specifically, a is called the *numerator*, and b is called the *denominator.*

EXAMPLE 1 Name the numerator and denominator for each fraction.

a. $\frac{3}{4}$ b. $\frac{a}{5}$ c. 7

Solution

a. The terms of the fraction $\frac{3}{4}$ are 3 and 4. The 3 is called the numerator, and the 4 is called the denominator.

b. The numerator of the fraction $\frac{a}{5}$ is a. The denominator is 5. Both a and 5 are called terms.

c. The number 7 may also be put in fraction form, because it can be written as $\frac{7}{1}$. In this case, 7 is the numerator and 1 is the denominator.

DEFINITION proper and improper fraction

A *proper fraction* is a fraction in which the numerator is less than the denominator. If the numerator is greater than or equal to the denominator, the fraction is called an *improper fraction.*

CLARIFICATION 1: The fractions $\frac{3}{4}$, $\frac{1}{8}$, and $\frac{9}{10}$ are all proper fractions, because in each case the numerator is less than the denominator.

CLARIFICATION 2: The numbers $\frac{9}{5}$, $\frac{10}{10}$, and 6 are all improper fractions, because in each case the numerator is greater than or equal to the denominator. (As we have seen, 6 can be written as $\frac{6}{1}$, in which case 6 is the numerator and 1 is the denominator.)

B Fractions on the Number Line

We can give meaning to the fraction $\frac{2}{3}$ by using a number line. If we take that part of the number line from 0 to 1 and divide it into *three equal parts*, we say that we have divided it into *thirds* (see Figure 2). Each of the three segments is $\frac{1}{3}$ (one third) of the whole segment from 0 to 1.

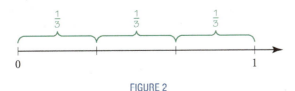

FIGURE 2

Practice Problems

1. Name the numerator and denominator for each fraction.

     a. $\frac{5}{6}$

     b. $\frac{1}{8}$

     c. $\frac{x}{3}$

Answers

1. **a.** Numerator: 5; denominator: 6
   b. Numerator: 1; denominator: 8
   c. Numerator: x; denominator: 3

Note   There are many ways to give meaning to fractions like $\frac{2}{3}$ other than by using the number line. One popular way is to think of cutting a pie into three equal pieces, as shown below. If you take two of the pieces, you have taken $\frac{2}{3}$ of the pie.

Two of these smaller segments together are $\frac{2}{3}$ (two thirds) of the whole segment. And three of them would be $\frac{3}{3}$ (three thirds), or the whole segment, as indicated in Figure 3.

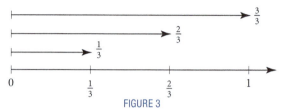

FIGURE 3

Let's do the same thing again with six and twelve equal divisions of the segment from 0 to 1 (see Figure 4).

The same point that we labeled with $\frac{1}{3}$ in Figure 3 is now labeled with $\frac{2}{6}$ and with $\frac{4}{12}$. It must be true then that

$$\frac{4}{12} = \frac{2}{6} = \frac{1}{3}$$

Although these three fractions look different, each names the same point on the number line, as shown in Figure 4. All three fractions have the same value, because they all represent the same number.

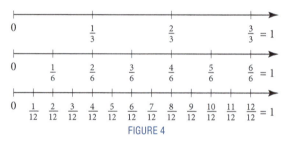

FIGURE 4

C Equivalent Fractions

> **DEFINITION** equivalent fractions
>
> Fractions that represent the same number are said to be *equivalent.* Equivalent fractions may look different, but they must have the same value.

It is apparent that every fraction has many different representations, each of which is equivalent to the original fraction. The next two properties give us a way of changing the terms of a fraction without changing its value.

> **PROPERTY** Multiplication Property for Fractions
>
> If a, b, and c are numbers and b and c are not 0, then it is always true that
>
> $$\frac{a}{b} = \frac{a \cdot c}{b \cdot c}$$
>
> *In words:* If the numerator and the denominator of a fraction are multiplied by the same nonzero number, the resulting fraction is equivalent to the original fraction.

2. Write $\frac{5}{6}$ as an equivalent fraction with a denominator of 30.

EXAMPLE 2   Write $\frac{3}{4}$ as an equivalent fraction with a denominator of 20.

Solution   The denominator of the original fraction is 4. The fraction we are trying to find must have a denominator of 20. We know that if we multiply 4 by 5, we get 20. The multiplication property for fractions indicates that we are free to multiply the denominator by 5 so long as we do the same to the numerator.

$$\frac{3}{4} = \frac{3 \cdot 5}{4 \cdot 5} = \frac{15}{20}$$

The fraction $\frac{15}{20}$ is equivalent to the fraction $\frac{3}{4}$.

Answer

2. $\frac{25}{30}$

EXAMPLE 3   Write $\frac{3}{4}$ as an equivalent fraction with a denominator of $12x$.

Solution   If we multiply 4 by $3x$, we will have $12x$.

$$\frac{3}{4} = \frac{3 \cdot 3x}{4 \cdot 3x} = \frac{9x}{12x}$$

PROPERTY **Division Property for Fractions**

If a, b, and c are integers and b and c are not 0, then it is always true that

$$\frac{a}{b} = \frac{a \div c}{b \div c}$$

In words: If the numerator and the denominator of a fraction are divided by the same nonzero number, the resulting fraction is equivalent to the original fraction.

EXAMPLE 4   Write $\frac{10}{12}$ as an equivalent fraction with a denominator of 6.

Solution   If we divide the original denominator 12 by 2, we obtain 6. The division property for fractions indicates that if we divide both the numerator and the denominator by 2, the resulting fraction will be equal to the original fraction.

$$\frac{10}{12} = \frac{10 \div 2}{12 \div 2} = \frac{5}{6}$$

You know from previous examples that the following is true about fractions containing negative numbers.

If a and b are numbers and b is not equal to 0, then

$$\frac{-a}{b} = \frac{a}{-b} = -\frac{a}{b} \quad \text{and} \quad \frac{-a}{-b} = \frac{a}{b}$$

EXAMPLE 5   Write the following as equivalent fractions without a negative sign.

a. $\frac{-1}{-5}$ **b.** $\frac{-2}{-3}$ **c.** $\frac{-a}{-b}$ **d.** $\frac{-5}{-y}$

Solution   Applying the rules for negative numbers, we have

a. $\frac{-1}{-5} = \frac{1}{5}$ **b.** $\frac{-2}{-3} = \frac{2}{3}$ **c.** $\frac{-a}{-b} = \frac{a}{b}$ **d.** $\frac{-5}{-y} = \frac{5}{y}$

EXAMPLE 6   Write the following as equivalent fractions by moving the negative sign in front of the fraction bar.

a. $\frac{-1}{4}$ **b.** $\frac{3}{-4}$ **c.** $\frac{-a}{b}$ **d.** $\frac{6}{-y}$

Solution   Applying our rules for negative numbers, we have

a. $\frac{-1}{4} = -\frac{1}{4}$ **b.** $\frac{3}{-4} = -\frac{3}{4}$ **c.** $\frac{-a}{b} = -\frac{a}{b}$ **d.** $\frac{6}{-y} = -\frac{6}{y}$

The Number 1 and Fractions

There are two situations involving fractions and the number 1 that occur frequently in mathematics. The first is when the denominator of a fraction is 1. In this case, if we let a represent any number, then

$$\frac{a}{1} = a$$

The second situation occurs when the numerator and the denominator of a fraction are the same nonzero number.

$$\frac{a}{a} = 1$$

---

3. Write $\frac{2}{3}$ as an equivalent fraction with a denominator of $15x$.

4. Write $\frac{20}{25}$ as an equivalent fraction with a denominator of 5.

5. Write the following as equivalent fractions without a negative sign.

  a. $\frac{-6}{-7}$

  b. $\frac{-11}{-3}$

  c. $\frac{-x}{-y}$

  d. $\frac{-a}{-4}$

6. Write the following as equivalent fractions by moving the negative sign in front of the fraction bar.

  a. $\frac{-2}{9}$

  b. $\frac{12}{-13}$

  c. $\frac{x}{-y}$

  d. $\frac{-8}{b}$

Answers

3. $\frac{10x}{15x}$

4. $\frac{4}{5}$

5. a. $\frac{6}{7}$ **b.** $\frac{11}{3}$ **c.** $\frac{x}{y}$ **d.** $\frac{a}{4}$

6. a $-\frac{2}{9}$ **b.** $-\frac{12}{13}$ **c.** $-\frac{x}{y}$ **d.** $-\frac{8}{b}$

7. Simplify each fraction.

a. $\frac{15}{1}$

b. $\frac{15}{15}$

c. $-\frac{45}{15}$

d. $-\frac{150}{15}$

8. Write each fraction as an equivalent fraction with denominator 36. Then write the original fractions in order from smallest to largest.

$\frac{5}{9}, \frac{1}{3}, \frac{5}{6}, \frac{5}{12}$

EXAMPLE 7   Simplify each fraction.

a. $\frac{24}{1}$ b. $\frac{24}{24}$ c. $-\frac{48}{24}$ d. $-\frac{72}{24}$

Solution   In each case, we divide the numerator by the denominator.

a. $\frac{24}{1} = 24$ b. $\frac{24}{24} = 1$ c. $-\frac{48}{24} = -2$ d. $-\frac{72}{24} = -3$

Comparing Fractions

We can compare fractions to see which is larger or smaller when they have the same denominator.

EXAMPLE 8   Write each fraction as an equivalent fraction with denominator 24. Then write them in order from smallest to largest.

$\frac{5}{8} \quad \frac{5}{6} \quad \frac{3}{4} \quad \frac{2}{3}$

Solution   We begin by writing each fraction as an equivalent fraction with denominator 24.

$\frac{5}{8} = \frac{15}{24} \quad \frac{5}{6} = \frac{20}{24} \quad \frac{3}{4} = \frac{18}{24} \quad \frac{2}{3} = \frac{16}{24}$

Now that they all have the same denominator, the smallest fraction is the one with the smallest numerator and the largest fraction is the one with the largest numerator. Writing them in order from smallest to largest we have

$\frac{15}{24} < \frac{16}{24} < \frac{18}{24} < \frac{20}{24}$

or

$\frac{5}{8} < \frac{2}{3} < \frac{3}{4} < \frac{5}{6}$

EXAMPLE 9   Write each fraction as an equivalent fraction with denominator 24.

$-\frac{1}{4} \quad -\frac{3}{8} \quad -\frac{5}{6} \quad -\frac{5}{8}$

Solution   We begin by writing each fraction as an equivalent fraction with denominator 24.

$-\frac{1}{4} = -\frac{6}{24} \quad -\frac{3}{8} = -\frac{9}{24} \quad -\frac{5}{6} = -\frac{20}{24} \quad -\frac{5}{8} = -\frac{15}{24}$

Now we can write them in order from smallest to largest.

$-\frac{20}{24} < -\frac{15}{24} < -\frac{9}{24} < -\frac{6}{24}$

or

$-\frac{5}{6} < -\frac{5}{8} < -\frac{3}{8} < -\frac{1}{4}$

GETTING READY FOR CLASS

After reading through the preceding section, respond in your own words and in complete sentences.

A. What is a fraction?

B. Which term in the fraction $\frac{7}{8}$ is the numerator?

C. Is the fraction $\frac{3}{9}$ a proper fraction?

D. What word do we use to describe fractions such as $\frac{1}{5}$ and $\frac{4}{20}$, which look different, but have the same value?

Answers

7. a. 15 **b.** 1 **c.** -3 **d.** -10

8. $\frac{1}{3} < \frac{5}{12} < \frac{5}{9} < \frac{5}{6}$

Vocabulary Review

Choose the correct words to fill in the blanks below.

numerator   equivalent   proper   fraction   denominator   improper

1. A _____ is any number that can be put in the form $\frac{a}{b}$, where a and b are numbers and b is not zero.

2. For the fraction $\frac{a}{b}$, the term a is called the _____, and the term b is called the

   _____.

3. A(n) _____ fraction is a fraction in which the numerator is less than the denominator.

4. A fraction is considered to be a(n) _____ fraction if the numerator is equal to or greater than the denominator.

5. _____ fractions are fractions that may look different but represent the same number.

Problems

A Name the numerator of each fraction.

1. $\frac{1}{3}$ 2. $\frac{1}{4}$ 3. $\frac{-2}{3}$ 4. $\frac{2}{4}$ 5. $\frac{x}{-8}$ 6. $\frac{y}{10}$ 7. $\frac{-a}{b}$ 8. $\frac{x}{-y}$

Name the denominator of each fraction.

9. $\frac{2}{5}$ 10. $\frac{3}{5}$ 11. 6   12. 2   13. $\frac{a}{-12}$ 14. $\frac{-b}{14}$

Complete the following tables.

15.

Numerator a	Denominator b	Fraction $\frac{a}{b}$
3	5	
-1		$-\frac{1}{7}$
	$-y$	$\frac{x}{y}$
$x+1$	x	

16.

Numerator a	Denominator b	Fraction $\frac{a}{b}$
2	9	
	-3	$-\frac{4}{3}$
-1		$\frac{1}{x}$
x		$\frac{x}{x+1}$

17. For the set of numbers $\left\{ \frac{3}{4}, \frac{6}{5}, \frac{12}{3}, \frac{1}{2}, \frac{9}{10}, \frac{20}{10} \right\}$, list all the proper fractions.

18. For the set of numbers $\left\{ \frac{1}{8}, \frac{7}{9}, \frac{6}{3}, \frac{18}{6}, \frac{3}{5}, \frac{9}{8} \right\}$, list all the improper fractions.

Indicate whether each of the following is *True* or *False*.

19. Every whole number greater than 1 can also be expressed as an improper fraction.

20. Some improper fractions are also proper fractions.

21. Adding the same number to the numerator and the denominator of a fraction will not change its value.

22. The fractions $\frac{3}{4}$ and $\frac{9}{16}$ are equivalent.

B Locate each of the following numbers on this number line.

23. $-\frac{3}{4}$

24. $-\frac{7}{8}$

25. $-\frac{15}{16}$

26. $-\frac{3}{8}$

27. $-\frac{1}{4}$

28. $-\frac{1}{16}$

29. $\frac{1}{2}$

30. $\frac{1}{4}$

31. $\frac{15}{16}$

32. $\frac{7}{8}$

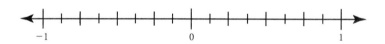

33. Write each fraction as an equivalent fraction with denominator 100. Then write the original fractions in order from smallest to largest.

$$\frac{3}{10} \quad \frac{1}{20} \quad \frac{4}{25} \quad \frac{2}{5}$$

34. Write each fraction as an equivalent fraction with denominator 30. Then write the original fractions in order from smallest to largest.

$$\frac{1}{15} \quad \frac{5}{6} \quad \frac{7}{10} \quad \frac{1}{2}$$

Write each of the following fractions as an equivalent fraction with a denominator of 6.

35. $\frac{2}{3}$

36. $\frac{1}{2}$

37. $-\frac{55}{66}$

38. $-\frac{65}{78}$

Write each of the following fractions as an equivalent fraction with a denominator of 12.

39. $\frac{2}{3}$

40. $\frac{5}{6}$

41. $-\frac{56}{84}$

42. $-\frac{143}{156}$

Write each fraction as an equivalent fraction with a denominator of 12x.

43. $\dfrac{1}{6}$ **44.** $\dfrac{3}{4}$ **45.** $\dfrac{1}{2}$ **46.** $\dfrac{2}{3}$

Write each number as an equivalent fraction with a denominator of 8.

47. 2 **48.** 1 **49.** -5 **50.** -8

51. One-fourth of the first circle below is shaded. Use the other three circles to show three other ways to shade one-fourth of the circle.

52. The six-sided figures below are hexagons. One-third of the first hexagon is shaded. Shade the other three hexagons to show three other ways to represent one-third.

C Simplify by dividing the numerator by the denominator.

53. $\dfrac{3}{1}$ **54.** $\dfrac{3}{3}$ **55.** $-\dfrac{6}{3}$ **56.** $-\dfrac{12}{3}$ **57.** $\dfrac{-37}{-1}$ **58.** $\dfrac{-37}{-37}$

Divide the numerator and the denominator of each of the following fractions by 2.

59. $\dfrac{6}{8}$ **60.** $\dfrac{10}{12}$ **61.** $-\dfrac{86}{94}$ **62.** $-\dfrac{106}{142}$

Divide the numerator and the denominator of each of the following fractions by 3.

63. $\dfrac{12}{9}$ **64.** $\dfrac{33}{27}$ **65.** $\dfrac{-39}{51}$ **66.** $\dfrac{57}{-69}$

67. For each square below, what fraction of the area is given by the shaded region?

a.   b.

c.   d.

68. For each square below, what fraction of the area is given by the shaded region?

a.   b.

c.   d.

Applying the Concepts

69. Sending E-mail The pie chart below shows the number of workers who responded to a survey about sending non-work-related e-mail from the office. 100 people responded to the survey. Use the pie chart to fill in the table.

Workers sending personal e-mail from the office

Never 16
More than ten times a day 5
One to five times a day 47
Five to ten times a day 32

How often workers send non-work-related e-mail from the office	Fraction of respondents saying yes
Never	
1 to 5 times a day	
5 to 10 times a day	
More than 10 times a day	

70. Surfing the Internet The pie chart below shows the number of workers who responded to a survey about viewing non-work-related sites during working hours. 100 people responded to the survey. Use the pie chart to fill in the table.

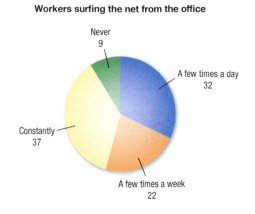

Workers surfing the net from the office

Never 9
A few times a day 32
Constantly 37
A few times a week 22

How often workers view non-work-related sites from the office	Fraction of respondents saying yes
Never	
A few times a week	
A few times a day	
Constantly	

71. Number of Children If there are 3 girls in a family with 5 children, then we say that $\frac{3}{5}$ of the children are girls. If there are 4 girls in a family with 5 children, what fraction of the children are girls?

72. Medical School If 3 out of every 7 people who apply to medical school actually get accepted, what fraction of the people who apply get accepted?

73. Number of Students Of the 43 people who started a math class meeting at 10:00 each morning, only 29 finished the class. What fraction of the people finished the class?

74. Number of Students In a class of 51 students, 23 are freshmen and 28 are juniors. What fraction of the students are freshmen?

75. Expenses If your monthly income is $1,791 and your house payment is $1,121, what fraction of your monthly income must go to pay your house payment?

76. Expenses If you spend $623 on food each month and your monthly income is $2,599, what fraction of your monthly income do you spend on food?

Use the chart on the right to answer questions 77 and 78.

77. Write the number of iPhones in Western Europe as an equivalent fraction with denominator 50.

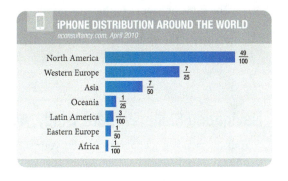

78. Write the number of iPhones in Oceania as an equivalent fraction with denominator 75.

Estimating

79. Which of the following fractions is closest to the number 0?

a. $\frac{1}{2}$ b. $\frac{1}{3}$ c. $\frac{1}{4}$ d. $\frac{1}{5}$

80. Which of the following fractions is closest to the number 1?

a. $\frac{1}{2}$ b. $\frac{1}{3}$ c. $\frac{1}{4}$ d. $\frac{1}{5}$

81. Which of the following fractions is closest to the number 0?

a. $\frac{1}{8}$ b. $\frac{3}{8}$ c. $\frac{5}{8}$ d. $\frac{7}{8}$

82. Which of the following fractions is closest to the number 1?

a. $\frac{1}{8}$ b. $\frac{3}{8}$ c. $\frac{5}{8}$ d. $\frac{7}{8}$

Getting Ready for the Next Section

Multiply.

83. $2 \cdot 2 \cdot 3 \cdot 3 \cdot 3$

84. $2^2 \cdot 3^3$

85. $2^2 \cdot 3 \cdot 5$

86. $2 \cdot 3^2 \cdot 5$

Divide.

87. $12 \div 3$

88. $15 \div 3$

89. $20 \div 4$

90. $24 \div 4$

91. $42 \div 6$

92. $72 \div 8$

93. $102 \div 2$

94. $105 \div 7$

Find the Mistake

Each sentence below contains a mistake. Circle the mistake and write the correct word(s) or numbers(s) on the line provided.

1. For the fraction $\frac{21}{7}$, the numerator is 7. _____

2. The fraction $\frac{90}{15}$ is considered a proper fraction. _____

3. The multiplication property for fractions states that if the numerator and the denominator of a fraction are multiplied by the same nonzero number, the resulting fraction is improper to the original fraction. _____

4. If we divide the numerator and denominator of the fraction $\frac{8}{12}$ by 4, then we get the equivalent fraction $\frac{4}{3}$. _____

Navigation Skills: Prepare, Study, Achieve

Your instructor is a vital resource for your success in this class. Make note of your instructor's office hours and utilize them regularly. Compile a resource list that you keep with your class materials. This list should contain your instructor's office hours and contact information (e.g., office phone number or e-mail), as well as classmates' contact information that you can utilize outside of class. Communicate often with your classmates about how the course is going for you and any questions you may have. Odds are that someone else has the same question and you may be able to work together to find the answer.

Prime Numbers, Factors, and Reducing to Lowest Terms

©iStockphoto.com/andreas_lamm

In 2006, Christian Schou earned the world record title for highest slackline walk at a height of 3,280 feet across a Norwegian fjord. A slackline is made of flat nylon webbing and anchored between two points, similar to a tightrope. But in contrast to a taut tightrope, the slackline is strung to allow it to stretch and bounce under the walker's feet.

Suppose the length of a walker's slackline is 30 feet. If the walker walks 15 feet to the middle of the rope, we can say the the walker has walked $\frac{15}{30}$ of the rope's length. We know that he is also standing in the middle of the rope, therefore, he has walked $\frac{1}{2}$ of the rope's length. The fraction $\frac{15}{30}$ is equivalent to the fraction $\frac{1}{2}$; that is, they both have the same value. The mathematical process we use to rewrite $\frac{15}{30}$ as $\frac{1}{2}$ is called *reducing to lowest terms*. Before we look at that process, we need to define some new terms.

A Prime and Composite Numbers

> **DEFINITION** prime number
>
> A *prime number* is any whole number greater than 1 that has exactly two divisors—itself and 1. (A number is a divisor of another number if it divides it without a remainder.)
>
> Prime numbers = {2, 3, 5, 7, 11, 13, 17, 19, 23, 29, 31, 37, . . . }

> **DEFINITION** composite number
>
> Any whole number greater than 1 that is not a prime number is called a *composite number*. A composite number always has at least one divisor other than itself and 1.

EXAMPLE 1   Identify each of the numbers below as either a prime number or a composite number. For those that are composite, give two divisors other than the number itself or 1.

a. 43   **b.** 12

Solution

a. 43 is a prime number, because the only numbers that divide it without a remainder are 43 and 1.

b. 12 is a composite number, because it can be written as $12 = 4 \cdot 3$, which means that 4 and 3 are divisors of 12. (These are not the only divisors of 12; other divisors are 1, 2, 6, and 12.)

**Video Examples
Section 2.2**

Note You may have already noticed that the word *divisor* as we are using it here means the same as the word *factor*. A divisor and a factor of a number are the same thing. A number can't be a divisor of another number without also being a factor of it.

Practice Problems

1. Identify each number as either prime or composite. For those that are composite, give two divisors other than the number itself or 1.
   a. 61
   b. 33

Answers
1. a. Prime   **b.** 3, 11

2. Factor 80 into a product of prime factors.

Every composite number can be written as the product of prime factors. Let's look at the composite number 108. We know we can write 108 as $2 \cdot 54$. The number 2 is a prime number, but 54 is not prime. Because 54 can be written as $2 \cdot 27$, we have

$$108 = 2 \cdot 54$$
$$= 2 \cdot 2 \cdot 27$$

Now the number 27 can be written as $3 \cdot 9$ or $3 \cdot 3 \cdot 3$ (because $9 = 3 \cdot 3$), so

$$108 = 2 \cdot 54$$
$$108 = 2 \cdot 2 \cdot 27$$
$$108 = 2 \cdot 2 \cdot 3 \cdot 9$$
$$108 = 2 \cdot 2 \cdot 3 \cdot 3 \cdot 3$$

This last line is the number 108 written as the product of prime factors. We can use exponents to rewrite the last line.

$$108 = 2^2 \cdot 3^3$$

EXAMPLE 2   Factor 60 into a product of prime factors.

Solution   We begin by writing 60 as $6 \cdot 10$ and continue factoring until all factors are prime numbers.

$$60 = 6 \cdot 10$$
$$= 2 \cdot 3 \cdot 2 \cdot 5$$
$$= 2^2 \cdot 3 \cdot 5$$

Notice that if we had started by writing 60 as $3 \cdot 20$, we would have achieved the same result:

$$60 = 3 \cdot 20$$
$$= 3 \cdot 2 \cdot 10$$
$$= 3 \cdot 2 \cdot 2 \cdot 5$$
$$= 2^2 \cdot 3 \cdot 5$$

B Reducing Fractions

We can use the method of factoring numbers into prime factors to help reduce fractions to lowest terms. Here is the definition for lowest terms:

DEFINITION lowest terms

A fraction is said to be in *lowest terms* if the numerator and the denominator have no factors in common other than the number 1.

Clarification 1: The fractions $\frac{1}{2}, \frac{1}{3}, \frac{2}{3}, \frac{1}{4}, \frac{3}{4}, \frac{1}{5}, \frac{2}{5}, \frac{3}{5}$, and $\frac{4}{5}$ are all in lowest terms, because in each case the numerator and the denominator have no factors other than 1 in common. That is, in each fraction, no number other than 1 divides both the numerator and the denominator exactly (without a remainder).

Clarification 2: The fraction $\frac{6}{8}$ is not written in lowest terms, because the numerator and the denominator are both divisible by 2. To write $\frac{6}{8}$ in lowest terms, we apply the division property for fractions and divide both the numerator and the denominator by 2.

$$\frac{6}{8} = \frac{6 \div 2}{8 \div 2} = \frac{3}{4}$$

The fraction $\frac{3}{4}$ is in lowest terms, because 3 and 4 have no factors in common except the number 1.

Reducing a fraction to lowest terms is simply a matter of dividing the numerator and the denominator by all the factors they have in common. We know from the division property for fractions that this will produce an equivalent fraction.

EXAMPLE 3 Reduce the fraction $\frac{12}{15}$ to lowest terms by first factoring the numerator and the denominator into prime factors, and then dividing both the numerator and the denominator by the factor they have in common.

Solution The numerator and the denominator factor as follows:

$$12 = 2 \cdot 2 \cdot 3 \quad \text{and} \quad 15 = 3 \cdot 5$$

The factor they have in common is 3. The division property for fractions tells us that we can divide both terms of a fraction by 3 to produce an equivalent fraction.

$$\frac{12}{15} = \frac{2 \cdot 2 \cdot 3}{3 \cdot 5} \qquad \text{Factor the numerator and the denominator completely.}$$

$$= \frac{2 \cdot 2 \cdot 3 \div 3}{3 \cdot 5 \div 3} \qquad \text{Divide by 3.}$$

$$= \frac{2 \cdot 2}{5} = \frac{4}{5}$$

The fraction $\frac{4}{5}$ is equivalent to $\frac{12}{15}$ and is in lowest terms, because the numerator and the denominator have no factors other than 1 in common.

We can shorten the work involved in reducing fractions to lowest terms by using a slash to indicate division. For example, we can write the above problem this way:

$$\frac{12}{15} = \frac{2 \cdot 2 \cdot 3}{3 \cdot 5} = \frac{4}{5}$$

So long as we understand that the slashes through the 3s indicate that we have divided both the numerator and the denominator by 3, we can use this notation.

EXAMPLE 4 Laura is having a party. She puts 4 six-packs of diet soda in a cooler for her guests. At the end of the party, she finds that only 4 sodas have been consumed. What fraction of the sodas are left? Write your answer in lowest terms.

Solution She had 4 six-packs of soda, which is $4(6) = 24$ sodas. Only 4 were consumed at the party, so 20 are left. The fraction of sodas left is

$$\frac{20}{24}$$

Factoring 20 and 24 completely and then dividing out both the factors they have in common gives us

$$\frac{20}{24} = \frac{2 \cdot 2 \cdot 5}{2 \cdot 2 \cdot 2 \cdot 3} = \frac{5}{6}$$

3. Reduce the fraction $\frac{30}{45}$ to lowest terms by using prime factors.

Note The slashes in Example 4 indicate that we have divided both the numerator and the denominator by $2 \cdot 2$, which is equal to 4. With some fractions it is apparent at the start what number divides the numerator and the denominator. For instance, you may have recognized that both 20 and 24 in Example 4 are divisible by 4. We can divide both terms by 4 without factoring first. The division property for fractions guarantees that dividing both terms of a fraction by 4 will produce an equivalent fraction.

$$\frac{20}{24} = \frac{20 \div 4}{24 \div 4} = \frac{5}{6}$$

4. Laura is baking cupcakes for her party. She bakes two dozen cupcakes, and six are eaten. What fraction of the cupcakes are left?

Answers

3. $\frac{2}{3}$

4. $\frac{3}{4}$

5. Reduce $\frac{10}{35}$ to lowest terms.

EXAMPLE 5   Reduce $\frac{6}{42}$ to lowest terms.

Solution   We begin by factoring both terms. We then divide through by any factors common to both terms.

$$\frac{6}{42} = \frac{2 \cdot 3}{2 \cdot 3 \cdot 7} = \frac{1}{7}$$

We must be careful in a problem like this to remember that the slashes indicate division. They are used to indicate that we have divided both the numerator and the denominator by $2 \cdot 3 = 6$. The result of dividing the numerator 6 by $2 \cdot 3$ is 1. It is a very common mistake to call the numerator 0 instead of 1 or to leave the numerator out of the answer.

6. Reduce $\frac{6}{30}$ to lowest terms.

EXAMPLE 6   Reduce $\frac{4}{40}$ to lowest terms.

$$\frac{4}{40} = \frac{2 \cdot 2 \cdot 1}{2 \cdot 2 \cdot 2 \cdot 5}$$

$$= \frac{1}{10}$$

7. Reduce $\frac{140}{30}$ to lowest terms.

EXAMPLE 7   Reduce $\frac{105}{30}$ to lowest terms.

$$\frac{105}{30} = \frac{3 \cdot 5 \cdot 7}{2 \cdot 3 \cdot 5}$$

$$= \frac{7}{2}$$

GETTING READY FOR CLASS

After reading through the preceding section, respond in your own words and in complete sentences.

A. What is a prime number?

B. Why is the number 22 a composite number?

C. Factor 120 into a product of prime factors.

D. How would you reduce a fraction to lowest terms?

Answers

5. $\frac{2}{7}$

6. $\frac{1}{5}$

7. $\frac{14}{3}$

Vocabulary Review

Choose the correct words to fill in the blanks below.

  prime   composite   remainder   lowest terms

1. A _____ number is any whole number greater than 1 that has exactly two divisors: 1 and the number itself.

2. A number is a divisor of another number if it divides it without a _____.

3. Any whole number greater than 1 that is not a prime number is called a _____ number.

4. A fraction is said to be in _____ if the numerator and the denominator have no factors in common other than the number 1.

Problems

A Identify each of the numbers below as either a prime number or a composite number. For those that are composite, give at least one divisor (factor) other than the number itself or the number 1.

1. 11   **2.** 23   **3.** 105   **4.** 41

5. 81   **6.** 50   **7.** 13   **8.** 219

Factor each of the following into a product of prime factors.

9. 12   **10.** 8   **11.** 81   **12.** 210

13. 215   **14.** 75   **15.** 15   **16.** 42

B Reduce each fraction to lowest terms.

17. $\dfrac{5}{10}$ **18.** $\dfrac{3}{6}$ **19.** $\dfrac{4}{6}$ **20.** $\dfrac{4}{10}$ **21.** $\dfrac{8}{10}$ **22.** $\dfrac{6}{10}$

23. $\dfrac{36}{20}$ **24.** $\dfrac{32}{12}$ **25.** $\dfrac{42}{66}$ **26.** $\dfrac{36}{60}$ **27.** $\dfrac{24}{40}$ **28.** $\dfrac{50}{75}$

29. $\dfrac{14}{98}$ **30.** $\dfrac{12}{84}$ **31.** $\dfrac{70}{90}$ **32.** $\dfrac{80}{90}$ **33.** $\dfrac{42}{30}$ **34.** $\dfrac{18}{90}$

35. $\dfrac{150}{210}$ **36.** $\dfrac{110}{70}$ **37.** $\dfrac{45}{75}$ **38.** $\dfrac{180}{108}$ **39.** $\dfrac{60}{36}$ **40.** $\dfrac{105}{30}$

41. $\dfrac{96}{108}$ **42.** $\dfrac{66}{84}$ **43.** $\dfrac{126}{165}$ **44.** $\dfrac{210}{462}$ **45.** $\dfrac{102}{114}$ **46.** $\dfrac{255}{285}$

47. $\dfrac{294}{693}$ **48.** $\dfrac{273}{385}$

49. Reduce each fraction to lowest terms.

    a. $\dfrac{6}{51}$ **b.** $\dfrac{6}{52}$ **c.** $\dfrac{6}{54}$ **d.** $\dfrac{6}{56}$ **e.** $\dfrac{6}{57}$

50. Reduce each fraction to lowest terms.

    a. $\dfrac{6}{42}$ **b.** $\dfrac{6}{44}$ **c.** $\dfrac{6}{45}$ **d.** $\dfrac{6}{46}$ **e.** $\dfrac{6}{48}$

51. Reduce each fraction to lowest terms.

    a. $\dfrac{2}{90}$ **b.** $\dfrac{3}{90}$ **c.** $\dfrac{5}{90}$ **d.** $\dfrac{6}{90}$ **e.** $\dfrac{9}{90}$

52. Reduce each fraction to lowest terms.

    a. $\dfrac{3}{105}$ **b.** $\dfrac{5}{105}$ **c.** $\dfrac{7}{105}$ **d.** $\dfrac{15}{105}$ **e.** $\dfrac{21}{105}$

53. The answer to each problem below is wrong. Give the correct answer.

    a. $\dfrac{5}{15} = \dfrac{5}{3 \cdot 5} = \dfrac{0}{3}$

    b. $\dfrac{5}{6} = \dfrac{3 + 2}{4 + 2} = \dfrac{0}{4}$

    c. $\dfrac{6}{30} = \dfrac{2 \cdot 3}{2 \cdot 3 \cdot 5} = 5$

54. The answer to each problem below is wrong. Give the correct answer.

    a. $\dfrac{10}{20} = \dfrac{7 + 3}{17 + 3} = \dfrac{7}{17}$

    b. $\dfrac{9}{36} = \dfrac{3 \cdot 3}{2 \cdot 2 \cdot 3 \cdot 3} = \dfrac{0}{4}$

    c. $\dfrac{4}{12} = \dfrac{2 \cdot 2}{2 \cdot 2 \cdot 3} = 3$

55. Which of the fractions $\frac{6}{8}$, $\frac{15}{20}$, $\frac{9}{16}$, and $\frac{21}{28}$ does not reduce to $\frac{3}{4}$?

56. Which of the fractions $\frac{4}{9}$, $\frac{10}{15}$, $\frac{8}{12}$, and $\frac{6}{12}$ do not reduce to $\frac{2}{3}$?

The number line below extends from 0 to 2, with the segment from 0 to 1 and the segment from 1 to 2 each divided into 8 equal parts. Locate each of the following numbers on this number line.

57. $\dfrac{1}{2}$, $\dfrac{2}{4}$, $\dfrac{4}{8}$, and $\dfrac{8}{16}$

58. $\dfrac{3}{2}$, $\dfrac{6}{4}$, $\dfrac{12}{8}$, and $\dfrac{24}{16}$

59. $\dfrac{5}{4}$, $\dfrac{10}{8}$, and $\dfrac{20}{16}$

60. $\dfrac{1}{4}$, $\dfrac{2}{8}$, and $\dfrac{4}{16}$

Applying the Concepts

61. Income A family's monthly income is $2,400, and they spend $600 each month on food. Write the amount they spend on food as a fraction of their monthly income in lowest terms.

62. Hours and Minutes There are 60 minutes in 1 hour. What fraction of an hour is 20 minutes? Write your answer in lowest terms.

63. Final Exam Suppose 33 people took the final exam in a math class. If 11 people got an A on the final exam, what fraction of the students did not get an A on the exam? Write your answer in lowest terms.

64. Income Tax A person making $21,000 a year pays $3,000 in income tax. What fraction of the person's income is paid as income tax? Write your answer in lowest terms.

Nutrition The nutrition labels below are from two different snack crackers. Use them to work Problems 65–70.

Cheez-It Crackers

Nutrition Facts
Serving Size 30 g. (About 27 crackers)
Servings Per Container: 9

Amount Per Serving

Calories 150   Calories from fat 70

% Daily Value*

Total Fat 8g	**12%**
Saturated Fat 2g	**10%**
Trans Fat 0g	
Polysaturated Fat 4g	
Monounsaturated Fat 2g	
Cholesterol 0mg	**0%**
Sodium 230mg	**10%**
Total Carbohydrate 17g	**6%**
Dietary Fiber less than 1g	**3%**
Sugars 0g	
Protein 3g	

Vitamin A 2%   •   Vitamin C 0%
Calcium 4%   •   Iron 6%
*Percent Daily Values are based on a 2,000 calorie diet

Goldfish Crackers

Nutrition Facts
Serving Size 55 pieces
Servings Per Container About 4

Amount Per Serving

Calories 140   Calories from fat 45

% Daily Value*

Total Fat 5g	**8%**
Saturated Fat 1g	**5%**
Cholesterol 5mg	**2%**
Sodium 250mg	**10%**
Total Carbohydrate 20g	**7%**
Dietary Fiber 1g	**4%**
Sugars 1g	
Protein 4g	

Vitamin A 0%   •   Vitamin C 0%
Calcium 4%   •   Iron 2%
*Percent Daily Values are based on a 2,000 calorie diet

65. What fraction of the calories in Cheez-It crackers comes from fat?

66. What fraction of the calories in Goldfish crackers comes from fat?

67. For Cheez-It crackers, what fraction of the total fat is from saturated fat?

68. For Goldfish crackers, what fraction of the total fat is from saturated fat?

69. What fraction of the total carbohydrates in Cheez-It crackers is from sugar?

70. What fraction of the total carbohydrates in Goldfish crackers is from sugar?

Getting Ready for the Next Section

Multiply.

71. $1 \cdot 3 \cdot 1$

72. $2 \cdot 4 \cdot 5$

73. $3 \cdot 5 \cdot 3$

74. $1 \cdot 4 \cdot 1$

75. $5 \cdot 5 \cdot 1$

76. $6 \cdot 6 \cdot 2$

Factor into prime factors.

77. 60   **78.** 72   **79.** $15 \cdot 4$ **80.** $8 \cdot 9$

Expand and multiply.

81. 3^2 **82.** 4^2 **83.** 5^2 **84.** 6^2

Improving Your Quantitative Literacy

85. Wimbledon The graphic shown here gives the most Wimbledon Men's Champions by country. Which of the following is the closest to the fraction of champions from France as compared to Australia?

a. $\dfrac{1}{10}$ **b.** $\dfrac{1}{3}$ **c.** $\dfrac{3}{4}$ **d.** $\dfrac{1}{2}$

WIMBLEDON - MOST MEN'S SINGLES CHAMPS
wimbledon.org

United States   33
British Isles   32
Australia   21
France   7
Sweden   7
Switzerland   6

Find the Mistake

Each sentence below contains a mistake. Circle the mistake and write the correct word(s) or numbers(s) on the line provided.

1. The number 30 is a prime number because it has 10 as a divisor. _____

2. The number 70 factored into a product of primes is $7 \cdot 10$. _____

3. When reducing the fraction $\dfrac{32}{48}$ to lowest terms, we divide out the common factors 2 and 3 to get $\dfrac{2}{3}$.

_____

4. Reducing the fraction $\dfrac{112}{14}$ to lowest terms gives us $\dfrac{7}{2}$. _____

Landmark Review: Checking Your Progress

Name the numerator and denominator for each fraction.

1. $\dfrac{3}{5}$ **2.** $\dfrac{1}{3}$ **3.** $\dfrac{7}{15}$ **4.** $\dfrac{4}{x}$

Write each of the following fractions as an equivalent fraction with a denominator of $8x$.

5. $\dfrac{1}{2}$ **6.** $\dfrac{3}{4}$ **7.** $\dfrac{1}{8}$ **8.** $\dfrac{5}{2}$

Reduce each fraction to lowest terms.

9. $\dfrac{17}{34}$ **10.** $\dfrac{15}{25}$ **11.** $\dfrac{48}{80}$ **12.** $\dfrac{135}{216}$ **13.** $\dfrac{68}{72}$ **14.** $\dfrac{93}{126}$

2.3 Multiplication with Fractions, and the Area of a Triangle

©iStockphoto.com/ Zinni-Online

OBJECTIVES

A Multiply fractions with no common factors.

B Multiply and simplify fractions.

C Apply the distributive property and find the area of a triangle.

KEY WORDS

simplify

area of a triangle

Once a year on Christmas Island, in Australia, millions of bright red crabs migrate from the rain forest for several miles to the beach. They spawn in the sea, where the eggs hatch almost immediately upon contact with the salt water. For days, the crabs swarm train tracks, highways, and other busy thoroughfares on the island. They even take over golf courses, and according to a special rule, a golfer must play a ball where it lies even if a crab knocks it to another spot.

Let's suppose a group of crabs are crawling down a railroad track when they hear a train coming. $\frac{3}{4}$ of the crabs scurry into the bushes on either side of the track, $\frac{1}{2}$ of the $\frac{3}{4}$ going to the right and $\frac{1}{2}$ to the left. What fraction of the original group of crabs crawls to the right of the track? This question can be answered by multiplying $\frac{1}{2}$ and $\frac{3}{4}$. Here is the problem written in symbols:

$$\frac{1}{2} \cdot \frac{3}{4} = \frac{3}{8}$$

If you analyze this example, you will discover that to multiply two fractions, we multiply the numerators and then multiply the denominators. We begin this section with the rule for multiplication of fractions.

Video Examples Section 2.3

A Multiplication of Fractions

RULE Product of Two Fractions

The product of two fractions is a fraction whose numerator is the product of the two numerators, and whose denominator is the product of the two denominators. We can write this rule in symbols as follows:

If a, b, c, and d represent any numbers and b and d are not zero, then

$$\frac{a}{b} \cdot \frac{c}{d} = \frac{a \cdot c}{b \cdot d}$$

EXAMPLE 1 Multiply: $\frac{3}{5} \cdot \frac{2}{7}$.

Solution Using our rule for multiplication, we multiply the numerators and multiply the denominators.

$$\frac{3}{5} \cdot \frac{2}{7} = \frac{3 \cdot 2}{5 \cdot 7} = \frac{6}{35}$$

The product of $\frac{3}{5}$ and $\frac{2}{7}$ is the fraction $\frac{6}{35}$. The numerator 6 is the product of 3 and 2, and the denominator 35 is the product of 5 and 7.

Practice Problems

1. Multiply: $\frac{5}{6} \cdot \frac{7}{8}$.

Answer

1. $\frac{35}{48}$

2. Multiply: $\frac{7}{12} \cdot 5$.

EXAMPLE 2   Multiply: $\frac{3}{8} \cdot 5$.

Solution   The number 5 can be written as $\frac{5}{1}$. That is, 5 can be considered a fraction with numerator 5 and denominator 1. Writing 5 this way enables us to apply the rule for multiplying fractions.

$$\frac{3}{8} \cdot 5 = \frac{3}{8} \cdot \frac{5}{1}$$

$$= \frac{3 \cdot 5}{8 \cdot 1}$$

$$= \frac{15}{8}$$

3. Multiply: $\frac{2}{3} \left(\frac{1}{3} \cdot \frac{5}{9} \right)$.

EXAMPLE 3   Multiply: $\frac{1}{2} \left(\frac{3}{4} \cdot \frac{1}{5} \right)$.

Solution   We find the product inside the parentheses first and then multiply the result by $\frac{1}{2}$.

$$\frac{1}{2} \left(\frac{3}{4} \cdot \frac{1}{5} \right) = \frac{1}{2} \left(\frac{3}{20} \right)$$

$$= \frac{1 \cdot 3}{2 \cdot 20}$$

$$= \frac{3}{40}$$

The properties of multiplication apply to fractions as well. That is, if a, b, and c are fractions, then

$$a \cdot b = b \cdot a \qquad \text{Multiplication with fractions is commutative.}$$

$$a \cdot (b \cdot c) = (a \cdot b) \cdot c \qquad \text{Multiplication with fractions is associative.}$$

To demonstrate the associative property for fractions, let's do Example 3 again, but this time we will apply the associative property first.

$$\frac{1}{2} \left(\frac{3}{4} \cdot \frac{1}{5} \right) = \left(\frac{1}{2} \cdot \frac{3}{4} \right) \cdot \frac{1}{5} \qquad \text{Associative property}$$

$$= \left(\frac{1 \cdot 3}{2 \cdot 4} \right) \cdot \frac{1}{5}$$

$$= \left(\frac{3}{8} \right) \cdot \frac{1}{5}$$

$$= \frac{3 \cdot 1}{8 \cdot 5}$$

$$= \frac{3}{40}$$

The result is identical to that of Example 3.

B Multiplying and Simplifying Fractions

The answers to all the examples so far in this section have been in lowest terms. Let's see what happens when we multiply two fractions to get a product that is not in lowest terms.

4. Multiply: $\frac{9}{16} \cdot \frac{4}{3}$.

EXAMPLE 4   Multiply: $\frac{15}{8} \cdot \frac{4}{9}$.

Solution   Multiplying the numerators and multiplying the denominators, we have

$$\frac{15}{8} \cdot \frac{4}{9} = \frac{15 \cdot 4}{8 \cdot 9}$$

$$= \frac{60}{72}$$

Answers

2. $\frac{35}{12}$

3. $\frac{10}{81}$

4. $\frac{3}{4}$

The product is $\frac{60}{72}$, which can be reduced to lowest terms by factoring 60 and 72 and then dividing out any factors they have in common.

$$\frac{60}{72} = \frac{\cancel{2} \cdot \cancel{2} \cdot \cancel{3} \cdot 5}{\cancel{2} \cdot \cancel{2} \cdot 2 \cdot \cancel{3} \cdot 3}$$

$$= \frac{5}{6}$$

We can actually save ourselves some time by factoring before we multiply. Here's how it is done:

$$\frac{15}{8} \cdot \frac{4}{9} = \frac{15 \cdot 4}{8 \cdot 9}$$

$$= \frac{(3 \cdot 5) \cdot (2 \cdot 2)}{(2 \cdot 2 \cdot 2) \cdot (3 \cdot 3)}$$

$$= \frac{\cancel{3} \cdot 5 \cdot \cancel{2} \cdot \cancel{2}}{\cancel{2} \cdot \cancel{2} \cdot 2 \cdot \cancel{3} \cdot 3}$$

$$= \frac{5}{6}$$

The result is the same in both cases. Reducing to lowest terms before we actually multiply takes less time. Here are some additional examples. Problems like these will be useful when we solve equations.

EXAMPLE 5 $\dfrac{9}{2} \cdot \dfrac{8}{18} = \dfrac{9 \cdot 8}{2 \cdot 18}$

$$= \frac{(3 \cdot 3) \cdot (2 \cdot 2 \cdot 2)}{2 \cdot (2 \cdot 3 \cdot 3)}$$

$$= \frac{\cancel{3} \cdot \cancel{3} \cdot \cancel{2} \cdot 2 \cdot 2}{\cancel{2} \cdot 2 \cdot \cancel{3} \cdot \cancel{3}}$$

$$= \frac{2}{1}$$

$$= 2$$

EXAMPLE 6 $\dfrac{2}{3} \cdot \dfrac{6}{5} \cdot \dfrac{5}{8} = \dfrac{2 \cdot 6 \cdot 5}{3 \cdot 5 \cdot 8}$

$$= \frac{2 \cdot (2 \cdot 3) \cdot 5}{3 \cdot 5 \cdot (2 \cdot 2 \cdot 2)}$$

$$= \frac{\cancel{2} \cdot \cancel{2} \cdot \cancel{3} \cdot \cancel{5}}{\cancel{3} \cdot \cancel{5} \cdot \cancel{2} \cdot 2 \cdot \cancel{2}}$$

$$= \frac{1}{2}$$

In Chapter 1, we did some work with exponents. We can extend our work with exponents to include fractions, as the following examples indicate.

EXAMPLE 7 $\left(\dfrac{3}{4}\right)^2 = \dfrac{3}{4}\left(\dfrac{3}{4}\right)$

$$= \frac{3 \cdot 3}{4 \cdot 4}$$

$$= \frac{9}{16}$$

5. Multiply: $\frac{9}{5} \cdot \frac{10}{21}$.

Note   Although $\frac{2}{1}$ is in lowest terms, it is still simpler to write the answer as just 2. We will always do this when the denominator is the number 1.

6. Multiply: $\frac{3}{4} \cdot \frac{4}{9} \cdot \frac{6}{7}$.

7. Simplify: $\left(\frac{5}{6}\right)^2$.

Answers

5. $\frac{6}{7}$

6. $\frac{2}{7}$

7. $\frac{25}{36}$

8. Multiply: $\left(\frac{4}{5}\right)^2 \cdot \frac{5}{6}$.

EXAMPLE 8 $\left(\frac{5}{6}\right)^2 \cdot \frac{1}{2} = \frac{5}{6} \cdot \frac{5}{6} \cdot \frac{1}{2}$

$$= \frac{5 \cdot 5 \cdot 1}{6 \cdot 6 \cdot 2}$$

$$= \frac{25}{72}$$

9. Simplify each expression.

   a. $\left(\frac{4}{5}\right)\left(-\frac{2}{5}\right)$

   b. $\left(-\frac{3}{5}\right)\left(-\frac{7}{10}\right)$

EXAMPLE 9   Simplify each expression.

   a. $\left(\frac{2}{3}\right)\left(-\frac{3}{5}\right)$ **b.** $\left(-\frac{7}{8}\right)\left(-\frac{5}{14}\right)$

Solution

   a. $\left(\frac{2}{3}\right)\left(-\frac{3}{5}\right) = -\frac{6}{15} = -\frac{2}{5}$ *The rule for multiplication also holds for fractions.*

   b. $\left(-\frac{7}{8}\right)\left(-\frac{5}{14}\right) = \frac{35}{112} = \frac{5}{16}$

The word *of* used in connection with fractions indicates multiplication. If we want to find $\frac{1}{2}$ of $\frac{2}{3}$, then what we do is multiply $\frac{1}{2}$ and $\frac{2}{3}$.

10. Find $\frac{3}{4}$ of $\frac{5}{6}$.

EXAMPLE 10   Find $\frac{1}{2}$ of $\frac{2}{3}$.

Solution   Knowing the word *of*, as used here, indicates multiplication, we have

$$\frac{1}{2} \text{ of } \frac{2}{3} = \frac{1}{2} \cdot \frac{2}{3}$$

$$= \frac{1 \cdot 2}{2 \cdot 3} = \frac{1}{3}$$

This seems to make sense. Logically, $\frac{1}{2}$ of $\frac{2}{3}$ should be $\frac{1}{3}$, as Figure 1 shows.

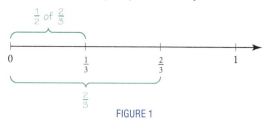

FIGURE 1

11. What is $\frac{7}{8}$ of 24?

Note   As you become familiar with multiplying fractions, you may notice shortcuts that reduce the number of steps in the problems. It's okay to use these shortcuts if you understand why they work and are consistently getting correct answers. If you are using shortcuts and not consistently getting correct answers, then go back to showing all the work until you completely understand the process.

EXAMPLE 11   What is $\frac{3}{4}$ of 12?

Solution   Again, the word *of* means multiply.

$$\frac{3}{4} \text{ of } 12 = \frac{3}{4}(12)$$

$$= \frac{3}{4}\left(\frac{12}{1}\right)$$

$$= \frac{3 \cdot 12}{4 \cdot 1}$$

$$= \frac{3 \cdot 2 \cdot 2 \cdot 3}{2 \cdot 2 \cdot 1}$$

$$= \frac{9}{1} = 9$$

Answers

8. $\frac{8}{15}$

9. a. $-\frac{4}{5}$ **b.** $\frac{21}{50}$

10. $\frac{5}{8}$

11. 21

C The Distributive Property and Area

The distributive property can also be applied to fractions. Recall that

$$a(b + c) = a \cdot b + a \cdot c \quad \text{or} \quad a(b - c) = a \cdot b - a \cdot c$$

The next example illustrates using the distributive property with fractions.

EXAMPLE 12   Apply the distributive property, then simplify.

a. $6\left(2 + \dfrac{1}{3}\right)$ **b.** $12\left(\dfrac{3}{4} - \dfrac{1}{2}\right)$ **c.** $-5\left(4 - \dfrac{2}{3}\right)$

Solution   Apply the distributive property to each expression.

a. $6\left(2 + \dfrac{1}{3}\right) = 6 \cdot 2 + 6 \cdot \dfrac{1}{3} = 12 + \dfrac{6}{3} = 12 + 2 = 14$

b. $12\left(\dfrac{3}{4} - \dfrac{1}{2}\right) = 12 \cdot \dfrac{3}{4} - 12 \cdot \dfrac{1}{2} = \dfrac{12}{1} \cdot \dfrac{3}{4} - \dfrac{12}{1} \cdot \dfrac{1}{2} = 9 - 6 = 3$

c. $-5\left(4 - \dfrac{2}{3}\right) = -5 \cdot 4 - (-5)\dfrac{2}{3} = -20 + \dfrac{10}{3} = -\dfrac{60}{3} + \dfrac{10}{3} = -\dfrac{50}{3}$

12. Apply the distributive property, then simplify.

a. $5\left(4 + \dfrac{3}{5}\right)$

b. $8\left(\dfrac{3}{4} - \dfrac{1}{2}\right)$

c. $-3\left(2 - \dfrac{1}{3}\right)$

FACTS FROM GEOMETRY   The Area of a Triangle

The formula for the area of a triangle is one application of multiplication with fractions. Figure 2 shows a triangle with base b and height h. Below the triangle is the formula for its area. As you can see, it is a product containing the fraction $\dfrac{1}{2}$.

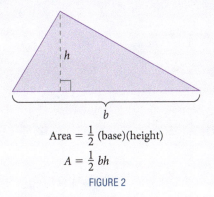

Area $= \dfrac{1}{2}$ (base)(height)

$A = \dfrac{1}{2} bh$

FIGURE 2

EXAMPLE 13   Find the area of the triangle in Figure 3.

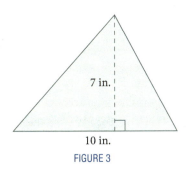

7 in.

10 in.

FIGURE 3

13. Find the area of a triangle with base 8 in. and height 11 in.

Solution   Applying the formula for the area of a triangle, we have

$$A = \dfrac{1}{2} bh = \dfrac{1}{2} \cdot 10 \cdot 7 = 5 \cdot 7 = 35 \text{ in}^2$$

Answers
12. a. 23   **b.** 2   **c.** -5
13. 44 in.²

EXERCISE SET 2.3

Problems

A Find each of the following products. (Multiply.)

1. $\frac{2}{3} \cdot \frac{4}{5}$

2. $\frac{5}{6} \cdot \frac{7}{4}$

3. $\frac{1}{2} \cdot \frac{7}{4}$

4. $\frac{3}{5} \cdot -\frac{4}{7}$

5. $\frac{5}{3} \cdot \frac{3}{5}$

6. $\frac{4}{7} \cdot -\frac{7}{4}$

7. $\frac{3}{4} \cdot 9$

8. $\frac{2}{3} \cdot -5$

9. $\frac{6}{7}\left(\frac{7}{6}\right)$

10. $-\frac{2}{9}\left(\frac{9}{2}\right)$

11. $\frac{1}{2} \cdot \frac{1}{3} \cdot \frac{1}{4}$

12. $\frac{2}{3} \cdot \frac{4}{5} \cdot \frac{1}{3}$

13. $\frac{2}{5} \cdot \frac{3}{5} \cdot \frac{4}{5}$

14. $\frac{1}{4} \cdot -\frac{3}{4} \cdot \frac{3}{4}$

15. $\frac{3}{2} \cdot \frac{5}{2} \cdot \frac{7}{2}$

16. $\frac{4}{3} \cdot -\frac{5}{3} \cdot \frac{7}{3}$

Complete the following tables.

17.

First Number x	Second Number y	Their Product xy
$\frac{1}{2}$	$\frac{2}{3}$	
$\frac{2}{3}$	$\frac{3}{4}$	
$\frac{3}{4}$	$\frac{4}{5}$	
$\frac{5}{a}$	$-\frac{a}{6}$	

18.

First Number x	Second Number y	Their Product xy
12	$\frac{1}{2}$	
12	$\frac{1}{3}$	
12	$\frac{1}{4}$	
12	$\frac{1}{6}$	

19.

First Number x	Second Number y	Their Product xy
$\frac{1}{2}$	30	
$\frac{1}{5}$	30	
$\frac{1}{6}$	30	
$\frac{1}{15}$	30	

20.

First Number x	Second Number y	Their Product xy
$\frac{1}{3}$	$\frac{3}{5}$	
$\frac{3}{5}$	$\frac{5}{7}$	
$\frac{5}{7}$	$\frac{7}{9}$	
$-\frac{7}{b}$	$\frac{b}{11}$	

B Multiply each of the following. Be sure all answers are written in lowest terms.

21. $\dfrac{9}{20} \cdot \dfrac{4}{3}$

22. $\dfrac{135}{16} \cdot -\dfrac{2}{45}$

23. $\dfrac{3}{4} \cdot 12$

24. $\dfrac{3}{4} \cdot 20$

25. $\dfrac{1}{3}\,(3)$

26. $\dfrac{1}{5}\,(5)$

27. $\dfrac{2}{5} \cdot -20$

28. $\dfrac{3}{5} \cdot 15$

29. $\dfrac{72}{35} \cdot -\dfrac{55}{108} \cdot \dfrac{7}{110}$

30. $\dfrac{32}{27} \cdot \dfrac{72}{49} \cdot \dfrac{1}{40}$

Expand and simplify each of the following.

31. $\left(\dfrac{2}{3}\right)^2$

32. $\left(\dfrac{3}{5}\right)^2$

33. $\left(\dfrac{3}{4}\right)^2$

34. $\left(\dfrac{2}{7}\right)^2$

35. $\left(\dfrac{1}{2}\right)^2$

36. $\left(-\dfrac{1}{3}\right)^2$

37. $\left(\dfrac{2}{3}\right)^3$

38. $\left(-\dfrac{3}{5}\right)^3$

39. $\left(\dfrac{3}{4}\right)^2 \cdot \dfrac{8}{9}$

40. $\left(\dfrac{5}{6}\right)^2 \cdot \dfrac{12}{15}$

41. $\left(\dfrac{1}{2}\right)^2 \left(\dfrac{3}{5}\right)^2$

42. $\left(\dfrac{3}{8}\right)^2 \left(-\dfrac{4}{3}\right)^2$

43. $\left(\dfrac{1}{2}\right)^2 \cdot 8 + \left(\dfrac{1}{3}\right)^2 \cdot 9$

44. $\left(\dfrac{2}{3}\right)^2 \cdot 9 + \left(\dfrac{1}{2}\right)^2 \cdot 4$

45. Find $\frac{3}{8}$ of 64.

46. Find $\frac{2}{3}$ of 18.

47. What is $\frac{1}{3}$ of the sum of 8 and 4?

48. What is $\frac{3}{5}$ of the sum of 8 and 7?

49. Find $\frac{1}{2}$ of $\frac{3}{4}$ of 24.

50. Find $\frac{3}{5}$ of $\frac{1}{3}$ of 15.

Find the mistakes in Problems 51–54. Correct the right-hand side of each one.

51. $\frac{1}{2} \cdot \frac{3}{5} = \frac{4}{10}$

52. $\frac{2}{7} \cdot \frac{3}{5} = \frac{5}{35}$

53. $-\frac{3}{4} \cdot -\frac{1}{3} = -\frac{1}{4}$

54. $-\frac{2}{5} \cdot -\frac{2}{3} = -\frac{4}{15}$

C Apply the distributive property, then simplify.

55. $4\left(3 + \frac{1}{2}\right)$

56. $4\left(2 - \frac{3}{4}\right)$

57. $-12\left(\frac{1}{2} + \frac{2}{3}\right)$

58. $-12\left(\frac{3}{4} - \frac{1}{6}\right)$

59. $9\left(\frac{2}{3} - \frac{1}{9}\right)$

60. $12\left(\frac{1}{2} - \frac{1}{3}\right)$

61. $-16\left(\frac{5}{8} - \frac{1}{4}\right)$

62. $-24\left(\frac{2}{3} - \frac{1}{6}\right)$

63. Find the area of the triangle with base 19 inches and height 14 inches.

64. Find the area of the triangle with base 13 inches and height 8 inches.

65. The base of a triangle is $\frac{4}{3}$ feet and the height is $\frac{2}{3}$ feet. Find the area.

66. The base of a triangle is $\frac{8}{7}$ feet and the height is $\frac{14}{5}$ feet. Find the area.

Find the area of each figure.

67.

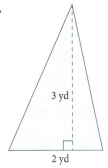

3 yd

2 yd

68.

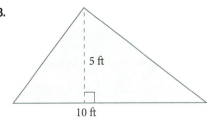

5 ft

10 ft

Applying the Concepts

Cal Poly Enrollment

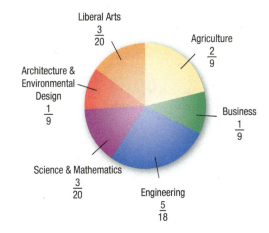

Liberal Arts
$\frac{3}{20}$

Agriculture
$\frac{2}{9}$

Architecture &
Environmental
Design
$\frac{1}{9}$

Business
$\frac{1}{9}$

Science & Mathematics
$\frac{3}{20}$

Engineering
$\frac{5}{18}$

Use the information in the pie chart to answer questions 69 and 70. Round to the nearest student.

69. Reading a Pie Chart If there are approximately 15,800 students attending Cal Poly, approximately how many of them are studying agriculture?

70. Reading a Pie Chart If there are exactly 15,828 students attending Cal Poly, how many of them are studying engineering? Round to the nearest student.

71. Hot Air Balloon Aerostar International makes a hot air balloon called the Rally 105 that has a volume of 105,400 cubic feet. Another balloon, the Rally 126, was designed with a volume that is approximately $\frac{6}{5}$ the volume of the Rally 105. Find the volume of the Rally 126 to the nearest hundred cubic feet.

72. Health Care According to the Department of Veteran's Affairs, approximately $\frac{4}{25}$ of military veterans have diabetes. If there approximately are 3.1 million veterans, how many have diabetes? Round number to the nearest ten thousand.

73. Bicycle Safety The National Safe Kids Campaign and Bell Sports sponsored a study that surveyed 8,159 children ages 5 to 14 who were riding bicycles. Approximately $\frac{2}{5}$ of the children were wearing helmets, and of those, only $\frac{13}{20}$ were wearing the helmets correctly. About how many of the children were wearing helmets correctly?

74. Bicycle Safety From the information in Problem 73, how many of the children surveyed do not wear helmets?

Geometric Sequences A geometric sequence is a sequence in which each term comes from the previous term by multiplying by the same number each time. For example, the sequence $1, \frac{1}{2}, \frac{1}{4}, \frac{1}{8}, \ldots$ is a geometric sequence in which each term is found by multiplying the previous term by $\frac{1}{2}$. By observing this fact, we know that the next term in the sequence will be $\frac{1}{8} \cdot \frac{1}{2} = \frac{1}{16}$.

Find the next number in each of the geometric sequences below.

75. $1, \frac{1}{3}, \frac{1}{9}, \ldots$

76. $1, \frac{1}{4}, \frac{1}{16}, \ldots$

77. $\frac{3}{2}, 1, \frac{2}{3}, \frac{4}{9}, \ldots$

78. $\frac{2}{3}, 1, \frac{3}{2}, \frac{9}{4}, \ldots$

Estimating For each problem below, mentally estimate if the answer will be closest to 0, 1, 2 or 3. Make your estimate without using pencil and paper or a calculator.

79. $\frac{11}{5} \cdot \frac{19}{20}$

80. $\frac{3}{5} \cdot \frac{1}{20}$

81. $\frac{16}{5} \cdot \frac{23}{24}$

82. $\frac{9}{8} \cdot \frac{31}{32}$

Getting Ready for the Next Section

In the next section we will do division with fractions. As you already know, division and multiplication are closely related. These review problems are intended to let you see more of the relationship between multiplication and division.

Perform the indicated operations.

83. $8 \div 4$

84. $8 \cdot \frac{1}{4}$

85. $15 \div 3$

86. $15 \cdot \frac{1}{3}$

87. $18 \div 6$

88. $18 \cdot \frac{1}{6}$

For each number below, find a number to multiply it by to obtain 1.

89. $\frac{3}{4}$

90. $\frac{9}{5}$

91. $\frac{1}{3}$

92. $\frac{1}{4}$

93. 7

94. 2

Find the Mistake

Each sentence below contains a mistake. Circle the mistake and write the correct word(s) or numbers(s) on the line provided.

1. To find the product of two fractions, multiply the numerators and put them over the largest denominator.

_____

2. To multiply $\frac{6}{7}$ by $\frac{12}{9}$, find the product of the numerators and divide it by the sum of the denominators to get $\frac{72}{63}$.

_____

3. Simplifying $\left(\frac{5}{6}\right)^2 \cdot \frac{8}{9}$ gives $\frac{80}{54} = \frac{40}{27}$. _____

4. The area of a triangle with a height of 14 inches and a base of 32 inches is 448 square inches. _____

OBJECTIVES

A Divide fractions and simplify.

B Use division in order of operations problems.

KEY WORDS

reciprocal

Image © Ben Barczi 2010

In 2010, a Peruvian inventor won the $200,000 grand prize for a competition that requested its entrants to suggest a new and ingenious way to save the earth. To battle global warming, the inventor proposed covering 173 acres of dry rocky land in the Andes Mountains with a mixture made from egg whites, lime, and water. The mixture creates a whitewash that when spread over the rocks would reflect sunlight and ideally reduce temperatures, thus slowing the melting of the area's glaciers.

Suppose a truck transports the whitewash in a 32-gallon tank up to the mountaintop. Let's say the inventor used a bucket that only held $\frac{3}{4}$ of a gallon to pour the whitewash onto the rocks. How many times would the inventor need to fill up his bucket to empty the tank? In order to answer this question, we need to learn more about how to divide with fractions. But before we define division with fractions, we must first introduce the idea of *reciprocals*. Look at the following multiplication problems:

$$\frac{3}{4} \cdot \frac{4}{3} = \frac{12}{12} = 1 \qquad \frac{7}{8} \cdot \frac{8}{7} = \frac{56}{56} = 1$$

In each case the product is 1. Whenever the product of two numbers is 1, we say the two numbers are reciprocals.

Video Examples
Section 2.4

A Dividing Fractions

> **DEFINITION** reciprocals
>
> Two numbers whose product is 1 are said to be *reciprocals.* In symbols, the reciprocal of $\frac{a}{b}$ is $\frac{b}{a}$, because
>
> $$\frac{a}{b} \cdot \frac{b}{a} = \frac{a \cdot b}{b \cdot a} = \frac{a \cdot b}{a \cdot b} = 1 \qquad (a \neq 0, b \neq 0)$$

Every number has a reciprocal except 0. The reason 0 does not have a reciprocal is because the product of *any* number with 0 is 0. It can never be 1. Reciprocals of whole numbers are fractions with 1 as the numerator. For example, the reciprocal of 5 is $\frac{1}{5}$, because

$$5 \cdot \frac{1}{5} = \frac{5}{1} \cdot \frac{1}{5} = \frac{5}{5} = 1$$

Table 1 lists some numbers and their reciprocals.

Table 1

Number	Reciprocal	Reason
$\frac{3}{4}$	$\frac{4}{3}$	Because $\frac{3}{4} \cdot \frac{4}{3} = \frac{12}{12} = 1$
$\frac{9}{5}$	$\frac{5}{9}$	Because $\frac{9}{5} \cdot \frac{5}{9} = \frac{45}{45} = 1$
$\frac{1}{3}$	3	Because $\frac{1}{3} \cdot 3 = \frac{1}{3} \cdot \frac{3}{1} = \frac{3}{3} = 1$
7	$\frac{1}{7}$	Because $7 \cdot \frac{1}{7} = \frac{7}{1} \cdot \frac{1}{7} = \frac{7}{7} = 1$

Division with fractions is accomplished by using reciprocals. More specifically, we can define division by a fraction to be the same as multiplication by its reciprocal. Here is the precise definition:

> **DEFINITION**   division by a fraction
>
> If a, b, c, and d are numbers and b, c, and d are all not equal to 0, then
> $$\frac{a}{b} \div \frac{c}{d} = \frac{a}{b} \cdot \frac{d}{c}$$

This definition states that dividing by the fraction $\frac{c}{d}$ is exactly the same as multiplying by its reciprocal $\frac{d}{c}$. Because we developed the rule for multiplying fractions in the last section, we do not need a new rule for division. We simply *replace the divisor by its reciprocal* and multiply. Here are some examples to illustrate the procedure:

EXAMPLE 1   Divide: $\frac{1}{2} \div \frac{1}{4}$.

Solution   The divisor is $\frac{1}{4}$, and its reciprocal is $\frac{4}{1}$. Applying the definition of division for fractions, we have

$$\frac{1}{2} \div \frac{1}{4} = \frac{1}{2} \cdot \frac{4}{1}$$
$$= \frac{1 \cdot 4}{2 \cdot 1}$$
$$= \frac{1 \cdot 2 \cdot 2}{2 \cdot 1}$$
$$= \frac{2}{1}$$
$$= 2$$

The quotient of $\frac{1}{2}$ and $\frac{1}{4}$ is 2. Or, $\frac{1}{4}$ "goes into" $\frac{1}{2}$ two times. Logically, our definition for division of fractions seems to be giving us answers that are consistent with what we know about fractions from previous experience. Because 2 times $\frac{1}{4}$ is $\frac{2}{4}$ or $\frac{1}{2}$, it seems logical that $\frac{1}{2}$ divided by $\frac{1}{4}$ should be 2.

EXAMPLE 2   Divide: $-\frac{3}{8} \div \frac{9}{4}$.

Solution   Dividing by $\frac{9}{4}$ is the same as multiplying by its reciprocal, which is $\frac{4}{9}$.

$$-\frac{3}{8} \div \frac{9}{4} = -\frac{3}{8} \cdot \frac{4}{9}$$
$$= -\frac{3 \cdot 2 \cdot 2}{2 \cdot 2 \cdot 2 \cdot 3 \cdot 3}$$
$$= -\frac{1}{6}$$

The quotient of $-\frac{3}{8}$ and $\frac{9}{4}$ is $-\frac{1}{6}$.

Note   Defining division to be the same as multiplication by the reciprocal does make sense. If we divide 6 by 2, we get 3. On the other hand, if we multiply 6 by $\frac{1}{2}$ (the reciprocal of 2), we also get 3. Whether we divide by 2 or multiply by $\frac{1}{2}$, we get the same result.

Practice Problems

1. Divide: $\frac{3}{4} \div \frac{1}{6}$.

2. Divide: $-\frac{8}{15} \div \frac{4}{5}$.

Answers

1. $\frac{9}{2}$

2. $-\frac{2}{3}$

EXAMPLE 3 Divide: $\frac{2}{3} \div 2$.

Solution The reciprocal of 2 is $\frac{1}{2}$. Applying the definition for division of fractions, we have

$$\frac{2}{3} \div 2 = \frac{2}{3} \cdot \frac{1}{2}$$

$$= \frac{2 \cdot 1}{3 \cdot 2}$$

$$= \frac{1}{3}$$

EXAMPLE 4 Divide: $2 \div \frac{1}{3}$.

Solution We replace $\frac{1}{3}$ by its reciprocal, which is 3, and multiply.

$$2 \div \frac{1}{3} = 2(3)$$

$$= 6$$

Here are some further examples of division with fractions. Notice in each case that the first step is the only new part of the process.

EXAMPLE 5 $\frac{4}{27} \div \frac{16}{9}$

Solution We replace $\frac{16}{9}$ by its reciprocal and multiply.

$$= \frac{4}{27} \cdot \frac{9}{16}$$

$$= \frac{4 \cdot 9}{3 \cdot 9 \cdot 4 \cdot 4}$$

$$= \frac{1}{12}$$

In Example 5 we did not factor the numerator and the denominator completely in order to reduce to lowest terms because, as you have probably already noticed, it is not necessary to do so. We need to factor only enough to show what numbers are common to the numerator and the denominator. If we factored completely in the second step, it would look like this:

$$= \frac{2 \cdot 2 \cdot 3 \cdot 3}{3 \cdot 3 \cdot 3 \cdot 2 \cdot 2 \cdot 2 \cdot 2}$$

$$= \frac{1}{12}$$

The result is the same in both cases. From now on, we will factor numerators and denominators only enough to show the factors we are dividing out.

EXAMPLE 6 Divide.

a. $\frac{16}{35} \div 8$ **b.** $27 \div \frac{3}{2}$

Solution

a. $\frac{16}{35} \div 8 = \frac{16}{35} \cdot \frac{1}{8}$

$$= \frac{2 \cdot 8 \cdot 1}{35 \cdot 8}$$

$$= \frac{2}{35}$$

3. Divide: $\frac{5}{6} \div 3$.

4. Divide: $4 \div \frac{2}{3}$.

5. Divide: $\frac{9}{16} \div \frac{9}{4}$.

6. Divide.

a. $\frac{15}{26} \div 5$

b. $18 \div \frac{9}{10}$

Answers

3. $\frac{5}{18}$

4. 6

5. $\frac{1}{4}$

6. a. $\frac{3}{26}$ **b.** 20

b. $27 \div \dfrac{3}{2} = 27 \cdot \dfrac{2}{3}$

$\qquad\qquad = \dfrac{3 \cdot 9 \cdot 2}{3}$

$\qquad\qquad = 18$

B Order of Operations

The next two examples combine what we have learned about division of fractions with the rule for order of operations.

7. The quotient of $\frac{2}{3}$ and $\frac{1}{6}$ is decreased by 2. What number results?

EXAMPLE 7   The quotient of $\frac{8}{3}$ and $\frac{1}{6}$ is increased by 5. What number results?

Solution   Translating to symbols, we have

$$\frac{8}{3} \div \frac{1}{6} + 5 = \left[\frac{8}{3} \cdot \frac{6}{1} \right] + 5$$

$$= 16 + 5$$

$$= 21$$

8. Simplify: $25 \div \left(\frac{5}{6}\right)^2 + 32 \div \left(\frac{4}{5}\right)^2$.

EXAMPLE 8   Simplify $32 \div \left(\frac{4}{3}\right)^2 + 75 \div \left(\frac{5}{2}\right)^2$.

Solution   According to the rule for order of operations, we must first evaluate the numbers with exponents, then we divide, and finally we add.

$$\left[32 \div \left(\frac{4}{3}\right)^2 \right] + \left[75 \div \left(\frac{5}{2}\right)^2 \right] = 32 \div \frac{16}{9} + 75 \div \frac{25}{4}$$

$$= 32 \cdot \frac{9}{16} + 75 \cdot \frac{4}{25}$$

$$= 18 + 12$$

$$= 30$$

9. Repeat Example 9 if they have 16 yards of material, and each blanket requires $\frac{2}{3}$ yard of material.

EXAMPLE 9   A 4-H Club is making blankets to keep their lambs clean at the county fair. If each blanket requires $\frac{3}{4}$ yard of material, how many blankets can they make from 9 yards of material?

Solution   To answer this question we must divide 9 by $\frac{3}{4}$.

$$9 \div \frac{3}{4} = 9 \cdot \frac{4}{3}$$

$$= 3 \cdot 4$$

$$= 12$$

They can make 12 blankets from the 9 yards of material.

GETTING READY FOR CLASS

After reading through the preceding section, respond in your own words and in complete sentences.

A. What do we call two numbers whose product is 1?

B. True or false? The quotient of $\frac{3}{5}$ and $\frac{3}{8}$ is the same as the product of $\frac{3}{5}$ and $\frac{8}{3}$.

C. How are multiplication and division of fractions related?

D. Dividing by $\frac{19}{9}$ is the same as multiplying by what number?

Answers
7. 2
8. 86
9. 24 blankets

Vocabulary Review

Choose the correct words to fill in the blanks below.

divisor   product   fraction   reciprocal

1. Two numbers whose _____ is 1 are said to be reciprocals.

2. In symbols, the _____ of $\frac{a}{b}$ is $\frac{b}{a}$ if a and b are not equal to zero.

3. To divide by a fraction, replace the _____ by its reciprocal and multiply.

4. In symbols, division by a _____ is written as $\frac{a}{b} \div \frac{c}{d} = \frac{a}{b} \cdot \frac{d}{c}$.

Problems

A Find the quotient in each case by replacing the divisor by its reciprocal and multiplying.

1. $\frac{3}{4} \div \frac{1}{5}$

2. $\frac{1}{3} \div \frac{1}{2}$

3. $\frac{2}{3} \div \frac{1}{2}$

4. $\frac{5}{8} \div \frac{1}{4}$

5. $6 \div \frac{2}{3}$

6. $8 \div -\frac{3}{4}$

7. $20 \div \frac{1}{10}$

8. $16 \div \frac{1}{8}$

9. $\frac{3}{4} \div 2$

10. $\frac{3}{5} \div 2$

11. $\frac{7}{8} \div -\frac{7}{8}$

12. $\frac{4}{3} \div \frac{4}{3}$

13. $\frac{7}{8} \div \frac{8}{7}$

14. $\frac{4}{3} \div \frac{3}{4}$

15. $\frac{9}{16} \div \frac{3}{4}$

16. $\frac{25}{36} \div \frac{5}{6}$

17. $\frac{25}{46} \div \frac{40}{69}$

18. $-\frac{25}{24} \div \frac{15}{36}$

19. $\frac{13}{28} \div \frac{39}{14}$

20. $\frac{28}{125} \div \frac{5}{2}$

21. $\frac{27}{196} \div \frac{9}{392}$

22. $\frac{16}{135} \div \frac{2}{45}$

23. $\frac{25}{18} \div 5$

24. $-\frac{30}{27} \div 6$

25. $6 \div \frac{4}{3}$

26. $-12 \div \frac{4}{3}$

27. $\frac{4}{3} \div 6$

28. $\frac{4}{3} \div 12$

29. $\frac{3}{4} \div \frac{1}{2} \cdot 6$

30. $12 \div \frac{6}{7} \cdot 7$

31. $\frac{2}{3} \cdot \frac{3}{4} \div \frac{5}{8}$

32. $4 \cdot -\frac{7}{6} \div 7$

33. $\frac{35}{10} \cdot \frac{80}{63} \div -\frac{16}{27}$

34. $\frac{20}{72} \cdot \frac{42}{18} \div \frac{20}{16}$

B Simplify each expression as much as possible using order of operations.

35. $10 \div \left(\dfrac{1}{2}\right)^2$

36. $12 \div \left(\dfrac{1}{4}\right)^2$

37. $\dfrac{18}{35} \div \left(\dfrac{6}{7}\right)^2$

38. $\dfrac{48}{55} \div \left(-\dfrac{8}{11}\right)^2$

39. $\dfrac{4}{5} \div \dfrac{1}{10} + 5$

40. $\dfrac{3}{8} \div \dfrac{1}{16} + 4$

41. $10 + \dfrac{11}{12} \div \dfrac{11}{24}$

42. $15 + \dfrac{13}{14} \div \dfrac{13}{42}$

43. $24 \div \left(\dfrac{2}{5}\right)^2 + 25 \div \left(\dfrac{5}{6}\right)^2$

44. $18 \div \left(\dfrac{3}{4}\right)^2 + 49 \div \left(\dfrac{7}{9}\right)^2$

45. $100 \div \left(\dfrac{5}{7}\right)^2 + 200 \div \left(\dfrac{2}{3}\right)^2$

46. $64 \div \left(\dfrac{8}{11}\right)^2 + 81 \div \left(\dfrac{9}{11}\right)^2$

47. What is the quotient of $\frac{3}{8}$ and $\frac{5}{8}$?

48. Find the quotient of $\frac{4}{5}$ and $\frac{16}{25}$.

49. If the quotient of 18 and $\frac{3}{5}$ is increased by 10, what number results?

50. If the quotient of 50 and $\frac{5}{3}$ is increased by 8, what number results?

51. Show that multiplying 3 by 5 is the same as dividing 3 by $\frac{1}{5}$.

52. Show that multiplying 8 by $\frac{1}{2}$ is the same as dividing 8 by 2.

Applying the Concepts

Although many of the application problems that follow involve division with fractions, some do not. Be sure to read the problems carefully.

53. Sewing If $\frac{6}{7}$ yard of material is needed to make a blanket, how many blankets can be made from 12 yards of material?

54. Manufacturing A clothing manufacturer is making scarves that require $\frac{3}{8}$ yard of material each. How many can be made from 27 yards of material?

55. Capacity Suppose a bag of candy holds exactly $\frac{1}{4}$ pound of candy. How many of these bags can be filled from 12 pounds of candy?

56. Capacity A certain size bottle holds exactly $\frac{4}{5}$ pint of liquid. How many of these bottles can be filled from a 20-pint container?

57. Cooking Audra is making cookies from a recipe that calls for $\frac{3}{4}$ teaspoon of oil. If the only measuring spoon she can find is a $\frac{1}{8}$ teaspoon, how many of these will she have to fill with oil in order to have a total of $\frac{3}{4}$ teaspoon of oil?

58. Cooking A cake recipe calls for $\frac{1}{2}$ cup of sugar. If the only measuring cup available is a $\frac{1}{8}$ cup, how many of these will have to be filled with sugar to make a total of $\frac{1}{2}$ cup of sugar?

59. Student Population If 14 of every 32 students attending Cuesta College are female, what fraction of the students is female? (Simplify your answer.)

60. Population If 27 of every 48 residents of a small town are male, what fraction of the population is male? (Simplify your answer.)

61. Student Population If 14 of every 32 students attending Cuesta College are female, and the total number of students at the school is 4,064, how many of the students are female?

62. Population If 27 of every 48 residents of a small town are male, and the total population of the town is 17,808, how many of the residents are male?

63. Cartons of Milk If a small carton of milk holds exactly $\frac{1}{2}$ pint, how many of the $\frac{1}{2}$-pint cartons can be filled from a 14-pint container?

64. Pieces of Pipe How many pieces of pipe that are $\frac{2}{3}$ foot long must be laid together to make a pipe 16 feet long?

Find the Mistake

Each sentence below contains a mistake. Circle the mistake and write the correct word(s) or numbers(s) on the line provided.

1. Two numbers whose quotient is 1 are said to be reciprocals. _____

2. Dividing the fraction $\frac{12}{7}$ by $\frac{4}{9}$, is equivalent to $\frac{7}{12} \cdot \frac{9}{4}$. _____

3. To work the problem $\frac{22}{5} \div \frac{10}{3}$, multiply the first fraction by its reciprocal. _____

4. The quotient of $\frac{14}{11}$ and $\frac{32}{6}$ is $\frac{224}{33}$. _____

Supplies Needed

A   Piece of graph paper
B   4 colored pens or pencils

Working With Fractions

This project will provide a visual approach for working with the multiplication and division of fractions. You will need a piece of graph paper and 4 colored pens or pencils.

1. On your graph paper, draw two squares each with a side length of 20 boxes. Fill in the boxes of your first square using your pens or pencils. You may choose any pattern as long as you use all four colors and fill in every box in your square.

2. Assign each color to a number (1-4).

    Color 1 _____

    Color 2 _____

    Color 3 _____

    Color 4 _____

3. Find the area of your first square. This will give you the total number of boxes in your square.

    Total area _____

4. Count the total number of boxes for each color. Enter these quantities in the second column of the table titled, "Number of Boxes."

Colors	Number of boxes	Fraction (not reduced)	Fraction (reduced)	Multiplied by $\frac{2}{3}$	Multiplied by $\frac{3}{4}$

5. In the third column, "Fraction (not reduced)," show the number of boxes for each color in the form of a fraction.

6. Reduce each fraction, if possible, and write the reduced fraction in the fourth column of the table.

7. Multiply each of the reduced fractions for Color 1 and Color 2 by $\frac{2}{3}$. Write these new fractions in the fifth column of the table.

8. Multiply each of the reduced fractions for Color 3 and Color 4 by $\frac{3}{4}$. Write these new fractions in the sixth column of the table.

9. Suppose the fractions in the fifth and sixth columns of your table now represent new quantities of boxes per color. Fill in your second square with these new quantities. Round to the nearest whole number. How many boxes are left over? Write this number as a fraction.

Definition of Fractions [2.1]

A fraction is any number that can be written in the form $\frac{a}{b}$, where a and b are numbers and b is not 0. The number a is called the *numerator*, and the number b is called the *denominator*.

EXAMPLES

1. Each of the following is a fraction:

$$\frac{1}{2}, \quad \frac{3}{4}, \quad \frac{8}{1}, \quad \frac{7}{3}$$

Properties of Fractions [2.1]

Multiplying the numerator and the denominator of a fraction by the same nonzero number will produce an equivalent fraction. The same is true for dividing the numerator and denominator by the same nonzero number. In symbols the properties look like this:

If a, b, and c are numbers and b and c are not 0, then

$$\text{Multiplication property for fractions} \quad \frac{a}{b} = \frac{a \cdot c}{b \cdot c}$$

$$\text{Division property for fraction} \quad \frac{a}{b} = \frac{a \div c}{b \div c}$$

2. Change $\frac{3}{4}$ to an equivalent fraction with denominator 12.

$$\frac{3}{4} = \frac{3 \cdot 3}{4 \cdot 3} = \frac{9}{12}$$

Fractions and the Number 1 [2.1]

If a represents any number, then

$$\frac{a}{1} = a \quad \text{and} \quad \frac{a}{a} = 1 \quad \text{(where } a \text{ is not 0)}$$

3. $\frac{5}{1} = 5, \frac{5}{5} = 1$

Reducing Fractions to Lowest Terms [2.2]

To reduce a fraction to lowest terms, factor the numerator and the denominator, and then divide both the numerator and denominator by any factors they have in common.

4.
$$\frac{90}{588} = \frac{2 \cdot 3 \cdot 3 \cdot 5}{2 \cdot 2 \cdot 3 \cdot 7 \cdot 7}$$
$$= \frac{3 \cdot 5}{2 \cdot 7 \cdot 7}$$
$$= \frac{15}{98}$$

Multiplying Fractions [2.3]

To multiply fractions, multiply numerators and multiply denominators.

5. $\frac{3}{5} \cdot \frac{4}{7} = \frac{3 \cdot 4}{5 \cdot 7} = \frac{12}{35}$

The Area of a Triangle [2.3]

The formula for the area of a triangle with base b and height h is

$$A = \frac{1}{2} bh$$

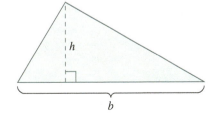

6. If the base of a triangle is 10 inches and the height is 7 inches, then the area is

$$A = \frac{1}{2} bh$$
$$= \frac{1}{2} \cdot 10 \cdot 7$$
$$= 5 \cdot 7$$
$$= 35 \text{ square inches}$$

Reciprocals [2.4]

7. The reciprocal of $\frac{2}{3}$ is $\frac{3}{2}$.

$$\frac{2}{3} \cdot \frac{3}{2} = 1$$

Any two numbers whose product is 1 are called *reciprocals*. The reciprocal of a is $\frac{1}{a}$ because their product is 1.

Division with Fractions [2.4]

8. $\frac{3}{8} \div \frac{1}{3} = \frac{3}{8} \cdot \frac{3}{1} = \frac{9}{8}$

To divide by a fraction, you must multiply by its reciprocal. That is, the quotient of two fractions is defined to be the product of the first fraction with the reciprocal of the second fraction (the divisor).

COMMON MISTAKE

1. A common mistake made with division of fractions occurs when we multiply by the reciprocal of the first fraction instead of the reciprocal of the divisor. For example,

$$\frac{2}{3} \div \frac{5}{6} \neq \frac{3}{2} \cdot \frac{5}{6}$$

Remember, we perform division by multiplying by the reciprocal of the divisor (the fraction to the right of the division symbol).

2. If the answer to a problem turns out to be a fraction, that fraction should always be written in lowest terms. It is a mistake not to reduce to lowest terms.

Reduce to lowest terms. [2.2]

1. $\dfrac{75}{105}$

2. $\dfrac{72}{192}$

3. $\dfrac{64}{208}$

4. $\dfrac{176}{330}$

Perform the indicated operations. Reduce all answers to lowest terms.[2.2, 2.3, 2.4]

5. $\dfrac{1}{3} \cdot \dfrac{6}{7}$

6. $\dfrac{2}{5} \cdot \dfrac{3}{8}$

7. $6 \cdot \dfrac{7}{12}$

8. $\dfrac{4}{15}\left(\dfrac{5}{8}\right)$

9. $-\dfrac{3}{7} \cdot \dfrac{4}{9}$

10. $-\dfrac{9}{4}\left(\dfrac{1}{3}\right)$

11. $-\dfrac{7}{4} \cdot \dfrac{8}{3} \cdot -\dfrac{9}{14}$

12. $\dfrac{6}{5} \cdot -\dfrac{15}{7} \cdot -\dfrac{3}{9}$

13. $\dfrac{7}{9} \div \dfrac{2}{3}$

14. $\dfrac{8}{5} \div \dfrac{2}{3}$

15. $-\dfrac{9}{16} \div \dfrac{3}{12}$

16. $\dfrac{4}{3} \div -16$

17. $15 \div \dfrac{3}{5}$

18. $\dfrac{24}{14} \div \dfrac{6}{7}$

Simplify each of the following as much as possible.
[2.2, 2.3, 2.4]

19. $\dfrac{8}{5} \cdot \dfrac{3}{4} \div \dfrac{9}{10}$

20. $\dfrac{12}{7} \div 3 \cdot \dfrac{5}{6}$

21. $\left(\dfrac{3}{14}\right)\left(\dfrac{4}{6}\right) \div \dfrac{1}{3}$

22. $\left(\dfrac{2}{3}\right)^3 \cdot \dfrac{9}{16}$

23. $\dfrac{4}{9} \div \dfrac{2}{7} \div \dfrac{1}{3}$

24. $\dfrac{8}{9} \div \left(\dfrac{4}{6}\right)^2$

25. $\dfrac{1}{6} \div \left(\dfrac{15}{18} \cdot -\dfrac{9}{25}\right)$

26. $\dfrac{5}{9} \cdot \dfrac{6}{7} \div -\dfrac{20}{14}$

27. $\left(\dfrac{4}{21}\right)\left(\dfrac{3}{8}\right) \div \dfrac{1}{6}$

28. $\left(\dfrac{3}{7} \div \dfrac{9}{14}\right)\left(\dfrac{6}{7}\right)$

29. Find the area of the triangle. [2.3]

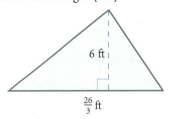

6 ft

$\dfrac{26}{3}$ ft

30. Sewing A dress requires $\dfrac{7}{4}$ yards of material to make. If you have 14 yards of material, how many dresses can you make? [2.4]

The illustration shows the most popular Netflix plans and the percentage of customers who choose them.

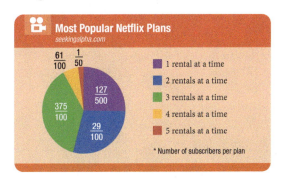

Most Popular Netflix Plans
seekingalpha.com

$\dfrac{61}{100}$ $\dfrac{1}{50}$

$\dfrac{127}{500}$

$\dfrac{375}{100}$

$\dfrac{29}{100}$

■ 1 rental at a time
■ 2 rentals at a time
■ 3 rentals at a time
■ 4 rentals at a time
■ 5 rentals at a time

* Number of subscribers per plan

31. Write the number of subscribers who choose 3 rentals at a time as a fraction in lowest terms.

32. Write the number of subscribers who choose 1 rental at a time as an equivalent fraction with a denominator of 1000.

Give the opposite and absolute value of each number.

1. $-\dfrac{4}{5}$

2. 27

Simplify.

3. $\left(\dfrac{4}{9}\right)^2$

4. $16 - (13 - 17)$

5. $4(2) - (19 - 15)$

6. $\dfrac{1}{2} \div \dfrac{1}{6}$

7. $\dfrac{4}{7} \cdot \dfrac{5}{8}$

8. $6^2 - 2^4$

9. $3 \div \dfrac{1}{5}$

10. $270 \div -15 \div 6$

11. $\dfrac{62}{12}$

12. $\dfrac{144}{240}$

13. 13^2

14. $(8 - 5) - (9 - 12)$

15. $3^2 - 4 \div 2$

16. $2,050(131)$

17. $\dfrac{4}{16} \div \dfrac{12}{18}$

18. $3 + 6^2 - 3 \cdot 4 + 7$

19. $15 \cdot \dfrac{2}{3}$

20. $\left(\dfrac{3}{5}\right)^2 \cdot \left(-\dfrac{1}{3}\right)^3$

21. $4,936 - 691$

22. $\left(\dfrac{7}{14}\right)\left(-\dfrac{8}{16}\right)\left(-\dfrac{4}{5}\right)$

23. $\left(-\dfrac{4}{7}\right)^2\left(\dfrac{14}{24}\right)$

24. $\left(\dfrac{2}{3}\right)\left(\dfrac{6}{16}\right) \div \dfrac{2}{3}$

25. $\dfrac{7(3) + 4(2 - 3)}{2(10 + 7)}$.

26. Find the product of $\dfrac{3}{5}$, $\dfrac{7}{15}$, and $\dfrac{5}{9}$.

27. Three times the difference of 5 and 1 is decreased by 4. What number results?

28. Reduce $\dfrac{15}{40}$ to lowest terms.

29. What is the quotient of $\dfrac{4}{5}$ and the product of $\dfrac{2}{3}$ and $\dfrac{6}{7}$

30. Aaron's checking account has a balance of $243. On Monday, he deposits $314. On Tuesday, he writes a check for $137 and Wednesday he writes a check for $38. What is his balance at the end of the day on Wednesday?

The chart shows the number of viewers for ABC's top primetime shows.

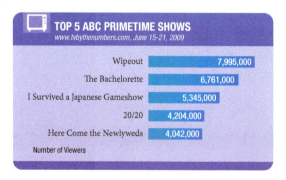

31. **TV Shows** How many viewers watched The Bachelorette and 20/20?

Reduce to lowest terms. [2.2]

1. $\dfrac{25}{125}$

2. $\dfrac{32}{128}$

3. $\dfrac{15}{70}$

4. $\dfrac{255}{340}$

Perform the indicated operations. Reduce all answers to lowest terms. [2.2, 2.3, 2.4]

5. $\dfrac{1}{4} \cdot \dfrac{3}{5}$

6. $\dfrac{2}{3} \cdot \dfrac{5}{7}$

7. $3 \cdot \dfrac{5}{8}$

8. $\dfrac{3}{16}\left(\dfrac{8}{9}\right)$

9. $\dfrac{3}{8} \cdot -\dfrac{12}{15}$

10. $\dfrac{7}{5}\left(-\dfrac{1}{14}\right)$

11. $\dfrac{1}{2} \cdot -\dfrac{3}{8} \cdot -\dfrac{4}{5}$

12. $\dfrac{7}{5} \cdot -\dfrac{12}{7} \cdot -\dfrac{1}{3}$

13. $\dfrac{8}{5} \div \dfrac{2}{5}$

14. $\dfrac{3}{8} \div \dfrac{4}{5}$

15. $\dfrac{8}{15} \div -\dfrac{2}{3}$

16. $\dfrac{3}{5} \div -9$

17. $16 \div \dfrac{8}{3}$

18. $\dfrac{36}{18} \div \dfrac{9}{2}$

Simplify each of the following as much as possible. [2.2, 2.3, 2.4]

19. $\dfrac{2}{3} \cdot \dfrac{1}{2} \div \dfrac{5}{6}$

20. $\dfrac{10}{7} \div 5 \cdot \dfrac{3}{4}$

21. $\left(\dfrac{4}{11}\right)\left(\dfrac{5}{8}\right) \div \dfrac{1}{4}$

22. $\left(\dfrac{3}{4}\right)^2 \cdot \dfrac{1}{3}$

23. $\dfrac{6}{11} \div \dfrac{3}{5} \div \dfrac{1}{2}$

24. $\dfrac{7}{8} \div \left(\dfrac{3}{2}\right)^2$

25. $-\dfrac{1}{5} \div \left(\dfrac{2}{3} \cdot \dfrac{4}{5}\right)$

26. $\dfrac{3}{8} \cdot -\dfrac{5}{9} \div \dfrac{5}{12}$

27. $\left(\dfrac{5}{18}\right)\left(\dfrac{4}{5}\right) \div \dfrac{10}{9}$

28. $\left(\dfrac{6}{5} \div \dfrac{9}{8}\right)\left(\dfrac{5}{8}\right)$

29. Find the area of the triangle. [2.3]

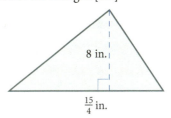

8 in.

$\dfrac{15}{4}$ in.

30. **Sewing** A dress requires $\dfrac{6}{5}$ yards of material to make. If you have 12 yards of material, how many dresses can you make? [2.4]

The illustration shows what new iPhone buyers had as their previous phone.

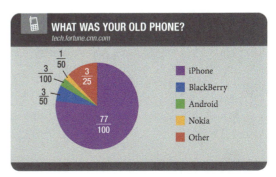

31. Write the fraction of iPhone buyers who had a Blackberry as their previous phone as an equivalent fraction with a denominator of 100.

32. Write the fraction of iPhone buyers who previously had a Nokia phone as an equivalent fraction with a denominator of 200.

Addition and Subtraction of Fractions, Mixed Numbers

3

Grand Canyon, Arizona
Image © 2010 DigitalGlobe
Image USDA Farm Service Agency

Google

The Grand Canyon in Arizona has some of the most breathtaking views in the world. Carved by the Colorado River over millions of years, the Grand Canyon now measures 277 miles in length. If you travel to the canyon's western rim, the Grand Canyon Skywalk is not to be missed. The U-shaped observation deck has a glass bottom and allows visitors to "walk the sky" some 4,000 feet above the Colorado River. Weighing 1.2 million pounds, the glass is the only thing separating patrons from the deep canyon below. Built with permission from the local Hualapai, a Native American tribe, the skywalk has hosted over 2 million tourists from 50 countries since its 2007 grand opening.

Inner Canyon Precipitation (in inches)

Jan	Feb	Mar	April	May	June	July	Aug	Sept	Oct	Nov	Dec
$\frac{17}{25}$	$\frac{3}{4}$	$\frac{79}{100}$	$\frac{47}{100}$	$\frac{9}{25}$	$\frac{21}{25}$	$\frac{21}{25}$	$1\frac{2}{5}$	$\frac{97}{100}$	$\frac{65}{100}$	$\frac{43}{100}$	$\frac{87}{100}$

To ensure an optimal viewing experience, it is important to check the weather forecast before traveling to the Grand Canyon. The table shows the average precipitation inside the canyon. Suppose you were planning to spend the months of August and September touring the region. How much precipitation can you expect during those two months? If you want to know whether to pack an umbrella you must know how to add fractions, which is one of the topics we will cover in this chapter.

A Add and subtract fractions with common denominators.

B Add and subtract fractions with unlike denominators.

least common denominator
(LCD)

Video Examples
Section 3.1

Note Most people who have done any work with adding fractions know that you add fractions that have the same denominator by adding their numerators, but not their denominators. However, most people don't know why this works. The reason why we add numerators but not denominators is because of the distributive property. That is what the discussion here is all about. If you really want to understand addition of fractions, pay close attention to this discussion.

3.1 Addition and Subtraction with Fractions

©iStockphoto.com/polygrafix

In Las Vegas, the Stratosphere's hotel and casino has opened another attraction for thrill-seekers. SkyJump is a death-defying controlled free-fall from the 108th floor of the hotel to the ground. Jump Package 1 includes the jump cost plus a DVD of the jump for $114.99. A jump without the DVD costs $99.99. Suppose you are part of a group of people that schedules an appointment for the SkyJump. $\frac{2}{7}$ of the group buys Jump Package 1 and $\frac{3}{7}$ buys the jump without the DVD. The remaining $\frac{1}{7}$ of the group decides not to jump and instead buys the "Chicken" shirt available for purchase at the SkyJump store. What fraction represents the amount of people in the group that actually jumped? To answer this question, we must be able to add fractions with a common denominator.

A Addition and Subtraction with Common Denominators

Adding and subtracting fractions is actually just another application of the distributive property. The distributive property looks like this:

$$a(b + c) = a(b) + a(c)$$

where a, b, and c may be whole numbers or fractions. We will want to apply this property to expressions like

$$\frac{2}{7} + \frac{3}{7}$$

But before we do, we must make one additional observation about fractions.

The fraction $\frac{2}{7}$ can be written as $2 \cdot \frac{1}{7}$, because

$$2 \cdot \frac{1}{7} = \frac{2}{1} \cdot \frac{1}{7} = \frac{2}{7}$$

Likewise, the fraction $\frac{3}{7}$ can be written as $3 \cdot \frac{1}{7}$, because

$$3 \cdot \frac{1}{7} = \frac{3}{1} \cdot \frac{1}{7} = \frac{3}{7}$$

In general, we can say that the fraction $\frac{a}{b}$ can always be written as $a \cdot \frac{1}{b}$, because

$$a \cdot \frac{1}{b} = \frac{a}{1} \cdot \frac{1}{b} = \frac{a}{b}$$

To add the fractions $\frac{2}{7}$ and $\frac{3}{7}$, we simply rewrite each of them as we have done above and apply the distributive property. Here is how it works:

$$\frac{2}{7} + \frac{3}{7} = 2 \cdot \frac{1}{7} + 3 \cdot \frac{1}{7} \qquad \text{\textit{Rewrite each fraction.}}$$

$$= (2 + 3) \cdot \frac{1}{7} \qquad \text{\textit{Apply the distributive property.}}$$

$$= 5 \cdot \frac{1}{7} \qquad \text{\textit{Add 2 and 3 to get 5.}}$$

$$= \frac{5}{7} \qquad \text{\textit{Rewrite }} 5 \cdot \frac{1}{7} \text{ \textit{as} } \frac{5}{7}.$$

We can visualize the process shown above by using circles that are divided into 7 equal parts.

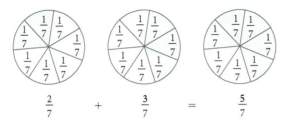

$$\frac{2}{7} \qquad + \qquad \frac{3}{7} \qquad = \qquad \frac{5}{7}$$

The fraction $\frac{5}{7}$ is the sum of $\frac{2}{7}$ and $\frac{3}{7}$. The steps and diagrams above show why we add numerators, *but do not add denominators*. Using this example as justification, we can write a rule for adding two fractions that have the same denominator.

> **RULE** Addition with Common Denominator
>
> To add two fractions that have the same denominator, we add their numerators to get the numerator of the answer. The denominator in the answer is the same denominator as in the original fractions.

What we have here is the sum of the numerators placed over the *common denominator*. In symbols, we have the following:

> **Addition and Subtraction of Fractions**
>
> If a, b, and c are numbers, and c is not equal to 0, then
> $$\frac{a}{c} + \frac{b}{c} = \frac{a+b}{c}$$
> This rule holds for subtraction as well. That is,
> $$\frac{a}{c} - \frac{b}{c} = \frac{a-b}{c}$$

EXAMPLE 1 Add or subtract.

a. $\frac{3}{8} + \frac{1}{8}$ **b.** $\frac{9}{5} - \frac{3}{5}$ **c.** $\frac{3}{7} + \frac{2}{7} + \frac{9}{7}$

Solution

a. $\frac{3}{8} + \frac{1}{8} = \frac{3+1}{8}$ Add numerators; keep the same denominator.

$\qquad\qquad = \frac{4}{8}$ The sum of 3 and 1 is 4.

$\qquad\qquad = \frac{1}{2}$ Reduce to lowest terms.

b. $\frac{9}{5} - \frac{3}{5} = \frac{9-3}{5}$ Subtract numerators; keep the same denominator.

$\qquad\qquad = \frac{6}{5}$ The difference of 9 and 3 is 6.

c. $\frac{3}{7} + \frac{2}{7} + \frac{9}{7} = \frac{3+2+9}{7}$

$\qquad\qquad\qquad = \frac{14}{7}$

$\qquad\qquad\qquad = 2$

Practice Problems

1. Add or subtract.

 a. $\frac{5}{12} + \frac{1}{12}$

 b. $\frac{3}{10} + \frac{7}{10}$

 c. $\frac{11}{6} - \frac{5}{6} - \frac{1}{6}$

Answers

1. **a.** $\frac{1}{2}$ **b.** 1 **c.** $\frac{5}{6}$

B Addition and Subtraction with Unlike Denominators

As Example 1 indicates, addition and subtraction are simple, straightforward processes when all the fractions have the same denominator. We will now turn our attention to the process of adding fractions that have different denominators. In order to get started, we need the following definition:

> **DEFINITION** least common denominator
>
> The *least common denominator* (LCD) for a set of denominators is the smallest number that is exactly divisible by each denominator. (Note that, in some books, the least common denominator is also called the *least common multiple*.)

In other words, all the denominators of the fractions involved in a problem must divide into the least common denominator exactly. That is, they divide it without leaving a remainder.

2. Find the LCD for the fractions $\frac{7}{15}$ and $\frac{9}{25}$.

EXAMPLE 2   Find the LCD for the fractions $\frac{5}{12}$ and $\frac{7}{18}$.

Solution   The least common denominator for the denominators 12 and 18 must be the smallest number divisible by both 12 and 18. We can factor 12 and 18 completely and then build the LCD from these factors. Factoring 12 and 18 completely gives us

$$12 = 2 \cdot 2 \cdot 3 \qquad 18 = 2 \cdot 3 \cdot 3$$

Now, if 12 is going to divide the LCD exactly, then the LCD must have factors of $2 \cdot 2 \cdot 3$. If 18 is to divide it exactly, it must have factors of $2 \cdot 3 \cdot 3$. We don't need to repeat the factors that 12 and 18 have in common.

$$\left.\begin{array}{l} 12 = 2 \cdot 2 \cdot 3 \\ 18 = 2 \cdot 3 \cdot 3 \end{array}\right\} \qquad \begin{array}{c} \textit{12 divides the LCD.} \\ \text{LCD} = 2 \cdot 2 \cdot 3 \cdot 3 = 36 \\ \textit{18 divides the LCD.} \end{array}$$

Note The ability to find least common denominators is very important in mathematics. The discussion here is a detailed explanation of how to find an LCD.

The LCD for 12 and 18 is 36. It is the smallest number that is divisible by both 12 and 18; 12 divides it exactly three times, and 18 divides it exactly two times.

We can visualize the results in Example 2 with the diagram below. It shows that 36 is the smallest number that both 12 and 18 divide evenly. As you can see, 12 divides 36 exactly 3 times, and 18 divides 36 exactly 2 times.

12	12	12

18	18

36

3. Add: $\frac{7}{15} + \frac{9}{25}$.

EXAMPLE 3   Add: $\frac{5}{12} + \frac{7}{18}$.

Solution   We can add fractions only when they have the same denominators. In Example 2, we found the LCD for $\frac{5}{12}$ and $\frac{7}{18}$ to be 36. We change $\frac{5}{12}$ and $\frac{7}{18}$ to equivalent fractions that have 36 for a denominator by applying the multiplication property for fractions.

$$\frac{5}{12} = \frac{5 \cdot 3}{12 \cdot 3} = \frac{15}{36}$$

$$\frac{7}{18} = \frac{7 \cdot 2}{18 \cdot 2} = \frac{14}{36}$$

The fraction $\frac{15}{36}$ is equivalent to $\frac{5}{12}$, because it was obtained by multiplying both the numerator and the denominator by 3. Likewise, $\frac{14}{36}$ is equivalent to $\frac{7}{18}$, because it was obtained by multiplying the numerator and the denominator by 2. All we have left to do is to add numerators.

$$\frac{15}{36} + \frac{14}{36} = \frac{29}{36}$$

The sum of $\frac{5}{12}$ and $\frac{7}{18}$ is the fraction $\frac{29}{36}$. Let's write the complete problem again step by step.

$$\frac{5}{12} + \frac{7}{18} = \frac{5 \cdot 3}{12 \cdot 3} + \frac{7 \cdot 2}{18 \cdot 2}$$ *Rewrite each fraction as an equivalent fraction with denominator 36.*

$$= \frac{15}{36} + \frac{14}{36}$$

$$= \frac{29}{36}$$ *Add numerators; keep the common denominator.*

EXAMPLE 4 Find the LCD for $\frac{3}{4}$ and $\frac{1}{6}$.

Solution We factor 4 and 6 into products of prime factors and build the LCD from these factors.

$$\left. \begin{array}{l} 4 = 2 \cdot 2 \\ 6 = 2 \cdot 3 \end{array} \right\} \quad \text{LCD} = 2 \cdot 2 \cdot 3 = 12$$

The LCD is 12. Both denominators divide it exactly; 4 divides 12 exactly 3 times, and 6 divides 12 exactly 2 times.

EXAMPLE 5 Add: $\frac{3}{4} + \frac{1}{6}$.

Solution In Example 4, we found that the LCD for these two fractions is 12. We begin by changing $\frac{3}{4}$ and $\frac{1}{6}$ to equivalent fractions with denominator 12.

$$\frac{3}{4} = \frac{3 \cdot 3}{4 \cdot 3} = \frac{9}{12}$$

$$\frac{1}{6} = \frac{1 \cdot 2}{6 \cdot 2} = \frac{2}{12}$$

The fraction $\frac{9}{12}$ is equal to the fraction $\frac{3}{4}$, because it was obtained by multiplying the numerator and the denominator of $\frac{3}{4}$ by 3. Likewise, $\frac{2}{12}$ is equivalent to $\frac{1}{6}$, because it was obtained by multiplying the numerator and the denominator of $\frac{1}{6}$ by 2. To complete the problem, we add numerators.

$$\frac{9}{12} + \frac{2}{12} = \frac{11}{12}$$

The sum of $\frac{3}{4}$ and $\frac{1}{6}$ is $\frac{11}{12}$. Here is how the complete problem looks:

$$\frac{3}{4} + \frac{1}{6} = \frac{3 \cdot 3}{4 \cdot 3} + \frac{1 \cdot 2}{6 \cdot 2}$$ *Rewrite each fraction as an equivalent fraction with denominator 12.*

$$= \frac{9}{12} + \frac{2}{12}$$

$$= \frac{11}{12}$$ *Add numerators; keep the same denominator.*

4. Find the LCD for $\frac{7}{12}$ and $\frac{3}{8}$.

5. Add: $\frac{7}{12} + \frac{3}{8}$.

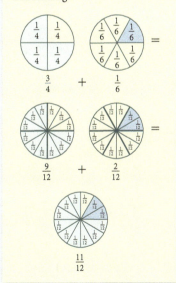

Note We can visualize the work in Example 5 using circles and shading.

6. Subtract: $\frac{3}{4} - \frac{5}{12}$.

EXAMPLE 6   Subtract: $\frac{7}{15} - \frac{3}{10}$.

Solution   Let's factor 15 and 10 completely and use these factors to build the LCD.

15 divides the LCD.

$$\left.\begin{array}{l} 15 = 3 \cdot 5 \\ 10 = 2 \cdot 5 \end{array}\right\} \text{LCD} = 2 \cdot 3 \cdot 5 = 30$$

10 divides the LCD.

Changing to equivalent fractions and subtracting, we have

$$\frac{7}{15} - \frac{3}{10} = \frac{7 \cdot 2}{15 \cdot 2} - \frac{3 \cdot 3}{10 \cdot 3} \quad \text{Rewrite as equivalent fractions with the LCD for the denominator.}$$

$$= \frac{14}{30} - \frac{9}{30}$$

$$= \frac{5}{30} \qquad \text{Subtract numerators; keep the LCD.}$$

$$= \frac{1}{6} \qquad \text{Reduce to lowest terms.}$$

As a summary of what we have done so far, and as a guide to working other problems, we now list the steps involved in adding and subtracting fractions with different denominators.

HOW TO To Add or Subtract Any Two Fractions

Step 1: Factor each denominator completely, and use the factors to build the LCD. (Remember, the LCD is the smallest number divisible by each of the denominators in the problem.)

Step 2: Rewrite each fraction as an equivalent fraction with the LCD. This is done by multiplying both the numerator and the denominator of the fraction in question by the appropriate whole number.

Step 3: Add or subtract the numerators of the fractions produced in Step 2. This is the numerator of the sum or difference. The denominator of the sum or difference is the LCD.

Step 4: Reduce the fraction produced in Step 3 to lowest terms if it is not already in lowest terms.

The idea behind adding or subtracting fractions is really very straight-forward. We can only add or subtract fractions that have the same denominators. If the fractions we are trying to add or subtract do not have the same denominators, we rewrite each of them as an equivalent fraction with the LCD for a denominator.

Here are some additional examples of sums and differences of fractions:

7. Subtract: $\frac{1}{2} - \frac{3}{10}$.

EXAMPLE 7   Subtract: $\frac{3}{5} - \frac{1}{6}$.

Solution   The LCD for 5 and 6 is their product, 30. We begin by rewriting each fraction with this common denominator.

$$\frac{3}{5} - \frac{1}{6} = \frac{3 \cdot 6}{5 \cdot 6} - \frac{1 \cdot 5}{6 \cdot 5}$$

$$= \frac{18}{30} - \frac{5}{30}$$

$$= \frac{13}{30}$$

Answers

6. $\frac{1}{3}$

7. $\frac{1}{5}$

EXAMPLE 8 Add: $\frac{1}{6} + \frac{1}{8} + \frac{1}{4}$.

Solution We begin by factoring the denominators completely and building the LCD from the factors that result.

$$
\left.\begin{array}{l}
6 = 2 \cdot 3 \\
8 = 2 \cdot 2 \cdot 2 \\
4 = 2 \cdot 2
\end{array}\right\}
\qquad
\begin{array}{c}
\textit{8 divides the LCD.} \\
\text{LCD} = 2 \cdot 2 \cdot 2 \cdot 3 = 24 \\
\textit{4 divides the LCD.} \quad \textit{6 divides the LCD.}
\end{array}
$$

We then change to equivalent fractions and add as usual.

$$\frac{1}{6} + \frac{1}{8} + \frac{1}{4} = \frac{1 \cdot 4}{6 \cdot 4} + \frac{1 \cdot 3}{8 \cdot 3} + \frac{1 \cdot 6}{4 \cdot 6} = \frac{4}{24} + \frac{3}{24} + \frac{6}{24} = \frac{13}{24}$$

8. Add: $\frac{1}{3} + \frac{3}{8} + \frac{1}{4}$.

EXAMPLE 9 Subtract: $3 - \frac{5}{6}$.

Solution The denominators are 1 (because $3 = \frac{3}{1}$) and 6. The smallest number divisible by both 1 and 6 is 6.

$$3 - \frac{5}{6} = \frac{3}{1} - \frac{5}{6} = \frac{3 \cdot 6}{1 \cdot 6} - \frac{5}{6} = \frac{18}{6} - \frac{5}{6} = \frac{13}{6}$$

9. Subtract: $4 - \frac{5}{7}$.

Here are three examples that involve addition with negative numbers.

EXAMPLE 10 Add: $\frac{3}{8} + \left(-\frac{1}{8}\right)$.

Solution We subtract absolute values. The answer will be positive, because $\frac{3}{8}$ is positive.

$$\frac{3}{8} + \left(-\frac{1}{8}\right) = \frac{2}{8}$$

$$= \frac{1}{4} \qquad \textit{Reduce to lowest terms.}$$

10. Add: $\frac{4}{5} + \left(-\frac{2}{5}\right)$.

EXAMPLE 11 Add: $\frac{1}{10} + \left(-\frac{4}{5}\right) + \left(-\frac{3}{20}\right)$.

Solution To begin, change each fraction to an equivalent fraction with an LCD of 20.

$$\frac{1}{10} + \left(-\frac{4}{5}\right) + \left(-\frac{3}{20}\right) = \frac{1 \cdot 2}{10 \cdot 2} + \left(-\frac{4 \cdot 4}{5 \cdot 4}\right) + \left(-\frac{3}{20}\right)$$

$$= \frac{2}{20} + \left(-\frac{16}{20}\right) + \left(-\frac{3}{20}\right)$$

$$= -\frac{14}{20} + \left(-\frac{3}{20}\right)$$

$$= -\frac{17}{20}$$

11. Add: $\frac{5}{12} + \left(-\frac{5}{6}\right) + \left(-\frac{5}{24}\right)$.

EXAMPLE 12 Find the difference of $-\frac{3}{5}$ and $\frac{2}{5}$.

Solution

$$-\frac{3}{5} - \frac{2}{5} = -\frac{3}{5} + \left(-\frac{2}{5}\right)$$

$$= -\frac{5}{5}$$

$$= -1$$

12. Find the difference of $-\frac{5}{8}$ and $\frac{3}{8}$.

Comparing Fractions

As we have shown previously, we can compare fractions to see which is larger or smaller when they have the same denominator. Now that we know how to find the LCD for a set of fractions, we can use the LCD to write equivalent fractions with the intention of comparing them.

13. Write the fractions from smallest to largest.

$$\frac{3}{4}, \frac{7}{12}, \frac{5}{8}, \frac{13}{16}$$

EXAMPLE 13   Find the LCD for the fractions below, then write each fraction as an equivalent fraction with the LCD for a denominator. Then write them in order from smallest to largest.

$$\frac{5}{8} \qquad \frac{5}{16} \qquad \frac{3}{4} \qquad \frac{1}{2}$$

Solution   The LCD for the four fractions is 16. We begin by writing each fraction as an equivalent fraction with denominator 16.

$$\frac{5}{8} = \frac{10}{16} \qquad \frac{5}{16} = \frac{5}{16} \qquad \frac{3}{4} = \frac{12}{16} \qquad \frac{1}{2} = \frac{8}{16}$$

Now that they all have the same denominator, the smallest fraction is the one with the smallest numerator, and the largest fraction is the one with the largest numerator. Writing them in order from smallest to largest we have

$$\frac{5}{16} < \frac{8}{16} < \frac{10}{16} < \frac{12}{16}$$

$$\frac{5}{16} < \frac{1}{2} < \frac{5}{8} < \frac{3}{4}$$

GETTING READY FOR CLASS

After reading through the preceding section, respond in your own words and in complete sentences.

A. When adding two fractions with the same denominators, we always add their _____, but we never add their _____.

B. What does the abbreviation LCD stand for?

C. What is the first step when finding the LCD for the fractions $\frac{5}{12}$ and $\frac{7}{18}$?

D. When adding fractions, what is the last step?

Answers

13. $\frac{7}{12} < \frac{5}{8} < \frac{3}{4} < \frac{13}{16}$

Vocabulary Review

The following is a list of steps for adding and subtracting fractions. Choose the correct words to fill in the blanks below.

numerators   least common denominator   lowest terms   equivalent

Step 1: Factor each denominator completely, and use the factors to build the _____.

Step 2: Rewrite each fraction as an _____ fraction with the LCD.

Step 3: Add or subtract the _____ of the fractions produced in Step 2.

Step 4: Reduce the fraction produced in Step 3 to _____ if it is not already.

Problems

A Find the following sums and differences, and reduce to lowest terms. (Add or subtract as indicated.)

1. $\dfrac{3}{6} + \dfrac{1}{6}$

2. $\dfrac{2}{5} + \dfrac{3}{5}$

3. $\dfrac{5}{8} - \dfrac{3}{8}$

4. $\dfrac{6}{7} - \dfrac{1}{7}$

5. $\dfrac{3}{4} - \dfrac{1}{4}$

6. $-\dfrac{7}{9} + \dfrac{4}{9}$

7. $\dfrac{2}{3} + \left(-\dfrac{1}{3}\right)$

8. $\dfrac{9}{8} - \dfrac{1}{8}$

9. $\dfrac{1}{4} + \dfrac{2}{4} + \dfrac{3}{4}$

10. $\dfrac{2}{5} + \dfrac{3}{5} + \dfrac{4}{5}$

11. $\dfrac{x+7}{2} + \left(-\dfrac{1}{2}\right)$

12. $\dfrac{x+5}{4} - \dfrac{3}{4}$

13. $\dfrac{1}{10} + \dfrac{3}{10} + \dfrac{4}{10}$

14. $\dfrac{3}{20} + \dfrac{1}{20} + \dfrac{4}{20}$

15. $\dfrac{1}{3} + \dfrac{4}{3} + \dfrac{5}{3}$

16. $\dfrac{5}{4} + \dfrac{4}{4} + \dfrac{3}{4}$

B Complete the following tables.

17.

First Number a	Second Number b	The Sum of a and b $a + b$
$\dfrac{1}{2}$	$\dfrac{1}{3}$	
$\dfrac{1}{3}$	$\dfrac{1}{4}$	
$\dfrac{1}{4}$	$\dfrac{1}{5}$	
$\dfrac{1}{5}$	$\dfrac{1}{6}$	

18.

First Number a	Second Number b	The Sum of a and b $a + b$
1	$\dfrac{1}{2}$	
1	$\dfrac{1}{3}$	
1	$\dfrac{1}{4}$	
1	$\dfrac{1}{5}$	

19.

First Number a	Second Number b	The Sum of a and b $a + b$
$\dfrac{1}{12}$	$\dfrac{1}{2}$	
$\dfrac{1}{12}$	$\dfrac{1}{3}$	
$\dfrac{1}{12}$	$\dfrac{1}{4}$	
$\dfrac{1}{12}$	$\dfrac{1}{6}$	

20.

First Number a	Second Number b	The Sum of a and b $a + b$
$\dfrac{1}{8}$	$\dfrac{1}{2}$	
$\dfrac{1}{8}$	$\dfrac{1}{4}$	
$\dfrac{1}{8}$	$\dfrac{1}{16}$	
$\dfrac{1}{8}$	$\dfrac{1}{24}$	

Find the LCD for each of the following, then use the methods developed in this section to add or subtract as indicated.

21. $\dfrac{4}{9} + \dfrac{1}{3}$

22. $\dfrac{1}{2} + \dfrac{1}{4}$

23. $2 + \dfrac{1}{3}$

24. $3 + \dfrac{1}{2}$

25. $\dfrac{3}{4} + 1$

26. $\dfrac{3}{4} + 2$

27. $\dfrac{1}{2} + \dfrac{2}{3}$

28. $\dfrac{1}{8} + \dfrac{3}{4}$

29. $\dfrac{1}{4} + \dfrac{1}{5}$

30. $\dfrac{1}{3} + \dfrac{1}{5}$

31. $\dfrac{1}{2} + \dfrac{1}{5}$

32. $\dfrac{1}{2} - \dfrac{1}{5}$

33. $\dfrac{5}{12} + \dfrac{3}{8}$

34. $\dfrac{9}{16} + \dfrac{7}{12}$

35. $\dfrac{8}{30} - \dfrac{1}{20}$

36. $\dfrac{9}{40} - \dfrac{1}{30}$

37. $\dfrac{3}{10} + \dfrac{1}{100}$

38. $\dfrac{9}{100} + \dfrac{7}{10}$

39. $\dfrac{10}{36} + \dfrac{9}{48}$

40. $\dfrac{12}{28} + \dfrac{9}{20}$

41. $\dfrac{17}{30} + \dfrac{11}{42}$

42. $\dfrac{19}{42} + \dfrac{13}{70}$

43. $\dfrac{25}{84} - \left(-\dfrac{41}{90} \right)$

44. $\dfrac{23}{70} + \dfrac{29}{84}$

45. $\dfrac{13}{126} - \dfrac{13}{180}$

46. $\dfrac{17}{84} + \left(-\dfrac{17}{90} \right)$

47. $\dfrac{3}{4} + \dfrac{1}{8} + \dfrac{5}{6}$

48. $\dfrac{3}{8} + \dfrac{2}{5} + \dfrac{1}{4}$

49. $\dfrac{3}{10} + \dfrac{5}{12} + \dfrac{1}{6}$

50. $\dfrac{5}{21} + \dfrac{1}{7} + \dfrac{3}{14}$

51. $\dfrac{1}{2} + \dfrac{1}{3} + \dfrac{1}{4} + \dfrac{1}{6}$

52. $\dfrac{1}{8} + \dfrac{1}{4} - \left(-\dfrac{1}{5} \right) + \dfrac{1}{10}$

53. $10 - \dfrac{2}{9}$

54. $9 - \dfrac{3}{5}$

55. $4 - \dfrac{2}{3}$

56. $5 + \left(-\dfrac{3}{4}\right)$

57. $\dfrac{1}{10} + \dfrac{4}{5} - \dfrac{3}{20}$

58. $\dfrac{1}{2} + \dfrac{3}{4} - \dfrac{5}{8}$

59. $\dfrac{1}{4} - \dfrac{1}{8} - \left(-\dfrac{1}{2}\right) - \dfrac{3}{8}$

60. $\dfrac{7}{8} + \left(-\dfrac{3}{4}\right) + \dfrac{5}{8} - \dfrac{1}{2}$

61. $-\dfrac{1}{6} - \dfrac{5}{6}$

62. $-\dfrac{4}{7} - \dfrac{3}{7}$

63. $-\dfrac{13}{70} - \dfrac{23}{42}$

64. $-\dfrac{17}{60} - \dfrac{17}{90}$

There are two ways to work the problems below. You can combine the fractions inside the parentheses first and then multiply, or you can apply the distributive property first, then add.

65. $15\left(\dfrac{2}{3} + \dfrac{3}{5}\right)$

66. $15\left(\dfrac{4}{5} - \dfrac{1}{3}\right)$

67. $4\left(\dfrac{1}{2} + \dfrac{1}{4}\right)$

68. $6\left(\dfrac{1}{3} + \dfrac{1}{2}\right)$

69. $4\left(3 - \dfrac{3}{4}\right)$

70. $6\left(5 - \dfrac{2}{3}\right)$

71. $9\left(\dfrac{1}{3} + \dfrac{1}{9}\right)$

72. $12\left(\dfrac{1}{3} + \dfrac{1}{4}\right)$

73. Write the fractions in order from smallest to largest.

$$\dfrac{3}{4} \qquad \dfrac{3}{8} \qquad \dfrac{1}{2} \qquad \dfrac{1}{4}$$

74. Write the fractions in order from smallest to largest.

$$\dfrac{1}{2} \qquad \dfrac{1}{6} \qquad \dfrac{1}{4} \qquad \dfrac{1}{3}$$

75. Find the sum of $\dfrac{3}{7}$, 2, and $\dfrac{1}{9}$.

76. Find the sum of 6, $\dfrac{6}{11}$, and 11.

77. Give the difference of $\dfrac{7}{8}$ and $\dfrac{1}{4}$.

78. Give the difference of $\dfrac{9}{10}$ and $\dfrac{1}{100}$.

Arithmetic Sequences An arithmetic sequence is a sequence in which each term comes from the previous term by adding the same number each time. For example, the sequence $1, \frac{3}{2}, 2, \frac{5}{2}, \ldots$ is an arithmetic sequence that starts with the number 1. Then each term after that is found by adding $\frac{1}{2}$ to the previous term. By observing this fact, we know that the next term in the sequence will be $\frac{5}{2} + \frac{1}{2} = \frac{6}{2} = 3$.

Find the next number in each arithmetic sequence below.

79. $1, \dfrac{4}{3}, \dfrac{5}{3}, 2, \ldots$

80. $1, \dfrac{5}{4}, \dfrac{3}{2}, \dfrac{7}{4}, \ldots$

81. $\dfrac{3}{2}, 2, \dfrac{5}{2}, \ldots$

82. $\dfrac{2}{3}, 1, \dfrac{4}{3}, \ldots$

Applying the Concepts

Some of the application problems below involve multiplication or division, while others involve addition or subtraction.

83. Capacity One carton of milk contains $\frac{1}{2}$ pint while another contains 4 pints. How much milk is contained in both cartons?

84. Baking A recipe calls for $\frac{2}{3}$ cup of flour and $\frac{3}{4}$ cup of sugar. What is the total amount of flour and sugar called for in the recipe?

85. Budget A family decides that they can spend $\frac{5}{8}$ of their monthly income on house payments. If their monthly income is $2,120, how much can they spend for house payments?

86. Savings A family saves $\frac{3}{16}$ of their income each month. If their monthly income is $1,264, how much do they save each month?

Reading a Pie Chart The pie chart below shows how the students at Cal Poly State University are distributed among the different schools at the university. Use the information in the pie chart to answer questions 79 and 80.

87. If the students in the Schools of Engineering and Business are combined, what fraction results?

88. What fraction of the university's students are enrolled in the Schools of Agriculture, Engineering, and Business combined?

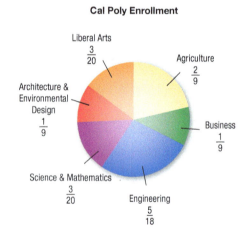

Cal Poly Enrollment

89. Final Exam Grades The table gives the fraction of students in a class of 40 that received grades of A, B, or C on the final exam. Fill in all the missing parts of the table.

Grade	Number of Students	Fraction of Students
A		$\frac{1}{8}$
B		$\frac{1}{5}$
C		$\frac{1}{2}$
Below C		
Total	40	1

90. Flu During a flu epidemic a company with 200 employees has $\frac{1}{10}$ of their employees call in sick on Monday and another $\frac{3}{10}$ call in sick on Tuesday. What is the total number of employees calling in sick during this 2-day period?

91. Subdivision A 6-acre piece of land is subdivided into $\frac{3}{5}$-acre lots. How many lots are there?

92. Cutting Wood A 12-foot piece of wood is cut into shelves. If each is $\frac{3}{4}$ foot in length, how many shelves are there?

Find the perimeter of each figure.

93.

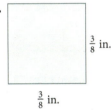

$\frac{3}{8}$ in.

$\frac{3}{8}$ in.

94.

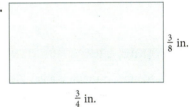

$\frac{3}{8}$ in.

$\frac{3}{4}$ in.

95.

$\frac{3}{10}$ ft

$\frac{3}{5}$ ft

96.

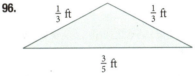

$\frac{1}{3}$ ft $\frac{1}{3}$ ft

$\frac{3}{5}$ ft

Getting Ready for the Next Section

Simplify.

97. $9 \cdot 6 + 5$

98. $4 \cdot 6 + 3$

99. Write 2 as a fraction with denominator 8.

100. Write 2 as a fraction with denominator 4.

101. Write 1 as a fraction with denominator 8.

102. Write 5 as a fraction with denominator 4.

Add.

103. $\frac{8}{4} + \frac{3}{4}$ **104.** $\frac{16}{8} + \frac{1}{8}$ **105.** $2 + \frac{1}{8}$ **106.** $2 + \frac{3}{4}$ **107.** $1 + \frac{1}{8}$ **108.** $5 + \frac{3}{4}$

Divide.

109. $11 \div 4$ **110.** $10 \div 3$ **111.** $208 \div 24$ **112.** $207 \div 26$

Find the Mistake

Each sentence below contains a mistake. Circle the mistake and write the correct word(s) or numbers(s) on the line provided.

1. The fractions $\frac{a}{c}$ and $\frac{b}{c}$ can be added to become $\frac{a+b}{c}$ because they have different denominators. _____

2. Subtracting $\frac{12}{21}$ from $\frac{18}{21}$, gives us $\frac{30}{21}$. _____

3. The least common denominator for a set of denominators is the smallest number that is exactly divisible by each numerator. _____

4. The LCD for the fractions $\frac{4}{6}$, $\frac{2}{8}$ and $\frac{3}{4}$ is 12. _____

Navigation Skills: Prepare, Study, Achieve

Completing homework assignments in full is a key piece to succeeding in this class. To do this effectively, you must pay special attention to each set of instructions. When you do your homework, you usually work a number of similar problems at a time. But the problems may vary on a test. It is very important to make a habit of paying attention to the instructions to elicit correct answers on a test. Secondly, to complete an assignment efficiently, you will need to memorize various definitions, properties, and formulas. Reading the definition in the book alone is not enough. There are many techniques for successful memorization. Here are a few:

- Spend some time rereading the definition.
- Say the definition out loud.
- Explain the definition to another person.
- Write the definition down on a separate sheets of notes.
- Create a mnemonic device using key words from the definition.
- Analyze how the definition applies to your homework problems.

The above suggestions are ways to engage your senses when memorizing an abstract concept. This will help anchor it in your memory. For instance, it is easier to remember explaining to your friend a difficult math formula, than it is to simply recall it from a single read of the chapter. Lastly, once you've completed an assignment, take any extra time you've allotted for studying to work more problems, and if you feel ready, read ahead and work problems you will encounter in the next section.

3.2 Mixed-Number Notation

OBJECTIVES

A Convert mixed numbers to improper fractions.

B Convert improper fractions to mixed numbers.

KEY WORDS

mixed numbers

Pogopalooza is an annual world championship for stunt pogo stick athletes. Pogo jumpers compete in various events, such as the most or the least jumps in a minute, the highest jump, and numerous exhibitions of acrobatic stunts. Imagine you are competing in the High Jump competition. The crowd cheers as you bounce a remarkable $8\frac{1}{4}$ feet into the air! Being a skilled mathematician, you realize that your score of $8\frac{1}{4}$ feet, or eight and one-fourth feet, is actually a *mixed number*. It is the sum of a whole number and a proper fraction. With mixed-number notation, we leave out the addition sign.

Mixed-Number Notation

Here are some further examples of mixed number notation:

$$2\frac{1}{8} = 2 + \frac{1}{8}, \quad 6\frac{5}{9} = 6 + \frac{5}{9}, \quad 11\frac{2}{3} = 11 + \frac{2}{3}$$

The notation used in writing mixed numbers (writing the whole number and the proper fraction next to each other) must always be interpreted as addition. It is a mistake to read $5\frac{3}{4}$ as meaning 5 times $\frac{3}{4}$. If we want to indicate multiplication, we must use parentheses or a multiplication symbol. That is,

$$5\frac{3}{4} \text{ is } \textbf{not} \text{ the same as } 5\left(\frac{3}{4}\right).$$

This implies addition.　These imply multiplication.

$$5\frac{3}{4} \text{ is } \textbf{not} \text{ the same as } 5 \cdot \frac{3}{4}.$$

**Video Examples
Section 3.2**

A Changing Mixed Numbers to Improper Fractions

To change a mixed number to an improper fraction, we write the mixed number with the + sign showing and then add the two numbers, as we did earlier.

EXAMPLE 1 Change $2\frac{3}{4}$ to an improper fraction.

Solution

$$2\frac{3}{4} = 2 + \frac{3}{4}$$ *Write the mixed number as a sum.*

$$= \frac{2}{1} + \frac{3}{4}$$ *Show that the denominator of 2 is 1.*

$$= \frac{4 \cdot 2}{4 \cdot 1} + \frac{3}{4}$$ *Multiply the numerator and the denominator of $\frac{2}{1}$ by 4 so both fractions will have the same denominator.*

Practice Problems

1. Change $4\frac{2}{3}$ to an improper fraction.

Answer

1. $\frac{14}{3}$

$$= \frac{8}{4} + \frac{3}{4}$$

$$= \frac{11}{4} \qquad \textit{Add the numerators; keep the common denominator.}$$

The mixed number $2\frac{3}{4}$ is equal to the improper fraction $\frac{11}{4}$. The diagram that follows further illustrates the equivalence of $2\frac{3}{4}$ and $\frac{11}{4}$.

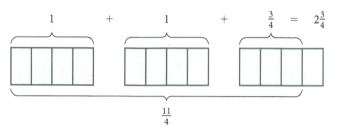

2. Change $3\frac{1}{5}$ to an improper fraction.

EXAMPLE 2   Change $2\frac{1}{8}$ to an improper fraction.

Solution $2\frac{1}{8} = 2 + \frac{1}{8}$ *Write as addition.*

$$= \frac{2}{1} + \frac{1}{8} \qquad \textit{Write the whole number 2 as a fraction.}$$

$$= \frac{8 \cdot 2}{8 \cdot 1} + \frac{1}{8} \qquad \textit{Change } \tfrac{2}{1} \textit{ to a fraction with denominator 8.}$$

$$= \frac{16}{8} + \frac{1}{8}$$

$$= \frac{17}{8} \qquad \textit{Add the numerators.}$$

If we look closely at Examples 1 and 2, we can see the following shortcut that will let us change a mixed number to an improper fraction without so many steps.

    Shortcut: To change a mixed number to an improper fraction, simply multiply the whole number by the denominator of the fraction, and add the result to the numerator of the fraction. The result is the numerator of the improper fraction we are looking for. The denominator is the same as the original denominator.

3. Change $5\frac{7}{8}$ to an improper fraction.

EXAMPLE 3   Use the shortcut to change $5\frac{3}{4}$ to an improper fraction.

Solution

    1. First, we multiply 4×5 to get 20.

    2. Next, we add 20 to 3 to get 23.

    3. The improper fraction equal to $5\frac{3}{4}$ is $\frac{23}{4}$.

Here is a diagram showing what we have done:

Step 1   Multiply $4 \times 5 = 20$.   *Step 2*

Step 2   Add $20 + 3 = 23$. $5\frac{3}{4}$

                   Step 1

Mathematically, our shortcut is written like this:

$$5\frac{3}{4} = \frac{(4 \cdot 5) + 3}{4} = \frac{20 + 3}{4} = \frac{23}{4} \qquad \textit{The result will always have the same denominator as the original mixed number.}$$

Answers

2. $\frac{16}{5}$

3. $\frac{47}{8}$

The shortcut shown in Example 3 works because the whole-number part of a mixed number can always be written with a denominator of 1. Therefore, the LCD for a whole number and fraction will always be the denominator of the fraction. That is why we multiply the whole number by the denominator of the fraction.

$$5\frac{3}{4} = 5 + \frac{3}{4} = \frac{5}{1} + \frac{3}{4} = \frac{4 \cdot 5}{4 \cdot 1} + \frac{3}{4} = \frac{4 \cdot 5 + 3}{4} = \frac{23}{4}$$

EXAMPLE 4 Change $6\frac{5}{9}$ to an improper fraction.

Solution Using the first method, we have

$$6\frac{5}{9} = 6 + \frac{5}{9} = \frac{6}{1} + \frac{5}{9} = \frac{9 \cdot 6}{9 \cdot 1} + \frac{5}{9} = \frac{54}{9} + \frac{5}{9} = \frac{59}{9}$$

Using the shortcut method, we have

$$6\frac{5}{9} = \frac{(9 \cdot 6) + 5}{9} = \frac{54 + 5}{9} = \frac{59}{9}$$

B Changing Improper Fractions to Mixed Numbers

To change an improper fraction to a mixed number, we divide the numerator by the denominator. The result is used to write the mixed number.

EXAMPLE 5 Change $\frac{11}{4}$ to a mixed number.

Solution Dividing 11 by 4 gives us

$$
\begin{array}{r}
2 \\
4\overline{)11} \\
-8 \\
\hline
3
\end{array}
$$

We see that 4 goes into 11 two times with 3 for a remainder. We write this result as

$$\frac{11}{4} = 2 + \frac{3}{4} = 2\frac{3}{4}$$

The improper fraction $\frac{11}{4}$ is equivalent to the mixed number $2\frac{3}{4}$.

One easy way to visualize the results in Example 5 is to imagine running 13 laps on a $\frac{1}{4}$ mile track. Your 13 laps are equivalent to $\frac{13}{4}$ miles. In miles, your laps are equal to 3 miles plus 1 quarter-mile, or $3\frac{1}{4}$ miles.

EXAMPLE 6 Write as a mixed number.

a. $\frac{10}{3}$ b. $\frac{208}{24}$

Solution

a.
$$
\begin{array}{r}
3 \\
3\overline{)10} \\
9 \\
\hline
1
\end{array}
$$
so $\frac{10}{3} = 3 + \frac{1}{3} = 3\frac{1}{3}$

b.
$$
\begin{array}{r}
8 \\
24\overline{)208} \\
192 \\
\hline
16
\end{array}
$$
so $\frac{208}{24} = 8 + \frac{16}{24} = 8 + \frac{2}{3} = 8\frac{2}{3}$

Reduce to lowest terms.

4. Change $8\frac{6}{7}$ to an improper fraction.

Calculator Note The sequence of keys to press on a calculator to obtain the numerator in Example 4 looks like this:

$$9 \times 6 + 5 =$$

5. Change $\frac{13}{3}$ to a mixed number.

Note This division process shows us how many ones are in $\frac{11}{4}$ and, when the ones are taken out, how many fourths are left.

6. Write as a mixed number.

a. $\frac{27}{4}$

b. $\frac{76}{8}$

Answers

4. $\frac{62}{7}$

5. $4\frac{1}{3}$

6. a. $6\frac{3}{4}$ **b.** $9\frac{1}{2}$

Long Division, Remainders, and Mixed Numbers

Mixed numbers give us another way of writing the answers to long division problems that contain remainders. Here is how we divide 1,690 by 67:

$$
\begin{array}{r}
25 \text{ R } 15 \\
67\overline{)1690} \\
\underline{134} \\
350 \\
\underline{335} \\
15
\end{array}
$$

The answer is 25 with a remainder of 15. Using mixed numbers, we can now write the answer as $25\frac{15}{67}$. That is,

$$\frac{1,690}{67} = 25\frac{15}{67}$$

The quotient of 1,690 and 67 is $25\frac{15}{67}$.

GETTING READY FOR CLASS

After reading through the preceding section, respond in your own words and in complete sentences.

A. What is a mixed number?

B. The expression $5\frac{3}{4}$ is equivalent to what addition problem?

C. The improper fraction $\frac{11}{4}$ is equivalent to what mixed number?

D. Why is $\frac{13}{5}$ an improper fraction, but $\frac{3}{5}$ is not an improper fraction?

Vocabulary Review

Choose the correct words to fill in the blanks below.

addition   denominator   proper fraction

improper fraction   remainder   mixed number

1. A mixed number is the sum of a whole number and a _____.

2. The notation used in writing mixed numbers is always interpreted as _____.

3. We leave out the addition sign when writing a number as a _____.

4. To change a mixed number to an _____, write the mixed number with the addition sign showing and then add the two numbers.

5. To change an improper fraction to a mixed number, divide the numerator by the _____.

6. An answer to a long division problem that contains a _____ can be rewritten as a mixed number.

Problems

A Change each mixed number to an improper fraction.

1. $4\frac{2}{3}$ **2.** $3\frac{5}{8}$ **3.** $5\frac{1}{4}$ **4.** $7\frac{1}{2}$ **5.** $1\frac{5}{8}$ **6.** $1\frac{6}{7}$

7. $15\frac{2}{3}$ **8.** $-\left(17\frac{3}{4}\right)$ **9.** $4\frac{20}{21}$ **10.** $-\left(5\frac{18}{19}\right)$ **11.** $12\frac{31}{33}$ **12.** $14\frac{29}{31}$

B Change each improper fraction to a mixed number.

13. $\frac{9}{8}$ **14.** $\frac{10}{9}$ **15.** $-\frac{19}{4}$ **16.** $\frac{23}{5}$ **17.** $\frac{29}{6}$ **18.** $\frac{7}{2}$

19. $\frac{13}{4}$ **20.** $\frac{41}{15}$ **21.** $\frac{109}{27}$ **22.** $\frac{319}{23}$ **23.** $-\frac{428}{15}$ **24.** $\frac{769}{27}$

Applying the Concepts

25. iPad Applications   The chart shows the number of each type of applications compatible with the iPad. Use the information to answer the following questions.

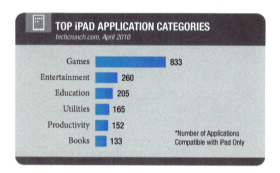

a. The number of game apps is what fraction of the education apps?

b. The number of productivity apps is what fraction of the entertainment apps?

26. NBA Finals   The chart shows the number of NBA Finals appearances by several teams. Use the information to answer the following questions.

a. The number of Lakers appearances is what fraction of the Bulls appearances?

b. The number of Pistons appearances is what fraction of the Celtics appearances?

27. Stocks   Suppose a stock is selling on a stock exchange for $5\frac{1}{4}$ dollars per share. If the price increases $\frac{3}{4}$ dollar per share, what is the new price of the stock?

28. Stocks   Suppose a stock is selling on a stock exchange for $5\frac{1}{4}$ dollars per share. If the price increases 2 dollars per share, what is the new price of the stock?

29. Height   If a man is 71 inches tall, then in feet his height is $5\frac{11}{12}$ feet. Change $5\frac{11}{12}$ to an improper fraction.

30. Height   If a woman is 63 inches tall, then her height in feet is $\frac{63}{12}$. Write $\frac{63}{12}$ as a mixed number.

31. Gasoline Prices   The price of unleaded gasoline is $305\frac{1}{5}$¢ per gallon. Write this number as an improper fraction.

32. Gasoline Prices   Suppose the price of gasoline is $308\frac{1}{5}$¢ if purchased with a credit card, but 5¢ less if purchased with cash. What is the cash price of the gasoline?

Getting Ready for the Next Section

Change to improper fractions.

33. $2\frac{3}{4}$ **34.** $3\frac{1}{5}$ **35.** $4\frac{5}{8}$ **36.** $1\frac{3}{5}$ **37.** $2\frac{4}{5}$ **38.** $5\frac{9}{10}$

Find the following products. (Multiply.)

39. $\frac{3}{8} \cdot \frac{3}{5}$ **40.** $\frac{11}{4} \cdot \frac{16}{5}$ **41.** $-\frac{2}{3}\left(\frac{9}{16}\right)$ **42.** $\frac{7}{10}\left(\frac{5}{21}\right)$

Find the quotients. (Divide.)

43. $\frac{4}{5} \div \frac{7}{8}$ **44.** $\frac{3}{4} \div \frac{1}{2}$ **45.** $-\frac{8}{5} \div \frac{14}{5}$ **46.** $\frac{59}{10} \div -2$

Improving Your Quantitative Literacy

Starbucks Stores The chart shows the number of Starbucks stores in each state. Use the information to answer the following questions.

47. According to the chart, which of the following statements is closer to the truth?

   a. There are five times as many stores in California as there are in Illinois.

   b. There are twice as many stores in Washington as there are in New York.

STARBUCKS STORES EVERYWHERE
www.statemaster.com

California	2,010
Texas	604
Washington	559
Illinois	412
New York	384
Florida	375
Colorado	322

48. According to the chart, which fraction best represents the fraction of stores in the chart that are described below. Choose between $\frac{1}{2}$, $\frac{1}{3}$, $\frac{1}{4}$, and $\frac{1}{5}$.

   a. Illinois as compared to California

   b. Colorado as compared to Texas

Find the Mistake

Each sentence below contains a mistake. Circle the mistake and write the correct word(s) or numbers(s) on the line provided.

1. For mixed-number notation, writing the whole number next to the fraction implies multiplication.

_____

2. A shortcut for changing a mixed number to an improper fraction is to multiply the whole number by the numerator of the fraction, and then add the result to the denominator of the fraction.

_____

3. Changing $6\frac{4}{5}$ to an improper fraction gives us $\frac{29}{5}$. _____

4. Writing $\frac{70}{12}$ as a mixed number gives us $6\frac{1}{6}$. _____

3.3 Multiplication and Division with Mixed Numbers

©iStockphoto.com/negaprion

Fire tornadoes are a difficult and dangerous element of firefighting. The visible part of the fire tornado is called the core, where hot ash and combustible gases ignite in flame and rise hundreds of feet to the sky. The outer layer of the tornado is the circulating air that fuels the core with fresh oxygen. As the fire tornado creeps across the land, it sets anything in its path ablaze, hurls debris into the air, and spins so fast it can knock down trees. The funnel of flames can create wind speeds of more than 100 miles per hour, which is as strong as a category 2 hurricane.

Suppose a fire tornado created a wind speed of 100 miles per hour. What if this speed was $2\frac{1}{2}$ times the wind speed outside the tornado? How fast is the wind speed outside the tornado? Recalling our work from the previous section, we know that the number $2\frac{1}{2}$ is a *mixed number*. To answer the question, we must be able to divide 100 by $2\frac{1}{2}$. Division with mixed numbers is one of the topics we will cover in this section.

A Multiplication with Mixed Numbers

Video Examples
Section 3.3

Practice Problems

1. Multiply: $1\frac{1}{4} \cdot 2\frac{3}{5}$.

EXAMPLE 1 Multiply: $2\frac{3}{4} \cdot 3\frac{1}{5}$.

Solution We begin by changing each mixed number to an improper fraction.

$$2\frac{3}{4} = \frac{11}{4} \quad \text{and} \quad 3\frac{1}{5} = \frac{16}{5}$$

Using the resulting improper fractions, we multiply as usual. (That is, we multiply numerators and multiply denominators.)

$$\frac{11}{4} \cdot \frac{16}{5} = \frac{11 \cdot 16}{4 \cdot 5}$$

$$= \frac{11 \cdot 4 \cdot 4}{4 \cdot 5}$$

$$= \frac{44}{5} \quad \text{or} \quad 8\frac{4}{5}$$

Note As you can see, once you have changed each mixed number to an improper fraction, you multiply the resulting fractions the same way you did in Section 2.3.

2. Multiply: $6 \cdot 2\frac{1}{8}$.

EXAMPLE 2 Multiply: $3 \cdot 4\frac{5}{8}$.

Solution Writing each number as an improper fraction, we have

$$3 = \frac{3}{1} \quad \text{and} \quad 4\frac{5}{8} = \frac{37}{8}$$

Answers

1. $3\frac{1}{4}$

2. $12\frac{3}{4}$

Chapter 3 Addition and Subtraction of Fractions, Mixed Numbers

The complete problem looks like this:

$$3 \cdot 4\frac{5}{8} = \frac{3}{1} \cdot \frac{37}{8}$$ *Change to improper fractions.*

$$= \frac{111}{8}$$ *Multiply numerators and multiply denominators.*

$$= 13\frac{7}{8}$$ *Write the answer as a mixed number.*

B Division with Mixed Numbers

Dividing mixed numbers also requires that we change all mixed numbers to improper fractions before we actually do the division.

EXAMPLE 3   Divide: $1\frac{3}{5} \div 2\frac{4}{5}$.

Solution   We begin by rewriting each mixed number as an improper fraction.

$$1\frac{3}{5} = \frac{8}{5} \quad \text{and} \quad 2\frac{4}{5} = \frac{14}{5}$$

We then divide using the same method we used in Section 2.4. We multiply by the reciprocal of the divisor. Here is the complete problem:

$$1\frac{3}{5} \div 2\frac{4}{5} = \frac{8}{5} \div \frac{14}{5}$$ *Change to improper fractions.*

$$= \frac{8}{5} \cdot \frac{5}{14}$$ *To divide by $\frac{14}{5}$, multiply by $\frac{5}{14}$.*

$$= \frac{8 \cdot 5}{5 \cdot 14}$$ *Multiply numerators and multiply denominators.*

$$= \frac{4 \cdot 2 \cdot 5}{5 \cdot 2 \cdot 7}$$ *Divide out factors common to the numerator and denominator.*

$$= \frac{4}{7}$$ *Answer in lowest terms.*

EXAMPLE 4   Divide: $5\frac{9}{10} \div 2$.

Solution   We change to improper fractions and proceed as usual.

$$5\frac{9}{10} \div 2 = \frac{59}{10} \div \frac{2}{1}$$ *Write each number as an improper fraction.*

$$= \frac{59}{10} \cdot \frac{1}{2}$$ *Write division as multiplication by the reciprocal.*

$$= \frac{59}{20}$$ *Multiply numerators and multiply denominators.*

$$= 2\frac{19}{20}$$ *Change to a mixed number.*

GETTING READY FOR CLASS

After reading through the preceding section, respond in your own words and in complete sentences.

A. What is the first step when multiplying or dividing mixed numbers?

B. What is the reciprocal of $2\frac{4}{5}$?

C. Dividing $5\frac{9}{10}$ by 2 is equivalent to multiplying $5\frac{9}{10}$ by what number?

D. Find $4\frac{5}{8}$ of 3.

3. Divide: $3\frac{3}{4} \div 2\frac{1}{4}$.

4. Divide: $8\frac{3}{4} \div 5$.

Answers

3. $1\frac{2}{3}$

4. $1\frac{3}{4}$

EXERCISE SET 3.3

Problems

Write your answers as proper fractions or mixed numbers, not as improper fractions.

A Find the following products. (Multiply.)

1. $3\frac{2}{5} \cdot 1\frac{1}{2}$

2. $2\frac{1}{3} \cdot 6\frac{3}{4}$

3. $5\frac{1}{8} \cdot 2\frac{2}{3}$

4. $1\frac{5}{6} \cdot 1\frac{4}{5}$

5. $2\frac{1}{10} \cdot 3\frac{3}{10}$

6. $4\frac{7}{10} \cdot 3\frac{1}{10}$

7. $1\frac{1}{4} \cdot 4\frac{2}{3}$

8. $3\frac{1}{2} \cdot 2\frac{1}{6}$

9. $2 \cdot 4\frac{7}{8}$

10. $-10 \cdot 1\frac{1}{4}$

11. $\frac{3}{5} \cdot 5\frac{1}{3}$

12. $\frac{2}{3} \cdot 4\frac{9}{10}$

13. $2\frac{1}{2} \cdot 3\frac{1}{3} \cdot 1\frac{1}{2}$

14. $3\frac{1}{5} \cdot 5\frac{1}{6} \cdot 1\frac{1}{8}$

15. $\frac{3}{4} \cdot 7 \cdot 1\frac{4}{5}$

16. $\frac{7}{8} \cdot 6 \cdot \left(-1\frac{5}{6}\right)$

B Find the following quotients. (Divide.)

17. $3\frac{1}{5} \div 4\frac{1}{2}$

18. $1\frac{4}{5} \div 2\frac{5}{6}$

19. $6\frac{1}{4} \div 3\frac{3}{4}$

20. $8\frac{2}{3} \div 4\frac{1}{3}$

21. $10 \div 2\frac{1}{2}$

22. $12 \div 3\frac{1}{6}$

23. $8\frac{3}{5} \div 2$

24. $-12\frac{6}{7} \div 3$

25. $\left(\frac{3}{4} \div 2\frac{1}{2}\right) \div 3$

26. $\frac{7}{8} \div \left(1\frac{1}{4} \div -4\right)$

27. $\left(8 \div 1\frac{1}{4}\right) \div -2$

28. $8 \div \left(1\frac{1}{4} \div 2\right)$

29. $2\frac{1}{2} \cdot \left(3\frac{2}{5} \div 4\right)$

30. $4\frac{3}{5} \cdot \left(2\frac{1}{4} \div 5\right)$

31. Find the product of $2\frac{1}{2}$ and 3.

32. Find the product of $\frac{1}{5}$ and $3\frac{2}{3}$.

33. What is the quotient of $2\frac{3}{4}$ and $3\frac{1}{4}$?

34. What is the quotient of $1\frac{1}{5}$ and $2\frac{2}{5}$?

....

Applying the Concepts

35. Cooking A certain recipe calls for $2\frac{3}{4}$ cups of sugar. If the recipe is to be doubled, how much sugar should be used?

36. Cooking If a recipe calls for $3\frac{1}{2}$ cups of flour, how much flour will be needed if the recipe is tripled?

37. Cooking If a recipe calls for $2\frac{1}{2}$ cups of sugar, how much sugar is needed to make $\frac{1}{3}$ of the recipe?

38. Cooking A recipe calls for $3\frac{1}{4}$ cups of flour. If Diane is using only half the recipe, how much flour should she use?

39. Number Problem Find $\frac{3}{4}$ of $1\frac{7}{9}$. (Remember that *of* means multiply.)

40. Number Problem Find $\frac{5}{6}$ of $2\frac{4}{15}$.

41. Cost of Gasoline If a gallon of gas costs $305\frac{1}{5}$ ¢, how much does 8 gallons cost?

42. Cost of Gasoline If a gallon of gas costs $308\frac{1}{5}$ ¢, how much does $\frac{1}{2}$ gallon cost?

43. Distance Traveled If a car can travel $32\frac{3}{4}$ miles on a gallon of gas, how far will it travel on 5 gallons of gas?

44. Distance Traveled If a new car can travel $20\frac{3}{10}$ miles on 1 gallon of gas, how far can it travel on $\frac{1}{2}$ gallon of gas?

45. Sewing If it takes $1\frac{1}{2}$ yards of material to make a pillow cover, how much material will it take to make 3 pillow covers?

46. Sewing If the material for the pillow covers in Problem 45 costs $2 a yard, how much will it cost for the material for the 3 pillow covers?

Find the area of each figure.

47.

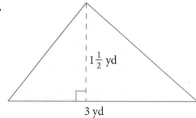

$1\frac{1}{2}$ yd

3 yd

48.

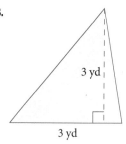

3 yd

3 yd

49. Write the numbers in order from smallest to largest.

$2\frac{1}{8}$ $\quad$ $\frac{5}{4}$ $\quad$ $\frac{3}{4}$ $\quad$ $1\frac{1}{2}$

50. Write the numbers in order from smallest to largest.

$1\frac{3}{8}$ $\quad$ $\frac{7}{8}$ $\quad$ $\frac{7}{4}$ $\quad$ $1\frac{11}{16}$

Nutrition The figure below shows nutrition labels for two different cans of corn.

Can 1

Nutrition Facts

Serving Size 1 cup
Servings Per Container About 2 ½

Amount Per Serving

Calories 133	Calories from fat 15

	% Daily Value*
Total Fat 3g	**3%**
Saturated Fat 1g	**1%**
Cholesterol 0mg	**0%**
Sodium 530mg	**22%**
Total Carbohydrate 30g	**10%**
Dietary Fiber 3g	**13%**
Sugars 4g	
Protein 4g	

Vitamin A 0%	•	Vitamin C 23%
Calcium 1%	•	Iron 8%

*Percent Daily Values are based on a 2,000 calorie diet

Can 2

Nutrition Facts

Serving Size 1 cup
Servings Per Container About 1 ½

Amount Per Serving

Calories 184	Calories from fat 10

	% Daily Value*
Total Fat 2g	**2%**
Saturated Fat 1g	**1%**
Cholesterol 0mg	**0%**
Sodium 730mg	**30%**
Total Carbohydrate 46g	**15%**
Dietary Fiber 3g	**12%**
Sugars 6g	
Protein 4g	

Vitamin A 0%	•	Vitamin C 20%
Calcium 1%	•	Iron 5%

*Percent Daily Values are based on a 2,000 calorie diet

51. Compare the total number of calories in the two cans of corn.

52. Compare the total amount of sugar in the two cans of corn.

53. Compare the total amount of sodium in the two cans of corn.

54. Compare the total amount of protein in the two cans of corn.

Getting Ready for the Next Section

55. Write as equivalent fractions with denominator 15.

     a. $\dfrac{2}{3}$ **b.** $\dfrac{1}{5}$ **c.** $\dfrac{3}{5}$ **d.** $\dfrac{1}{3}$

56. Write as equivalent fractions with denominator 12.

     a. $\dfrac{3}{4}$ **b.** $\dfrac{1}{3}$ **c.** $\dfrac{5}{6}$ **d.** $\dfrac{1}{4}$

57. Write as equivalent fractions with denominator 20.

     a. $\dfrac{1}{4}$ **b.** $\dfrac{3}{5}$ **c.** $\dfrac{9}{10}$ **d.** $\dfrac{1}{10}$

58. Write as equivalent fractions with denominator 24.

     a. $\dfrac{3}{4}$ **b.** $\dfrac{7}{8}$ **c.** $\dfrac{5}{8}$ **d.** $\dfrac{3}{8}$

Add or subtract the following fractions, as indicated.

59. $\dfrac{2}{3} + \dfrac{1}{5}$

60. $\dfrac{3}{4} + \dfrac{5}{6}$

61. $\dfrac{2}{3} + \dfrac{8}{9}$

62. $\dfrac{1}{4} + \dfrac{3}{5} + \dfrac{9}{10}$

63. $\dfrac{9}{10} - \dfrac{3}{10}$

64. $\dfrac{7}{10} - \dfrac{3}{5}$

65. $\dfrac{5}{12} - \dfrac{1}{6}$

66. $\dfrac{2}{3} - \dfrac{1}{4} - \dfrac{1}{6}$

One Step Further

To find the square of a mixed number, we first change the mixed number to an improper fraction, and then we square the result. For example, $\left(2\dfrac{1}{2}\right)^2 = \left(\dfrac{5}{2}\right)^2 = \dfrac{25}{4}$. If we are asked to write our answer as a mixed number, we write it as $6\dfrac{1}{4}$.

Find each of the following squares, and write your answers as mixed numbers.

67. $\left(1\dfrac{1}{2}\right)^2$

68. $\left(3\dfrac{1}{2}\right)^2$

69. $\left(1\dfrac{3}{4}\right)^2$

70. $\left(2\dfrac{3}{4}\right)^2$

71. The length of one side of a square is $1\dfrac{1}{2}$ ft. Write the area of the square as a mixed number.

72. The length of one side of a square is $2\dfrac{1}{2}$ ft. Write the area of the square as a mixed number.

73. The length of one side of a cube is $1\dfrac{1}{2}$ ft. Write the volume of the cube as a mixed number.

74. The length of one side of a cube is $1\dfrac{1}{3}$ ft. Write the volume of the cube as a mixed number.

Find the Mistake

Each sentence below contains a mistake. Circle the mistake and write the correct word(s) or numbers(s) on the line provided.

1. To multiply two mixed numbers, multiply the whole numbers and multiply the fractions.

   _____

2. Multiplying $4\frac{3}{8}$ and $9\frac{2}{7}$ gives us the mixed number $45\frac{1}{3}$. _____

3. To divide the mixed numbers $12\frac{2}{5}$ and $3\frac{12}{5}$, change the mixed numbers to improper fractions and then multiply numerators and denominators. _____

4. The answer to the division problem $3\frac{9}{14} \div 2$ written as a mixed number is $\frac{51}{28}$.

   _____

Landmark Review: Checking Your Progress

Find the following sums and differences and reduce to lowest terms.

1. $\frac{7}{10} + \frac{3}{10}$

2. $\frac{2}{5} - \frac{1}{5}$

3. $\frac{2}{3} + \frac{3}{5}$

4. $\frac{1}{2} + \frac{3}{4}$

5. $\frac{3}{5} + \frac{2}{3} - \frac{4}{7}$

6. $\frac{4+x}{3} - \frac{2}{3}$

7. Find the sum of $\frac{2}{3}$, $\frac{1}{5}$, and $\frac{1}{2}$.

8. Find the difference of $\frac{7}{10}$ and $\frac{3}{5}$.

Change each mixed number to an improper fraction.

9. $3\frac{5}{8}$

10. $4\frac{2}{3}$

11. $10\frac{1}{2}$

12. $1\frac{1}{4}$

Change each improper fraction to a mixed number.

13. $\frac{14}{3}$

14. $\frac{23}{5}$

15. $\frac{7}{2}$

16. $\frac{42}{17}$

Perform the indicated operations.

17. $4\frac{1}{4} \cdot 5\frac{1}{2}$

18. $3\frac{1}{3} \cdot 2\frac{5}{6}$

19. $5\frac{2}{3} \cdot 6\frac{5}{8}$

20. $5\frac{1}{4} \div 4\frac{3}{8}$

21. $3\frac{7}{10} \div 1\frac{3}{5}$

22. $4\frac{5}{8} \div 2\frac{2}{3}$

Dock diving is a sport where dogs compete to see who can jump the highest or the farthest off the end of a dock into a body of water. Many of the dogs' handlers use a chase object to lure the dog as far as possible off the dock. The handler commands the dog to begin a running start and then throws the chase object out over the water. The dog leaps after the object, and once he splashes into the water, an event official measures his jump distance. The official measures the dog's final distance from the end of the dock to the point where the base of the dog's tail first hit the water.

In this section, we will work addition and subtraction problems that involve mixed numbers. Let's say that the top two dogs jumped 21 feet $10\frac{1}{4}$ inches and 21 feet $11\frac{7}{8}$ inches. How much further did the latter dog jump? To solve this problem, we need to know how to subtract $10\frac{1}{4}$ from $11\frac{7}{8}$. After reading this section, you will be more comfortable working with mixed numbers such as these.

The notation we use for mixed numbers is especially useful for addition and subtraction. When adding and subtracting mixed numbers, we will assume you recall from section 3.1 how to go about finding a least common denominator (LCD).

Video Examples
Section 3.4

A Addition with Mixed Numbers

EXAMPLE 1 Add: $3\frac{2}{3} + 4\frac{1}{5}$.

Solution We begin by writing each mixed number showing the + sign. We then apply the commutative and associative properties to rearrange the order and grouping.

$$3\frac{2}{3} + 4\frac{1}{5} = 3 + \frac{2}{3} + 4 + \frac{1}{5} \qquad \text{Expand each number to show the + sign.}$$

$$= 3 + 4 + \frac{2}{3} + \frac{1}{5} \qquad \text{Commutative property}$$

$$= (3 + 4) + \left(\frac{2}{3} + \frac{1}{5}\right) \qquad \text{Associative property}$$

$$= 7 + \left(\frac{5 \cdot 2}{5 \cdot 3} + \frac{3 \cdot 1}{3 \cdot 5}\right) \qquad \text{Add } 3 + 4 = 7; \text{ then multiply to get the LCD.}$$

$$= 7 + \left(\frac{10}{15} + \frac{3}{15}\right) \qquad \text{Write each fraction with the LCD.}$$

$$= 7 + \frac{13}{15} \qquad \text{Add the numerators.}$$

$$= 7\frac{13}{15} \qquad \text{Write the answer in mixed-number notation.}$$

Practice Problems
1. Add: $2\frac{1}{3} + 3\frac{3}{4}$.

Answer

1. $6\frac{1}{12}$

As you can see, we obtain our result by adding the whole-number parts $(3 + 4 = 7)$ and the fraction parts $\left(\frac{2}{3} + \frac{1}{5} = \frac{13}{15}\right)$ of each mixed number. Knowing this, we can save ourselves some writing by doing the same problem in columns.

$$3\frac{2}{3} = 3\frac{2\cdot5}{3\cdot5} = 3\frac{10}{15}$$

$$+4\frac{1}{5} = 4\frac{1\cdot3}{5\cdot3} = 4\frac{3}{15} \qquad \text{\textit{Write each fraction with LCD 15}}$$

$$\overline{\hspace{4cm}} $$

$$7\frac{13}{15} \qquad \text{\textit{Add whole numbers, then add fractions.}}$$

The second method shown above requires less writing and lends itself to mixed-number notation. We will use this method for the rest of this section.

2. Redo Practice Problem 1 using method 2.

EXAMPLE 2   Add: $5\frac{3}{4} + 9\frac{5}{6}$.

Solution   The LCD for 4 and 6 is 12. Writing the mixed numbers in a column and then adding looks like this:

$$5\frac{3}{4} = 5\frac{3\cdot3}{4\cdot3} = 5\frac{9}{12}$$

$$+9\frac{5}{6} = 9\frac{5\cdot2}{6\cdot2} = 9\frac{10}{12}$$

$$\overline{\hspace{4cm}}$$

$$14\frac{19}{12}$$

The fraction part of the answer is an improper fraction. We rewrite it as a whole number and a proper fraction.

$$14\frac{19}{12} = 14 + \frac{19}{12} \qquad \text{\textit{Write the mixed number with a + sign.}}$$

$$= 14 + 1\frac{7}{12} \qquad \text{\textit{Write } \frac{19}{12} \text{ as a mixed number.}}$$

$$= 15\frac{7}{12} \qquad \text{\textit{Add 14 and 1.}}$$

Note   Once you see how to change from a whole number and an improper fraction to a whole number and a proper fraction, you will be able to do this step without showing any work.

3. Add: $4\frac{3}{4} + 5\frac{3}{8}$.

EXAMPLE 3   Add: $5\frac{2}{3} + 6\frac{8}{9}$.

Solution

$$5\frac{2}{3} = 5\frac{2\cdot3}{3\cdot3} = 5\frac{6}{9}$$

$$+6\frac{8}{9} = 6\frac{8}{9} \quad . \quad = 6\frac{8}{9}$$

$$\overline{\hspace{4cm}}$$

$$11\frac{14}{9} = 12\frac{5}{9}$$

The last step involves writing $\frac{14}{9}$ as $1\frac{5}{9}$ and then adding 11 and 1 to get 12.

4. Add: $2\frac{2}{3} + 3\frac{1}{4} + 5\frac{5}{6}$.

EXAMPLE 4   Add: $3\frac{1}{4} + 2\frac{3}{5} + 1\frac{9}{10}$.

Solution   The LCD is 20. We rewrite each fraction as an equivalent fraction with denominator 20 and add.

$$3\frac{1}{4} = 3\frac{1\cdot5}{4\cdot5} = 3\frac{5}{20}$$

$$2\frac{3}{5} = 2\frac{3\cdot4}{5\cdot4} = 2\frac{12}{20}$$

$$+1\frac{9}{10} = 1\frac{9\cdot2}{10\cdot2} = 1\frac{18}{20}$$

$$\overline{\hspace{5cm}}$$

$$6\frac{35}{20} = 7\frac{15}{20} = 7\frac{3}{4} \qquad \text{\textit{Reduce to lowest terms.}}$$

$$\frac{35}{20} = 1\frac{15}{20} \qquad \text{\textit{Change to a mixed number.}}$$

Answers

2. $6\frac{1}{12}$

3. $10\frac{1}{8}$

4. $11\frac{3}{4}$

We should note here that we could have worked each of the first four examples in this section by first changing each mixed number to an improper fraction and then adding as we did earlier in this chapter. To illustrate, if we were to work Example 4 this way, it would look like this:

$$3\frac{1}{4} + 2\frac{3}{5} + 1\frac{9}{10} = \frac{13}{4} + \frac{13}{5} + \frac{19}{10} \qquad \text{Change to improper fractions.}$$

$$= \frac{13 \cdot 5}{4 \cdot 5} + \frac{13 \cdot 4}{5 \cdot 4} + \frac{19 \cdot 2}{10 \cdot 2} \qquad \text{LCD is 20.}$$

$$= \frac{65}{20} + \frac{52}{20} + \frac{38}{20} \qquad \text{Equivalent fractions}$$

$$= \frac{155}{20} \qquad \text{Add numerators.}$$

$$= 7\frac{15}{20} = 7\frac{3}{4} \qquad \text{Change to a mixed number, and reduce.}$$

As you can see, the result is the same as the result we obtained in Example 4.

There are advantages to both methods. The method just shown works well when the whole-number parts of the mixed numbers are small. The vertical method shown in Examples 1–4 works well when the whole-number parts of the mixed numbers are large.

B Subtraction with Mixed Numbers

Subtraction with mixed numbers is very similar to addition with mixed numbers.

EXAMPLE 5 Subtract: $3\frac{9}{10} - 1\frac{3}{10}$.

Solution Because the denominators are the same, we simply subtract the whole numbers and subtract the fractions.

$$\begin{array}{r} 3\frac{9}{10} \\ - 1\frac{3}{10} \\ \hline 2\frac{6}{10} = 2\frac{3}{5} \end{array} \qquad \text{Reduce to lowest terms.}$$

5. Subtract: $4\frac{5}{8} - 1\frac{1}{8}$.

An easy way to visualize the results in Example 5 is to imagine 3 dollar bills and 9 dimes in your pocket. If you spend 1 dollar and 3 dimes, you will have 2 dollars and 6 dimes left.

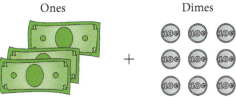

Ones   Dimes

EXAMPLE 6 Subtract: $12\frac{7}{10} - 8\frac{3}{5}$.

6. Subtract: $8\frac{1}{2} - 3\frac{1}{6}$.

Solution The common denominator is 10. We must rewrite $8\frac{3}{5}$ as an equivalent fraction with denominator 10.

$$\begin{array}{r} 12\frac{7}{10} = 12\frac{7}{10} = 12\frac{7}{10} \\ - 8\frac{3}{5} = -8\frac{3 \cdot 2}{5 \cdot 2} = -8\frac{6}{10} \\ \hline 4\frac{1}{10} \end{array}$$

7. Subtract: $12 - 8\frac{5}{9}$.

> *Note* Convince yourself that 10 is the same as $9\frac{7}{7}$. The reason we choose to write the 1 we borrowed as $\frac{7}{7}$ is that the fraction we eventually subtracted from $\frac{7}{7}$ was $\frac{2}{7}$. Both fractions must have the same denominator, 7, so that we can subtract.

8. Subtract: $12\frac{1}{6} - 5\frac{5}{6}$.

9. Subtract: $9\frac{1}{3} - 3\frac{3}{4}$.

EXAMPLE 7   Subtract: $10 - 5\frac{2}{7}$.

Solution   In order to have a fraction from which to subtract $\frac{2}{7}$, we borrow 1 from 10 and rewrite the 1 we borrow as $\frac{7}{7}$. The process looks like this:

$$
\begin{array}{r}
10 = \quad 9\frac{7}{7} \quad \longleftarrow \text{We rewrite 10 as } 9 + 1, \text{ which is } 9 + \frac{7}{7} = 9\frac{7}{7}. \\
- \ 5\frac{2}{7} = -5\frac{2}{7} \quad \text{Then we can subtract as usual.} \\
\hline
4\frac{5}{7}
\end{array}
$$

EXAMPLE 8   Subtract: $8\frac{1}{4} - 3\frac{3}{4}$.

Solution   Because $\frac{3}{4}$ is larger than $\frac{1}{4}$, we again need to borrow 1 from the whole number. The 1 that we borrow from the 8 is rewritten as $\frac{4}{4}$, because 4 is the denominator of both fractions.

$$
\begin{array}{r}
8\frac{1}{4} = \quad 7\frac{5}{4} \quad \longleftarrow \text{Borrow 1 in the form } \frac{4}{4}; \text{ then } \frac{4}{4} + \frac{1}{4} = \frac{5}{4}. \\
- \ 3\frac{3}{4} = -3\frac{3}{4} \\
\hline
4\frac{2}{4} = 4\frac{1}{2} \quad \text{Reduce to lowest terms.}
\end{array}
$$

EXAMPLE 9   Subtract: $4\frac{3}{4} - 1\frac{5}{6}$.

Solution   This is about as complicated as it gets with subtraction of mixed numbers. We begin by rewriting each fraction with the common denominator 12.

$$
\begin{array}{r}
4\frac{3}{4} = \quad 4\frac{3 \cdot 3}{4 \cdot 3} = \quad 4\frac{9}{12} \\
- \ 1\frac{5}{6} = -1\frac{5 \cdot 2}{6 \cdot 2} = -1\frac{10}{12} \\
\hline
\end{array}
$$

Because $\frac{10}{12}$ is larger than $\frac{9}{12}$, we must borrow 1 from 4 in the form $\frac{12}{12}$ before we subtract.

$$
\begin{array}{r}
4\frac{9}{12} = \quad 3\frac{21}{12} \quad \longleftarrow 4 = 3 + 1 = 3 + \frac{12}{12}, \text{ so } 4\frac{9}{12} = \left(3 + \frac{12}{12}\right) + \frac{9}{12} \\
- \ 1\frac{10}{12} = -1\frac{10}{12} \qquad\qquad\qquad = 3 + \left(\frac{12}{12} + \frac{9}{12}\right) \\
\hline
2\frac{11}{12} \qquad\qquad\qquad\qquad = 3 + \frac{21}{12} \\
= 3\frac{21}{12}
\end{array}
$$

GETTING READY FOR CLASS

After reading through the preceding section, respond in your own words and in complete sentences.

A. Is it necessary to "borrow" when subtracting $1\frac{3}{10}$ from $3\frac{9}{10}$?

B. To subtract $1\frac{2}{7}$ from 10, it is necessary to rewrite 10 as what mixed number?

C. To subtract $11\frac{20}{30}$ from $15\frac{3}{30}$, it is necessary to rewrite $15\frac{3}{30}$ as what mixed number?

D. Rewrite $14\frac{19}{12}$ so that the fraction part is a proper fraction instead of an improper fraction.

Answers

7. $3\frac{4}{9}$

8. $6\frac{1}{3}$

9. $5\frac{7}{12}$

Vocabulary Review

Choose the correct words to fill in the blanks below.

proper   columns   addition sign   improper   LCD   borrow

1. To add mixed numbers, rewrite each mixed number showing the _____, and then apply the commutative and associative properties.

2. Adding mixed numbers in _____ requires less writing than the method that uses the addition sign and grouping.

3. When adding mixed numbers in columns, (1) add the whole numbers, (2) write each fraction with the _____, and (3) add the fractions.

4. If your answer to a mixed-number addition problem is a(n) _____ fraction, rewrite it as a whole number and a _____ fraction.

5. When subtracting mixed numbers, if the fraction in the second mixed number is larger than the fraction in the first mixed number, _____ 1 from the whole number in the first mixed number.

Problems

A B Add and subtract the following mixed numbers as indicated.

1. $2\frac{1}{5} + 3\frac{3}{5}$

2. $8\frac{2}{9} + 1\frac{5}{9}$

3. $4\frac{3}{10} + 8\frac{1}{10}$

4. $5\frac{2}{7} + 3\frac{3}{7}$

5. $6\frac{8}{9} - 3\frac{4}{9}$

6. $12\frac{5}{12} + \left(-7\frac{1}{12}\right)$

7. $9\frac{1}{6} + 2\frac{5}{6}$

8. $9\frac{1}{4} - \left(-5\frac{3}{4}\right)$

9. $3\frac{5}{8} - 2\frac{1}{4}$

10. $7\frac{9}{10} - 6\frac{3}{5}$

11. $11\frac{1}{3} + 2\frac{5}{6}$

12. $1\frac{5}{8} + 2\frac{1}{2}$

13. $7\frac{5}{12} - 3\frac{1}{3}$

14. $7\frac{3}{4} - 3\frac{5}{12}$

15. $6\frac{1}{3} - 4\frac{1}{4}$

16. $5\frac{4}{5} - 3\frac{1}{3}$

17. $10\frac{5}{6} + 15\frac{3}{4}$

18. $11\frac{7}{8} - \left(-9\frac{1}{6}\right)$

19. $18\frac{1}{8} + \left(-6\frac{3}{4}\right)$

20. $10\frac{1}{3} - 4\frac{1}{6}$

21. $5\frac{2}{3}$
   $+ 6\frac{1}{3}$

22. $8\frac{5}{6}$
   $+ 9\frac{5}{6}$

23. $10\frac{13}{16}$
   $- 8\frac{5}{16}$

24. $17\frac{7}{12}$
   $- 9\frac{5}{12}$

25. $6\frac{1}{2}$
   $+ 2\frac{5}{14}$

26. $9\frac{11}{12}$
   $+ 4\frac{1}{6}$

27. $1\frac{5}{8}$
   $+ 1\frac{3}{4}$

28. $7\frac{6}{7}$
   $+ 2\frac{3}{14}$

29. $4\frac{2}{3}$
   $+ 5\frac{3}{5}$

30. $9\frac{4}{9}$
   $+ 1\frac{1}{6}$

31. $5\frac{4}{10}$
   $- 3\frac{1}{3}$

32. $12\frac{7}{8}$
   $- 3\frac{5}{6}$

A Find the following sums. (Add.)

33. $1\frac{1}{4} + 2\frac{3}{4} + 5$

34. $6 + 5\frac{3}{5} + 8\frac{2}{5}$

35. $7\frac{1}{10} + 8\frac{3}{10} + 2\frac{7}{10}$

36. $5\frac{2}{7} + 8\frac{1}{7} + 3\frac{5}{7}$

37. $\frac{3}{4} + 8\frac{1}{4} + 5$

38. $\frac{5}{8} + 1\frac{1}{8} + 7$

39. $3\frac{1}{2} + 8\frac{1}{3} + 5\frac{1}{6}$

40. $4\frac{1}{5} + 7\frac{1}{3} + 8\frac{1}{15}$

41. $8\frac{2}{3}$
   $9\frac{1}{8}$
   $+ 6\frac{1}{4}$

42. $7\frac{3}{5}$
   $8\frac{2}{3}$
   $+ 1\frac{1}{5}$

43. $6\frac{1}{7}$
   $9\frac{3}{14}$
   $+ 12\frac{1}{2}$

44. $1\frac{5}{6}$
   $2\frac{3}{4}$
   $+ 5\frac{1}{2}$

45. $10\frac{1}{20}$
   $11\frac{4}{5}$
   $+ 15\frac{3}{10}$

46. $18\frac{7}{12}$
   $19\frac{3}{16}$
   $+ 10\frac{2}{3}$

47. $10\frac{3}{4}$
   $12\frac{5}{6}$
   $+ 9\frac{5}{8}$

48. $4\frac{5}{9}$
   $9\frac{2}{3}$
   $+ 8\frac{5}{6}$

B The following problems all involve the concept of borrowing. Subtract in each case.

49. $8 - 1\frac{3}{4}$

50. $5 - 3\frac{1}{3}$

51. $15 - 5\frac{3}{10}$

52. $24 - 10\frac{5}{12}$

53. $8\frac{1}{4} - 2\frac{3}{4}$

54. $12\frac{3}{10} - 5\frac{7}{10}$

55. $9\frac{1}{3} - 8\frac{2}{3}$

56. $7\frac{1}{6} - 6\frac{5}{6}$

57. $4\frac{1}{4} - 2\frac{1}{3}$

58. $6\frac{1}{5} - 1\frac{2}{3}$

59. $9\frac{2}{3} - 5\frac{3}{4}$

60. $12\frac{5}{6} - 8\frac{7}{8}$

61. $16\frac{3}{4} - 10\frac{4}{5}$

62. $18\frac{5}{12} - 9\frac{3}{4}$

63. $10\frac{3}{10} - 4\frac{4}{5}$

64. $9\frac{4}{7} - 7\frac{2}{3}$

65. $13\frac{1}{6} - 12\frac{5}{8}$

66. $21\frac{2}{5} - 20\frac{5}{6}$

67. $19\frac{1}{4} - 8\frac{5}{6}$

68. $22\frac{7}{10} - 18\frac{4}{5}$

69. Find the difference between $6\frac{1}{5}$ and $2\frac{7}{10}$.

70. Give the difference between $5\frac{1}{3}$ and $1\frac{5}{6}$.

71. Find the sum of $3\frac{1}{8}$ and $2\frac{3}{5}$.

72. Find the sum of $1\frac{5}{6}$ and $3\frac{4}{9}$.

Applying the Concepts

73. Building Two pieces of molding $5\frac{7}{8}$ inches and $6\frac{3}{8}$ inches long are placed end to end. What is the total length of the two pieces of molding together?

74. Jogging A jogger runs $2\frac{1}{2}$ miles on Monday, $3\frac{1}{4}$ miles on Tuesday, and $2\frac{2}{5}$ miles on Wednesday. What is the jogger's total mileage for this 3-day period?

75. Horse Racing If a racehorse runs at both Churchill Downs and Keeneland racetracks, she will run $1\frac{3}{8}$ miles at Churchill Downs, and $1\frac{1}{2}$ miles at Keeneland. How much further will she run at Keeneland?

76. Triple Crown The three races that constitute the Triple Crown in horse racing are shown in the table.

  a. Write the distances in order from smallest to largest.

Race	Distance (miles)
Kentucky Derby	$1\frac{1}{4}$
Preakness Stakes	$1\frac{3}{16}$
Belmont Stakes	$1\frac{1}{2}$

Source: ESPN.com

  b. How much longer is the Belmont Stakes race than the Preakness Stakes?

77. Length of Jeans A pair of jeans is $32\frac{1}{2}$ inches long. How long are the jeans after they have been washed if they shrink $1\frac{1}{3}$ inches?

78. Manufacturing A clothing manufacturer has two rolls of cloth. One roll is $35\frac{1}{2}$ yards, and the other is $62\frac{5}{8}$ yards. What is the total number of yards in the two rolls?

Area and Perimeter The diagrams below show the dimensions of playing fields for the National Football League (NFL), the Canadian Football League, and arena football.

Football Fields

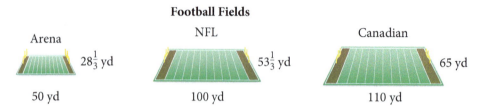

Arena	NFL	Canadian
$28\frac{1}{3}$ yd	$53\frac{1}{3}$ yd	65 yd
50 yd	100 yd	110 yd

79. Find the perimeter of each football field.

80. Find the area of each football field.

Stock Prices In March 1995, rumors that Michael Jordan would return to basketball sent stock prices for the companies whose products he endorses higher. The table below gives some of the details of those increases. Use the table to work Problems 81-84.

81. a. Find the difference in the price of Nike stock between March 13 and March 8.

  b. If you owned 100 shares of Nike stock, how much more are the 100 shares worth on March 13 than on March 8?

82. a. Find the difference in price of General Mills stock between March 13 and March 8.

  b. If you owned 1,000 shares of General Mills stock on March 8, how much more would they be worth on March 13?

Stock Prices for Companies with Michael Jordan Endorsements

Company	Product Endorsed	Stock Price (Dollars) 3/8/95	Stock Price (Dollars) 3/13/95
Nike	Air Jordans	$74\frac{7}{8}$	$77\frac{3}{8}$
Quaker Oats	Gatorade	$32\frac{1}{4}$	$32\frac{5}{8}$
General Mills	Wheaties	$60\frac{1}{2}$	$63\frac{3}{8}$
McDonald's		$32\frac{7}{8}$	$34\frac{3}{8}$

83. If you owned 200 shares of McDonald's stock on March 8, how much more would they be worth on March 13?

84. If you owned 100 shares of McDonald's stock on March 8, how much more would they be worth on March 13?

Getting Ready for the Next Section

Multiply or divide as indicated.

85. $\dfrac{11}{8} \cdot \dfrac{29}{8}$

86. $\dfrac{3}{4} \div \dfrac{5}{6}$

87. $\dfrac{7}{6} \cdot \dfrac{12}{7}$

88. $10\dfrac{1}{3} \div 8\dfrac{2}{3}$

Combine.

89. $\dfrac{3}{4} + \dfrac{5}{8}$

90. $\dfrac{1}{2} + \dfrac{2}{3}$

91. $2\dfrac{3}{8} + 1\dfrac{1}{4}$

92. $3\dfrac{2}{3} + 4\dfrac{1}{3}$

Improving Your Quantitative Literacy

93. Horse Racing A column on horse racing in the *Daily News* in Los Angeles reported that the horse Action This Day ran 3 furlongs in $35\frac{1}{5}$ seconds and another horse, Halfbridled, went two-fifths of a second faster. How many seconds did it take Halfbridled to run 3 furlongs?

Find the Mistake

Each sentence below contains a mistake. Circle the mistake and write the correct word(s) or numbers(s) on the line provided.

1. To begin adding $3\frac{9}{14}$ and $5\frac{1}{3}$, write each mixed number with the addition sign and then apply the commutative and associative properties, such that $\left(3 + \frac{2}{5}\right) + \left(5 + \frac{1}{3}\right)$. _____

2. The final answer for the problem $5\frac{2}{3} + 6\frac{7}{9}$ is $11\frac{13}{9}$. _____

3. The first step when subtracting $8 - 3\frac{2}{7} = 4\frac{5}{7}$, is to borrow 1 from 8 in the form of $\frac{8}{8}$. _____

4. The work for the subtraction problem $5\frac{2}{3} - 2\frac{4}{5}$ is shown below. Circle the mistake and write the correct work on the lines provided.

$$5\frac{2}{3} = 5\frac{2 \cdot 5}{3 \cdot 5} = 5\frac{10}{15} = 5\left(\frac{10}{15} + \frac{15}{15}\right) = 5\frac{25}{15}$$ _____

$$-\quad 2\frac{4}{5} = 2\frac{4 \cdot 3}{5 \cdot 3} = 2\frac{12}{15} \qquad \rightarrow \qquad 2\frac{12}{15}$$ _____

$$3\frac{13}{15}$$ _____

**Video Examples
Section 3.5**

Practice Problems

1. Simplify the expression.

$$4 + \left(1\frac{1}{3}\right)\left(4\frac{1}{2}\right)$$

Each year, the coastal waters of Qingdao, China are infiltrated with a vibrant green algae bloom that covers more than 150 square miles. The algae thrive in high temperatures and in polluted waters caused by runoff from agriculture land and fish farms. The bloom sucks oxygen out of the water, threatening the local marine life. As it washes onshore and dries in the hot sun, it gives off a noxious odor that smells like rotten eggs. Soldiers and other volunteers work tirelessly to clear the water and the beaches of the algae before the bloom drastically damages the local ecosystem.

Suppose you have volunteered to help clean up a portion of the algae-affected coast. The area of the beach you'll be responsible for has a width of $5\frac{1}{4}$ feet and a length of $10\frac{2}{3}$ feet. You also need to clean part of the water that measures 20 square feet. You wonder how large the total area is for which you are responsible. The total area is given by the following expression:

$$20 + \left(5\frac{1}{4}\right)\left(10\frac{2}{3}\right)$$

In this section, we will learn how to simplify expressions, such as the one above.

A Order of Operations

Let's use the order of operations to simplify expressions that contain both fractions and mixed numbers.

EXAMPLE 1   Simplify the expression: $5 + \left(2\frac{1}{2}\right)\left(3\frac{2}{3}\right)$.

Solution   The rule for order of operations indicates that we should multiply $2\frac{1}{2}$ times $3\frac{2}{3}$ and then add 5 to the result.

$$5 + \left(2\frac{1}{2}\right)\left(3\frac{2}{3}\right) = 5 + \left(\frac{5}{2}\right)\left(\frac{11}{3}\right)$$
Change the mixed numbers to improper fractions.

$$= 5 + \frac{55}{6}$$
Multiply the improper fractions.

$$= \frac{30}{6} + \frac{55}{6}$$
Write 5 as $\frac{30}{6}$ so both numbers have the same denominator.

$$= \frac{85}{6}$$
Add fractions by adding their numerators.

$$= 14\frac{1}{6}$$
Write the answer as a mixed number.

Answer

1. 10

EXAMPLE 2 Simplify: $\left(\frac{3}{4} + \frac{5}{8} \right)\left(2\frac{3}{8} + 1\frac{1}{4} \right)$.

Solution We begin by combining the numbers inside the parentheses.

$$\frac{3}{4} + \frac{5}{8} = \frac{3 \cdot 2}{4 \cdot 2} + \frac{5}{8} \qquad \text{and} \qquad 2\frac{3}{8} = \quad 2\frac{3}{8} \quad = \quad 2\frac{3}{8}$$

$$= \frac{6}{8} + \frac{5}{8} \qquad\qquad\qquad \underline{+ 1\frac{1}{4} = + 1\frac{1 \cdot 2}{4 \cdot 2} = + 1\frac{2}{8}}$$

$$= \frac{11}{8} \qquad\qquad\qquad\qquad\qquad 3\frac{5}{8}$$

Now that we have combined the expressions inside the parentheses, we can complete the problem by multiplying the results.

$$\left(\frac{3}{4} + \frac{5}{8} \right)\left(2\frac{3}{8} + 1\frac{1}{4} \right) = \left(\frac{11}{8} \right)\left(3\frac{5}{8} \right)$$

$$= \frac{11}{8} \cdot \frac{29}{8} \qquad \textit{Change } 3\tfrac{5}{8} \textit{ to an improper fraction.}$$

$$= \frac{319}{64} \qquad \textit{Multiply fractions.}$$

$$= 4\frac{63}{64} \qquad \textit{Write the answer as a mixed number.}$$

EXAMPLE 3 Simplify: $\frac{3}{5} + \frac{1}{2}\left(3\frac{2}{3} + 4\frac{1}{3} \right)^2$.

Solution We begin by combining the expressions inside the parentheses.

$$\frac{3}{5} + \frac{1}{2}\left(3\frac{2}{3} + 4\frac{1}{3} \right)^2 = \frac{3}{5} + \frac{1}{2}\,(8)^2 \quad \textit{The sum inside the parentheses is 8.}$$

$$= \frac{3}{5} + \frac{1}{2}\,(64) \quad \textit{The square of 8 is 64.}$$

$$= \frac{3}{5} + 32 \qquad \tfrac{1}{2} \textit{ of 64 is 32.}$$

$$= 32\frac{3}{5} \qquad \textit{The result is a mixed number.}$$

B Complex Fractions

DEFINITION complex fraction

A *complex fraction* is a fraction in which the numerator and/or the denominator are themselves fractions or combinations of fractions.

Each of the following is a complex fraction:

$$\frac{\frac{3}{4}}{\frac{5}{6}}, \qquad \frac{3 + \frac{1}{2}}{2 - \frac{3}{4}}, \qquad \frac{\frac{1}{2} + \frac{2}{3}}{\frac{3}{4} - \frac{1}{6}}$$

2. Simplify: $\left(\frac{2}{3} + \frac{5}{6} \right)\left(3\frac{1}{4} - 1\frac{1}{2} \right)$.

3. Simplify: $3\frac{5}{6} + \frac{1}{3}\left(4\frac{1}{3} + 2\frac{2}{3} \right)^2$.

Answers

2. $2\frac{5}{8}$

3. $20\frac{1}{6}$

4. Simplify: $\dfrac{\frac{2}{3}}{\frac{5}{6}}$.

EXAMPLE 4   Simplify: $\dfrac{\frac{3}{4}}{\frac{5}{6}}$.

Solution   This is actually the same as the problem $\frac{3}{4} \div \frac{5}{6}$, because the bar between $\frac{3}{4}$ and $\frac{5}{6}$ indicates division. Therefore, it must be true that

$$\frac{\frac{3}{4}}{\frac{5}{6}} = \frac{3}{4} \div \frac{5}{6}$$
$$= \frac{3 \cdot 6}{4 \cdot 5}$$
$$= \frac{18}{20}$$
$$= \frac{9}{10}$$

As you can see, we continue to use properties we have developed previously when we encounter new situations. In Example 4, we use the fact that division by a number and multiplication by its reciprocal produce the same result. We are taking a new problem, simplifying a complex fraction, and thinking of it in terms of a problem we have done previously, division by a fraction.

5. Simplify: $\dfrac{\frac{1}{4} + \frac{1}{3}}{\frac{1}{2} - \frac{1}{3}}$.

Note We are going to simplify this complex fraction by two different methods. This is the first method.

EXAMPLE 5   Simplify: $\dfrac{\frac{1}{2} + \frac{2}{3}}{\frac{3}{4} - \frac{1}{6}}$.

Solution   Let's decide to call the numerator of this complex fraction the *top* of the fraction and its denominator the *bottom* of the complex fraction. It will be less confusing if we name them this way. The LCD for all the denominators on the top and bottom is 12, so we can multiply the top and bottom of this complex fraction by 12 and be sure all the denominators will divide it exactly. This will leave us with only whole numbers on the top and bottom.

$$\frac{\frac{1}{2} + \frac{2}{3}}{\frac{3}{4} - \frac{1}{6}} = \frac{12\left(\frac{1}{2} + \frac{2}{3}\right)}{12\left(\frac{3}{4} - \frac{1}{6}\right)} \quad \text{Multiply the top and bottom by the LCD.}$$

$$= \frac{12 \cdot \frac{1}{2} + 12 \cdot \frac{2}{3}}{12 \cdot \frac{3}{4} - 12 \cdot \frac{1}{6}} \quad \text{Distributive property}$$

$$= \frac{6 + 8}{9 - 2} \quad \text{Multiply each fraction by 12.}$$

$$= \frac{14}{7} \quad \text{Add on the top and subtract on the bottom.}$$

$$= 2 \quad \text{Reduce to lowest terms.}$$

The problem can be worked in another way also. We can simplify the top and bottom of the complex fraction separately. Simplifying the top, we have

$$\frac{1}{2} + \frac{2}{3} = \frac{1 \cdot 3}{2 \cdot 3} + \frac{2 \cdot 2}{3 \cdot 2} = \frac{3}{6} + \frac{4}{6} = \frac{7}{6}$$

Simplifying the bottom, we have

$$\frac{3}{4} - \frac{1}{6} = \frac{3 \cdot 3}{4 \cdot 3} - \frac{1 \cdot 2}{6 \cdot 2} = \frac{9}{12} - \frac{2}{12} = \frac{7}{12}$$

Answers

4. $\frac{4}{5}$

5. $3\frac{1}{2}$

We now write the original complex fraction again using the simplified expressions for the top and bottom. Then we proceed as we did in Example 4.

$$\frac{\frac{1}{2} + \frac{2}{3}}{\frac{3}{4} - \frac{1}{6}} = \frac{\frac{7}{6}}{\frac{7}{12}}$$

$$= \frac{7}{6} \div \frac{7}{12} \qquad \text{The divisor is } \frac{7}{12}.$$

$$= \frac{7}{6} \cdot \frac{12}{7} \qquad \text{Replace } \frac{7}{12} \text{ by its reciprocal and multiply.}$$

$$= \frac{7 \cdot 2 \cdot 6}{6 \cdot 7} \qquad \text{Divide out common factors.}$$

$$= 2$$

Note The fraction bar that separates the numerator of the complex fraction from its denominator works like parentheses. If we were to rewrite this problem without it, we would write it like this:

$$\left(\frac{1}{2} + \frac{2}{3}\right) \div \left(\frac{3}{4} - \frac{1}{6}\right)$$

That is why we simplify the top and bottom of the complex fraction separately and then divide.

EXAMPLE 6   Simplify: $\dfrac{3 + \frac{1}{2}}{2 - \frac{3}{4}}$.

6. Simplify: $\dfrac{4 - \frac{1}{3}}{2 + \frac{1}{2}}$.

Solution   The simplest approach here is to multiply both the top and bottom by the LCD for all fractions, which is 4.

$$\frac{3 + \frac{1}{2}}{2 - \frac{3}{4}} = \frac{4\left(3 + \frac{1}{2}\right)}{4\left(2 - \frac{3}{4}\right)} \qquad \text{Multiply the top and bottom by 4.}$$

$$= \frac{4 \cdot 3 + 4 \cdot \frac{1}{2}}{4 \cdot 2 - 4 \cdot \frac{3}{4}} \qquad \text{Distributive property}$$

$$= \frac{12 + 2}{8 - 3} \qquad \text{Multiply each number by 4.}$$

$$= \frac{14}{5} \qquad \text{Add on the top and subtract on the bottom.}$$

$$= 2\frac{4}{5}$$

EXAMPLE 7   Simplify: $\dfrac{10\frac{1}{3}}{8\frac{2}{3}}$.

7. Simplify: $\dfrac{8\frac{3}{4}}{5\frac{1}{2}}$.

Solution   The simplest way to simplify this complex fraction is to think of it as a division problem.

$$\frac{10\frac{1}{3}}{8\frac{2}{3}} = 10\frac{1}{3} \div 8\frac{2}{3} \qquad \text{Write with a } \div \text{ symbol.}$$

$$= \frac{31}{3} \div \frac{26}{3} \qquad \text{Change to improper fractions.}$$

$$= \frac{31}{3} \cdot \frac{3}{26} \qquad \text{Write in terms of multiplication.}$$

$$= \frac{31 \cdot 3}{3 \cdot 26} \qquad \text{Divide out the common factor 3.}$$

$$= \frac{31}{26} = 1\frac{5}{26} \qquad \text{Answer as a mixed number.}$$

Answers

6. $1\frac{7}{15}$

7. $1\frac{13}{22}$

GETTING READY FOR CLASS

After reading through the preceding section, respond in your own words and in complete sentences.

A. What is a complex fraction?

B. Rewrite $\dfrac{\frac{5}{6}}{\frac{1}{3}}$ as a multiplication problem.

C. True or false? The rules for order of operations tell us to work inside parentheses first.

D. True or false? We find the LCD when we add or subtract fractions, but not when we multiply them. Explain.

Vocabulary Review

Choose the correct words to fill in the blanks below.

LCD   simplify   complex fraction

whole numbers   bottom   top

1. A _____ is a fraction in which the numerator and/or the denominator are themselves fractions or combinations of fractions.

2. We call the numerator of a complex fraction the _____, and the denominator the _____.

3. One method to simplify a complex fraction is to find the _____ of all the denominators in the top and bottom of the complex fraction.

4. Multiplying the top and bottom of a complex fraction by the LCD will leave only _____ on the top and the bottom.

5. A second method to _____ a complex fraction is to simplify the top and bottom of the complex fraction separately.

Problems

A Use the rule for order of operations to simplify each of the following.

1. $3 + \left(1\frac{1}{2}\right)\left(2\frac{2}{3}\right)$

2. $7 - \left(1\frac{3}{5}\right)\left(2\frac{1}{2}\right)$

3. $8 - \left(\frac{6}{11}\right)\left(1\frac{5}{6}\right)$

4. $10 + \left(2\frac{4}{5}\right)\left(\frac{5}{7}\right)$

5. $\frac{2}{3}\left(1\frac{1}{2}\right) + \frac{3}{4}\left(1\frac{1}{3}\right)$

6. $\frac{2}{5}\left(2\frac{1}{2}\right) + \frac{5}{8}\left(3\frac{1}{5}\right)$

7. $2\left(1\frac{1}{2}\right) + 5\left(6\frac{2}{5}\right)$

8. $4\left(5\frac{3}{4}\right) + 6\left(3\frac{5}{6}\right)$

9. $\left(\frac{3}{5} + \frac{1}{10}\right)\left[\frac{1}{2} - \left(-\frac{3}{4}\right)\right]$

10. $\left(\frac{2}{9} + \frac{1}{3}\right)\left(\frac{1}{5} + \frac{1}{10}\right)$

11. $\left(2 + \frac{2}{3}\right)\left(3 + \frac{1}{8}\right)$

12. $\left(3 - \frac{3}{4}\right)\left(3 + \frac{1}{3}\right)$

13. $\left(1 + \frac{5}{6}\right)\left(1 - \frac{5}{6}\right)$

14. $\left(2 - \frac{1}{4}\right)\left(2 + \frac{1}{4}\right)$

15. $\frac{2}{3} + \frac{1}{3}\left(2\frac{1}{2} + \frac{1}{2}\right)^2$

16. $\dfrac{3}{5} + \dfrac{1}{4}\left(2\dfrac{1}{2} - \dfrac{1}{2}\right)^{3}$

17. $2\dfrac{3}{8} + \dfrac{1}{2}\left(\dfrac{1}{3} + \dfrac{5}{3}\right)^{3}$

18. $8\dfrac{2}{3} + \dfrac{1}{3}\left(\dfrac{8}{5} + \dfrac{7}{5}\right)^{2}$

19. $2\left(\dfrac{1}{2} + \dfrac{1}{3}\right) + 3\left(\dfrac{2}{3} + \dfrac{1}{4}\right)$

20. $5\left(\dfrac{1}{5} + \dfrac{3}{10}\right) + 2\left(\dfrac{1}{10} + \dfrac{1}{2}\right)$

B Simplify each complex fraction as much as possible.

21. $\dfrac{\frac{2}{3}}{\frac{3}{4}}$

22. $\dfrac{\frac{5}{6}}{\frac{3}{12}}$

23. $\dfrac{\frac{2}{3}}{\frac{4}{3}}$

24. $\dfrac{\frac{7}{9}}{\frac{5}{9}}$

25. $\dfrac{\frac{11}{20}}{\frac{5}{10}}$

26. $\dfrac{\frac{9}{16}}{\frac{3}{4}}$

27. $\dfrac{\frac{1}{2} + \frac{1}{3}}{\frac{1}{2} - \frac{1}{3}}$

28. $\dfrac{\frac{1}{4} + \frac{1}{5}}{\frac{1}{4} - \frac{1}{5}}$

29. $\dfrac{\frac{5}{8} - \frac{1}{4}}{\frac{1}{8} + \frac{1}{2}}$

30. $\dfrac{\frac{3}{4} + \frac{1}{3}}{\frac{2}{3} + \frac{1}{6}}$

31. $\dfrac{\frac{9}{20} - \frac{1}{10}}{\frac{1}{10} + \frac{9}{20}}$

32. $\dfrac{\frac{1}{2} + \frac{2}{3}}{\frac{3}{4} + \frac{5}{6}}$

33. $\dfrac{1 + \frac{2}{3}}{1 - \frac{2}{3}}$

34. $\dfrac{5 - \frac{3}{4}}{2 + \frac{3}{4}}$

35. $\dfrac{2 + \frac{5}{6}}{5 - \frac{1}{3}}$

36. $\dfrac{9 - \frac{11}{5}}{3 + \frac{13}{10}}$

37. $\dfrac{3 + \frac{5}{6}}{1 + \frac{5}{3}}$

38. $\dfrac{10 + \frac{9}{10}}{5 + \frac{4}{5}}$

39. $\dfrac{\frac{1}{3} + \frac{3}{4}}{2 - \frac{1}{6}}$

40. $\dfrac{3 + \frac{5}{2}}{\frac{5}{6} + \frac{1}{4}}$

41. $\dfrac{\frac{5}{6}}{3 + \frac{2}{3}}$

42. $\dfrac{9 - \frac{3}{2}}{\frac{7}{4}}$

43. $\dfrac{8 + \frac{3}{4}}{\frac{5}{8}}$

44. $\dfrac{\frac{5}{6}}{3 - \frac{1}{3}}$

Simplify each of the following complex fractions.

45. $\dfrac{2\frac{1}{2} + \frac{1}{2}}{3\frac{3}{5} - \frac{2}{5}}$

46. $\dfrac{5\frac{3}{8} + \frac{5}{8}}{4\frac{1}{4} + 1\frac{3}{4}}$

47. $\dfrac{2 + 1\frac{2}{3}}{3\frac{5}{6} - 1}$

48. $\dfrac{5 + 8\frac{3}{5}}{2\frac{3}{10} + 4}$

49. $\dfrac{3\frac{1}{4} - 2\frac{1}{2}}{5\frac{3}{4} + 1\frac{1}{2}}$

50. $\dfrac{9\frac{3}{8} + 2\frac{5}{8}}{6\frac{1}{2} + 7\frac{1}{2}}$

51. $\dfrac{3\frac{1}{4} + 5\frac{1}{6}}{2\frac{1}{3} + 3\frac{1}{4}}$

52. $\dfrac{8\frac{5}{6} + 1\frac{2}{3}}{7\frac{1}{3} + 2\frac{1}{4}}$

53. $\dfrac{6\frac{2}{3} + 7\frac{3}{4}}{8\frac{1}{2} + 9\frac{7}{8}}$

54. $\dfrac{3\frac{4}{5} - 1\frac{9}{10}}{6\frac{5}{6} - 2\frac{3}{4}}$

55. $\dfrac{4\frac{1}{3} - 1\frac{2}{3}}{8\frac{1}{4} - 5\frac{1}{2}}$

56. $\dfrac{3\frac{1}{2} + 2\frac{1}{6}}{2 - \frac{5}{6}}$

57. What is twice the sum of $2\frac{1}{5}$ and $\frac{3}{6}$?

58. Find 3 times the difference of $1\frac{7}{9}$ and $\frac{2}{9}$.

59. Add $5\frac{1}{4}$ to the sum of $\frac{3}{4}$ and 2.

60. Subtract $\frac{7}{8}$ from the product of 2 and $3\frac{1}{2}$.

Applying the Concepts

61. **Manufacturing** A dress manufacturer usually buys two rolls of cloth, one of $32\frac{1}{2}$ yards and the other of $25\frac{1}{3}$ yards, to fill his weekly orders. If his orders double one week, how much of the cloth should he order? (Give the total yardage.)

62. **Body Temperature** Suppose your normal body temperature is $98\frac{3}{5}°$ Fahrenheit. If your temperature goes up $3\frac{1}{5}°$ on Monday and then down $1\frac{4}{5}°$ on Tuesday, what is your temperature on Tuesday?

Find the Mistake

Each sentence below contains a mistake. Circle the mistake and write the correct word(s) or numbers(s) on the line provided.

1. The first step to solving the problem $\frac{1}{4} - \left(2\frac{3}{8} - 1\frac{5}{8}\right)^2$ is to rewrite the problem so it looks like $\frac{1}{4} + \left(-2\frac{3}{8} + -1\frac{5}{8}\right)^2$.

_____

2. To simplify $5\frac{2}{3} + \left(10\frac{1}{3} \cdot \frac{2}{3}\right)$ you must first change $10\frac{1}{3}$ into the reciprocal $\frac{31}{3}$ before multiplying by $\frac{2}{3}$.

_____

3. A complex fraction is a fraction in which a fraction or combination of fractions appear in the denominator of the original fraction. _____

4. To simplify the complex fraction $\dfrac{\frac{1}{6} + \frac{2}{3}}{\frac{5}{6} + \frac{5}{12}}$, multiply the top and bottom of the fraction by 6. _____

Fractions and Word Frequency

Choose the first paragraph from one of the five sections in this chapter. Disregarding any numbers or fractions, count the total number of words in the paragraph. Now count the words grouped by the number of letters each one contains. For example, find the total number of 1-letter words in the paragraph. Do the same for 2-letter words, 3-letter words, and so on. Group any words 9 letters or greater into a single category. Create a table with this information.

Now using the information from your table, add a new column that represents each category of words as a fraction and reduce.

Answer the following questions using your reduced fractions.

1. What fraction represents the total number of words that contain two to five letters?

2. What fraction represents the total number of words that contain six to eight letters?

3. What fraction represents the difference between the category with the most words and the category with the least words?

Note: Gathering this information is a way to analyze the frequency of words in a paragraph. We will work more with frequency later in the book.

Least Common Denominator (LCD) [3.1]

The *least common denominator* (LCD) for a set of denominators is the smallest number that is exactly divisible by each denominator.

Addition and Subtraction of Fractions [3.1]

To add (or subtract) two fractions with a common denominator, add (or subtract) numerators and use the common denominator. *In symbols*: If a, b, and c are numbers with c not equal to 0, then

$$\frac{a}{c} + \frac{b}{c} = \frac{a+b}{c} \quad \text{and} \quad \frac{a}{c} - \frac{b}{c} = \frac{a-b}{c}$$

EXAMPLES

$$1.\ \frac{1}{8} + \frac{3}{8} = \frac{1+3}{8}$$
$$= \frac{4}{8}$$
$$= \frac{1}{2}$$

Additional Facts about Fractions

1. In some books, fractions are called *rational numbers*.

2. Every whole number can be written as a fraction with a denominator of 1.

3. The commutative, associative, and distributive properties are true for fractions.

4. The word *of* as used in the expression "$\frac{2}{3}$ *of* 12" indicates that we are to multiply $\frac{2}{3}$ and 12.

5. Two fractions with the same value are called *equivalent fractions*.

Mixed-Number Notation [3.2]

A mixed number is the sum of a whole number and a fraction. The $+$ sign is not shown when we write mixed numbers; it is implied. The mixed number $4\frac{2}{3}$ is actually the sum $4 + \frac{2}{3}$.

Changing Mixed Numbers to Improper Fractions [3.2]

To change a mixed number to an improper fraction, we write the mixed number showing the $+$ sign and add as usual. The result is the same if we multiply the denominator of the fraction by the whole number and add what we get to the numerator of the fraction, putting this result over the denominator of the fraction.

$$2.\ 4\frac{2}{3} = \frac{3 \cdot 4 + 2}{3} = \frac{14}{3}$$

↑ Mixed number   ↑ Improper fraction

Changing an Improper Fraction to a Mixed Number [3.2]

To change an improper fraction to a mixed number, divide the denominator into the numerator. The quotient is the whole-number part of the mixed number. The fraction part is the remainder over the divisor.

3. Change $\frac{14}{3}$ to a mixed number.

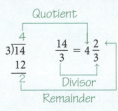

Multiplication and Division with Mixed Numbers [3.3]

4. $2\frac{1}{3} \cdot 1\frac{3}{4} = \frac{7}{3} \cdot \frac{7}{4} = \frac{49}{12} = 4\frac{1}{12}$

To multiply or divide two mixed numbers, change each to an improper fraction and multiply or divide as usual.

Addition and Subtraction with Mixed Numbers [3.4]

5.
$$\begin{aligned}3\frac{4}{9} &= 3\frac{4}{9} & = 3\frac{4}{9}\\ +2\frac{2}{3} &= 2\frac{2\cdot3}{3\cdot3} & = 2\frac{6}{9}\\ & & 5\frac{10}{9} = 6\frac{1}{9}\end{aligned}$$

Common denominator │ Add fractions. Add whole numbers.

To add or subtract two mixed numbers, add or subtract the whole-number parts and the fraction parts separately. This is best done with the numbers written in columns.

Borrowing in Subtraction with Mixed Numbers [3.4]

6.
$$\begin{aligned}4\frac{1}{3} &= & 4\frac{2}{6} &= & 3\frac{8}{6}\\ -1\frac{5}{6} &= & -1\frac{5}{6} &= & -1\frac{5}{6}\\ & & & & 2\frac{3}{6} = 2\frac{1}{2}\end{aligned}$$

It is sometimes necessary to borrow when doing subtraction with mixed numbers. We always change to a common denominator before we actually borrow.

Complex Fractions [3.5]

7.
$$\frac{4+\frac{1}{3}}{2-\frac{5}{6}} = \frac{6\left(4+\frac{1}{3}\right)}{6\left(2-\frac{5}{6}\right)}$$

$$= \frac{6\cdot4+6\cdot\frac{1}{3}}{6\cdot2-6\cdot\frac{5}{6}}$$

$$= \frac{24+2}{12-5}$$

$$= \frac{26}{7} = 3\frac{5}{7}$$

A fraction that contains a fraction in its numerator or denominator is called a *complex fraction*.

COMMON MISTAKE

1. The most common mistake when working with fractions occurs when we try to add two fractions without using a common denominator. For example,

   $$\frac{2}{3} + \frac{4}{5} \neq \frac{2+4}{3+5}$$

   If the two fractions we are trying to add don't have the same denominators, then we *must* rewrite each one as an equivalent fraction with a common denominator. *We never add denominators when adding fractions.*
   Note We do *not* need a common denominator when multiplying fractions.

2. A common mistake when working with mixed numbers is to confuse mixed-number notation for multiplication of fractions. The notation $3\frac{2}{5}$ does *not* mean 3 *times* $\frac{2}{5}$. It means 3 *plus* $\frac{2}{5}$.

3. Another mistake occurs when multiplying mixed numbers. The mistake occurs when we don't change the mixed number to an improper fraction before multiplying and instead try to multiply the whole numbers and fractions separately.

   $$2\frac{1}{2} \cdot 3\frac{1}{3} = (2 \cdot 3) + \left(\frac{1}{2} \cdot \frac{1}{3}\right) \qquad \text{Mistake}$$

   $$= 6 + \frac{1}{6}$$

   $$= 6\frac{1}{6}$$

   Remember, the correct way to multiply mixed numbers is to first change to improper fractions and then multiply numerators and multiply denominators. This is correct:

   $$2\frac{1}{2} \cdot 3\frac{1}{3} = \frac{5}{2} \cdot \frac{10}{3} = \frac{50}{6} = 8\frac{2}{6} = 8\frac{1}{3} \qquad \text{Correct}$$

Perform the indicated operations. Reduce all answers to lowest terms. [3.1, 3.2]

1. $\dfrac{7}{9} + \dfrac{5}{9}$

2. $\dfrac{7}{12} - \dfrac{1}{12}$

3. $4 + \dfrac{4}{5}$

4. $\dfrac{3}{8} - \left(-\dfrac{1}{4}\right)$

5. $\dfrac{1}{2} + \dfrac{7}{12} + \dfrac{3}{20}$

6. $9\dfrac{2}{3} + 2\dfrac{3}{5}$

7. $6\dfrac{2}{5} - 4\dfrac{3}{10}$

8. $9 + \left(-\dfrac{3}{5}\right)$

9. $\dfrac{5}{8} - \dfrac{3}{16}$

10. $\dfrac{7}{8} + \dfrac{3}{4} + 4$

11. $\dfrac{9}{24} - \dfrac{1}{8}$

12. $2\dfrac{5}{6} + \left(-3\dfrac{1}{3}\right)$

13. $12\dfrac{5}{7} - \dfrac{3}{14}$

14. $\dfrac{4}{9} - \dfrac{5}{18}$

15. $5\dfrac{7}{12} - 1\dfrac{3}{8}$

16. $10\dfrac{9}{16} - \left(-\dfrac{7}{8}\right)$

17. $11\dfrac{2}{3} + 5\dfrac{5}{6}$

18. $4\dfrac{1}{4} + 3\dfrac{5}{8} - 1\dfrac{7}{12}$

19. Sewing A dress that is $24\dfrac{3}{8}$ inches long is shortened by $4\dfrac{3}{4}$ inches. What is the new length of the dress? [3.1]

20. Find the perimeter of the triangle below. [3.1]

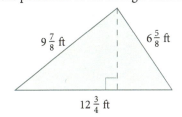

$9\dfrac{7}{8}$ ft $6\dfrac{5}{8}$ ft

$12\dfrac{3}{4}$ ft

Simplify each of the following as much as possible. [3.3, 3.4, 3.5]

21. $7 + 3\left(2\dfrac{4}{5}\right)$

22. $\left(5\dfrac{7}{12} - \dfrac{1}{3}\right)\left(3 - 2\dfrac{5}{8}\right)$

23. $\dfrac{\dfrac{5}{16} - \dfrac{1}{4}}{\dfrac{1}{2} - \dfrac{3}{8}}$

24. $8 - 2\left(2\dfrac{5}{8}\right)$

25. $\left(5\dfrac{3}{8} - 2\dfrac{1}{4}\right)\left(6\dfrac{7}{12} - 2\dfrac{1}{3}\right)$

26. $\left(\dfrac{5}{4}\right)^2 - \dfrac{17}{24}$

27. $\left(\dfrac{2}{9}\right)\left(3\dfrac{3}{5}\right) - \dfrac{3}{10}$

28. $\dfrac{4\dfrac{3}{16}}{2\dfrac{7}{8}}$

29. $\dfrac{\dfrac{7}{12}}{6\dfrac{3}{4}}$

30. $\dfrac{4\dfrac{5}{8} - 2\dfrac{7}{12}}{5\dfrac{1}{3} - 1\dfrac{1}{4}}$

The chart shows iPhone distribution around the world. Use the information to answer the following questions.

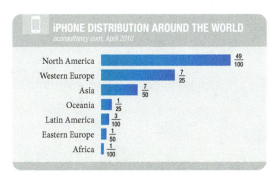

31. What is the fraction of iPhones found in Western Europe and Africa combined?

32. What is the fraction of iPhones found in Asia and Oceania combined?

Reduce to lowest terms.

1. $\dfrac{27}{36}$

2. $\dfrac{30}{45}$

3. $\dfrac{51}{68}$

4. $\dfrac{135}{216}$

Simplify.

5. $\dfrac{7}{12} - \dfrac{5}{12}$

6. $23\dfrac{9}{14} - 7\dfrac{13}{21}$

7. $\dfrac{4}{9} \cdot \dfrac{12}{20}$

8. $7^2 - 3^3$

9. $6 \div \dfrac{1}{2}$

10. $512 \div 16 \div 8$

11. $\dfrac{46}{13}$

12. $\dfrac{170}{306}$

13. 14^2

14. $(9 + 13) - (8 - 4)$

15. $\dfrac{2}{5} + \dfrac{1}{3}$

16. $2{,}074(304)$

17. $\dfrac{4}{16} \div \dfrac{3}{8}$

18. $8 + 4^2 - 4 \cdot 5 + 7$

19. $16 \cdot \dfrac{3}{4}$

20. $\left(\dfrac{6}{5}\right)^2 \cdot \left(\dfrac{2}{3}\right)^3$

21. $7{,}482 - 594$

22. $\dfrac{4\frac{2}{3}}{3\frac{1}{9}}$

23. $\left(\dfrac{3}{8} + \dfrac{7}{24}\right) + \left(-\dfrac{13}{48}\right)$

24. $4 + \dfrac{4}{9} \div \dfrac{2}{3}$

25. Find the sum of $\dfrac{3}{8}$, $\dfrac{1}{4}$, and $\dfrac{2}{3}$.

26. Find the sum of $\dfrac{5}{6}$, $\dfrac{3}{4}$, and $\dfrac{1}{5}$.

27. Write the fraction $\dfrac{7}{15}$ as an equivalent fraction with a denominator of $60x$.

28. Reduce $\dfrac{39}{143}$ to lowest terms.

29. Add $\dfrac{2}{3}$ to half of $\dfrac{6}{7}$.

30. Three times the difference of 8 and 3 is decreased by 3. What number results?

The chart shows the number of Twitter followers for some celebrities. Use the information to answer the following questions.

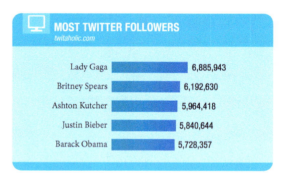

31. How many followers do Ashton Kutcher and Lady Gaga have together?

32. How many more followers does Britney Spears have than Justin Bieber?

Perform the indicated operations. Reduce all answers to lowest terms. [3.1, 3.2]

1. $\dfrac{3}{8} + \dfrac{1}{8}$

2. $\dfrac{9}{14} - \dfrac{3}{14}$

3. $5 + \dfrac{5}{8}$

4. $\dfrac{5}{6} - \left(-\dfrac{2}{3}\right)$

5. $\dfrac{3}{8} + \dfrac{7}{12} + \dfrac{1}{3}$

6. $5\dfrac{1}{5} + 4\dfrac{3}{4}$

7. $4\dfrac{1}{3} - 2\dfrac{7}{9}$

8. $4 + \left(-\dfrac{7}{8}\right)$

9. $\dfrac{4}{7} - \dfrac{3}{14}$

10. $\dfrac{5}{6} + \dfrac{2}{3} + 2$

11. $\dfrac{7}{18} - \dfrac{1}{6}$

12. $1\dfrac{5}{8} + \left(-2\dfrac{1}{4}\right)$

13. $5\dfrac{3}{4} - \dfrac{7}{12}$

14. $\dfrac{3}{4} - \dfrac{7}{16}$

15. $3\dfrac{5}{6} - 2\dfrac{2}{3}$

16. $9\dfrac{1}{4} - \left(-\dfrac{7}{12}\right)$

17. $9\dfrac{5}{6} + 6\dfrac{11}{18}$

18. $3\dfrac{5}{8} + 4\dfrac{1}{6} - 2\dfrac{4}{9}$

19. **Sewing** A dress that is $18\dfrac{5}{8}$ inches long is shortened by $2\dfrac{3}{4}$ inches. What is the new length of the dress? [3.1]

20. Find the perimeter of the triangle below. [3.1]

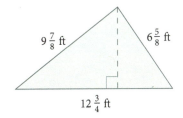

$9\dfrac{7}{8}$ ft $6\dfrac{5}{8}$ ft

$12\dfrac{3}{4}$ ft

Simplify each of the following as much as possible. [3.3, 3.4, 3.5]

21. $5 + 2\left(4\dfrac{3}{4}\right)$

22. $\left(4\dfrac{3}{4} + \dfrac{1}{2}\right)\left(6\dfrac{5}{6} - 4\dfrac{1}{3}\right)$

23. $\dfrac{\dfrac{1}{12} + \dfrac{3}{4}}{\dfrac{5}{6} - \dfrac{1}{3}}$

24. $6 - 3\left(1\dfrac{4}{5}\right)$

25. $\left(7\dfrac{1}{8} - \dfrac{3}{4}\right)\left(3\dfrac{5}{6} + \dfrac{1}{3}\right)$

26. $\left(\dfrac{4}{3}\right)^2 - \dfrac{5}{6}$

27. $\left(\dfrac{1}{3}\right)\left(2\dfrac{1}{4}\right) - \dfrac{3}{8}$

28. $\dfrac{2\dfrac{4}{15}}{1\dfrac{3}{5}}$

29. $\dfrac{\dfrac{9}{16}}{5\dfrac{1}{4}}$

30. $\dfrac{3\dfrac{1}{6} + \dfrac{1}{3}}{5\dfrac{5}{12} - 2\dfrac{1}{6}}$

The chart shows the fraction of computer users who are using each platform. Use the information to answer the following questions. Write your answers as fractions and explain what the fractions mean.

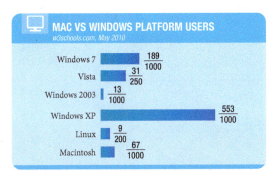

MAC VS WINDOWS PLATFORM USERS
w3schools.com, May 2010

Platform	Fraction
Windows 7	$\dfrac{189}{1000}$
Vista	$\dfrac{31}{250}$
Windows 2003	$\dfrac{13}{1000}$
Windows XP	$\dfrac{553}{1000}$
Linux	$\dfrac{9}{200}$
Macintosh	$\dfrac{67}{1000}$

31. What fraction of users report using Windows 7 and Vista?

32. What fraction of users report using Macintosh and Linux platforms combined?

Decimals

El Giza, Egypt
Data SIO, NOAA, U.S. Navy, NGA, GEBCO
Image © 2010 GeoEye, Image © 2010 DigitalGlobe
© 2010 Cnes/Spot Image

Google

The ancient Egyptian culture required that the body of a ruler be housed in a large tomb; the more impressive the structure, the smoother that ruler's journey into the afterlife would be. Thus explains the spectacular architectural wonders that are the Egyptian Pyramids. The largest of Egypt's famous pyramids, the Great Pyramid of Giza is believed to be the tomb of King Khufu and is recognized as one of the seven wonders of the ancient world. Khufu began construction on his pyramid soon after taking his throne, completing it approximately 20 years later around 2560 BC.

The current height of the Great Pyramid measures 138.8 meters. Historians believe that erosion has shortened the pyramid throughout its 4,000-year history, and that its original height was 146.5 meters. Suppose you wanted to calculate how much volume was lost due to this erosion. You would need to know that each side s of the pyramid measures 230.4 meters in length and that the area of the base B of a pyramid can be found using the formula

$$B = s \cdot s$$

You would also need to know that volume V of a pyramid can be found with the formula

$$V = \frac{1}{3} B \cdot h$$

where h is the height of the pyramid at its apex. If you wanted to find the volume of the Great Pyramid when it was built and its volume now, you will need to know how to multiply and subtract numbers containing decimals. Decimal numbers are what we will explore in this chapter.

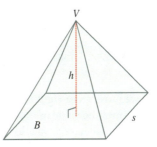

OBJECTIVES

A Write decimals in expanded form and with words or as mixed numbers.

B Round decimals to given place values.

KEY WORDS

decimals

place value

Video Examples
Section 4.1

Image © sxc.hu, shho, 2009

A person's handwriting is extremely difficult to mimic. To make copying signatures of the world's elite businessmen and women even more impossible, a company based in Germany has produced a pen with plant DNA embedded in the ink. Each pen sold has a distinctively different type of DNA, making words written with the pen distinguishable from words written with other pens. A forensic test is needed to view the plant DNA, but uniqueness of each pen gives peace of mind that a forgery is less likely to happen.

Each pen containing plant DNA sells for $15,850. If this pen sold in San Luis Obispo, California, the sales tax would be an additional $1,386.88. In this chapter, we will focus our attention on decimals. Anyone who has used money in the United States has worked with decimals already. For example, the sales tax on the pen contains a decimal:

$1386.88

└──— Decimal point

What is interesting and useful about decimals is their relationship to fractions and to powers of ten. The work we have done up to now—especially our work with fractions—can be used to develop the properties of decimal numbers.

A Decimal Notation and Place Value

Previously you have used the idea of place value for the digits in a whole number. The name and the place value of each of the first seven columns in our number system is shown here:

Millions Column	Hundred Thousands Column	Ten Thousands Column	Thousands Column	Hundreds Column	Tens Column	Ones Column
1,000,000	100,000	10,000	1,000	100	10	1

As we move from right to left, we multiply by 10 each time. The value of each column is 10 times the value of the column on its right, with the rightmost column being 1. Up until now we have always looked at place value as increasing by a factor of 10 each time we move one column to the left.

Ten Thousands		Thousands		Hundreds		Tens		Ones
10,000	←	1,000	←	100	←	10	←	1
	Multiply by 10		Multiply by 10		Multiply by 10		Multiply by 10	

Chapter 4 Decimals

To understand the idea behind decimal numbers, we notice that moving in the opposite direction, from left to right, we *divide* by 10 each time.

Ten Thousands		Thousands		Hundreds		Tens		Ones
10,000	→	1,000	→	100	→	10	→	1
	Divide by 10		Divide by 10		Divide by 10		Divide by 10	

If we keep going to the right, the next column will have to be

$$1 \div 10 = \frac{1}{10} \qquad \text{Tenths}$$

The next one after that will be

$$\frac{1}{10} \div 10 = \frac{1}{10} \cdot \frac{1}{10} = \frac{1}{100} \qquad \text{Hundredths}$$

After that, we have

$$\frac{1}{100} \div 10 = \frac{1}{100} \cdot \frac{1}{10} = \frac{1}{1,000} \qquad \text{Thousandths}$$

We could continue this pattern as long as we wanted. We simply divide by 10 to move one column to the right. (And remember, dividing by 10 gives the same result as multiplying by $\frac{1}{10}$.)

To show where the ones column is, we use a *decimal point* between the ones column and the tenths column.

Thousands	Hundreds	Tens	Ones		Tenths	Hundredths	Thousandths	Ten Thousandths	Hundred Thousandths
1,000	100	10	1	↑ Decimal point	$\frac{1}{10}$	$\frac{1}{100}$	$\frac{1}{1,000}$	$\frac{1}{10,000}$	$\frac{1}{100,000}$

Note Because the digits to the right of the decimal point have fractional place values, numbers with digits to the right of the decimal point are called *decimal fractions*. In this book, we will also call them *decimal numbers*, or simply *decimals* for short.

The ones column can be thought of as the middle column, with columns larger than 1 to the left and columns smaller than 1 to the right. The first column to the right of the ones column is the tenths column, the next column to the right is the hundredths column, the next is the thousandths column, and so on. The decimal point is always written between the ones column and the tenths column.

We can use the place value of decimal fractions to write them in expanded form.

EXAMPLE 1 Write 423.576 in expanded form.

Solution $423.576 = 400 + 20 + 3 + \dfrac{5}{10} + \dfrac{7}{100} + \dfrac{6}{1,000}$

EXAMPLE 2 Write each number in words.

a. 0.4   **b.** 0.04   **c.** 0.004

Solution

a. 0.4 is "four tenths."

b. 0.04 is "four hundredths."

c. 0.004 is "four thousandths."

Practice Problems

1. Write 18.439 in expanded form.

2. Write each number in words.
   a. 0.1
   b. 0.01
   c. 0.001

Answers

1. $10 + 8 + \frac{4}{10} + \frac{3}{100} + \frac{9}{1000}$

2. **a.** One tenth
   b. One hundredth
   c. One thousandth

3. Write each number in words.
   a. 4.2
   b. 4.12
   c. 4.012

4. Write 10.805 in words.

5. Write 14.0836 in words.

6. Write each number as a fraction or mixed number. Do not reduce to lowest terms.
   a. 0.08

   b. 4.206

   c. 12.5814

When a decimal number contains digits to the left of the decimal point, we use the word "and" to indicate where the decimal point is when writing the number in words.

EXAMPLE 3   Write each number in words.
a. 5.4   **b.** 5.04   **c.** 5.004
Solution
a. 5.4 is "five and four tenths."
b. 5.04 is "five and four hundredths."
c. 5.004 is "five and four thousandths."

EXAMPLE 4   Write 3.64 in words.
Solution   The number 3.64 is read "three and sixty-four hundredths." The place values of the digits are as follows:

We read the decimal part as "sixty-four hundredths" because

$$6 \text{ tenths} + 4 \text{ hundredths} = \frac{6}{10} + \frac{4}{100} = \frac{60}{100} + \frac{4}{100} = \frac{64}{100}$$

EXAMPLE 5   Write 25.4936 in words.
Solution   Using the idea given in Example 4, we write 25.4936 in words as "twenty-five and four thousand, nine hundred thirty-six ten thousandths."

In order to understand addition and subtraction of decimals in the next section, we need to be able to convert decimal numbers to fractions or mixed numbers.

EXAMPLE 6   Write each number as a fraction or a mixed number. Do not reduce to lowest terms.
a. 0.004   **b.** 3.64   **c.** 25.4936
Solution
a. Because 0.004 is 4 thousandths, we write

$$0.004 = \frac{4}{1{,}000}$$

Three digits after the decimal point   Three zeros

b. Looking over the work in Example 4, we can write

$$3.64 = 3\frac{64}{100}$$

Two digits after the decimal point   Two zeros

c. From the way in which we wrote 25.4936 in words in Example 5, we have

$$25.4936 = 25\frac{4936}{10{,}000}$$

Four digits after the decimal point   Four zeros

Answers
3. **a.** Four and two tenths
   b. Four and twelve hundredths
   c. Four and twelve thousandths
4. Ten and eight hundred five thousandths.
5. Fourteen and eight hundred thirty-six ten thousandths.
6. **a.** $\frac{8}{100}$ **b.** $4\frac{206}{1000}$ **c.** $12\frac{5814}{10{,}000}$

B Rounding Decimal Numbers

The rule for rounding decimal numbers is similar to the rule for rounding whole numbers. If the digit in the column to the right of the one we are rounding to is 5 or more, we add 1 to the digit in the column we are rounding to; otherwise, we leave it alone. We then replace all digits to the right of the column we are rounding to with zeros if they are to the left of the decimal point; otherwise, we simply delete them. Table 1 illustrates the procedure.

Table 1

Rounded to the Nearest

Number	Whole Number	Tenth	Hundredth
24.785	25	24.8	24.79
2.3914	2	2.4	2.39
0.98243	1	1.0	0.98
14.0942	14	14.1	14.09
0.545	1	0.5	0.55

EXAMPLE 7   Round 9,235.492 to the nearest hundred.

Solution   The number next to the hundreds column is 3, which is less than 5. We change all digits to the right to 0, and we can drop all digits to the right of the decimal point, so we write

   9,200

7. Round 8,456.085 to the nearest hundred.

EXAMPLE 8   Round 0.00346 to the nearest ten thousandth.

Solution   Because the number to the right of the ten thousandths column is more than 5, we add 1 to the 4 and get

   0.0035

8. Round 8,456.085 to the nearest hundredth.

GETTING READY FOR CLASS

After reading through the preceding section, respond in your own words and in complete sentences.

A. Write 754.326 in expanded form.

B. Write $400 + 70 + 5 + \frac{1}{10} + \frac{3}{100} + \frac{7}{1,000}$ in decimal form.

C. Write seventy-two and three tenths in decimal form.

D. How many places to the right of the decimal point is the hundredths column?

Answers

7. 8,500
8. 8,456.09

EXERCISE SET 4.1

Vocabulary Review

Choose the correct words to fill in the blanks below.

hundredths	left	decimal point	fractional
tenths	right	thousandths	place value

1. Digits to the right of a decimal point have _____ place values.

2. A _____ is used to separate the ones column and the tenths column.

3. In the decimal number 0.036, the 6 is in the _____ column.

4. In the decimal number 4.169, the 1 is in the _____ column.

5. In the decimal number 10.0977, the 9 is in the _____ column.

6. We use _____ of decimal fractions to write them in expanded form.

7. When a decimal number contains digits to the _____ of the decimal point, we use the word *and* to indicate where the decimal point is when writing the number in words.

8. When rounding decimal numbers, if the digit in the column to the _____ of the one we are rounding to is 5 or more, we add 1 to the digit in the column to which we are rounding.

Problems

A Write out the name of each number in words.

1. 0.3   **2.** 0.03   **3.** 0.015   **4.** 0.0015

5. 3.4   **6.** 2.04   **7.** 52.7   **8.** 46.8

Write each number as a fraction or a mixed number. Do not reduce your answers.

9. 405.36   **10.** 362.78   **11.** 9.009   **12.** 60.06

13. 1.234   **14.** 12.045   **15.** 0.00305   **16.** 2.00106

Give the place value of the 5 in each of the following numbers.

17. 458.327   **18.** 327.458   **19.** 29.52   **20.** 25.92   **21.** 0.00375

22. 0.00532   **23.** 275.01   **24.** 0.356   **25.** 539.76   **26.** 0.123456

Write each of the following as a decimal number.

27. Fifty-five hundredths

28. Two hundred thirty-five ten thousandths

29. Six and nine tenths

30. Forty-five thousand and six hundred twenty-one thousandths

31. Eleven and eleven hundredths

32. Twenty-six thousand, two hundred forty-five and sixteen hundredths

33. One hundred and two hundredths

34. Seventy-five and seventy-five hundred thousandths

35. Three thousand and three thousandths

36. One thousand, one hundred eleven and one hundred eleven thousandths

B Complete the following table.

	Rounded to the Nearest			
Number	Whole Number	Tenth	Hundredth	Thousandth
37. 47.5479				
38. 100.9256				
39. 0.8175				
40. 29.9876				
41. 0.1562				
42. 128.9115				
43. 2,789.3241				
44. 0.8743				
45. 99.9999				
46. 71.7634				

Applying the Concepts

47. Penny Weight If you have a penny dated anytime from 1959 through 1982, its original weight was 3.11 grams. If the penny has a date of 1983 or later, the original weight was 2.5 grams. Write the two weights in words.

48. 100 Meters The chart shows some close finishes at the 2010 Winter Olympics in Vancouver. Use the information to answer the following questions.

a. What is the place value of the 6 in the finish of the Men's Slalom?

b. Write the the difference in times for the Defago/Svindal finish in words.

49. Speed of Light The speed of light is 186,282.3976 miles per second. Round this number to the nearest hundredth.

50. Halley's Comet Halley's comet was seen from the earth during 1986. It will be another 76.1 years before it returns. Write 76.1 in words.

51. Nutrition A 50-gram egg contains 0.15 milligram of riboflavin. Write 0.15 in words.

52. Nutrition One medium banana contains 0.64 milligram of Vitamin B6. Write 0.64 in words.

53. Gasoline Prices The bar chart below was created from a survey by the U.S. Department of Energy's Energy Information Administration during four weeks in 2011. It gives the average price of regular gasoline for the United States on each Monday throughout the four week period. Use the information in the chart to fill in the table.

Price of 1 Gallon of Regular Gasoline	
Date	**Price (Dollars)**
3/21/11	
3/28/11	
4/4/11	
4/11/11	

54. Speed and Time The bar chart below was created from data given by *Car and Driver* magazine. It gives the minimum time in seconds for a Toyota Echo to reach various speeds from a complete stop. Use the information in the chart to fill in the table.

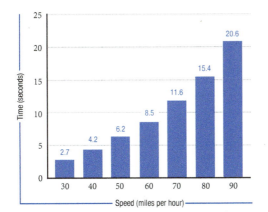

Speed (miles per hour)	Time (seconds)
30	
40	
50	
60	
70	
80	
90	

For each pair of numbers, place the correct symbol, < or >, between the numbers.

55. a. 0.02   0.2

  b. 0.3   0.032

56. a. 0.45   0.5

  b. 0.5   0.56

57. Write the following numbers in order from smallest to largest.

  0.02   0.05   0.025   0.052   0.005   0.002

58. Write the following numbers in order from smallest to largest.

  0.2   0.02   0.4   0.04   0.42   0.24

59. Which of the following numbers will round to 7.5?

  7.451   7.449   7.54   7.56

60. Which of the following numbers will round to 3.2?

  3.14999   3.24999   3.279   3.16111

Change each decimal to a fraction, and then reduce to lowest terms.

61. 0.25   **62.** 0.75   **63.** 0.125   **64.** 0.375

65. 0.625   **66.** 0.0625   **67.** 0.875   **68.** 0.1875

Estimating   For each pair of numbers, choose the number that is closest to 10.

69. 9.9 and 9.99   **70.** 8.5 and 8.05   **71.** 10.5 and 10.05   **72.** 10.9 and 10.99

Estimating   For each pair of numbers, choose the number that is closest to 0.

73. 0.5 and 0.05   **74.** 0.10 and 0.05   **75.** 0.01 and 0.02   **76.** 0.1 and 0.01

Getting Ready for the Next Section

In the next section we will do addition and subtraction with decimals. To understand the process of addition and subtraction, we need to understand the process of addition and subtraction with mixed numbers.

Find each of the following sums and differences. (Add or subtract.)

77. $4\frac{3}{10} + 2\frac{1}{100}$

78. $5\frac{35}{100} + 2\frac{3}{10}$

79. $8\frac{5}{10} - 2\frac{4}{100}$

80. $6\frac{3}{100} - 2\frac{125}{1,000}$

81. $5\frac{1}{10} + 6\frac{2}{100} + 7\frac{3}{1,000}$

82. $4\frac{27}{100} + 6\frac{3}{10} + 7\frac{123}{1,000}$

Find the Mistake

Each sentence below contains a mistake. Circle the mistake and write the correct word(s) or numbers(s) on the line provided.

1. To move a place value from the tens column to the hundreds column, you must divide by ten. _____

2. The decimal 0.09 can be written as the fraction $\frac{9}{1,000}$. _____

3. The decimal 142.9643 written as a mixed number is $142\frac{9,643}{1,000}$. _____

4. Rounding the decimal 0.06479 to the nearest tenth gives us 0.065. _____

Navigation Skills: Prepare, Study, Achieve

At the end of the previous chapter, we suggested ways to engage your senses when memorizing an abstract concept. These techniques will help anchor it in your memory. For instance, it is easier to remember explaining to your friend a difficult math formula, than it is to simply recall it from a single read of the chapter. You can engage your senses when you are studying alone, but it is also helpful to get together with fellow classmates to study as a group. Make sure to keep your group size small to make sure each person gets the attention he or she needs. During these sessions, talk openly about the concepts you are studying and the difficulties you may be having. Here are a few examples to make your group study sessions as effective as possible:

- Choose a regular time and place to meet.
- Attend every meeting and pull your weight. Showing up and helping others will, in turn, help you in this course.
- Make flashcards of definitions, properties, and formulas and quiz each other.
- Recite definitions or formulas out loud.
- Take turns explaining concepts to the group.
- Discuss how concepts learned in previous chapters are incorporated into the concepts you are currently studying.
- Role play where one person is the instructor verbally guiding another person as the student through a problem.
- Predict test questions and quiz each other.

©iStockphoto.com/masta4650

The chart shows the top finishing times for the women's 400-meter race during the 2009 World Track and Field Championship in Berlin. In order to analyze the different finishing times, it is important that you are able to add and subtract decimals, and that is what we will cover in this section.

Runner	Time (seconds)
Sanya Richards (USA)	49.00
Shericka Williams (Jamaica)	49.32
Antonia Krivoshapka (Russia)	49.71
Novlene Williams-Mills (Jamaica)	49.77

A Addition of Decimals

Suppose you are earning $8.50 an hour and you receive a raise of $1.25 an hour. Your new hourly rate of pay is

$$\begin{array}{r} \$8.50 \\ + \ \$1.25 \\ \hline \$9.75 \end{array}$$

To add the two rates of pay, we align the decimal points, and then add in columns.

To see why this is true in general, we can use mixed-number notation.

$$8.50 = 8\frac{50}{100}$$

$$+ \ 1.25 = 1\frac{25}{100}$$

$$9\frac{75}{100} = 9.75$$

We can visualize the mathematics above by thinking in terms of money.

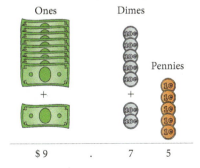

Video Examples
Section 4.2

1. Add by first changing to fractions: $16.47 + 5.9 + 45.842$.

EXAMPLE 1   Add by first changing to fractions: $25.43 + 2.897 + 379.6$.

Solution   We first change each decimal to a mixed number. We then write each fraction using the least common denominator and add as usual.

$$25.43 = 25\frac{43}{100} \;=\; 25\frac{430}{1,000}$$

$$2.897 = 2\frac{897}{1,000} \;=\; 2\frac{897}{1,000}$$

$$+\; 379.6 = 379\frac{6}{10} \;=\; 379\frac{600}{1,000}$$

$$\overline{\phantom{+\; 379.6 = 379\frac{6}{10} \;=\;}}\;406\frac{1,927}{1,000} = 407\frac{927}{1,000} = 407.927$$

Again, the result is the same if we just line up the decimal points and add as if we were adding whole numbers.

$$\begin{array}{r} 25.430 \\ 2.897 \\ +\;379.600 \\ \hline 407.927 \\ \uparrow \end{array}$$

Notice that we can fill in zeros on the right to help keep the numbers in the correct columns. Doing this does not change the value of any of the numbers.

Note: The decimal point in the answer is directly below the decimal points in the problem.

B Subtraction of Decimals

The same thing would happen if we were to subtract two decimal numbers. We can use these facts to write a rule for addition and subtraction of decimal numbers.

> **RULE** Addition (or Subtraction) of Decimal Numbers
>
> To add (or subtract) decimal numbers, we line up the decimal points and add (or subtract) as usual. The decimal point in the result is written directly below the decimal points in the problem.

We will use this rule for the rest of the examples in this section.

2. Subtract: $42.809 - 13.658$.

EXAMPLE 2   Subtract: $39.812 - 14.236$.

Solution   We write the numbers vertically, with the decimal points lined up, and subtract as usual.

$$\begin{array}{r} 39.812 \\ -\;14.236 \\ \hline 25.576 \end{array}$$

3. Add: $9 + 1.05 + 8.7 + 3.86$.

EXAMPLE 3   Add: $8 + 0.002 + 3.1 + 0.04$.

Solution   To make sure we keep the digits in the correct columns, we can write zeros to the right of the rightmost digits.

$$8 = 8.000$$
$$3.1 = 3.100$$
$$0.04 = 0.040$$

Writing the extra zeros here is really equivalent to finding a common denominator for the fractional parts of the original four numbers—now we have a thousandths column in all the numbers.

This doesn't change the value of any of the numbers, and it makes our task easier. Now we have

$$\begin{array}{r} 8.000 \\ 0.002 \\ 3.100 \\ + \ 0.040 \\ \hline 11.142 \end{array}$$

EXAMPLE 4   Subtract: $5.9 - 3.0814$.

Solution   In this case, it is very helpful to write 5.9 as 5.9000, since we will have to borrow in order to subtract.

$$\begin{array}{r} 5.9000 \\ - \ 3.0814 \\ \hline 2.8186 \end{array}$$

EXAMPLE 5   Subtract 3.09 from the sum of 9 and 5.472.

Solution   Writing the problem in symbols, we have

$$(9 + 5.472) - 3.09 = 14.472 - 3.09$$
$$= 11.382$$

EXAMPLE 6   Add: $-4.75 + (-2.25)$.

Solution   Because both signs are negative, we add absolute values. The answer will be negative.

$$-4.75 + (-2.25) = -7.00$$

EXAMPLE 7   Add: $3.42 + (-6.89)$.

Solution   The signs are different, so we subtract the smaller absolute value from the larger absolute value. The answer will be negative, because 6.89 is larger than 3.42 and the sign in front of 6.89 is $-$.

$$3.42 + (-6.89) = -3.47$$

> ### GETTING READY FOR CLASS
>
> *After reading through the preceding section, respond in your own words and in complete sentences.*
>
> **A.** When adding numbers with decimals, why is it important to line up the decimal points?
>
> **B.** Write 379.6 in mixed-number notation.
>
> **C.** Why do we line up the decimals when we add or subtract vertically?
>
> **D.** How many quarters does the decimal 0.75 represent?

4. Subtract: $3.7 - 1.9034$.

5. Subtract 8.92 from the sum of 12 and 5.01.

6. Add: $-3.5 + (-1.25)$.

7. Add: $6.58 + (-10.47)$.

Answers
4. 1.7966
5. 8.09
6. -4.75
7. -3.89

EXERCISE SET 4.2

Problems

A Find each of the following sums. (Add.)

1. $2.91 + 3.28$

2. $8.97 + 2.04$

3. $0.04 + 0.31 + 0.78$

4. $0.06 + 0.92 + 0.65$

5. $-3.89 + 2.4$

6. $7.65 + 3.8$

7. $4.532 + 1.81 + 2.7$

8. $9.679 + 3.49 + 6.5$

9. $-0.081 + (-5) + (-2.94)$

10. $0.396 + 7 + 3.96$

11. $5.0003 + 6.78 + 0.004$

12. $27.0179 + 7.89 + 0.009$

13.
$$\begin{array}{r} 7.123 \\ 8.12 \\ + 9.1 \\ \hline \end{array}$$

14.
$$\begin{array}{r} 5.432 \\ 4.32 \\ + 3.2 \\ \hline \end{array}$$

15.
$$\begin{array}{r} 9.001 \\ 8.01 \\ + 7.1 \\ \hline \end{array}$$

16.
$$\begin{array}{r} 6.003 \\ 5.02 \\ + 4.1 \\ \hline \end{array}$$

17.
$$\begin{array}{r} 89.7854 \\ 3.4 \\ 65.35 \\ + 100.006 \\ \hline \end{array}$$

18.
$$\begin{array}{r} 57.4698 \\ 9.89 \\ 32.032 \\ + 572.0079 \\ \hline \end{array}$$

19.
$$\begin{array}{r} 543.21 \\ + 123.45 \\ \hline \end{array}$$

20.
$$\begin{array}{r} 987.654 \\ + 456.789 \\ \hline \end{array}$$

B Find each of the following differences. (Subtract.)

21. $99.34 - 88.23$

22. $47.69 - 36.58$

23. $5.97 - 2.4$

24. $9.87 - 1.04$

25. $6.3 - 2.08$

26. $7.5 - 3.04$

27. $-28.96 - (-149.37)$

28. $796.45 - 32.68$

29. $45 - 0.067$

30. $48 - 0.075$

31. $8 - 0.327$

32. $-0.962 - (-12)$

33. $765.432 - 234.567$

34. $654.321 - 123.456$

35. $100.42 - 56.87$

36. $12 - 1.93$

37. $-4.082 - (-10)$

38. $20 - 5.86$

Subtract.

39. $\begin{array}{r} 34.07 \\ -\ 6.18 \\ \hline \end{array}$

40. $\begin{array}{r} 25.008 \\ -\ 3.119 \\ \hline \end{array}$

41. $\begin{array}{r} 40.04 \\ -\ 4.4 \\ \hline \end{array}$

42. $\begin{array}{r} 50.05 \\ -\ 5.5 \\ \hline \end{array}$

43. $\begin{array}{r} 768.436 \\ -\ 356.998 \\ \hline \end{array}$

44. $\begin{array}{r} 495.237 \\ -\ 247.668 \\ \hline \end{array}$

Add and subtract as indicated.

45. $(7.8 - 4.3) + 2.5$

46. $(8.3 - 1.2) + 3.4$

47. $7.8 - (4.3 + 2.5)$

48. $8.3 - (1.2 + 3.4)$

49. $(9.7 - 5.2) - 1.4$

50. $(7.8 - 3.2) - 1.5$

51. $9.7 - (5.2 - 1.4)$

52. $7.8 - (3.2 - 1.5)$

53. $5.9 - (4.03 - 2.3)$

54. $10 - (3 - 1.06)$

55. $12.2 - (9.1 + 0.3)$

56. $15.6 - (4.9 + 3.87)$

57. $-9.01 - 2.4$

58. $-8.23 - 5.4$

59. $-0.89 - 1.01$

60. $-0.42 - 2.04$

61. $-3.4 - 5.6 - 8.5$

62. $-2.1 - 3.1 - 4.1$

63. Subtract 5 from the sum of 8.2 and 0.072.

64. Subtract 8 from the sum of 9.37 and 2.5.

65. What number is added to 0.035 to obtain 4.036?

66. What number is added to 0.043 to obtain 6.054?

Applying the Concepts

67. Shopping A family buying school clothes for their two children spends $25.37 at one store, $39.41 at another, and $52.04 at a third store. What is the total amount spent at the three stores?

68. Expenses A 4-H Club member is raising a lamb to take to the county fair. If she spent $75 for the lamb, $25.60 for feed, and $35.89 for shearing tools, what was the total cost of the project?

69. Take-Home Pay A college professor making $2,105.96 per month has deducted from her check $311.93 for federal income tax, $158.21 for retirement, and $64.72 for state income tax. How much does the professor take home after the deductions have been taken from her monthly income?

70. Take-Home Pay A cook making $1,504.75 a month has deductions of $157.32 for federal income tax, $58.52 for Social Security, and $45.12 for state income tax. How much does the cook take home after the deductions have been taken from his check?

71. Rectangle The logo on a business letter is rectangular. The rectangle has a width of 0.84 inches and a length of 1.41 inches. Find the perimeter.

72. Rectangle A small sticky note is a rectangle. It has a width of 21.4 millimeters and a length of 35.8 millimeters. Find the perimeter.

73. Change A person buys $4.57 worth of candy. If he pays for the candy with a $10 bill, how much change should he receive?

74. Checking Account A checking account contains $342.38. If checks are written for $25.04, $36.71, and $210, how much money is left in the account?

Check No.	Date	Description of Transaction	Payment/Debit (-)	Deposit/Credit (+)	Balance
1630	7/3	Deposit		$342 38	$342 38
	7/5	Albertsons	$25 04		
	7/5	Rite Aid	$36 71		
	7/7	Gas Company	$210 00		?

Downhill Skiers The chart shows the times for several downhill skiers in the 2010 Vancouver Olympics. Use the information to answer the following questions.

75. How much faster was Elisabeth Goergl than Andrea Fischbacher?

76. How much faster was Lindsey Vonn than Julia Mancuso?

77. Mac Users The chart shows the increase in the number of Mac users each year. Use it to answer the following questions.

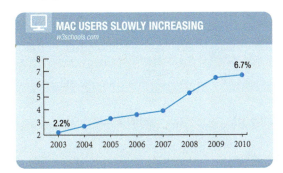

a. Was the increase in users during 2007 more or less than 5% of total users?

b. Was the increase in users during 2009 more or less than 5%?

c. Estimate the percent of total users increase from 2007 to 2009.

78. Movie Tickets The chart shows the increase in movie ticket prices. Use the information to answer the following questions.

a. Were movie ticket prices above $6 in 2001?

b. Were ticket prices above $6 in 2007?

c. Were ticket prices in 2008 below $7?

79. Geometry A rectangle has a perimeter of 9.5 inches. If the length is 2.75 inches, find the width.

80. Geometry A rectangle has a perimeter of 11 inches. If the width is 2.5 inches, find the length.

81. Change Suppose you eat dinner in a restaurant and the bill comes to $16.76. If you give the cashier a $20 bill and a penny, how much change should you receive? List the bills and coins you should receive for change.

82. Change Suppose you buy some tools at the hardware store and the bill comes to $37.87. If you give the cashier two $20 bills and 2 pennies, how much change should you receive? List the bills and coins you should receive for change.

Sequences Find the next number in each sequence.

83. 2.5, 2.75, 3, . . .

84. 3.125, 3.375, 3.625, . . .

Getting Ready for the Next Section

To understand how to multiply decimals, we need to understand multiplication with whole numbers, fractions, and mixed numbers. The following problems review these concepts.

85. $\dfrac{1}{10} \cdot \dfrac{3}{10}$

86. $\dfrac{5}{10} \cdot \dfrac{6}{10}$

87. $\dfrac{3}{100} \cdot \dfrac{17}{100}$

88. $\dfrac{7}{100} \cdot \dfrac{31}{100}$

89. $5\left(\dfrac{3}{10}\right)$

90. $7 \cdot \dfrac{7}{10}$

91. $56 \cdot 25$

92. $39(48)$

93. $\dfrac{5}{10} \cdot \dfrac{3}{10}$

94. $\dfrac{5}{100} \cdot \dfrac{3}{1,000}$

95. $2\dfrac{1}{10} \cdot \dfrac{7}{100}$

96. $3\dfrac{5}{10} \cdot \dfrac{4}{100}$

97. $305(436)$

98. $403(522)$

99. $5(420 + 3)$

100. $3(550 + 2)$

Improving Your Quantitative Literacy

101. Facebook Users The chart shows the number of active Facebook users over several years. Use the information to answer the following questions.

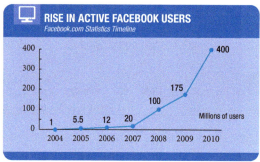

a. How many more users were there in 2007 than in 2006?

b. Between which years was the increase the most?

c. If the trend from 2009 to 2010 continues, how many users will there be in 2011?

Find the Mistake

Each sentence below contains a mistake. Circle the mistake and write the correct word(s) or numbers(s) on the line provided.

1. To add 32.69 and 4.837, align the rightmost digit of each number and add in columns. _____

2. To add $0.004 + 5.06 + 32$ by first changing each decimal to a fraction would give us the problem $\dfrac{4}{1,000} + \dfrac{5}{600} + 32$.

_____

3. When subtracting $8.7 - 2.0163$, we make sure to keep the digits in the correct columns by writing 8.7 as 8.0007. _____

4. Subtracting 4.367 from the sum of 12.1 and 0.036 gives us 11.3333. _____

4.3 Multiplication with Decimals

OBJECTIVES

A Multiply with decimals.

B Use order of operations with decimals.

The mystery of the roving rocks in Racetrack Playa of Death Valley, California has been puzzling scientists for decades. Water from winter rains wash down the slopes of the surrounding mountains and form a small lake in the Playa. During the summer, the lake evaporates and the muddy bed dries up, cracking into a mosaic of hexagonal clay tiles. Scientists have been trying to discover what causes large stones to move across the lakebed and leave behind tracks. One hypothesis suggests that as night temperatures fall below freezing, ice forms around each tile. Frequent winds up to 90 miles per hour can move this ice, allowing for a conveyor belt of sorts to move the rocks. To date, no one has ever witnessed the rocks moving, but scientists believe they may move at half the jogging speed of a human. Let's suppose a person is jogging at 8.5 miles per hour. If a rock in Racetrack Playa was moving at half the jogger's rate, it would be moving at 4.25 miles per hour. Therefore, it must be true that

$$\frac{1}{2} \text{ of } 8.5 \text{ is } 4.25$$

But because $\frac{1}{2}$ can be written as 0.5 and *of* translates to multiply, we can write this problem again as

$$0.5 \times 8.5 = 4.25$$

If we were to ignore the decimal points in this problem and simply multiply 5 and 85, the result would be 425.

**Video Examples
Section 4.3**

A Multiplication with Decimals

Multiplication with decimal numbers is similar to multiplication with whole numbers. The difference lies in deciding where to place the decimal point in the answer. To find out how this is done, we can use fraction notation.

EXAMPLE 1   Change each decimal to a fraction and multiply.

$$0.5 \times 0.3$$

Solution   Changing each decimal to a fraction and multiplying, we have

$0.5 \times 0.3 = \dfrac{5}{10} \times \dfrac{3}{10}$ *Change to fractions.*

$ = \dfrac{15}{100}$ *Multiply numerators and multiply denominators.*

$ = 0.15$ *Write the answer in decimal form.*

The result is 0.15, which has two digits to the right of the decimal point.

Practice Problems

1. Change each decimal to a fraction and multiply.

$$0.4 \times 0.08$$

Note To indicate multiplication we are using a × sign here instead of a dot so we won't confuse the decimal points with the multiplication symbol.

Answer

1. 0.032

What we want to do now is find a shortcut that will allow us to multiply decimals without first having to change each decimal number to a fraction. Let's look at another example.

EXAMPLE 2   Change each decimal to a fraction and multiply: 0.05×0.003.

Solution

$$0.05 \times 0.003 = \frac{5}{100} \times \frac{3}{1,000} \qquad \text{Change to fractions.}$$

$$= \frac{15}{100,000} \qquad \text{Multiply numerators and multiply denominators.}$$

$$= 0.00015 \qquad \text{Write the answer in decimal form.}$$

The result is 0.00015, which has a total of five digits to the right of the decimal point.

Looking over these first two examples, we can see that the digits in the result are just what we would get if we simply forgot about the decimal points and multiplied; that is, $3 \times 5 = 15$. The decimal point in the result is placed so that the total number of digits to its right is the same as the total number of digits to the right of both decimal points in the original two numbers. The reason this is true becomes clear when we look at the denominators after we have changed from decimals to fractions.

EXAMPLE 3   Multiply: 2.1×0.07.

Solution

$$2.1 \times 0.07 = 2\frac{1}{10} \times \frac{7}{100} \qquad \text{Change to fractions.}$$

$$= \frac{21}{10} \times \frac{7}{100}$$

$$= \frac{147}{1,000} \qquad \text{Multiply numerators and multiply denominators.}$$

$$= 0.147 \qquad \text{Write the answer as a decimal.}$$

Again, the digits in the answer come from multiplying $21 \times 7 = 147$. The decimal point is placed so that there are three digits to its right, because that is the total number of digits to the right of the decimal points in 2.1 and 0.07.

We summarize this discussion with the following rule:

RULE   Multiplication with Decimal Numbers

To multiply two decimal numbers, follow these steps:

1. Multiply as you would if the decimal points were not there.

2. Place the decimal point in the answer so that the number of digits to its right is equal to the total number of digits to the right of the decimal points in the original two numbers in the problem.

EXAMPLE 4   How many digits will be to the right of the decimal point in the following product?

$$2.987 \times 24.82$$

Solution   There are three digits to the right of the decimal point in 2.987 and two digits to the right in 24.82. Therefore, there will be $3 + 2 = 5$ digits to the right of the decimal point in their product.

---

2. Change each decimal to a fraction and multiply.

0.08×0.009.

3. Multiply: 3.2×0.09.

4. How many digits will be to the right of the decimal point in the following product?

4.632×0.0008

Answers

2. 0.00072
3. 0.288
4. Seven

EXAMPLE 5 Multiply: 3.05×4.36.

Solution We can set this up as if it were a multiplication problem with whole numbers. We multiply and then place the decimal point in the correct position in the answer.

$$
\begin{array}{r}
3.05 \\
\times\ 4.36 \\
\hline
1830 \\
915\ \ \\
+\ 12\,20\ \ \ \\
\hline
13.2980 \\
\end{array}
$$

⟵ 2 digits to the right of decimal point
⟵ 2 digits to the right of decimal point

The decimal point is placed so that there are $2 + 2 = 4$ digits to its right.

5. Multiply: 2.16×3.05.

As you can see, multiplying decimal numbers is just like multiplying whole numbers, except that we must place the decimal point in the result in the correct position.

Estimating

Look back to Example 5. We could have placed the decimal point in the answer by rounding the two numbers to the nearest whole number and then multiplying them. Because 3.05 rounds to 3 and 4.36 rounds to 4, and the product of 3 and 4 is 12, we estimate that the answer to 3.05×4.36 will be close to 12. Then, we place the decimal point in the product 132980 between the 3 and the 2 in order to make it into a number close to 12.

EXAMPLE 6 Estimate the answer to each of the following products.

a. 29.4×8.2 **b.** 68.5×172 **c.** $(6.32)^2$

Solution

a. Because 29.4 is approximately 30 and 8.2 is approximately 8, we estimate this product to be about $30 \times 8 = 240$. (If we were to multiply 29.4 and 8.2, we would find the product to be exactly 241.08.)

b. Rounding 68.5 to 70 and 172 to 170, we estimate this product to be $70 \times 170 = 11,900$. (The exact answer is 11,782.) Note here that we do not always round the numbers to the nearest whole number when making estimates. The idea is to round to numbers that will be easy to multiply.

c. Because 6.32 is approximately 6 and $6^2 = 36$, we estimate our answer to be close to 36. (The actual answer is 39.9424.)

6. Estimate the answer to each of the following products.
a. 39.6×9.1
b. 58.7×141.3
c. $(9.84)^2$

B Order of Operations with Decimals

We can use the rule for order of operations to simplify expressions involving decimal numbers and addition, subtraction, and multiplication.

EXAMPLE 7 Perform the indicated operations: $0.05(4.2 + 0.03)$.

Solution We begin by adding inside the parentheses.

$$0.05(4.2 + 0.03) = 0.05(4.23) \qquad \text{Add.}$$
$$= 0.2115 \qquad \text{Multiply.}$$

Notice that we could also have used the distributive property first, and the result would be unchanged.

$$0.05(4.2 + 0.03) = 0.05(4.2) + 0.05(0.03) \qquad \text{Distributive property}$$
$$= 0.210 + 0.0015 \qquad \text{Multiply.}$$
$$= 0.2115 \qquad \text{Add.}$$

7. Perform the indicated operations.
$$0.07\,(3.9 + 6.05)$$

Answers
5. 6.588
6. a. 360 **b.** 8,400 **c.** 100
7. 0.6965

8. Simplify: $9.5 + 11(2.3)^2$.

EXAMPLE 8   Simplify: $4.8 + 12(3.2)^2$.

Solution   According to the rule for order of operations, we must first evaluate the number with an exponent, then multiply, and finally add.

$$4.8 + 12(3.2)^2 = 4.8 + 12(10.24) \qquad (3.2)^2 = 10.24$$
$$= 4.8 + 122.88 \qquad \text{Multiply.}$$
$$= 127.68 \qquad \text{Add.}$$

9. Simplify each expression.
   a. $(-4)(1.2)$
   b. $(-0.2)(-0.5)$

EXAMPLE 9   Simplify each expression.

  a. $(-5)(3.4)$ **b.** $(-0.4)(-0.8)$

Solution

  a. $(-5)(3.4) = -17.0$ *The rule for multiplication also holds for decimals.*
  b. $(-0.4)(-0.8) = 0.32$

Applications

10. How much will Sally from Example 10 make if she works 48 hours in one week?

Note To estimate the answer to Example 10 before doing the actual calculations, we would do the following:

$6(40) + 9(6) = 240 + 54 = 294$

EXAMPLE 10   Sally earns \$6.32 for each of the first 36 hours she works in one week, and \$9.48 in overtime pay for each additional hour she works in the same week. How much money will she make if she works 42 hours in one week?

Solution   The difference between 42 and 36 is 6 hours of overtime pay. The total amount of money she will make is

    Pay for the first   *Pay for the*
      36 hours   *next 6 hours*

$$6.32(36) + 9.48(6) = 227.52 + 56.88$$
$$= 284.40$$

She will make \$284.40 for working 42 hours in one week.

GETTING READY FOR CLASS

After reading through the preceding section, respond in your own words and in complete sentences.

A. If you multiply 34.76 and 0.072, how many digits will be to the right of the decimal point in your answer?

B. To simplify the expression $0.053(9) + 67.42$, what would be the first step according to the rule for order of operations?

C. What is the purpose of estimating?

D. What are some applications of decimals that we use in our everyday lives?

Answers
8. 67.69
9. a. -4.8 **b.** 0.10
10. \$341.28

Vocabulary Review

Choose the correct words to fill in the blanks in the paragraph below.

      answer   decimal points   multiply   digits

To multiply two decimal numbers, first _____ as you would if the decimal points were not there. Then place the decimal point in the _____ so that the number of _____ to its right is equal to the total number of digits to the right of the _____ in the original two numbers in the problem.

Problems

A Find each of the following products. (Multiply.)

1.   0.7
      $\times\,0.4$

2.   0.8
      $\times\,0.3$

3.   0.07
      $\times\,0.4$

4.   0.8
      $\times\,0.03$

5.   0.03
      $\times\,0.09$

6.   0.07
      $\times\,0.002$

7. 2.6(0.3)

8. 8.9(0.2)

9.   0.9
      $\times\,0.88$

10.   0.8
      $\times\,0.99$

11.   3.12
      $\times\,0.005$

12.   4.69
      $\times\,0.006$

13.   4.003
      $\times\,6.07$

14.   7.0001
      $\times\quad3.04$

15. 5(0.006)

16. 7(0.005)

17.   75.14
      $\times\quad2.5$

18.   963.8
      $\times\,0.24$

19.   0.1
      $\times\,0.02$

20.   0.3
      $\times\,0.02$

21. 2.796(10)

22. 97.531(100)

23.   0.0043
      $\times\quad100$

24.   12.345
      $\times\,1{,}000$

25.   49.94
      $\times\,1{,}000$

26.   157.02
      $\times\,10{,}000$

27.   987.654
      $\times\,10{,}000$

28.   1.23
      $\times\,100{,}000$

B Perform the following operations according to the rule for order of operations.

29. $2.1(3.5 - 2.6)$

30. $5.4(9.9 - 6.6)$

31. $0.05(0.02 + 0.03)$

32. $-0.04(0.07 + 0.09)$

33. $2.02(0.03 + 2.5)$

34. $-4.04(0.05 + 6.6)$

35. $(2.1 + 0.03)(3.4 + 0.05)$

36. $(9.2 + 0.01)(3.5 + 0.03)$

37. $(2.1 - 0.1)(2.1 + 0.1)$

38. $(9.6 - 0.5)(9.6 + 0.5)$

39. $3.08 - 0.2(5 + 0.03)$

40. $4.09 + 0.5(6 + 0.02)$

41. $4.23 - 5(0.04 + 0.09)$

42. $7.89 - 2(0.31 + 0.76)$

43. $2.5 + 10(4.3)^2$

44. $3.6 + 15(2.1)^2$

45. $100(1 + 0.08)^2$

46. $500(1 + 0.12)^2$

47. $(1.5)^2 + (2.5)^2 + (3.5)^2$

48. $(1.1)^2 + (2.1)^2 + (3.1)^2$

Applying the Concepts

Solve each of the following word problems. Note that not all of the problems are solved by simply multiplying the numbers in the problems. Many of the problems involve addition and subtraction as well as multiplication.

49. Number Problem What is the product of 6 and the sum of 0.001 and 0.02?

50. Number Problem Find the product of 8 and the sum of 0.03 and 0.002.

51. Number Problem What does multiplying a decimal number by 100 do to the decimal point?

52. Number Problem What does multiplying a decimal number by 1,000 do to the decimal point?

53. **Home Mortgage** On a certain home mortgage, there is a monthly payment of $9.66 for every $1,000 that is borrowed. What is the monthly payment on this type of loan if $143,000 is borrowed?

54. **Caffeine Content** If 1 cup of regular coffee contains 105 milligrams of caffeine, how much caffeine is contained in 3.5 cups of coffee?

55. **Long-Distance Charges** If a phone company charges $0.45 for the first minute and $0.35 for each additional minute for a long-distance call, how much will a 20-minute long-distance call cost?

56. **Price of Gasoline** If gasoline costs $3.05 per gallon when you pay with a credit card, but $0.06 per gallon less if you pay with cash, how much do you save by filling up a 12-gallon tank and paying for it with cash?

57. **Car Rental** Suppose it costs $15 per day and $0.12 per mile to rent a car. What is the total bill if a car is rented for 2 days and is driven 120 miles?

58. **Car Rental** Suppose it costs $20 per day and $0.08 per mile to rent a car. What is the total bill if the car is rented for 2 days and is driven 120 miles?

59. **Wages** A man earns $5.92 for each of the first 36 hours he works in one week and $8.88 in overtime pay for each additional hour he works in the same week. How much money will he make if he works 45 hours in one week?

60. **Wages** A student earns $8.56 for each of the first 40 hours she works in one week and $12.84 in overtime pay for each additional hour she works in the same week. How much money will she make if she works 44 hours in one week?

61. **Rectangle** A rectangle has a width of 33.5 millimeters and a length of 254 millimeters. Find the area.

62. **Rectangle** A rectangle has a width of 2.56 inches and a length of 6.14 inches. Find the area rounded to the nearest hundredth.

63. **Rectangle** The logo on a business letter is rectangular. The rectangle has a width of 0.84 inches and a length of 1.41 inches. Find the area rounded to the nearest hundredth.

64. **Rectangle** A small sticky note is a rectangle. It has a width of 21.4 millimeters and a length of 35.8 millimeters. Find the area.

Getting Ready for the Next Section

To get ready for the next section, which covers division with decimals, we will review multiplication and division with whole numbers and fractions.

Perform the following operations.

65. $3,758 \div 2$

66. $9,900 \div 22$

67. $50,032 \div 33$

68. $90,902 \div 5$

69. $20\overline{)5,960}$

70. $30\overline{)4,620}$

71. 4×8.7

72. 5×6.7

73. 27×1.848

74. 35×32.54

75. $38\overline{)31,350}$

76. $25\overline{)377,800}$

Improving Your Quantitative Literacy

77. Containment System Holding tanks for hazardous liquids are often surrounded by containment tanks that will hold the hazardous liquid if the main tank begins to leak. We see that the center tank has a height of 16 feet and a radius of 6 feet. The outside containment tank has a height of 4 feet and a radius of 8 feet. The formula to calculate the volume of a tank is $V = \pi r^2 h$. Use 3.14 as an estimate for π. If the center tank is full of heating fuel and develops a leak at the bottom, will the containment tank be able to hold all the heating fuel that leaks out?

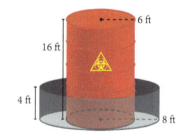

Find the Mistake

Each sentence below contains a mistake. Circle the mistake and write the correct word(s) or numbers(s) on the line provided.

1. To multiply 18.05 by 3.5, multiply as if the numbers were whole numbers and then place the decimal in the answer with two digits to its right. _____

2. To estimate the answer for 24.9×7.3, round 24.9 to 20 and 7.3 to 7. _____

3. To simplify $(8.43 + 1.002) - (0.05)(3.2)$, first subtract the product of 0.05 and 3.2 from 1.002 before adding 8.43.

_____

4. Lucy pays \$1.52 a pound for the first three pounds of candy she buys at a candy store, and pays \$3.27 for each additional pound. To find how much she will pay if she buys 5.2 pounds of candy, we must solve the problem $1.52(3) + 3.27(5.2)$.

_____

4.4 Division with Decimals

A coffee maker company announced the development of their new home espresso machine that scans your fingerprint to take an order. The new device can distinguish between six different housemates based on a scan of each fingerprint and its matching preferences (e.g., strength, froth, etc.). The current estimated selling price for the espresso machine is $3,200.

Suppose you and your three roommates have decided to purchase one of these espresso machines selling for $3,199.99. If you decide to split the bill equally, how much does each person owe? To find out, you will have to divide 3199.99 by 4. In this section, we will find out how to do division with any combination of decimal numbers.

A Dividing Decimal Numbers (Exact Answers)

EXAMPLE 1   Divide: $5{,}974 \div 20$.

Solution

```
         298
    20)5974
       40↓
       ───
       197
       180↓
       ───
       174
       160
       ───
        14
```

In the past, we have written this answer as $298\frac{14}{20}$ or, after reducing the fraction, $298\frac{7}{10}$. Because $\frac{7}{10}$ can be written as 0.7, we could also write our answer as 298.7. This last form of our answer is exactly the same result we obtain if we write 5,974 as 5,974.0 and continue the division until we have no remainder. Here is how it looks:

```
        298.7
    20)5974.0
       40↓
       ───
       197
       180↓
       ───
       174
       160↓
       ───
       140
       140
       ───
         0
```

Notice that we place the decimal point in the answer directly above the decimal point in the problem.

Video Examples Section 4.4

Practice Problems

1. Divide: $4{,}870 \div 50$.

> *Note* We can estimate the answer to Example 1 by rounding 5,974 to 6,000 and dividing by 20:
> $$\frac{6{,}000}{20} = 300$$

Answer

1. 97.4

2. Divide: $56.9 \div 5$.

> **Note**   We never need to make a mistake with division, because we can always check our results with multiplication.

Let's try another division problem. This time one of the numbers in the problem will be a decimal.

EXAMPLE 2   Divide: $34.8 \div 4$.

Solution   We can use the ideas from Example 1 and divide as usual. The decimal point in the answer will be placed directly above the decimal point in the problem.

$$
\begin{array}{r}
8.7 \\
4\overline{)34.8} \\
32 \\
\hline
2\,8 \\
2\,8 \\
\hline
0
\end{array}
\qquad
\begin{array}{r}
Check: \quad 8.7 \\
\times\ \ 4 \\
\hline
34.8
\end{array}
$$

The answer is 8.7.

We can use these facts to write a rule for dividing decimal numbers.

> **RULE**   Division with Decimal Numbers
>
> To divide a decimal by a whole number, we do the usual long division as if there were no decimal point involved. The decimal point in the answer is placed directly above the decimal point in the problem.

Here are some more examples to illustrate the procedure:

3. Divide: $54.632 \div 25$.

EXAMPLE 3   Divide: $49.896 \div 27$.

Solution

$$
\begin{array}{r}
1.848 \\
27\overline{)49.896} \\
27 \\
\hline
22\,8 \\
21\,6 \\
\hline
1\,29 \\
1\,08 \\
\hline
216 \\
216 \\
\hline
0
\end{array}
\qquad
\begin{array}{r}
Check: \quad 1.848 \\
\times\ \ 27 \\
\hline
12\,936 \\
36\,96 \\
\hline
49.896
\end{array}
$$

We can write as many zeros as we choose after the rightmost digit in a decimal number without changing the value of the number. For example,

$$6.91 = 6.910 = 6.9100 = 6.91000$$

There are times when this can be very useful, as Example 4 shows.

4. Divide: $1{,}105.8 \div 15$.

EXAMPLE 4   Divide: $1{,}138.9 \div 35$.

Solution

$$
\begin{array}{r}
32.54 \\
35\overline{)1138.90} \\
105 \\
\hline
88 \\
70 \\
\hline
18\,9 \\
17\,5 \\
\hline
1\,40 \\
1\,40 \\
\hline
0
\end{array}
\qquad
\begin{array}{r}
Check: \quad 32.54 \\
\times\ \ \ \ 35 \\
\hline
162\,70 \\
976\,2 \\
\hline
1{,}138.90
\end{array}
$$

Write 0 after the 9. It doesn't change the original number, but it gives us another digit to bring down.

Answers

2. 11.38
3. 2.18528
4. 73.72

Until now we have considered only division by whole numbers. Extending division to include division by decimal numbers is a matter of knowing what to do about the decimal point in the divisor.

EXAMPLE 5   Divide: $31.35 \div 3.8$.

Solution   In fraction form, this problem is equivalent to

$$\frac{31.35}{3.8}$$

If we want to write the divisor as a whole number, we can multiply the numerator and the denominator of this fraction by 10:

$$\frac{31.35 \times 10}{3.8 \times 10} = \frac{313.5}{38}$$

So, since this fraction is equivalent to the original fraction, our original division problem is equivalent to

$$\begin{array}{r} 8.25 \\ 38\overline{)313.50} \\ \underline{304} \\ 95 \\ \underline{76} \\ 190 \\ \underline{190} \\ 0 \end{array}$$ *Put 0 after the last digit.*

We can summarize division with decimal numbers by listing the following points, as illustrated in the first five examples.

Summary of Division with Decimals

1. We divide decimal numbers by the same process used to divide whole numbers. The decimal point in the answer is placed directly above the decimal point in the dividend.

2. We are free to write as many zeros after the last digit in a decimal number as we need.

3. If the divisor is a decimal, we can change it to a whole number by moving the decimal point to the right as many places as necessary so long as we move the decimal point in the dividend the same number of places.

B Dividing Decimal Numbers (Rounded Answers)

EXAMPLE 6   Divide, and round the answer to the nearest hundredth.

$$0.3778 \div 0.25$$

Solution   First, we move the decimal point two places to the right.

$$0.25\overline{)37.78}$$

5. Divide: $46.354 \div 4.9$.

Note We do not always use the rules for rounding numbers to make estimates. For example, to estimate the answer in Example 5, $31.35 \div 3.8$, we can get a rough estimate of the answer by reasoning that 3.8 is close to 4 and 31.35 is close to 32. Therefore, our answer will be approximately $32 \div 4 = 8$.

6. Divide, and round the answer to the nearest hundredth.

$$0.3849 \div 0.49$$

Note Moving the decimal point two places in both the divisor and the dividend is justified like this:

$$\frac{0.3778 \times 100}{0.25 \times 100} = \frac{37.78}{25}$$

Answers

5. 9.46
6. 0.79

Then we divide, using long division.

$$
\begin{array}{r}
1.5112 \\
25\overline{)37.7800} \\
\underline{25} \\
12\,7 \\
\underline{12\,5} \\
28 \\
\underline{25} \\
30 \\
\underline{25} \\
50 \\
\underline{50} \\
0
\end{array}
$$

Rounding to the nearest hundredth, we have 1.51. We actually did not need to have this many digits to round to the hundredths column. We could have stopped at the thousandths column and rounded off.

7. Divide, and round to the nearest tenth.

$19 \div 0.07$

EXAMPLE 7   Divide and round to the nearest tenth: $17 \div 0.03$.

Solution   Because we are rounding to the nearest tenth, we will continue dividing until we have a digit in the hundredths column. We don't have to go any further to round to the tenths column.

$$
\begin{array}{r}
566.66 \\
0.03\overline{)17.00.00} \\
\underline{15} \\
2\,0 \\
\underline{1\,8} \\
20 \\
\underline{18} \\
20 \\
\underline{18} \\
20 \\
\underline{18} \\
2
\end{array}
$$

Rounding to the nearest tenth, we have 566.7.

Applications

8. If you earn $8.30 per hour and receive a paycheck for $311.25 before deductions, how many hours did you work?

EXAMPLE 8   If a man earning $5.26 an hour receives a paycheck for $170.95 before deductions, how many hours did he work?

Solution   To find the number of hours the man worked, we divide $170.95 by $5.26.

$$
\begin{array}{r}
32.5 \\
5.26\overline{)170.95.0} \\
\underline{157\,8} \\
13\,15 \\
\underline{10\,52} \\
2\,63\,0 \\
\underline{2\,63\,0} \\
0
\end{array}
$$

The man worked 32.5 hours.

EXAMPLE 9 A telephone company charges $0.43 for the first minute and then $0.33 for each additional minute for a long-distance call. If a long-distance call costs $3.07, how many minutes was the call?

Solution To solve this problem we need to find the number of additional minutes for the call. To do so, we first subtract the cost of the first minute from the total cost, and then we divide the result by the cost of each additional minute. Without showing the actual arithmetic involved, the solution looks like this:

The number of additional minutes $= \dfrac{3.07 - 0.43}{0.33} = \dfrac{2.64}{0.33} = 8$

Total cost of the call — Cost of the first minute

Cost of each additional minute

The call was 9 minutes long. (The number 8 is the number of additional minutes past the first minute.)

9. Repeat Example 9 if the long distance call costs $13.63.

DESCRIPTIVE STATISTICS Grade Point Average

I have always been surprised by the number of my students who have difficulty calculating their grade point average (GPA). During her first semester in college, my daughter, Amy, earned the following grades:

Class	Units	Grade
Algebra	5	B
Chemistry	4	C
English	3	A
History	3	B

When her grades arrived in the mail, she told me she had a 3.0 grade point average, because the A and C grades averaged to a B. I told her that her GPA was a little less than a 3.0. What do you think? Can you calculate her GPA? If not, you will be able to after you finish this section.

When you calculate your grade point average (GPA), you are calculating what is called a *weighted average*. To calculate your grade point average, you must first calculate the number of grade points you have earned in each class that you have completed. The number of grade points for a class is the product of the number of units the class is worth times the value of the grade received. The table below shows the value that is assigned to each grade.

Grade	Value
A	4
B	3
C	2
D	1
F	0

If you earn a B in a 4-unit class, you earn $4 \times 3 = 12$ grade points. A grade of C in the same class gives you $4 \times 2 = 8$ grade points. To find your grade point average for one term (a semester or quarter), you must add your grade points and divide that total by the number of units. Round your answer to the nearest hundredth.

Answer

9. 41 minutes

10. Calculate the grade-point
    average.

Class	Units	Grade
Algebra	4	A
Biology	3	B
English	3	B
History	3	C
PE	1	A

EXAMPLE 10   Calculate Amy's grade point average using the previous information.

Solution   We begin by creating two more columns: one for the value of each grade (4 for an A, 3 for a B, 2 for a C, 1 for a D, and 0 for an F), and another for the grade points earned for each class. To fill in the grade points column, we multiply the number of units by the value of the grade.

Class	Units	Grade	Value	Grade Points
Algebra	5	B	3	$5 \times 3 = 15$
Chemistry	4	C	2	$4 \times 2 = 8$
English	3	A	4	$3 \times 4 = 12$
History	3	B	3	$3 \times 3 = 9$
Total Units	15			Total Grade Points: 44

To find her grade point average, we divide 44 by 15 and round (if necessary) to the nearest hundredth.

$$\text{Grade point average} = \frac{44}{15} = 2.93$$

GETTING READY FOR CLASS

After reading through the preceding section, respond in your own words and in complete sentences.

A. The answer to the division problem in Example 1 is $298\frac{14}{20}$. Write this number in decimal notation.

B. In Example 4, we place a 0 at the end of a number without changing the value of the number. Why is the placement of this 0 helpful?

C. The expression $0.3778 \div 0.25$ is equivalent to the expression $37.78 \div 25$ because each number was multiplied by what?

D. Briefly explain how to divide with decimals.

Answer
10. 3.14

Vocabulary Review

Choose the correct words to fill in the blanks below.

divisor   above   long division   right   last

1. To divide decimal numbers, use _____ as if the decimal point was not there.

2. In a long division problem that involves a decimal point, write the decimal point in the answer directly _____ the decimal point in the problem.

3. Writing zeros after the _____ digit in a decimal number will not change the value.

4. If the _____ is a decimal, we can change it to a whole number by moving the decimal point to the _____ as many places as necessary so long as we move the decimal point in the dividend the same number of places.

Problems

A Perform each of the following divisions.

1. $394 \div 20$
2. $486 \div 30$
3. $248 \div 40$
4. $372 \div 80$

5. $5\overline{)26}$
6. $8\overline{)36}$
7. $25\overline{)276}$
8. $50\overline{)276}$

9. $28.8 \div 6$
10. $15.5 \div 5$
11. $77.6 \div 8$
12. $31.48 \div 4$

13. $35\overline{)92.05}$
14. $26\overline{)146.38}$
15. $45\overline{)190.8}$
16. $55\overline{)342.1}$

17. $86.7 \div 34$
18. $411.4 \div 44$
19. $29.7 \div 22$
20. $488.4 \div 88$

21. $4.5\overline{)29.25}$
22. $3.3\overline{)21.978}$
23. $0.11\overline{)1.089}$
24. $0.75\overline{)2.40}$

25. $2.3\overline{)0.115}$
26. $6.6\overline{)0.198}$
27. $0.012\overline{)1.068}$
28. $0.052\overline{)0.23712}$

29. $1.1\overline{)2.42}$
30. $2.2\overline{)7.26}$
31. $0.014\overline{)0.0644}$
32. $0.38\overline{)9.652}$

B Carry out each of the following divisions only so far as needed to round the results to the nearest hundredth.

33. $26\overline{)35}$

34. $18\overline{)47}$

35. $3.3\overline{)56}$

36. $4.4\overline{)75}$

37. $0.1234 \div 0.5$

38. $-0.543 \div 2.1$

39. $19 \div 7$

40. $16 \div (-6)$

41. $0.059\overline{)0.69}$

42. $0.048\overline{)0.49}$

43. $1.99 \div 0.5$

44. $0.99 \div 0.5$

45. $2.99 \div 0.5$

46. $3.99 \div (-0.5)$

47. $-3.82 \div 0.9$

48. $1.79 \div 0.08$

Applying the Concepts

49. Hot Air Balloon Since the pilot of a hot air balloon can only control the balloon's altitude, he relies on the winds for travel. To ride on the jet streams, a hot air balloon must rise as high as 12 kilometers. Convert this to miles by dividing by 1.61. Round your answer to the nearest tenth of a mile.

50. Hot Air Balloon December and January are the best times for traveling in a hot-air balloon because the jet streams in the Northern Hemisphere are the strongest. They reach speeds of 400 kilometers per hour. Convert this to miles per hour by dividing by 1.61. Round to the nearest whole number.

51. Women's Golf The table below gives the top five money earners for the Ladies' Professional Golf Association (LPGA) in 2010. Fill in the last column of the table by finding the average earning per tournament for each golfer. Round your answers to the nearest ten dollars.

Rank	Name	Number of Tournaments	Total Earnings	Average per Tournament
1.	Na Yeon Choi	23	$1,871,165.50	
2.	Jiyai Shin	19	$1,783,127.00	
3.	Cristie Kerr	21	$1,601,551.75	
4.	Yani Tseng	19	$1,573,529.00	
5.	Suzann Pettersen	19	$1,557,174.50	

52. Men's Golf The table below gives the top five earners for the men's Professional Golf Association (PGA) in 2010. Fill in the last column of the table by finding the average earnings per tournament for each golfer. Round your answers to the nearest hundred dollars.

Rank	Name	Number of Tournaments	Total Earnings	Average per Tournament
1.	Matt Kuchar	26	$4,910,477	
2.	Jim Furyk	21	$4,809,622	
3.	Ernie Els	20	$4,558,861	
4.	Dustin Johnson	23	$4,473,122	
5.	Steve Stricker	19	$4,190,235	

53. Wages If a woman earns $33.90 for working 6 hours, how much does she earn per hour?

54. Wages How many hours does a person making $6.78 per hour have to work in order to earn $257.64?

55. Gas Mileage If a car travels 336 miles on 15 gallons of gas, how far will the car travel on 1 gallon of gas?

56. Gas Mileage If a car travels 392 miles on 16 gallons of gas, how far will the car travel on 1 gallon of gas?

57. Wages Suppose a woman earns $6.78 an hour for the first 36 hours she works in a week and then $10.17 an hour in overtime pay for each additional hour she works in the same week. If she makes $294.93 in one week, how many hours did she work overtime?

58. Wages Suppose a woman makes $286.08 in one week. If she is paid $5.96 an hour for the first 36 hours she works and then $8.94 an hour in overtime pay for each additional hour she works in the same week, how many hours did she work overtime that week?

59. Phone Bill Suppose a telephone company charges $0.41 for the first minute and then $0.32 for each additional minute for a long-distance call. If a long-distance call costs $2.33, how many minutes was the call?

60. Phone Bill Suppose a telephone company charges $0.45 for each of the first three minutes and then $0.29 for each additional minute for a long-distance call. If a long-distance call costs $3.67, how many minutes was the call?

Grade Point Average The following grades were earned by Steve during his first term in college. Use these data to answer Problems 61–64.

61. Calculate Steve's GPA.

Class	Units	Grade
Basic mathematics	3	A
Health	2	B
History	3	B
English	3	C
Chemistry	4	C

62. If his grade in chemistry had been a B instead of a C, by how much would his GPA have increased?

63. If his grade in health had been a C instead of a B, by how much would his grade point average have dropped?

64. If his grades in both English and chemistry had been Bs, what would his GPA have been?

Calculator Problems   Work each of the following problems on your calculator. If rounding is necessary, round to the nearest hundred thousandth.

65. $7 \div 9$

66. $11 \div 13$

67. $243 \div 0.791$

68. $67.8 \div 37.92$

69. $0.0503 \div 0.0709$

70. $429.87 \div 16.925$

Getting Ready for the Next Section

In the next section, we will consider the relationship between fractions and decimals in more detail. The problems below review some of the material that is necessary to make a successful start in the next section.

Reduce to lowest terms.

71. $\dfrac{75}{100}$

72. $\dfrac{220}{1,000}$

73. $\dfrac{12}{18}$

74. $\dfrac{15}{30}$

75. $\dfrac{75}{200}$

76. $\dfrac{220}{2,000}$

77. $\dfrac{38}{100}$

78. $\dfrac{75}{1,000}$

Write each fraction as an equivalent fraction with denominator 100.

79. $\dfrac{3}{5}$

80. $\dfrac{1}{2}$

81. $\dfrac{5}{1}$

82. $\dfrac{17}{20}$

Write each fraction as an equivalent fraction with denominator 15.

83. $\dfrac{4}{5}$

84. $\dfrac{2}{3}$

85. $\dfrac{4}{1}$

86. $\dfrac{2}{1}$

87. $\dfrac{6}{5}$

88. $\dfrac{7}{3}$

Divide.

89. $3 \div 4$

90. $3 \div 5$

91. $7 \div 8$

92. $3 \div 8$

Find the Mistake

Each sentence below contains a mistake. Circle the mistake and write the correct word(s) or numbers(s) on the line provided.

1. The answer to the problem 25)70.75 will have a decimal point placed with four digits to its right. _____

2. To work the problem 27.468 ÷ 8.4, multiply 8.4 by 10 and then divide. _____

3. To divide 0.6778 by 0.54, multiply both numbers by 10 to move the decimal point two places to the right. _____

4. Samantha earns $10.16 an hour as a cashier. She received a paycheck for $309.88. To find out how many hours she worked, you must solve the problem 10.16 ÷ 309.88. _____

Landmark Review: Checking Your Progress

Write each of the following in words.

1. 1.15　　　　**2.** 45.08　　　　**3.** 0.005　　　　**4.** 245.157

Write each of the following as a decimal number.

5. Sixty-seven ten thousandths　　**6.** Five and six tenths　　**7.** Twenty-three and fourteen thousandths　　**8.** Two thousand thirteen and fifteen hundredths

Find each of the following sums and differences.

9. 24.13 + 4.15　　**10.** 6.000014 + 3.15　　**11.** 100.00001 + 24.1583　　**12.** 5.387 + 6.412

13. 8.3 − 5.2　　**14.** 14.2 − 7.13　　**15.** 27.57 − 14.24　　**16.** 92.42 − 14.05

Perform each of the following operations.

17. 4.735(10)　　**18.** 0.075(0.03)　　**19.** 1.4 ÷ 0.07　　**20.** 0.24 ÷ 0.6

Perform the following operations according to the rule for order of operations.

21. 4.3(3.8 − 2.6)　　**22.** (2.85 − 1.7)(5.67 + 4.2)　　**23.** 5.5 + 2.2(14 − 12.5)　　**24.** $(1.3)^2 + (5.1)^2 + (2.4)^2$

KEY WORDS

convert

Video Examples
Section 4.5

Practice Problems

1. Write $\frac{5}{8}$ as a decimal.

2. Write $\frac{5}{7}$ as a decimal rounded to the thousandths place.

Answers
1. 0.625
2. 0.714

4.5 Fractions and Decimals

Image © sxc.hu, Thoursie, 2010

If you are shopping for clothes and a store has a sale advertising $\frac{1}{3}$ off the regular price, how much can you expect to pay for a pair of pants that normally sells for $31.95? If the sale price of the pants is $22.30, have they really been marked down by $\frac{1}{3}$? To answer questions like these, we need to know how to solve problems that involve fractions and decimals together.

We begin this section by showing how to convert back and forth between fractions and decimals.

A Converting Fractions to Decimals

You may recall that the notation we use for fractions can be interpreted as implying division. That is, the fraction $\frac{3}{4}$ can be thought of as meaning "3 divided by 4." We can use this idea to convert fractions to decimals.

EXAMPLE 1 Write $\frac{3}{4}$ as a decimal.

Solution Dividing 3 by 4, we have

$$
\begin{array}{r}
.75 \\
4\overline{)3.00} \\
\underline{28} \\
20 \\
\underline{20} \\
0
\end{array}
$$

The fraction $\frac{3}{4}$ is equal to the decimal 0.75.

EXAMPLE 2 Write $\frac{7}{12}$ as a decimal rounded to the thousandths place.

Solution Because we want the decimal to be rounded to the thousandths place, we divide to the ten thousandths place and round off to the thousandths place.

$$
\begin{array}{r}
.5833 \\
12\overline{)7.0000} \\
\underline{60} \\
100 \\
\underline{96} \\
40 \\
\underline{36} \\
40 \\
\underline{36} \\
4
\end{array}
$$

Rounding off to the thousandths place, we have 0.583. Because $\frac{7}{12}$ is not exactly the same as 0.583, we write

$$\frac{7}{12} \approx 0.583$$

where the symbol $\approx$ is read "is approximately equal to."

If we wrote more zeros after 0.583 in Example 2, the pattern of 3s would continue for as many places as we choose to divide. When we get a sequence of digits that repeat like this, 0.58333 . . . , we can indicate the repetition by writing

$$0.58\overline{3}$$ *The bar over the 3 indicates that the 3 repeats from there on.*

EXAMPLE 3   Write $\frac{3}{11}$ as a decimal.

Solution   Dividing 3 by 11, we have

```
        .272727
  11)3.000000
     22
     ──
      80
      77
      ──
       30
       22
       ──
        80
        77
        ──
         30
         22
         ──
          80
          77
          ──
           3
```

No matter how long we continue the division, the remainder will never be 0, and the pattern will continue. We write the decimal form of $\frac{3}{11}$ as $0.\overline{27}$, where

$$0.\overline{27} = 0.272727 \ldots$$ *The dots mean "and so on."*

Converting Decimals to Fractions

To convert decimals to fractions, we take advantage of the place values we assigned to the digits to the right of the decimal point.

EXAMPLE 4   Write 0.38 as a fraction in lowest terms.

Solution   0.38 is 38 hundredths, or

$$0.38 = \frac{38}{100}$$

$$= \frac{19}{50}$$ *Divide the numerator and the denominator by 2 to reduce to lowest terms.*

The decimal 0.38 is equal to the fraction $\frac{19}{50}$.

We could check our work here by converting $\frac{19}{50}$ back to a decimal. We do this by dividing 19 by 50. That is,

```
        .38
  50)19.00
     15 0
     ────
       4 00
       4 00
       ────
          0
```

3. Write $\frac{5}{6}$ as a decimal.

Note   The bar over the 2 and the 7 in $0.\overline{27}$ is used to indicate that the pattern repeats itself indefinitely.

4. Write 0.76 as a fraction in lowest terms.

Answers

3. $0.8\overline{3}$

4. $\frac{19}{25}$

5. Convert 0.045 to a fraction.

EXAMPLE 5   Convert 0.075 to a fraction.

Solution   We have 75 thousandths, or

$$0.075 = \frac{75}{1,000}$$

$$= \frac{3}{40} \qquad \text{\textit{Divide the numerator and the denominator by}}$$
$$\text{\textit{25 to reduce to lowest terms.}}$$

6. Write 12.08 as a mixed number.

EXAMPLE 6   Write 15.6 as a mixed number.

Solution   Converting 0.6 to a fraction, we have

$$0.6 = \frac{6}{10} = \frac{3}{5} \qquad \text{\textit{Reduce to lowest terms.}}$$

Since $0.6 = \frac{3}{5}$, we have $15.6 = 15\frac{3}{5}$.

B Problems Containing Both Fractions and Decimals

We continue this section by working some problems that involve both fractions and decimals.

7. Simplify: $\frac{12}{25}$ (1.41 − 0.56).

EXAMPLE 7   Simplify: $\frac{19}{50}$ (1.32 + 0.48).

Solution   In Example 4, we found that $0.38 = \frac{19}{50}$. Therefore, we can rewrite the problem as

$$\frac{19}{50}(1.32 + 0.48) = 0.38(1.32 + 0.48) \quad \text{\textit{Convert all numbers to decimals.}}$$

$$= 0.38(1.80) \qquad \text{\textit{Add: 1.32 + 0.48.}}$$

$$= 0.684 \qquad \text{\textit{Multiply: 0.38 × 1.80.}}$$

8. Simplify: $\frac{3}{4} + \frac{2}{3}$ (0.66).

EXAMPLE 8   Simplify: $\frac{1}{2} + (0.75)\left(\frac{2}{5}\right)$.

Solution   We could do this problem one of two different ways. First, we could convert all fractions to decimals and then simplify.

$$\frac{1}{2} + (0.75)\left(\frac{2}{5}\right) = 0.5 + 0.75(0.4) \quad \text{\textit{Convert to decimals.}}$$

$$= 0.5 + 0.300 \qquad \text{\textit{Multiply: 0.75 × 0.4.}}$$

$$= 0.8 \qquad \text{\textit{Add.}}$$

Or, we could convert 0.75 to $\frac{3}{4}$ and then simplify.

$$\frac{1}{2} + (0.75)\left(\frac{2}{5}\right) = \frac{1}{2} + \frac{3}{4}\left(\frac{2}{5}\right) \quad \text{\textit{Convert decimals to fractions.}}$$

$$= \frac{1}{2} + \frac{3}{10} \qquad \text{\textit{Multiply: } \frac{3}{4} \times \frac{2}{5}.}$$

$$= \frac{5}{10} + \frac{3}{10} \qquad \text{\textit{The common denominator is 10.}}$$

$$= \frac{8}{10} \qquad \text{\textit{Add numerators.}}$$

$$= \frac{4}{5} \qquad \text{\textit{Reduce to lowest terms.}}$$

The answers are equivalent. That is, $0.8 = \frac{8}{10} = \frac{4}{5}$. Either method can be used with problems of this type.

EXAMPLE 9 Simplify: $\left(\frac{1}{2}\right)^3 (2.4) + \left(\frac{1}{4}\right)^2 (3.2)$.

Solution This expression can be simplified without any conversions between fractions and decimals. To begin, we evaluate all numbers that contain exponents. Then we multiply. After that, we add.

$$\left(\frac{1}{2}\right)^3 (2.4) + \left(\frac{1}{4}\right)^2 (3.2) = \frac{1}{8}(2.4) + \frac{1}{16}(3.2) \quad \text{Evaluate exponents.}$$

$$= 0.3 + 0.2 \quad \text{Multiply by } \frac{1}{8} \text{ and } \frac{1}{16}.$$

$$= 0.5 \quad \text{Add.}$$

Applications

EXAMPLE 10 If a shirt that normally sells for \$27.99 is on sale for $\frac{1}{3}$ off, what is the sale price of the shirt?

Solution To find out how much the shirt is marked down, we must find $\frac{1}{3}$ of 27.99. That is, we multiply $\frac{1}{3}$ and 27.99, which is the same as dividing 27.99 by 3.

$$\frac{1}{3}(27.99) = \frac{27.99}{3} = 9.33$$

The shirt is marked down \$9.33. The sale price is \$9.33 less than the original price.

$$\text{Sale price} = 27.99 - 9.33 = 18.66$$

The sale price is \$18.66. We also could have solved this problem by simply multiplying the original price by $\frac{2}{3}$, since, if the shirt is marked $\frac{1}{3}$ off, then the sale price must be $\frac{2}{3}$ of the original price. Multiplying by $\frac{2}{3}$ is the same as dividing by 3 and then multiplying by 2. The answer would be the same.

GETTING READY FOR CLASS

After reading through the preceding section, respond in your own words and in complete sentences.

A. To convert fractions to decimals, do we multiply or divide the numerator by the denominator?

B. The decimal 0.13 is equivalent to what fraction?

C. Write 36 thousandths in decimal form and in fraction form.

D. Explain how to write the fraction $\frac{84}{1,000}$ in lowest terms.

9. Simplify: $\left(\frac{1}{3}\right)^2 (0.81) + \left(\frac{1}{2}\right)^3 (1.2)$.

10. A pair of jeans that normally sell for \$38.60 are on sale for $\frac{1}{4}$ off. What is the sale price of the jeans?

Answers
9. 0.24
10. \$28.95

EXERCISE SET 4.5

Problems

A Each circle below is divided into 8 equal parts. The number below each circle indicates what fraction of the circle is shaded. Convert each fraction to a decimal.

1.

$\dfrac{1}{8}$

2.

$\dfrac{3}{8}$

3.

$\dfrac{5}{8}$

4.

$\dfrac{7}{8}$

Complete the following tables by converting each fraction to a decimal.

5.

Fraction	$\frac{1}{5}$	$\frac{2}{5}$	$\frac{3}{5}$	$\frac{4}{5}$	$\frac{5}{5}$
Decimal					

6.

Fraction	$\frac{1}{6}$	$\frac{2}{6}$	$\frac{3}{6}$	$\frac{4}{6}$	$\frac{5}{6}$	$\frac{6}{6}$
Decimal						

Convert each of the following fractions to a decimal.

7. $\dfrac{1}{2}$

8. $-\dfrac{12}{25}$

9. $\dfrac{14}{25}$

10. $\dfrac{14}{32}$

11. $\dfrac{18}{32}$

12. $\dfrac{9}{16}$

13. $-\dfrac{13}{16}$

14. $\dfrac{13}{8}$

Write each fraction as a decimal rounded to the hundredths place.

15. $\dfrac{12}{13}$

16. $\dfrac{17}{19}$

17. $\dfrac{3}{11}$

18. $-\dfrac{5}{11}$

19. $\dfrac{2}{23}$

20. $-\dfrac{3}{28}$

21. $\dfrac{12}{43}$

22. $\dfrac{15}{51}$

Complete the following table by converting each decimal to a fraction.

23.

Decimal	0.125	0.250	0.375	0.500	0.625	0.750	0.875
Fraction							

24.

Decimal	0.1	0.2	0.3	0.4	0.5	0.6	0.7	0.8	0.9
Fraction									

Write each decimal as a fraction in lowest terms.

25. 0.15 **26.** 0.45 **27.** 0.08 **28.** 0.06 **29.** 0.375 **30.** 0.475

Write each decimal as a mixed number.

31. 5.6 **32.** 8.4 **33.** 5.06 **34.** 8.04 **35.** 1.22 **36.** 2.11

B Simplify each of the following as much as possible, and write all answers as decimals.

37. $\frac{1}{2}(2.3 + 2.5)$ **38.** $\frac{3}{4}(1.8 + 7.6)$ **39.** $\dfrac{1.99}{\frac{1}{2}}$ **40.** $\dfrac{2.99}{\frac{1}{2}}$

41. $3.4 - \frac{1}{2}(0.76)$ **42.** $6.7 - \frac{1}{5}(0.45)$ **43.** $\frac{2}{5}(0.3) + \frac{3}{5}(0.3)$ **44.** $\frac{1}{8}(0.7) + \frac{3}{8}(0.7)$

45. $6\left(\frac{3}{5}\right)(0.02)$ **46.** $8\left(\frac{4}{5}\right)(0.03)$ **47.** $\frac{5}{8} + 0.35\left(\frac{1}{2}\right)$ **48.** $\frac{7}{8} + 0.45\left(\frac{3}{4}\right)$

49. $\left(\frac{1}{3}\right)^2(5.4) + \left(\frac{1}{2}\right)^3(3.2)$ **50.** $\left(\frac{1}{5}\right)^2(7.5) + \left(\frac{1}{4}\right)^2(6.4)$ **51.** $(0.25)^2 + \left(\frac{1}{4}\right)^2(3)$ **52.** $(0.75)^2 + \left(\frac{1}{4}\right)^2(7)$

Applying the Concepts

53. Price of Beef If each pound of beef costs $2.59, how much does $3\frac{1}{4}$ pounds cost?

54. Price of Gasoline What does it cost to fill a $15\frac{1}{2}$-gallon gas tank if the gasoline is priced at 305.2¢ per gallon? Convert your answer to dollars.

55. Sale Price A dress that costs $57.99 is on sale for $\frac{1}{3}$ off. What is the sale price of the dress?

56. Sale Price A suit that normally sells for $121 is on sale for $\frac{1}{4}$ off. What is the sale price of the suit?

57. Perimeter of the Sierpinski Triangle The diagram shows one stage of what is known as the Sierpinski triangle. Each triangle in the diagram has three equal sides. The large triangle is made up of 4 smaller triangles. If each side of the large triangle is 2 inches, and each side of the smaller triangles is 1 inch, what is the perimeter of the shaded region?

58. Perimeter of the Sierpinski Triangle The diagram shows another stage of the Sierpinski triangle. Each triangle in the diagram has three equal sides. The largest triangle is made up of a number of smaller triangles. If each side of the large triangle is 2 inches, and each side of the smallest triangles is 0.5 inch, what is the perimeter of the shaded region?

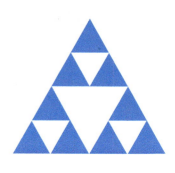

59. Average Gain in Stock Price The table below shows the amount of gain each day of one week in 2011 for the price of Verizon stock. Complete the table by converting each fraction to a decimal and rounding to the nearest hundredth if necessary.

Change in Stock Price		
Day	Gain ($)	As a Decimal ($) (to the nearest hundredth)
Monday	$\frac{3}{5}$	
Tuesday	$\frac{1}{2}$	
Wednesday	$\frac{1}{25}$	
Thursday	$\frac{1}{5}$	
Friday	$\frac{1}{10}$	

60. Average Gain in Stock Price The table below shows the amount of gain each day of one week in 2011 for the price of AT&T stock. Complete the table by converting each fraction to a decimal and rounding to the nearest hundredth, if necessary.

Change in Stock Price		
Day	Gain	As a Decimal ($) (to the nearest hundredth)
Monday	$\frac{3}{10}$	
Tuesday	$\frac{3}{50}$	
Wednesday	$\frac{2}{25}$	
Thursday	$\frac{1}{10}$	
Friday	0	

61. Nutrition If 1 ounce of ground beef contains 50.75 calories and 1 ounce of halibut contains 27.5 calories, what is the difference in calories between a $4\frac{1}{2}$-ounce serving of ground beef and a $4\frac{1}{2}$-ounce serving of halibut?

62. Nutrition If a 1-ounce serving of baked potato contains 48.3 calories and a 1-ounce serving of chicken contains 24.6 calories, how many calories are in a meal of $5\frac{1}{4}$ ounces of chicken and a $3\frac{1}{3}$-ounce baked potato?

Taxi Ride The Texas Junior College Teachers Association annual conference was held in Austin. At that time, a taxi ride in Austin was $1.25 for the first $\frac{1}{5}$ of a mile and $0.25 for each additional $\frac{1}{5}$ of a mile. The charge for a taxi to wait is $12.00 per hour. Use this information for Problems 63 through 66.

63. If the distance from one of the convention hotels to the airport is 7.5 miles, how much will it cost to take a taxi from that hotel to the airport?

64. If you were to tip the driver of the taxi in Problem 63 $1.50, how much would it cost to take a taxi from the hotel to the airport?

65. Suppose the distance from one of the hotels to one of the western dance clubs in Austin is 12.4 miles. If the fare meter in the taxi gives the charge for that trip as $16.50, is the meter working correctly?

66. Suppose that the distance from a hotel to the airport is 8.2 miles, and the ride takes 20 minutes. Is it more expensive to take a taxi to the airport or to just sit in the taxi?

Getting Ready for the Next Section

Expand and simplify.

67. 6^2

68. 8^2

69. 5^2

70. 10^2

71. 5^3

72. 2^5

73. 3^2

74. 2^3

75. $\left(\frac{1}{3}\right)^4$

76. $\left(\frac{3}{4}\right)^3$

77. $\left(\frac{5}{6}\right)^2$

78. $\left(\frac{3}{5}\right)^3$

79. $(0.5)^2$

80. $(0.1)^3$

81. $(1.2)^2$

82. $(2.1)^2$

83. $3^2 + 4^2$

84. $5^2 + 12^2$

85. $6^2 + 8^2$

86. $2^2 + 3^2$

Find the Mistake

Each sentence below contains a mistake. Circle the mistake and write the correct word(s) or numbers(s) on the line provided.

1. The correct way to write $\frac{6}{11}$ as a decimal is 0.54545454.... _____

2. Writing 14.3 as a fraction gives us $14 \cdot \frac{3}{10} = \frac{42}{10} = \frac{21}{5}$. _____

3. The simplified answer to the problem $\frac{12}{45(0.256 + 0.14)}$ contains both fractions and decimals. _____

4. Simplifying the problem $\left(\frac{3}{2}\right)(0.5) + \left(\frac{1}{2}\right)^2(6.7)$ by first converting all decimals to fractions gives us $\left(\frac{3}{2}\right)\left(\frac{1}{2}\right) + \left(\frac{1}{2}\right)^2\left(\frac{67}{100}\right)$.

_____

4.6 Square Roots and The Pythagorean Theorem

Photo courtesy of Megan Silva

KEY WORDS

square root

radical sign

perfect squares

irrational number

Pythagorean theorem

spiral of roots

Video Examples
Section 4.6

Imagine you are a seasoned backpacker and decide to hike through Evolution Valley in Kings Canyon National Park, California. You hike directly east along one leg of a trail for 6 miles, then turn north and hike for 8 more miles. If you were to draw a straight line from where you began to where you are now, what would be the distance? There is a formula in geometry that gives the distance d:

$$d = \sqrt{6^2 + 8^2}$$

where $\sqrt{}$ is called the square root symbol. If we simplify what is under the square root symbol, we have

$$d = \sqrt{36 + 64}$$
$$= \sqrt{100}$$

The expression $\sqrt{100}$ stands for the number we square to get 100. Because $10 \cdot 10 = 100$, that number is 10. Therefore, the straight-line distance from the starting point of your hike to where you are now is 10 miles. We'll discuss square roots in detail this section.

A Square Roots

Previously, we did some work with exponents. In particular, we spent some time finding squares of numbers. For example, we considered expressions like this:

$$5^2 = 5 \cdot 5 = 25$$
$$7^2 = 7 \cdot 7 = 49$$
$$x^2 = x \cdot x$$

We say that "the square of 5 is 25" and "the square of 7 is 49." To square a number, we multiply it by itself. When we ask for the *square root* of a given number, we want to know what number we square in order to obtain the given number. We say that the square root of 49 is 7, because 7 is the number we square to get 49. Likewise, the square root of 25 is 5, because $5^2 = 25$. The symbol we use to denote square root is $\sqrt{}$, which is also called a *radical sign*. Here is the precise definition of square root:

DEFINITION square root

The *square root* of a positive number a, written $\sqrt{a}$, is the number we square to get a.

If $\sqrt{a} = b$ then $b^2 = a$.

Note The square root we are describing here is actually the principal or positive square root. There is another square root that is a negative number. We won't see it in this book, but, if you go on to take an algebra course, you will see it there.

We list some common square roots in Table 1.

Table 1

Statement	In Words	Reason
$\sqrt{0} = 0$	The square root of 0 is 0	Because $0^2 = 0$
$\sqrt{1} = 1$	The square root of 1 is 1	Because $1^2 = 1$
$\sqrt{4} = 2$	The square root of 4 is 2	Because $2^2 = 4$
$\sqrt{9} = 3$	The square root of 9 is 3	Because $3^2 = 9$
$\sqrt{16} = 4$	The square root of 16 is 4	Because $4^2 = 16$
$\sqrt{25} = 5$	The square root of 25 is 5	Because $5^2 = 25$

Numbers like 1, 9, and 25, whose square roots are whole numbers, are called *perfect squares*. To find the square root of a perfect square, we look for the whole number that is squared to get the perfect square. The following examples involve square roots of perfect squares.

EXAMPLE 1   Simplify: $7\sqrt{64}$.

Solution   The expression $7\sqrt{64}$ means 7 times $\sqrt{64}$. To simplify this expression, we write $\sqrt{64}$ as 8 and multiply.

$$7\sqrt{64} = 7 \cdot 8 = 56$$

We know $\sqrt{64} = 8$, because $8^2 = 64$.

EXAMPLE 2   Simplify: $\sqrt{9} + \sqrt{16}$.

Solution   We write $\sqrt{9}$ as 3 and $\sqrt{16}$ as 4. Then we add.

$$\sqrt{9} + \sqrt{16} = 3 + 4 = 7$$

EXAMPLE 3   Simplify: $\sqrt{\dfrac{25}{81}}$.

Solution   We are looking for the number we square to get $\frac{25}{81}$. We know that when we multiply two fractions, we multiply the numerators and multiply the denominators. Because $5 \cdot 5 = 25$ and $9 \cdot 9 = 81$, the square root of $\frac{25}{81}$ must be $\frac{5}{9}$.

$$\sqrt{\frac{25}{81}} = \frac{5}{9} \quad \text{because} \quad \left(\frac{5}{9}\right)^2 = \frac{5}{9} \cdot \frac{5}{9} = \frac{25}{81}$$

In Examples 4–6, we simplify each expression as much as possible.

EXAMPLE 4   Simplify: $12\sqrt{25} = 12 \cdot 5 = 60$.

EXAMPLE 5   Simplify: $\sqrt{100} - \sqrt{36} = 10 - 6 = 4$.

EXAMPLE 6   Simplify: $\sqrt{\dfrac{49}{121}} = \dfrac{7}{11}$ because $\left(\dfrac{7}{11}\right)^2 = \dfrac{7}{11} \cdot \dfrac{7}{11} = \dfrac{49}{121}$.

Practice Problems

1. Simplify: $5\sqrt{81}$.

2. Simplify: $\sqrt{25} + \sqrt{36}$.

3. Simplify: $\sqrt{\frac{36}{49}}$.

4. Simplify: $25\sqrt{9}$.

5. Simplify: $\sqrt{144} - \sqrt{36}$.

6. Simplify: $\sqrt{\frac{81}{144}}$.

Answers
1. 45
2. 11
3. $\frac{6}{7}$
4. 75
5. 6
6. $\frac{3}{4}$

So far in this section we have been concerned only with square roots of perfect squares. The next question is, "What about square roots of numbers that are not perfect squares, like $\sqrt{7}$, for example?" We know that

$$\sqrt{4} = 2 \quad \text{and} \quad \sqrt{9} = 3$$

And because 7 is between 4 and 9, $\sqrt{7}$ should be between $\sqrt{4}$ and $\sqrt{9}$. That is, $\sqrt{7}$ should be between 2 and 3. But what is it exactly? The answer is, we cannot write it exactly in decimal or fraction form. Because of this, it is called an *irrational number*. We can approximate it with a decimal, but we can never write it exactly with a decimal. Table 2 gives some decimal approximations for $\sqrt{7}$. The decimal approximations were obtained by using a calculator. We could continue the list to any accuracy we desired. However, we would never reach a number in decimal form whose square was exactly 7.

Table 2

Approximations for the Square Root of 7		
Accurate to the nearest	The square root of 7 is	Check by squaring
Tenth	$\sqrt{7} = 2.6$	$(2.6)^2 = 6.76$
Hundredth	$\sqrt{7} = 2.65$	$(2.65)^2 = 7.0225$
Thousandth	$\sqrt{7} = 2.646$	$(2.646)^2 = 7.001316$
Ten thousandth	$\sqrt{7} = 2.6458$	$(2.6458)^2 = 7.00025764$

B Using Calculators to Find Square Roots

7. Give a decimal approximation for the expression $6\sqrt{10}$ that is accurate to the nearest thousandth.

EXAMPLE 7   Give a decimal approximation for the expression $5\sqrt{12}$ that is accurate to the nearest ten thousandth.

Solution   Let's agree not to round to the nearest ten thousandth until we have first done all the calculations. Using a calculator, we find $\sqrt{12} \approx 3.4641016$. Therefore,

$$5\sqrt{12} \approx 5(3.4641016) \qquad \text{\textit{$\sqrt{12}$ on calculator}}$$
$$= 17.320508 \qquad \text{\textit{Multiply.}}$$
$$= 17.3205 \qquad \text{\textit{Round to the nearest ten thousandth.}}$$

8. Approximate $\sqrt{245} - \sqrt{131}$ to the nearest thousandth.

EXAMPLE 8   Approximate $\sqrt{301} + \sqrt{137}$ to the nearest hundredth.

Solution   Using a calculator to approximate the square roots, we have

$$\sqrt{301} + \sqrt{137} \approx 17.349352 + 11.704700 = 29.054052$$

To the nearest hundredth, the answer is 29.05.

9. Approximate $\sqrt{\dfrac{9}{26}}$ to the nearest ten thousandth.

EXAMPLE 9   Approximate $\sqrt{\dfrac{7}{11}}$ to the nearest thousandth.

Solution   Because we are using calculators, we first change $\frac{7}{11}$ to a decimal and then find the square root.

$$\sqrt{\frac{7}{11}} \approx \sqrt{0.6363636} \approx 0.7977240$$

To the nearest thousandth, the answer is 0.798.

Answers
7. 18.974
8. 4.207
9. 0.5883

C The Pythagorean Theorem

FACTS FROM GEOMETRY Pythagorean Theorem

A *right triangle* is a triangle that contains a 90° (or right) angle. The longest side in a right triangle is called the *hypotenuse*, and we use the letter c to denote it. The two shorter sides are denoted by the letters a and b. The Pythagorean theorem states that the hypotenuse is the square root of the sum of the squares of the two shorter sides. Here is the previous statement in symbols:

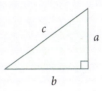

$$c = \sqrt{a^2 + b^2}$$

EXAMPLE 10 Find the length of the hypotenuse in each right triangle.

a.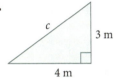

b.

Solution We apply the formula given above.

a. When $a = 3$ and $b = 4$,

$$c = \sqrt{3^2 + 4^2}$$
$$= \sqrt{9 + 16}$$
$$= \sqrt{25}$$
$$= 5 \text{ meters}$$

b. When $a = 5$ and $b = 7$,

$$c = \sqrt{5^2 + 7^2}$$
$$= \sqrt{25 + 49}$$
$$= \sqrt{74}$$
$$\approx 8.60 \text{ inches}$$

In part a, the solution is a whole number, whereas in part b, we must use a calculator to get 8.60 as an approximation to $\sqrt{74}$.

EXAMPLE 11 A ladder is leaning against a barn to reach a hay loft 6-feet above the barn floor. If the bottom of the ladder is 8 feet from the wall, how long is the ladder?

Solution A picture of the situation is shown in Figure 2. We let c denote the length of the ladder. Applying the Pythagorean theorem, we have

$$c = \sqrt{6^2 + 8^2}$$
$$= \sqrt{36 + 64}$$
$$= \sqrt{100}$$
$$= 10 \text{ feet}$$

The ladder is 10 feet long.

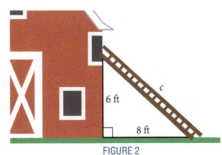

FIGURE 2

10. Find the length of the hypotenuse in a right triangle with the following leg lengths.
   a. 6 ft, 8 ft
   b. 9 m, 14 m

11. A rope is tied from the top of a 12-foot pole to the ground 5 feet from the base of the pole. If there is no slack in the rope, how long is the rope?

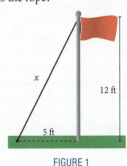

FIGURE 1

Answers
10. a. 10 ft   **b.** 16.64 m
11. 13 ft

The diagram showed here is called the *spiral of roots*. It is constructed using the Pythagorean theorem and it gives us a way to visualize positive square roots. The table below gives us the decimal equivalents (some of which are approximations) of the first 10 square roots in the spiral. The line graph can be constructed from the table or from the spiral.

Approximate length of diagonals

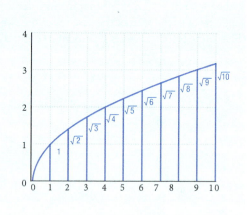

Number	Positive Square Root
1	1
2	1.41
3	1.73
4	2
5	2.24
6	2.45
7	2.65
8	2.83
9	3
10	3.16

To visualize the square roots of the counting numbers, we can construct the spiral of roots another way. To begin, we draw two line segments, each of length 1, at right angles to each other. Then we use the Pythagorean theorem to find the length of the diagonal. Figure 3 illustrates:

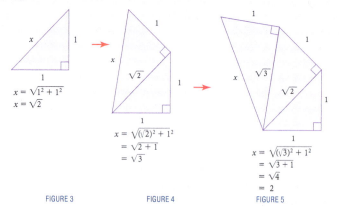

FIGURE 3   FIGURE 4   FIGURE 5

Next, we construct a second triangle by connecting a line segment of length 1 to the end of the first diagonal so that the angle formed is a right angle. We find the length of the second diagonal using the Pythagorean theorem. Figure 4 illustrates this procedure. As we continue to draw new triangles by connecting line segments of length 1 to the end of each previous diagonal, so that the angle formed is a right angle, the spiral of roots begins to appear (see Figure 5).

GETTING READY FOR CLASS

After reading through the preceding section, respond in your own words and in complete sentences.

A. Which number is larger, the square of 10 or the square root of 10?

B. Give a definition for the square root of a number.

C. What two numbers will the square root of 20 fall between?

D. What is the Pythagorean theorem?

Vocabulary Review

Choose the correct words to fill in the blanks below.

    spiral of roots   hypotenuse   irrational   radical sign
    Pythagorean theorem   square root   whole numbers   right

1. The _____ of a positive number a is the number we square to get a.
2. The symbol we use to denote a square root is called a _____.
3. Numbers whose square roots are _____ are called perfect squares.
4. An _____ number, such as $\sqrt{7}$, is a number that can be approximated with a decimal number but can never be written exactly as a decimal or fraction.
5. A _____ triangle is a triangle that contains a 90 degree angle.
6. The longest side of a right triangle is called the _____.
7. The _____ states that the hypotenuse is the square root of the sum of the squares of the two shorter sides.
8. The _____ is constructed using the Pythagorean theorem and is a way to visualize positive square roots of the counting numbers.

Problems

A Find each of the following square roots without using a calculator.

1. $\sqrt{64}$　　2. $\sqrt{100}$　　3. $\sqrt{81}$　　4. $\sqrt{49}$

5. $\sqrt{36}$　　6. $\sqrt{144}$　　7. $\sqrt{25}$　　8. $\sqrt{169}$

Simplify each of the following expressions without using a calculator.

9. $3\sqrt{25}$　　10. $9\sqrt{49}$　　11. $6\sqrt{64}$　　12. $11\sqrt{100}$

13. $15\sqrt{9}$　　14. $8\sqrt{36}$　　15. $16\sqrt{9}$　　16. $9\sqrt{16}$

17. $\sqrt{49}+\sqrt{64}$　　18. $\sqrt{1}+\sqrt{0}$　　19. $\sqrt{16}-\sqrt{9}$　　20. $\sqrt{25}-\sqrt{4}$

21. $3\sqrt{25}+9\sqrt{49}$　　22. $6\sqrt{64}+11\sqrt{100}$　　23. $15\sqrt{9}-9\sqrt{16}$　　24. $7\sqrt{49}-2\sqrt{4}$

25. $\sqrt{\dfrac{16}{49}}$

26. $\sqrt{\dfrac{100}{121}}$

27. $\sqrt{\dfrac{36}{64}}$

28. $\sqrt{\dfrac{81}{144}}$

29. $\sqrt{\dfrac{4}{9}}$

30. $\sqrt{\dfrac{25}{16}}$

31. $\sqrt{\dfrac{1}{16}}$

32. $\sqrt{\dfrac{1}{100}}$

Indicate whether each of the statements in Problems 33–36 is *True* or *False*.

33. $\sqrt{4} + \sqrt{9} = \sqrt{4 + 9}$

34. $\sqrt{\dfrac{16}{25}} = \dfrac{\sqrt{16}}{\sqrt{25}}$

35. $\sqrt{25 \cdot 9} = \sqrt{25} \cdot \sqrt{9}$

36. $\sqrt{100} - \sqrt{36} = \sqrt{100 - 36}$

B Calculator Problems

Use a calculator to work problems 37 through 56.

Approximate each of the following square roots to the nearest ten thousandth.

37. $\sqrt{1.25}$

38. $\sqrt{12.5}$

39. $\sqrt{125}$

40. $\sqrt{1250}$

Approximate each of the following expressions to the nearest hundredth.

41. $2\sqrt{3}$

42. $3\sqrt{2}$

43. $5\sqrt{5}$

44. $5\sqrt{3}$

45. $\dfrac{\sqrt{3}}{3}$

46. $\dfrac{\sqrt{2}}{2}$

47. $\sqrt{\dfrac{1}{3}}$

48. $\sqrt{\dfrac{1}{2}}$

Approximate each of the following expressions to the nearest thousandth.

49. $\sqrt{12} + \sqrt{75}$ **50.** $\sqrt{18} + \sqrt{50}$ **51.** $\sqrt{87}$ **52.** $\sqrt{68}$

53. $2\sqrt{3} + 5\sqrt{3}$ **54.** $3\sqrt{2} + 5\sqrt{2}$ **55.** $7\sqrt{3}$ **56.** $8\sqrt{2}$

C Find the length of the hypotenuse in each right triangle. Round to the nearest hundredth, if rounding is necessary.

57.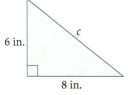

58.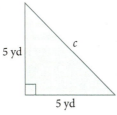

59.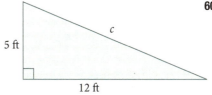

60.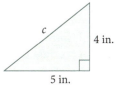

61.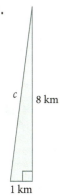

62.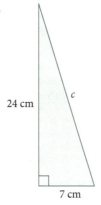

63.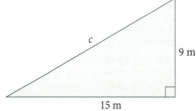

64.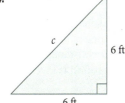

Applying the Concepts

65. Geometry One end of a wire is attached to the top of a 24-foot pole; the other end of the wire is anchored to the ground 18 feet from the bottom of the pole. If the pole makes an angle of 90° with the ground, find the length of the wire.

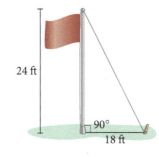

66. Geometry Two children are trying to cross a stream. They want to use a log that goes from one bank to the other. If the left bank is 5 feet higher than the right bank and the stream is 12 feet wide, how long must a log be to just barely reach?

67. Geometry A ladder is leaning against the top of a 15-foot wall. If the bottom of the ladder is 20 feet from the wall, how long is the ladder?

68. Geometry A wire from the top of a 24-foot pole is fastened to the ground by a stake that is 10 feet from the bottom of the pole. How long is the wire?

69. Spiral of Roots Construct your own spiral of roots by using a ruler. Draw the first triangle by using two 1-inch lines. The first diagonal will have a length of $\sqrt{2}$ inches. Each new triangle will be formed by drawing a 1-inch line segment at the end of the previous diagonal so that the angle formed is 90°. Draw your spiral until you have at least six right triangles.

70. Spiral of Roots Construct a spiral of roots by using line segments of length 2 inches. The length of the first diagonal will be $2\sqrt{2}$ inches. The length of the second diagonal will be $2\sqrt{3}$ inches. What will be the length of the third diagonal?

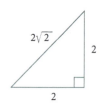

Use a calculator to work problems 71 and 72.

71. Lighthouse Problem The higher you are above the ground, the farther you can see. If your view is unobstructed, then the distance in miles that you can see from h feet above the ground is given by the formula

$$d = \sqrt{\frac{3h}{2}}$$

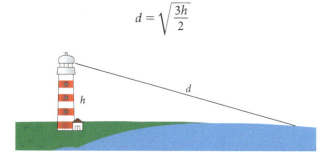

The following figure shows a lighthouse with a door and windows at various heights. The preceding formula can be used to find the distance to the ocean horizon from these heights. Use the formula and a calculator to complete the following table. Round your answers to the nearest whole number.

Height h (feet)	Distance d (miles)
10	
50	
90	
130	
170	
190	

72. Pendulum Problem The time (in seconds) it takes for the pendulum on a clock to swing through one complete cycle is given by the formula

$$T = \frac{11}{7}\sqrt{\frac{L}{2}}$$

where L is the length (in feet) of the pendulum. Use this formula and a calculator to complete the following table. Round your answers to the nearest hundredth.

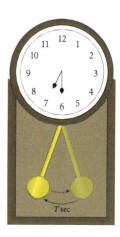

Length L (feet)	Time T (seconds)
1	
2	
3	
4	
5	
6	

Improving Your Quantitative Literacy

73. Millionaire Households The chart shows the number of millionaire households over a ten year period. Use the information to answer the following questions.

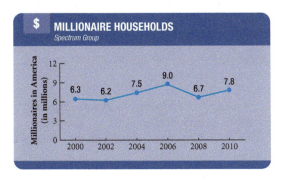

a. If the trend from 2008 to 2010 continues, how many millionaire households will there be in 2012?

b. Estimate the number of millionaire households in 2005.

Find the Mistake

Each sentence below contains a mistake. Circle the mistake and write the correct word(s) or numbers(s) on the line provided.

1. The square root of a positive number x is the number we square to get $\sqrt{x}$. _____

2. The square root of 225 is 15, and can be written in symbols as $\sqrt{15} = 225$. _____

3. Simplifying the radical $\sqrt{\frac{196}{36}}$ gives us $\frac{98}{36}$ because $\left(\frac{98}{36}\right)^2 = \frac{98}{36} + \frac{98}{36} = \frac{196}{36}$. _____

4. The Pythagorean theorem states that $b = \sqrt{a^2 + c^2}$. _____

Celebrating Pi

Since 1988, math enthusiasts around the world have spent March 14 celebrating the irrational number pi by eating pies and having discussions of the number's significance. Then in 2009, Congress passed a non-binding resolution that recognizes March 14 as National Pi Day. Pi, written using the Greek symbol π, represents the ratio of a circle's circumference to its diameter. Rounded to the nearest hundredth, π can be written as 3.14, which explains why National Pi Day occurs on the fourteenth day of the third month each year. However, since π is an irrational number, its digits will infinitely continue without repeating. Here are the first ten digits of π:

$$3.141592653$$

One of the activities often found on Pi Day is that of writing pi-kus. A pi-ku, the name derived from a short-form Japanese poem called haiku, is a short poem that uses the digits of pi to represent the syllables in each line. To write a pi-ku for the 3-digit approximation of pi, the first line of the poem contains three syllables, the second line one syllable, and the final line four syllables. For instance,

Today is	←	3 syllables
a	←	1 syllable
great day for math!	←	4 syllables

1. Write a pi-ku using the 3-digit approximation of π.

_____	←	3 syllables
_____	←	1 syllable
_____	←	4 syllables

2. Another style of writing known as Basic Pilish uses the digits of pi to represent the number of letters in each word. Here's an example:

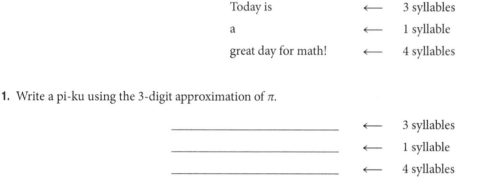

But I gaze a while patiently at grassy hills and seven gorillas gallivant merrily...
3. 1 4 1 5 9 2 6 5 3 5 8 9 7

Break into groups. Each group will be assigned a different sequential set of pi's digits.

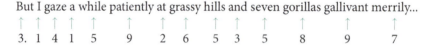

Group 1: 3.14159265358979...
Group 2: ...323846264338327...
Group 3: ...950288419716939...
Group 4: ...937510582097494...
Group 5: ...459230781640628...

For each digit, come up with a word that has the same number of letters as the digit. If the digit is 0, use a ten-letter word. The words should flow into sentences and tell a story. Punctuation may appear but is not counted as a digit. Each group should share its story with the class in the order the digits appear in pi (e.g., Group 1 shares first, then Group 2, and so on).

Hint: Use the stories as a mnemonic device to recite the digits of pi.

Chapter 4 Summary

Place Value [4.1]

The place values for the first five places to the right of the decimal point are

Decimal Point	Tenths	Hundredths	Thousandths	Ten Thousandths	Hundred Thousandths
.	$\frac{1}{10}$	$\frac{1}{100}$	$\frac{1}{1,000}$	$\frac{1}{10,000}$	$\frac{1}{100,000}$

Rounding Decimals [4.1]

If the digit in the column to the right of the one we are rounding to is 5 or more, we add 1 to the digit in the column we are rounding to; otherwise, we leave it alone. We then replace all digits to the right of the column we are rounding to with zeros if they are to the left of the decimal point; otherwise, we simply delete them.

Addition and Subtraction with Decimals [4.2]

To add (or subtract) decimal numbers, we align the decimal points and add (or subtract) as if we were adding (or subtracting) whole numbers. The decimal point in the answer goes directly below the decimal points in the problem.

Multiplication with Decimals [4.3]

To multiply two decimal numbers, we multiply as if the decimal points were not there. The decimal point in the product has as many digits to the right as there are total digits to the right of the decimal points in the two original numbers.

Division with Decimals [4.4]

To begin a division problem with decimals, we make sure that the divisor is a whole number. If it is not, we move the decimal point in the divisor to the right as many places as it takes to make it a whole number. We must then be sure to move the decimal point in the dividend the same number of places to the right. Once the divisor is a whole number, we divide as usual. The decimal point in the answer is placed directly above the decimal point in the dividend.

Changing Fractions to Decimals [4.5]

6. $\frac{4}{15} = 0.2\overline{6}$ because

$$
\begin{array}{r}
.266 \\
15\overline{)4.000} \\
\underline{3\,0} \\
1\,00 \\
\underline{90} \\
100 \\
\underline{90} \\
10
\end{array}
$$

To change a fraction to a decimal, we divide the numerator by the denominator.

Changing Decimals to Fractions [4.5]

7. $0.781 = \dfrac{781}{1,000}$

To change a decimal to a fraction, we write the digits to the right of the decimal point over the appropriate power of 10.

Square Roots [4.6]

8. $\sqrt{49} = 7$ because
$7^2 = 7 \cdot 7 = 49$

The square root of a positive number a, written $\sqrt{a}$, is the number we square to get a.

Pythagorean Theorem [4.6]

In any right triangle, the length of the longest side (the hypotenuse) is equal to the square root of the sum of the squares of the two shorter sides.

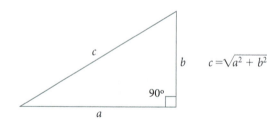

1. Write the decimal number 6.302 in words. [4.1]

2. Give the place value of the 6 in the number 23.4263. [4.1]

3. Write twenty-three and five thousand, six ten thousandths as a decimal number. [4.1]

4. Round 72.1950 to the nearest hundredth. [4.1]

Perform the following operations. Round to the nearest thousandth if necessary. [4.2, 4.3, 4.4]

5. $11 + 0.1 + 0.92$

6. $14.002 - 6.098$

7. $1.8(9.03)$

8. $11.913 \div 4.8$

9. $8.1 + 6.49$

10. $14.83 - 6.938$

11. $0.9(3.1)(-1.1)$

12. $12.364 \div 4$

13. A person purchases $7.23 worth of goods at a drugstore. If a $10 bill is used to pay for the purchases, how much change is received? [4.2]

14. If coffee sells for $5.29 per pound, how much will 4.5 pounds of coffee cost? [4.3]

15. If a person earns $540 for working 80 hours, what is the person's hourly wage? [4.4]

16. Write $\frac{17}{20}$ as a decimal. [4.5]

17. Write 0.62 as a fraction in lowest terms. [4.5]

18. $6.8(3.9 + 0.37)$

19. $8.7 - 5(0.23)$

20. $46.918 - 6(4.92 + 0.086)$ 21. $\frac{4}{3}(0.36) - \frac{7}{5}(0.3)$

22. $16.3 - 6(3.07 - 4.3)$ 23. $\frac{1}{5}(0.38) + 7(9.1 - 2.7)$

Find each of the following square roots without using a calculator. [4.6]

24. $\sqrt{4}$ 25. $\sqrt{121}$

Simplify each of the following without using a calculator. [4.6]

26. $4\sqrt{81}$ 27. $5\sqrt{25}$

Find the length of the hypotenuse in each right triangle. Round to the nearest hundredth if necessary. [4.6]

28.

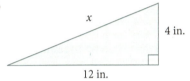

29.

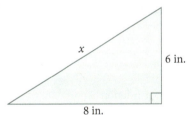

The diagram shows the annual sales for different brands of cookies. Use the information to answer the following questions.

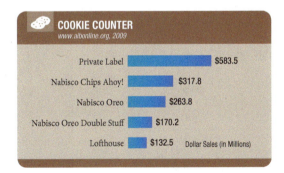

30. What are the average monthly sales for Nabisco Chips Ahoy! cookies?

31. What are the average monthly sales for Lofthouse and Nabisco Oreo Double Stuff cookies combined?

Simplify.

1. $4,832 - (-459)$ **2.** $620 + (-476)$

3. $93(190)$ **4.** $4.390 - 1.57$

5. $6\overline{)921.41}$ **6.** $\dfrac{3}{7} + \dfrac{5}{6}$

7. $9.1(11.03)$ **8.** $1,178 \div 19$

9. $\dfrac{9}{14} \div \dfrac{6}{35}$

10. Twice the sum of 3 and 4 decreased by 14 results in what number?

11. Change $\frac{81}{5}$ to a mixed number.

12. Change $3\frac{7}{8}$ to an improper fraction.

13. Find the product of $3\frac{5}{8}$ and 7.

14. Change each decimal into a fraction.

Decimal	Fraction
0.125	
0.250	
0.375	
0.500	
0.625	
0.750	
0.875	
1	

15. Give the quotient of 72 and 18.

16. Translate into symbols: Four times the sum of two and three is twenty.

17. Translate into symbols: Seven times the sum of six and three is sixty-three.

18. Reduce $\dfrac{75}{130}$.

19. True or False? Adding the same number to the numerator and denominator of a fraction produces an equivalent fraction.

Simplify.

20. $48(3)^2 - 9(5)^2$ **21.** $\dfrac{3 + 7(5)}{10 + 9}$

22. $20\left(\dfrac{1}{4}\right) + 8\left(\dfrac{3}{8}\right)$ **23.** $\dfrac{3}{5}(0.65) + \dfrac{9}{10}(0.7)$

24. $\left(\dfrac{1}{3}\right)^3 + \left(\dfrac{1}{9}\right)^2$ **25.** $\left(5\dfrac{1}{8} - \dfrac{1}{3}\right)\left(3\dfrac{1}{4} + \dfrac{7}{8}\right)$

26. Suppose the temperature at 5 a.m. is 50 °F . If the temperature rises 40° by 2 p.m., then decreases 4° by 6 p.m., what is the temperature?

27. Recipe A muffin recipe calls for $4\frac{3}{4}$ cups of flour. If the recipe is doubled, how many cups of flour will be needed?

28. Hourly Wage If you earn $288.75 for working 35 hours, what is your hourly wage?

1. Write the decimal number 11.819 in words. [4.1]

2. Give the place value of the 8 in the number 61.8276. [4.1]

3. Write seventy-three and forty-six ten thousandths as a decimal number. [4.1]

4. Round 100.9052 to the nearest hundredth. [4.1]

Perform the following operations. [4.2, 4.3, 4.4]

5. $6 + 0.8 + 0.22$

6. $28.332 - 16.608$

7. $6.9(2.40)$

8. $96.4768 \div 16.52$

9. $6.8 + 3.3$

10. $16.47 - 8.58$

11. $0.5(3.7)(-1.8)$

12. $11.616 \div 52.8$

13. A person purchases $11.39 worth of goods at a drugstore. If a $10 bill and a $5 bill were used to pay for the purchases, how much change is received? [4.2]

14. If coffee sells for $3.29 per pound, how much will 5.5 pounds of coffee cost? [4.3]

15. If a person earns $489 for working 60 hours, what is the person's hourly wage? [4.4]

16. Write $\frac{17}{25}$ as a decimal. [4.5]

17. Write 0.38 as a fraction in lowest terms. [4.5]

18. $6.8(3.7 + 0.08)$

19. $6.1 - 4(0.93)$

20. $41.901 - 7(3.11 + 0.462)$

21. $\frac{7}{8}(0.11) + \frac{5}{6}(0.45)$

22. $23.4 - 8(6.01 - 4.2)$

23. $\frac{1}{5}(0.17) + 8(6.13 - 2.8)$

Find each of the following square roots without using a calculator. [4.6]

24. $\sqrt{100}$

25. $\sqrt{9}$

Simplify each of the following without using a calculator. [4.6]

26. $3\sqrt{25}$

27. $10\sqrt{9}$

Find the length of the hypotenuse in each right triangle. Round to the nearest hundredth if necessary. [4.6]

28.

29.

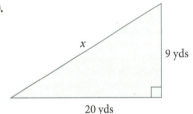

The diagram shows the annual sales for different magazines. Use the information to answer the following questions.

30. What is the average monthly sales for People Magazine?

31. What are the average monthly sales for US Weekly and Time Magazine combined?

Ratio and Proportion

5

Los Angeles, California
© 2010 Google

The Arroyo Seco Parkway, now called the Pasadena Freeway, in Los Angeles, California is significant in the history of American transportation. It is the first limited-access freeway that utilized diamond or cloverleaf ramps, which were a big departure from the traditional easy-access parkways found in the East. The two-car family was becoming a reality in Southern California by the end of World War I, therefore, adding more traffic congestion. The new freeway became necessary to help ease this congestion. When it opened in 1940, the freeway linked downtown Los Angeles with Pasadena. Its relatively high speeds allowed more rapid commutes and made getting around in 1940s LA a much quicker process. It was also unique in that art and landscaping were incorporated into the general construction plan to enhance the driving experience.

The freeway's expansion in 1953 to what was the nation's first four-lane interchange just north of downtown Los Angeles connected the Arroyo Seco Parkway to the expanding LA freeway network. This was a big step toward the modern freeway systems found in most large cities today. It is still part of the system that allows Angelinos to travel over 300 million miles every day.

Suppose you were to use the Pasadena Freeway during a 6 hour driving trip. If you traveled 270 miles during that time, how far would you go if you drove for 10 hours? Solving problems like this requires that you know how to set up and solve proportions, which is one of the topics we will cover in this chapter.

OBJECTIVES

A Write ratios as fractions.

KEY WORDS

ratio

Video Examples
Section 5.1

If a human faced off against an elephant in a hot dog eating contest, who would win? Each July, three brave humans sign up to find out firsthand. Three elephants, Susie, Minnie, and Bunny, from Ringling Bros. and Barnum & Bailey Circus stand behind a table on which a tall pile of hot dog buns sits. A few feet away, three competitive eating contestants stand in front of their own table outfitted with stacks of buns and large cups of water in which they can dunk the buns to help their cause. When the announcer yells "Go!", the two teams begin devouring the buns. They have 6 minutes to eat as many as possible. The 2010 final tally: 41 dozen buns for the elephants, 15 dozen for the humans. In mathematical terms, the ratio of hot dog buns eaten by elephants to those eaten by humans is 41 to 15. The *ratio* of two numbers is a way of comparing them which we will discuss in this section.

A Writing Ratios as Fractions

If we say that the ratio of two numbers is 2 to 1, then the first number is twice as large as the second number. For example, if there are 10 men and 5 women enrolled in a math class, then the ratio of men to women is 10 to 5. Because 10 is twice as large as 5, we can also say that the ratio of men to women is 2 to 1.

We can define the ratio of two numbers in terms of fractions.

> **DEFINITION** ratio
>
> The *ratio* of two numbers is a fraction, where the first number in the ratio is the numerator and the second number in the ratio is the denominator.
>
> *In symbols:* If a and b are any two numbers ($b \neq 0$), then the ratio of a to b is $\frac{a}{b}$.

We handle ratios the same way we handle fractions. For example, when we said that the ratio of 10 men to 5 women was the same as the ratio 2 to 1, we were actually saying

$$\frac{10}{5} = \frac{2}{1} \qquad \textit{Reduce to lowest terms.}$$

Because we have already studied fractions in detail, much of the introductory material on ratios will seem like review.

EXAMPLE 1 Express the ratio of 16 to 48 as a fraction in lowest terms.

Solution Because the ratio is 16 to 48, the numerator of the fraction is 16 and the denominator is 48.

$$\frac{16}{48} = \frac{1}{3}$$

Notice that the first number in the ratio becomes the numerator of the fraction, and the second number in the ratio becomes the denominator.

Practice Problems

1. Express the ratio of 15 to 25 as a fraction in lowest terms.

Answer

1. $\frac{3}{5}$

EXAMPLE 2   Write the ratio of $\frac{2}{3}$ to $\frac{4}{9}$ as a fraction in lowest terms.

Solution   We begin by writing the ratio of $\frac{2}{3}$ to $\frac{4}{9}$ as a complex fraction. The numerator is $\frac{2}{3}$, and the denominator is $\frac{4}{9}$. Then we simplify.

$$\frac{\frac{2}{3}}{\frac{4}{9}} = \frac{2}{3} \cdot \frac{9}{4} \quad \text{Division by } \tfrac{4}{9} \text{ is the same as multiplication by } \tfrac{9}{4}.$$

$$= \frac{18}{12} \quad \text{Multiply.}$$

$$= \frac{3}{2} \quad \text{Reduce to lowest terms.}$$

EXAMPLE 3   Write the ratio of 0.08 to 0.12 as a fraction in lowest terms.

Solution   When the ratio is in reduced form, it is customary to write it with whole numbers and not decimals. For this reason, we multiply the numerator and the denominator of the ratio by 100 to clear it of decimals. Then we reduce to lowest terms.

$$\frac{0.08}{0.12} = \frac{0.08 \times 100}{0.12 \times 100} \quad \text{Multiply the numerator and the denominator by 100 to clear the ratio of decimals.}$$

$$= \frac{8}{12} \quad \text{Multiply.}$$

$$= \frac{2}{3} \quad \text{Reduce to lowest terms.}$$

Table 1 shows several more ratios and their fractional equivalents. Notice that in each case the fraction has been reduced to lowest terms. Also, the ratio that contains decimals has been rewritten as a fraction that does not contain decimals.

Table 1

Ratio	Fraction	Fraction in lowest terms
25 to 35	$\frac{25}{35}$	$\frac{5}{7}$
35 to 25	$\frac{35}{25}$	$\frac{7}{5}$
8 to 2	$\frac{8}{2}$	$\frac{4}{1}$ We can also write this as just 4.
$\frac{1}{4}$ to $\frac{3}{4}$	$\frac{\frac{1}{4}}{\frac{3}{4}}$	$\frac{1}{3}$ because $\frac{\frac{1}{4}}{\frac{3}{4}} = \frac{1}{4} \cdot \frac{4}{3} = \frac{1}{3}$
0.6 to 1.7	$\frac{0.6}{1.7}$	$\frac{6}{17}$ because $\frac{0.6 \times 10}{1.7 \times 10} = \frac{6}{17}$

EXAMPLE 4   During a game, a basketball player makes 12 out of the 18 free throws he attempts. Write the ratio of the number of free throws he makes to the number of free throws he attempts as a fraction in lowest terms.

Solution   Because he makes 12 out of 18, we want the ratio 12 to 18, or

$$\frac{12}{18} = \frac{2}{3}$$

Because the ratio is 2 to 3, we can say that, in this particular game, he made 2 out of every 3 free throws he attempted.

2. Write the ratio of $\frac{3}{4}$ to $\frac{5}{8}$ as a fraction in lowest terms.

3. Write the ratio of 0.16 to 0.20 as a fraction in lowest terms.

Note   Another symbol used to denote ratio is the colon (:). The ratio of, say, 5 to 4 can be written as 5:4. Although we will not use it here, this notation is fairly common.

4. During a game, a baseball pitcher throws 50 strikes out of 80 pitches. Write the ratio of the strikes to total pitches as a fraction in lowest terms.

Answers

2. $\frac{6}{5}$

3. $\frac{4}{5}$

4. $\frac{5}{8}$

5. A solution of radiator fluid contains 2 pints antifreeze and 8 pints water. Find the ratio of antifreeze to water, water to antifreeze, antifreeze to total solution, and water to total solution.

EXAMPLE 5   A solution of alcohol and water contains 15 milliliters of water and 5 milliliters of alcohol. Find the ratio of alcohol to water, water to alcohol, water to total solution, and alcohol to total solution. Write each ratio as a fraction and reduce to lowest terms.

Solution   There are 5 milliliters of alcohol and 15 milliliters of water, so there are 20 milliliters of solution (alcohol + water). The ratios are as follows:

The ratio of alcohol to water is 5 to 15, or

$$\frac{5}{15} = \frac{1}{3} \qquad \text{In lowest terms}$$

The ratio of water to alcohol is 15 to 5, or

$$\frac{15}{1} = \frac{3}{1} \qquad \text{In lowest terms}$$

The ratio of water to total solution is 15 to 20, or

$$\frac{15}{20} = \frac{3}{4} \qquad \text{In lowest terms}$$

The ratio of alcohol to total solution is 5 to 20, or

$$\frac{5}{20} = \frac{1}{4} \qquad \text{In lowest terms}$$

6. Using the information from Example 6, find the ratio of the price of the medium pizza to the large pizza. Then change the ratio to a decimal rounded to the nearest hundredth.

EXAMPLE 6   Suppose a pizza restaurant advertised the following prices for their deep-dish pizza. Use the information to find the ratio of the cost of the large cheese pizza to the medium cheese pizza. Then change the ratio to a decimal rounded to the nearest tenth.

Size	Price
Medium 12" cheese	$12.00
Large 15" cheese	$16.50

Solution   The ratio of the large pizza to the medium is

$$\frac{16.5}{12.0} = \frac{16.5 \times 10}{12.0 \times 10} \qquad \text{Multiply the numerator and denominator by 10 to clear the ratio of decimals.}$$

$$= \frac{165}{120}$$

$$= \frac{11}{8} \qquad \text{Reduce to lowest terms.}$$

To convert to a decimal, we divide 11 by 8 and round to the nearest tenth.

$$\frac{11}{8} \approx 1.4$$

GETTING READY FOR CLASS

After reading through the preceding section, respond in your own words and in complete sentences.

A. In your own words, write a definition for the ratio of two numbers.

B. What does a ratio compare?

C. What are some different ways of using mathematics to write the ratio of a to b?

D. When will the ratio of two numbers be a complex fraction?

Answers

5. $\frac{1}{4}, \frac{4}{1}, \frac{1}{5}, \frac{4}{5}$

6. $\frac{8}{11} \approx 0.73$

Vocabulary Review

Choose the correct words to fill in the blanks below.

numerator   colon   complex   denominator   ratio   fraction

1. The ratio of two numbers is a _____, where the first number in the ratio is the numerator and second number in the ratio is the denominator.

2. When writing the ratio 6 to 8 as a fraction, 6 is the _____ and 8 is the _____.

3. If a and b are any two numbers, then the _____ of a to b is $\frac{a}{b}$.

4. A symbol used to denote a ratio is the _____.

5. The ratio of $\frac{1}{4}$ to $\frac{2}{3}$ will appear as a _____ fraction.

Problems

A Write each of the following ratios as a fraction in lowest terms. None of the answers should contain decimals.

1. 8 to 6

2. 6 to 8

3. 64 to 12

4. 12 to 64

5. 100 to 250

6. 250 to 100

7. 13 to 26

8. 36 to 18

9. $\frac{3}{4}$ to $\frac{1}{4}$

10. $\frac{5}{8}$ to $\frac{3}{8}$

11. $\frac{7}{3}$ to $\frac{6}{3}$

12. $\frac{9}{5}$ to $\frac{11}{5}$

13. $\frac{6}{5}$ to $\frac{6}{7}$

14. $\frac{5}{3}$ to $\frac{5}{8}$

15. $2\frac{1}{2}$ to $3\frac{1}{2}$

16. $5\frac{1}{4}$ to $1\frac{3}{4}$

17. $2\frac{2}{3}$ to $\frac{5}{3}$

18. $\frac{1}{2}$ to $3\frac{1}{2}$

19. 0.05 to 0.15

20. 0.21 to 0.03

21. 0.3 to 3

22. 0.5 to 10

23. 1.2 to 10

24. 6.4 to 0.8

Use the figures to answer the following questions.

25. a. What is the ratio of shaded squares to nonshaded squares?

b. What is the ratio of shaded squares to total squares?

c. What is the ratio of nonshaded squares to total squares?

26. a. What is the ratio of shaded squares to nonshaded squares?

b. What is the ratio of shaded squares to total squares?

c. What is the ratio of nonshaded squares to total squares?

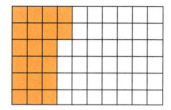

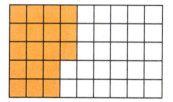

Applying the Concepts

27. Family Budget A family of four budgeted the amounts shown below for some of their monthly bills.

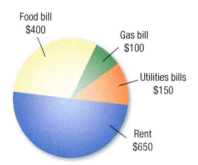

a. What is the ratio of the rent to the food bill?

b. What is the ratio of the gas bill to the food bill?

c. What is the ratio of the utilities bills to the food bill?

d. What is the ratio of the rent to the utilities bills?

28. Nutrition One cup of breakfast cereal was found to contain the nutrients shown here.

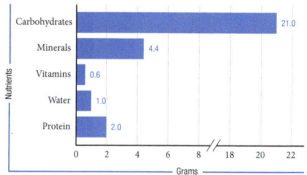

a. Find the ratio of water to protein.

b. Find the ratio of carbohydrates to protein.

c. Find the ratio of vitamins to minerals.

d. Find the ratio of protein to vitamins and minerals.

29. Pizza Prices The price of several menu items from a pizza restaurant are shown in the table. Find the ratio of the prices for the following items.

a. Large pizza to garlic bread

b. Buffalo wings to medium pizza

c. Garlic bread to buffalo wings

d. Garlic bread to medium pizza

Item	Price
Large pizza with pepperoni	$18.25
Medium pizza with pepperoni	$12.75
Buffalo wings	$6.50
Garlic Bread	$3.50

30. Profit and Revenue The following bar chart shows the profit and revenue of the Baby Steps Shoe Company each quarter for one year.

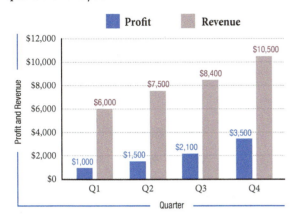

Find the ratio of revenue to profit for each of the following quarters. Write your answers in lowest terms.

a. Q1   **b.** Q2   **c.** Q3   **d.** Q4

e. Find the ratio of revenue to profit for the entire year.

31. Geometry In the diagram below, AC represents the length of the line segment that starts at A and ends at C. From the diagram we see that $AC = 8$.

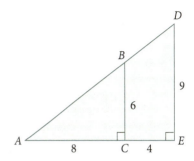

a. Find the ratio of BC to AC.

b. What is the length AE?

c. Find the ratio of DE to AE.

Calculator Problems

Write each of the following ratios as a fraction, and then use a calculator to change the fraction to a decimal. Round all decimal answers to the nearest hundredth. Do not reduce fractions.

Number of Students The total number of students attending a community college in the Midwest is 4,722. Of these students, 2,314 are male and 2,408 are female.

32. Give the ratio of males to females as a fraction and as a decimal.

33. Give the ratio of females to males as a fraction and as a decimal.

34. Give the ratio of males to total number of students as a fraction and as a decimal.

35. Give the ratio of total number of students to females as a fraction and as a decimal.

Getting Ready for the Next Section

The following problems review material from a previous section. Reviewing these problems will help you with the next section.

Write as a decimal.

36. $\dfrac{90}{5}$ **37.** $\dfrac{120}{3}$ **38.** $\dfrac{125}{2}$ **39.** $\dfrac{2}{10}$ **40.** $\dfrac{1.23}{2}$

41. $\dfrac{1.39}{2}$ **42.** $\dfrac{88}{0.5}$ **43.** $\dfrac{1.99}{0.5}$ **44.** $\dfrac{46}{0.25}$ **45.** $\dfrac{9}{0.25}$

Divide. Round answers to the nearest thousandth.

46. $0.48 \div 5.5$ **47.** $0.75 \div 11.5$ **48.** $2.19 \div 46$ **49.** $1.25 \div 50$

Improving Your Quantitative Literacy

50. Stock Market One method of comparing stocks on the stock market is the price to earnings ratio, or P/E.

$$P/E = \frac{\text{Current Stock Price}}{\text{Earnings per Share}}$$

Most stocks have a P/E between 25 and 40. A stock with a P/E of less than 25 may be undervalued, while a stock with a P/E greater than 40 may be overvalued. Fill in the P/E for each stock listed in the table below. Based on your results, are any of the stocks undervalued?

Stock	Price	Earnings Per Share	P/E
IBM	146.05	$6.35	
AOL	139.69	$0.61	
DIS	30.03	$0.91	
KM	15.64	$0.68	
GE	90.75	$2.75	
TOY	19.92	$1.66	

Find the Mistake

Each sentence below contains a mistake. Circle the mistake and write the correct word(s) or numbers(s) on the line provided.

1. Writing the ratio of $\frac{2}{5}$ to $\frac{3}{8}$ is the same as writing $\frac{2}{5} \cdot \frac{3}{8}$. _____

2. The ratio of 6 to 24 expressed in lowest terms is 4. _____

3. To write the ratio of 0.04 to 0.20 as a fraction in lowest terms, you must first multiply 0.20 by 10.

_____

4. A cleaning solution of bleach and water contains 100 milliliters of bleach and 150 milliliters of water. To find the ratio of water to the whole solution in lowest terms, you must write the ratio as $\frac{150}{100}$. _____

Navigation Skills: Prepare, Study, Achieve

Preparation is another key component for success in this course. What does it mean to be prepared for this class? Before you come to class, make sure to do the following:

- Complete in full the homework from the previous section.
- Read the upcoming section.
- Answer the Getting Ready for Class questions.
- Rework example problems, and work practice problems.
- Make a point to attend every class session and arrive on time.
- Prepare a list of questions to ask your instructor or fellow students that will help you work through difficult problems.

You shouldn't expect to master a new topic the first time you read about it or learn about it in class. Mastering mathematics takes a lot of practice, so make the commitment to come to class prepared and practice, practice, practice.

5.2 Rates and Unit Pricing

©iStockphoto.com/yellowcrestmedia

OBJECTIVES

A Convert ratio data to rates.

B Calculate unit prices.

KEY WORDS

rate

unit pricing

**Video Examples
Section 5.2**

A Rates

Rocket Man Bob Maddox plans to strap himself to four rockets and blast himself into space. While he's raising money to build these rockets, he spends his free time building small pulse jet engines for bicycles. For power, the engines pulse a remarkable 70 times per second. A single engine on a standard bicycle can propel the rider to a top speed of 50 miles per hour in 7 seconds. A bicycle with two engines attached can reach a top speed of 73 miles per hour.

We will begin this section with a discussion of rates. Whenever a ratio compares two quantities that have different units (and neither unit can be converted to the other), then the ratio is called a rate. For example, if it were possible for Bob Maddox to ride his single engine bicycle 150 miles for 3 hours, then his average rate of speed expressed as the ratio of miles to hours would be

$$\frac{150 \text{ miles}}{3 \text{ hours}} = \frac{50 \text{ miles}}{1 \text{ hour}}$$

Divide the numerator and the denominator by 3 to reduce to lowest terms.

The ratio $\frac{50 \text{ miles}}{1 \text{ hour}}$ can be expressed as

$$50 \frac{\text{miles}}{\text{hour}} \quad \text{or} \quad 50 \text{ miles/hour} \quad \text{or} \quad 50 \text{ miles per hour}$$

A rate is expressed in simplest form when the numerical part of the denominator is 1. To accomplish this we use division.

EXAMPLE 1 A train travels 125 miles in 2 hours. What is the train's rate in miles per hour?

Solution The ratio of miles to hours is

$$\frac{125 \text{ miles}}{2 \text{ hours}} = 62.5 \frac{\text{miles}}{\text{hours}} \quad \text{Divide 125 by 2.}$$

$$= 62.5 \text{ miles per hour}$$

If the train travels 125 miles in 2 hours, then its average rate of speed is 62.5 miles per hour.

EXAMPLE 2 A car travels 90 miles on 5 gallons of gas. Give the ratio of miles to gallons as a rate in miles per gallon.

Solution The ratio of miles to gallons is

$$\frac{90 \text{ miles}}{5 \text{ gallons}} = 18 \frac{\text{miles}}{\text{gallon}} \quad \text{Divide 90 by 5.}$$

$$= 18 \text{ miles/gallon}$$

The gas mileage of the car is 18 miles per gallon.

Practice Problems

1. A car travels 238 miles in 3.5 hours. What is the car's rate in miles per hour.

2. A car travels 357 miles on 8.5 gallons of gas. Give the ratio of miles to gallons as a rate.

Answers

1. 68 miles per hour
2. 42 miles per gallon

B Unit Pricing

One kind of rate that is very common is *unit pricing*. Unit pricing is the ratio of price to quantity when the quantity is one unit. Suppose a 1-liter bottle of a certain soft drink costs $1.19, whereas a 2-liter bottle of the same drink costs $1.39. Which is the better buy? That is, which has the lower price per liter?

$$\frac{\$1.19}{1 \text{ liter}} = \$1.19 \text{ per liter}$$

$$\frac{\$1.39}{2 \text{ liters}} = \$0.695 \text{ per liter}$$

The unit price for the 1-liter bottle is $1.19 per liter, whereas the unit price for the 2-liter bottle is 69.5¢ per liter. The 2-liter bottle is a better buy.

EXAMPLE 3   A supermarket sells low-fat milk in three different containers at the following prices:

1 gallon	$\frac{1}{2}$ gallon	1 quart (1 quart = $\frac{1}{4}$ gallon)
$3.59	$1.99	$1.29

Give the unit price in dollars per gallon for each one.

Solution   Because 1 quart = $\frac{1}{4}$ gallon, we have

$$\text{1-gallon container} \quad \frac{\$3.59}{1 \text{ gallon}} = \$3.59 \text{ per gallon}$$

$$\tfrac{1}{2}\text{-gallon container} \quad \frac{\$1.99}{\frac{1}{2} \text{ gallon}} = \frac{\$1.99}{0.5 \text{ gallon}} = \$3.98 \text{ per gallon}$$

$$\text{1-quart container} \quad \frac{\$1.29}{1 \text{ quart}} = \frac{\$1.29}{0.25 \text{ gallon}} = \$5.16 \text{ per gallon}$$

The 1-gallon container has the lowest unit price, whereas the 1-quart container has the highest unit price.

GETTING READY FOR CLASS

After reading through the preceding section, respond in your own words and in complete sentences.

A. In your own words, explain what a rate is.

B. When is a rate written in simplest terms?

C. What is unit pricing?

D. Give some examples of rates not found in your textbook.

---

3. A store sells paper towels for $2.57 (4-pack) and $5.99 (10-pack). Find the unit price for each package to determine the cheapest unit price.

Answers

3. 4-pack: $0.64 per roll;
10-pack: $0.60 per roll (cheapest)

Problems

A Solve each of the following word problems.

1. Miles/Hour A car travels 220 miles in 4 hours. What is the rate of the car in miles per hour?

2. Miles/Hour A train travels 360 miles in 5 hours. What is the rate of the train in miles per hour?

3. Kilometers/Hour It takes a car 3 hours to travel 252 kilometers. What is the rate in kilometers per hour?

4. Kilometers/Hour In 6 hours an airplane travels 4,200 kilometers. What is the rate of the airplane in kilometers per hour?

5. Gallons/Second The flow of water from a water faucet can fill a 3-gallon container in 15 seconds. Give the ratio of gallons to seconds as a rate in gallons per second.

6. Gallons/Minute A 225-gallon drum is filled in 3 minutes. What is the rate in gallons per minute?

7. Liters/Minute A gas tank which can hold a total of 56 liters contains only 8 liters of gas when the driver stops to refuel. If it takes 4 minutes to fill up the tank, what is the rate in liters per minute?

8. Liters/Hour The gas tank on a car holds 60 liters of gas. At the beginning of a 6-hour trip, the tank is full. At the end of the trip, it contains only 12 liters. What is the rate at which the car uses gas in liters per hour?

9. Miles/Gallon A car travels 95 miles on 5 gallons of gas. Give the ratio of miles to gallons as a rate in miles per gallon.

10. Miles/Gallon On a 384-mile trip, an economy car uses 8 gallons of gas. Give this as a rate in miles per gallon.

11. **Miles/Liter** The gas tank on a car has a capacity of 75 liters. On a full tank of gas, the car travels 325 miles. What is the gas mileage in miles per liter?

12. **Miles/Liter** A car pulling a trailer can travel 105 miles on 70 liters of gas. What is the gas mileage in miles per liter?

B

13. **Cents/Ounce** A 6-ounce can of frozen orange juice costs 96¢. Give the unit price in cents per ounce.

14. **Cents/Liter** A 2-liter bottle of root beer costs $1.25. Give the unit price in cents per liter.

15. **Cents/Ounce** A 20-ounce package of frozen peas is priced at 99¢. Give the unit price in cents per ounce.

16. **Dollars/Pound** A 4-pound bag of cat food costs $8.12. Give the unit price in dollars per pound.

17. **Best Buy** Find the unit price in cents per diaper for each of the brands shown below. Round to the nearest tenth of a cent. Which is the better buy?

Dry Baby	*Happy Baby*
36 Diapers, $12.49	38 Diapers, $11.99

18. **Best Buy** Find the unit price in cents per pill for each of the brands shown below. Round to the nearest tenth of a cent. Which is the better buy?

Relief	*New Life*
100 Pills, $5.99	225 Pills, $13.96

19. **Carbon Footprint** A car produces 38.5 tons of CO_2 over a 5-year period. Find the tons of CO_2 produced per year.

20. **Pounds/Gallon** A car uses 5 gallons of gas on a trip and produces 101 pounds of carbon dioxide. Find the amount of CO_2 per gallon produced by the car.

21. **Cents/Day** If a 15-day supply of vitamins costs $1.62, what is the price in cents per day?

22. **Miles/Hour** A car travels 675.4 miles in $12\frac{1}{2}$ hours. Give the rate in miles per hour to the nearest hundredth.

23. **Miles/Gallon** A truck's odometer reads 15,208.3 at the beginning of a trip and 15,336.7 at the end of the trip. If the trip takes 13.8 gallons of gas, what is the gas mileage in miles per gallon? (Round to the nearest tenth.)

24. **Miles/Hour** At the beginning of a trip, the odometer on a car read 32,567.2 miles. At the end of the trip, it read 32,741.8 miles. If the trip took $4\frac{1}{4}$ hours, what was the rate of the car in miles per hour to the nearest tenth?

Hourly Wages Jane has a job at the local department store. The graph shows how much Jane earns for working 8 hours per day for 5 days.

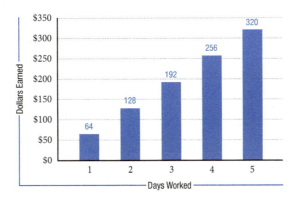

25. What is Jane's daily rate of pay?

26. What is her weekly rate of pay? (Assume she works 5 days per week.)

27. What is her annual rate of pay? (Assume she works 50 weeks per year.)

28. What is her hourly rate of pay? (Assume she works 8 hours per day.)

Getting Ready for the Next Section

Solve each equation by finding a number to replace n with that will make the equation a true statement.

29. $2 \cdot n = 12$

30. $3 \cdot n = 27$

31. $6 \cdot n = 24$

32. $8 \cdot n = 16$

33. $20 = 5 \cdot n$

34. $35 = 7 \cdot n$

35. $650 = 10 \cdot n$

36. $630 = 7 \cdot n$

Improving Your Quantitative Literacy

37. Unit Pricing An article in *USA Today* reported that the makers of Wisk liquid detergent cut the size of its popular midsize jug from 100 ounces (3.125 quarts) to 80 ounces (2.5 quarts). At the same time it lowered the price from $6.99 to $5.75. Fill in the table below and use your results to decide which of the two sizes is the better buy.

Wisk laundry detergent

	Old	New
Size	100 ounces	80 ounces
Container cost	$6.99	$5.75
Price per quart		

Find the Mistake

Each sentence below contains a mistake. Circle the mistake and write the correct word(s) or numbers(s) on the line provided.

1. The rate in miles per hour for a plane traveling 3000 miles in 6 hours is 50 miles per hour. _____

2. If a runner can run 16 miles in 2 hours, then her ratio of miles to hours is $\frac{1}{8}$ miles per hour. _____

3. If a supermarket sells a package of 20 cookies for $4.27, then the unit price for each cookie is $4.68. _____

4. A supermarket sells 10 packages of oatmeal for $5.33. A wholesale store sells the same oatmeal for $8.86 for 24 packages of oatmeal. Given the information, we find that the supermarket has the lowest unit price. _____

5.3 Solving Equations by Division

A new robot named EMILY (Emergency Integrated Lifesaving Lanyard) patrols dangerous ocean waters off the coast of Malibu, California. Currently, a lifeguard controls EMILY by remote. However, developers will soon equip the robot with sonar and an autonomous system able to detect swimmers in distress without the help of a lifeguard on shore. EMILY's electric-powered impeller drives the robot through even the roughest surf at 28 miles per hour, a speed 6 times faster than that of a human. A swimmer in distress can use EMILY as a flotation device until further help arrives, or EMILY can tow the swimmer back to shore.

Using the information about EMILY's speed, we can write the following equation:

$$28 = 6x$$

where x is a human's swimming speed. In this section, we will learn how to solve equations by division, which will help us solve the above equation for x and determine a human's swimming speed.

Previously, we solved equations like $3 \cdot n = 12$ by finding a number with which to replace n that would make the equation a true statement. The solution for the equation $3 \cdot n = 12$ is $n = 4$, because

$$\text{when} \rightarrow \quad n = 4$$
$$\text{the equation} \rightarrow \quad 3 \cdot n = 12$$
$$\text{becomes} \rightarrow \quad 3 \cdot 4 = 12$$
$$12 = 12 \quad \text{A true statement}$$

The problem with this method of solving equations is that we have to guess at the solution and then check it in the equation to see if it works. In this section, we will develop a method of solving equations like $3 \cdot n = 12$ that does not require any guessing.

A Solving Equations by Division

Earlier, we simplified expressions such as

$$\frac{2 \cdot 2 \cdot 3 \cdot 5 \cdot 7}{2 \cdot 5}$$

by dividing out any factors common to the numerator and the denominator.

$$\frac{2 \cdot 2 \cdot 3 \cdot 5 \cdot 7}{2 \cdot 5} = 2 \cdot 3 \cdot 7 = 42$$

Video Examples
Section 5.3

Image © Katherine Heistand Shields, 2010

The same process works with expressions that have variables for some of their factors. For example, the expression

$$\frac{2 \cdot n \cdot 7 \cdot 11}{n \cdot 11}$$

can be simplified by dividing out the factors common to the numerator and the denominator—namely, n and 11:

$$\frac{2 \cdot n \cdot 7 \cdot 11}{n \cdot 11} = 2 \cdot 7 = 14$$

Practice Problems

1. Divide the expression $8 \cdot n$ by 8.

EXAMPLE 1   Divide the expression $5 \cdot n$ by 5.

Solution   Applying the method above, we have

$$5 \cdot n \text{ divided by 5 is } \frac{5 \cdot n}{5} = n$$

If you are having trouble understanding this process because there is a variable involved, consider what happens when we divide 6 by 2 and when we divide 6 by 3. Because $6 = 2 \cdot 3$, when we divide by 2 we get 3.

$$\frac{6}{2} = \frac{2 \cdot 3}{2} = 3$$

When we divide by 3, we get 2.

$$\frac{6}{3} = \frac{2 \cdot 3}{3} = 2$$

2. Divide $5 \cdot w$ by 5.

EXAMPLE 2   Divide $7 \cdot y$ by 7.

Solution   Dividing by 7, we have

$$7 \cdot y \text{ divided by 7 is } \frac{7 \cdot y}{7} = y$$

We can use division to solve equations such as $3 \cdot n = 12$. Notice that the left side of the equation is $3 \cdot n$. The equation is solved when we have just n, instead of $3 \cdot n$, on the left side and a number on the right side. That is, we have solved the equation when we have rewritten it as

$$n = \text{a number}$$

We can accomplish this by dividing *both* sides of the equation by 3.

$$\frac{3 \cdot n}{3} = \frac{12}{3} \quad \text{Divide both sides by 3.}$$

$$n = 4$$

Because 12 divided by 3 is 4, the solution to the equation is $n = 4$, which we know to be correct from our discussion at the beginning of this section. Notice that it would be incorrect to divide just the left side by 3 and not the right side also. It is important to remember that whenever we divide one side of an equation by a number, we must also divide the other side by the same number.

Note   The choice of the letter we use for the variable is not important. The process works just as well with y as it does with n. The letters used for variables in equations are most often the letters $a, n, x, y,$ or z.

Note   In a later chapter of this book, we will devote a lot of time to solving equations. For now, we are concerned only with equations that can be solved by division.

3. Solve the equation $5 \cdot a = 45$ for a by dividing both sides by 5.

EXAMPLE 3   Solve the equation $7 \cdot y = 42$ for y by dividing both sides by 7.

Solution   Dividing both sides by 7, we have

$$\frac{7 \cdot y}{7} = \frac{42}{7}$$

$$y = 6$$

We can check our solution by replacing y with 6 in the original equation.

$$\text{When} \rightarrow \qquad y = 6$$
$$\text{the equation} \rightarrow \quad 7 \cdot y = 42$$
$$\text{becomes} \rightarrow \qquad 7 \cdot 6 = 42$$
$$42 = 42 \qquad \text{A true statement}$$

Answers

1. n

2. w

3. $a = 9$

EXAMPLE 4   Solve $30 = 5 \cdot a$ for a.

Solution   Our method of solving equations by division works regardless of which side the variable is on. In this case, the right side is $5 \cdot a$, and we would like it to be just a. Dividing both sides by 5, we have

$$\frac{30}{5} = \frac{5 \cdot a}{5}$$

$$6 = a$$

The solution is $a = 6$. (If 6 is a, then a is 6.)

We can write our solutions as improper fractions, mixed numbers, or decimals. Let's agree to write our answers as either whole numbers, proper fractions, or mixed numbers unless otherwise stated.

> **GETTING READY FOR CLASS**
>
> *After reading through the preceding section, respond in your own words and in complete sentences.*
>
> **A.** In your own words, explain what a solution to an equation is.
>
> **B.** What number results when you simplify $\frac{2 \cdot n \cdot 7 \cdot 11}{n \cdot 11}$?
>
> **C.** What is the result of dividing $7 \cdot y$ by 7?
>
> **D.** Explain how division is used to solve the equation $30 = 5 \cdot a$.

4. Solve for b: $32 = 4 \cdot b$.

Answer
4. $b = 8$

EXERCISE SET 5.3

Vocabulary Review

Choose the correct words to fill in the blanks below.

variable   equals sign   equation   division

1. To solve an _____, find a number with which to replace the _____ that would make the equation a true statement.

2. Solving equations by _____ eliminates the need to guess the value that the variable represents.

3. We have solved an equation when the answer has a variable isolated on one side of the _____ and a number on the other side.

Problems

A Simplify each of the following expressions by dividing out any factors common to the numerator and the denominator and then simplifying the result.

1. $\dfrac{3 \cdot 5 \cdot 5 \cdot 7}{3 \cdot 5}$

2. $\dfrac{2 \cdot 2 \cdot 3 \cdot 5 \cdot 7}{2 \cdot 5 \cdot 7}$

3. $\dfrac{2 \cdot n \cdot 3 \cdot 3 \cdot 5}{n \cdot 5}$

4. $\dfrac{3 \cdot 5 \cdot n \cdot 7 \cdot 7}{3 \cdot n \cdot 7}$

5. $\dfrac{2 \cdot 2 \cdot n \cdot 7 \cdot 11}{2 \cdot n \cdot 11}$

6. $\dfrac{3 \cdot n \cdot 7 \cdot 13 \cdot 17}{n \cdot 13 \cdot 17}$

7. $\dfrac{9 \cdot n}{9}$

8. $\dfrac{8 \cdot a}{8}$

9. $\dfrac{4 \cdot y}{4}$

10. $\dfrac{7 \cdot x}{7}$

11. $\dfrac{12 \cdot b}{12}$

12. $\dfrac{9 \cdot w}{9}$

Solve each of the following equations by dividing both sides by the appropriate number. Be sure to show the division in each case.

13. $4 \cdot n = 8$

14. $2 \cdot n = 8$

15. $5 \cdot x = 35$

16. $7 \cdot x = 35$

17. $3 \cdot y = 21$

18. $7 \cdot y = 21$

19. $6 \cdot n = 48$

20. $16 \cdot n = 48$

21. $5 \cdot a = 40$

22. $10 \cdot a = 40$

23. $3 \cdot x = 6$

24. $8 \cdot x = 40$

25. $2 \cdot y = 2$

26. $2 \cdot y = 12$

27. $3 \cdot a = 18$

28. $4 \cdot a = 4$

29. $5 \cdot n = 25$

30. $9 \cdot n = 18$

31. $6 = 2 \cdot x$

32. $56 = 7 \cdot x$

33. $42 = 6 \cdot n$

34. $30 = 5 \cdot n$

35. $4 = 4 \cdot y$

36. $90 = 9 \cdot y$

37. $63 = 7 \cdot y$

38. $3 = 3 \cdot y$

39. $2 \cdot n = 7$

40. $4 \cdot n = 10$

41. $6 \cdot x = 21$

42. $7 \cdot x = 8$

43. $5 \cdot a = 12$

44. $8 \cdot a = 13$

45. $4 = 7 \cdot y$

46. $3 = 9 \cdot y$

47. $10 = 13 \cdot y$

48. $9 = 11 \cdot y$

49. $12 \cdot x = 30$

50. $16 \cdot x = 56$

51. $21 = 14 \cdot n$

52. $48 = 20 \cdot n$

Getting Ready for the Next Section

Reduce.

53. $\dfrac{6}{8}$

54. $\dfrac{17}{34}$

55. $\dfrac{16}{24}$

56. $\dfrac{5}{20}$

Multiply.

57. $3(0.4)$

58. $2(0.15)$

59. $\dfrac{2}{3} \cdot 6$

60. $\dfrac{3}{4} \cdot 16$

Divide.

61. $65 \div 10$

62. $30 \div 4$

63. $1.5 \div 3$

64. $1.2 \div 8$

Find the Mistake

Each sentence below contains a mistake. Circle the mistake and write the correct word(s) or numbers(s) on the line provided.

1. To simplify $\dfrac{3 \cdot a \cdot 8 \cdot 11}{a \cdot 11}$, divide out a and 11 to get $264a$. _____

2. Dividing $6 \cdot z$ by 6 gives us 1. _____

3. Solving the equation $6 \cdot a = 48$ for a gives us $a = 48$. _____

4. Solving the equation $36 = w \cdot 12$ for w gives us $w = \dfrac{12}{36} = \dfrac{1}{3}$. _____

Landmark Review: Checking Your Progress

Write each of the following ratios as a fraction in lowest terms.

1. 5 to 6   **2.** 10 to 1   **3.** 0.6 to 6   **4.** 0.25 to 0.75

Solve each of the following word problems.

5. A car travels 210 miles in 3 hours. What is the rate of the car in miles per hour?

6. A 5-gallon bucket can be filled from a faucet in 40 seconds. What is the rate in gallons per second?

7. A car travels 80 miles on 5 gallons of gas. Give the rate in miles per gallon.

8. An 8-ounce can of corn costs $1.16. Give the unit price in cents per ounce?

Solve each of the following equations by division.

9. $5 \cdot x = 25$ **10.** $2 \cdot n = 8$ **11.** $4 \cdot y = 16$ **12.** $3 \cdot a = 9$

13. $7 \cdot b = 14$ **14.** $8 \cdot x = 24$ **15.** $100 \cdot y = 10$ **16.** $16 \cdot x = 4$

5.4 Proportions

A Solve proportions.

In June 2010, the Children's Museum of Indianapolis had the world's largest Etch-a-Sketch on display. The large toy measured 8 feet tall, compared to the original toy's measurements of 7 inches wide by 9 inches tall. When the screen needed to be cleared, instead of shaking it like the original, the display was designed to flip upside down and then return upright again.

We can compare the measurements of the museum's Etch-a-Sketch with the original using the concept of proportions. Let's allow x to equal the width of the museum's Etch-a-Sketch. Then we can put the measurements of both toys into two ratios and set them equal to each other to solve for x.

$$\frac{x}{8} = \frac{7}{9}$$

After reading this section, you will be able to find the value for x. First, let's explore the definition of proportion.

KEY WORDS

proportion

extremes

means

A Solving Proportions Using the Fundamental Property

In this section, we will solve problems using proportions. As you will see later in this chapter, proportions can model a number of everyday applications.

> **DEFINITION** proportion
>
> A statement that two ratios are equal is called a **proportion**. If $\frac{a}{b}$ and $\frac{c}{d}$ are two equal ratios, then the statement
>
> $$\frac{a}{b} = \frac{c}{d}$$
>
> is called a proportion.

Each of the four numbers in a proportion is called a *term* of the proportion. We number the terms of a proportion as follows:

$$\text{First term} \longrightarrow \frac{a}{b} = \frac{c}{d} \longleftarrow \text{Third term}$$
$$\text{Second term} \longrightarrow \qquad\qquad \longleftarrow \text{Fourth term}$$

The first and fourth terms of a proportion are called the *extremes*, and the second and third terms of a proportion are called the *means*.

$$\text{Means} \longrightarrow \frac{a}{b} = \frac{c}{d} \longleftarrow \text{Extremes}$$

Video Examples
Section 5.4

Practice Problems

1. In the proportion $\frac{5}{8} = \frac{15}{24}$, name the four terms, the means, and the extremes.

EXAMPLE 1   In the proportion $\frac{3}{4} = \frac{6}{8}$, name the four terms, the means, and the extremes.

Solution   The terms are numbered as follows:

First term = 3	Third term = 6
Second term = 4	Fourth term = 8

The means are 4 and 6; the extremes are 3 and 8.

The only additional thing we need to know about proportions is the following property.

> **PROPERTY**   Fundamental Property of Proportions
>
> In any proportion, the product of the extremes is equal to the product of the means. In symbols, it looks like this:
>
> $$\text{If } \frac{a}{b} = \frac{c}{d} \quad \text{then} \quad ad = bc \quad \text{for } b \neq 0 \text{ and } d \neq 0.$$

2. Verify the fundamental property of proportions for the following proportions.
   a. $\frac{9}{10} = \frac{36}{40}$

   b. $\frac{2}{3} = \frac{24}{36}$

EXAMPLE 2   Verify the fundamental property of proportions for the following proportions.

   a. $\dfrac{3}{4} = \dfrac{6}{8}$ b. $\dfrac{17}{34} = \dfrac{1}{2}$

Solution   We verify the fundamental property by finding the product of the means and the product of the extremes in each case.

Proportion	Product of the Means	Product of the Extremes
a. $\frac{3}{4} = \frac{6}{8}$	$4 \cdot 6 = 24$	$3 \cdot 8 = 24$
b. $\frac{17}{34} = \frac{1}{2}$	$34 \cdot 1 = 34$	$17 \cdot 2 = 34$

For each proportion the product of the means is equal to the product of the extremes.

We can use the fundamental property of proportions, along with a property we encountered in the last section, to solve an equation that has the form of a proportion.

A Note on Multiplication   Previously, we have used a multiplication dot to indicate multiplication, both with whole numbers and with variables. A more compact form for multiplication involving variables is simply to leave out the dot.
That is, $5 \cdot y = 5y$ and $10 \cdot x \cdot y = 10xy$.

3. Solve $\frac{3}{10} = \frac{12}{x}$ for x.

> *Note*   In some of these problems you will be able to see what the solution is just by looking the problem over. In those cases it is still best to show all the work involved in solving the proportion. It is good practice for the more difficult problems.

EXAMPLE 3   Solve for x.

$$\frac{2}{3} = \frac{4}{x}$$

Solution   Applying the fundamental property of proportions, we have

If $\dfrac{2}{3} = \dfrac{4}{x}$

then $2 \cdot x = 3 \cdot 4$ *The product of the extremes equals the product of the means.*

$2x = 12$ *Multiply.*

Answers

1. First term = 5, second term = 8, third term = 15, fourth term = 24; means: 8 and 15; extremes: 5 and 24.
2. a. $9 \cdot 40 = 360$; $10 \cdot 36 = 360$
   b. $2 \cdot 36 = 72$; $3 \cdot 24 = 72$
3. $x = 40$

The result is an equation. We know that we can divide both sides of an equation by the same nonzero number to find the solution to the equation. In this case, we divide both sides by 2 to solve for x.

$$2x = 12$$

$$\frac{2x}{2} = \frac{12}{2} \qquad \text{Divide both sides by 2.}$$

$$x = 6 \qquad \text{Simplify each side.}$$

The solution is 6. We can check our work by using the fundamental property of proportions.

$$\frac{2}{3} \bowtie \frac{4}{6}$$

$$\underset{\substack{\text{Product of} \\ \text{the means}}}{\underline{12}} = \underset{\substack{\text{Product of} \\ \text{the extremes}}}{\underline{12}}$$

Because the product of the means and the product of the extremes are equal, our work is correct.

EXAMPLE 4 Solve $\frac{5}{y} = \frac{10}{13}$ for y.

Solution We apply the fundamental property and solve as we did in Example 3.

$$\text{If} \qquad \frac{5}{y} = \frac{10}{13}$$

$$\text{then} \qquad 5 \cdot 13 = y \cdot 10 \qquad \text{The product of the extremes equals the product of the means.}$$

$$65 = 10y \qquad \text{Multiply } 5 \cdot 13.$$

$$\frac{65}{10} = \frac{10y}{10} \qquad \text{Divide both sides by 10.}$$

$$6.5 = y \qquad 65 \div 10 = 6.5$$

The solution is 6.5. We could check our result by substituting 6.5 for y in the original proportion and then finding the product of the means and the product of the extremes.

EXAMPLE 5 Find n if $\frac{n}{3} = \frac{0.4}{8}$.

Solution We proceed as we did in the previous two examples.

$$\text{If} \qquad \frac{n}{3} = \frac{0.4}{8}$$

$$\text{then} \qquad n \cdot 8 = 3(0.4) \qquad \text{The product of the extremes equals the product of the means.}$$

$$8n = 1.2 \qquad 3(0.4) = 1.2$$

$$\frac{8n}{8} = \frac{1.2}{8} \qquad \text{Divide both sides by 8.}$$

$$n = 0.15 \qquad 1.2 \div 8 = 0.15$$

The missing term is 0.15.

4. Solve $\frac{7}{y} = \frac{14}{30}$ for y.

5. Find n if $\frac{n}{8} = \frac{0.5}{5}$.

Answers

4. $y = 15$

5. $n = 0.8$

6. Solve $\dfrac{\frac{3}{4}}{6} = \dfrac{x}{10}$ for x.

EXAMPLE 6 Solve $\dfrac{\frac{2}{3}}{5} = \dfrac{x}{6}$ for x.

Solution We begin by multiplying the means and multiplying the extremes.

$$\text{If} \qquad \dfrac{\frac{2}{3}}{5} = \dfrac{x}{6}$$

$$\text{then} \qquad \dfrac{2}{3} \cdot 6 = 5 \cdot x \qquad \textit{The product of the extremes equals the}$$
$$\textit{product of the means.}$$

$$4 = 5 \cdot x \qquad \tfrac{2}{3} \cdot 6 = 4$$

$$\dfrac{4}{5} = \dfrac{5 \cdot x}{5} \qquad \textit{Divide both sides by 5.}$$

$$\dfrac{4}{5} = x$$

The missing term is $\frac{4}{5}$, or 0.8.

7. Solve $\dfrac{a}{20} = 4$.

EXAMPLE 7 Solve $\dfrac{b}{15} = 2$.

Solution Since the number 2 can be written as the ratio of 2 to 1, we can write this equation as a proportion, and then solve as we have in the examples above.

$$\dfrac{b}{15} = 2$$

$$\dfrac{b}{15} = \dfrac{2}{1} \qquad \textit{Write 2 as a ratio.}$$

$$b \cdot 1 = 15 \cdot 2 \qquad \textit{Product of the extremes equals the product of the means.}$$

$$b = 30$$

The procedure for finding a missing term in a proportion is always the same. We first apply the fundamental property of proportions to find the product of the extremes and the product of the means. Then we solve the resulting equation.

> **GETTING READY FOR CLASS**
>
> *After reading through the preceding section, respond in your own words and in complete sentences.*
>
> **A.** In your own words, give a definition of a proportion.
>
> **B.** In the proportion $\frac{2}{5} = \frac{4}{x}$, name the means and the extremes.
>
> **C.** State the fundamental property of proportions in words and in symbols.
>
> **D.** For the proportion $\frac{2}{5} = \frac{4}{x}$, find the product of the means and the product of the extremes.

Answers

6. $x = \frac{5}{4}$ or 1.25

7. $a = 80$

Vocabulary Review

Choose the correct words to fill in the blanks below.

| product | proportion | term | fundamental property of proportions |
| first | second | third | fourth |

1. If $\frac{a}{b}$ and $\frac{c}{d}$ are two equal ratios, then the statement $\frac{a}{b} = \frac{c}{d}$ is called a _____.

2. Each of the four numbers in a proportion is call a _____.

3. The _____ and _____ terms of a proportion are called the extremes.

4. The _____ and _____ terms of a proportion are called the means.

5. The fundamental property of proportions states that the _____ of the extremes is equal to the product of the means.

6. To find a missing term in a proportion, first apply the _____ and then solve the resulting equation.

Problems

A For each of the following proportions, name the means, name the extremes, and show that the product of the means is equal to the product of the extremes.

1. $\frac{1}{3} = \frac{5}{15}$

2. $\frac{6}{12} = \frac{1}{2}$

3. $\frac{10}{25} = \frac{2}{5}$

4. $\frac{5}{8} = \frac{10}{16}$

5. $\dfrac{\frac{1}{3}}{\frac{1}{2}} = \frac{4}{6}$

6. $\dfrac{2}{\frac{1}{4}} = \dfrac{4}{\frac{1}{2}}$

7. $\frac{0.5}{5} = \frac{1}{10}$

8. $\frac{0.3}{1.2} = \frac{1}{4}$

Find the missing term in each of the following proportions. Set up each problem like the examples in this section. For problems 30–36, write your answers in decimal form. For the other problems, write your answers as fractions in lowest terms.

9. $\dfrac{3}{5} = \dfrac{6}{x}$

10. $\dfrac{3}{8} = \dfrac{9}{x}$

11. $\dfrac{1}{y} = \dfrac{5}{12}$

12. $\dfrac{2}{y} = \dfrac{6}{10}$

13. $\dfrac{x}{4} = \dfrac{3}{8}$

14. $\dfrac{x}{5} = \dfrac{7}{10}$

15. $\dfrac{5}{9} = \dfrac{x}{2}$

16. $\dfrac{3}{7} = \dfrac{x}{3}$

17. $\dfrac{3}{7} = \dfrac{3}{x}$

18. $\dfrac{2}{9} = \dfrac{2}{x}$

19. $\dfrac{x}{2} = 7$

20. $\dfrac{x}{3} = 10$

21. $\dfrac{\frac{1}{2}}{y} = \dfrac{\frac{1}{3}}{12}$

22. $\dfrac{\frac{2}{3}}{y} = \dfrac{\frac{1}{3}}{5}$

23. $\dfrac{n}{12} = \dfrac{\frac{1}{4}}{\frac{1}{2}}$

24. $\dfrac{n}{10} = \dfrac{\frac{3}{5}}{\frac{3}{8}}$

25. $\dfrac{10}{20} = \dfrac{20}{n}$

26. $\dfrac{8}{4} = \dfrac{4}{n}$

27. $\dfrac{x}{10} = \dfrac{10}{2}$

28. $\dfrac{x}{12} = \dfrac{12}{48}$

29. $\dfrac{y}{12} = 9$

30. $\dfrac{y}{16} = 0.75$

31. $\dfrac{0.4}{1.2} = \dfrac{1}{x}$

32. $\dfrac{5}{0.5} = \dfrac{20}{x}$

33. $\dfrac{0.3}{0.18} = \dfrac{n}{0.6}$

34. $\dfrac{0.01}{0.1} = \dfrac{n}{10}$

35. $\dfrac{0.5}{x} = \dfrac{1.4}{0.7}$

36. $\dfrac{0.3}{x} = \dfrac{2.4}{0.8}$

37. $\dfrac{168}{324} = \dfrac{56}{x}$

38. $\dfrac{280}{530} = \dfrac{112}{x}$

39. $\dfrac{429}{y} = \dfrac{858}{130}$

40. $\dfrac{573}{y} = \dfrac{2,292}{316}$

41. $\dfrac{n}{39} = \dfrac{533}{507}$

42. $\dfrac{n}{47} = \dfrac{1,003}{799}$

43. $\dfrac{756}{903} = \dfrac{x}{129}$

44. $\dfrac{321}{1,128} = \dfrac{x}{376}$

Getting Ready for the Next Section

Divide.

45. $360 \div 18$

46. $2{,}700 \div 6$

47. $3{,}300 \div 11$

48. $1{,}440 \div 24$

Multiply.

49. $3.5(85)$

50. $4.75(105)$

51. $4.2(12)$

52. $1.25(34)$

Solve each equation.

53. $\dfrac{x}{10} = \dfrac{270}{6}$

54. $\dfrac{x}{45} = \dfrac{8}{18}$

55. $\dfrac{x}{25} = \dfrac{4}{20}$

56. $\dfrac{x}{3.5} = \dfrac{85}{1}$

Find the Mistake

Each sentence below contains a mistake. Circle the mistake and write the correct word(s) or numbers(s) on the line provided.

1. A statement that two proportions are equal is called a ratio. _____

2. For the proportion $\dfrac{5}{6} = \dfrac{10}{x}$, the means are 5 and x. _____

3. To solve $\dfrac{7}{10} = \dfrac{n}{0.2}$, set the product of the first and third terms equal to the product of second and fourth terms.

_____

4. Solving the proportion $\dfrac{8}{5} = \dfrac{n}{\frac{3}{10}}$ gives us $n = \dfrac{3}{16}$. _____

A Solve application problems using proportions.

5.5 Applications of Proportions

Image © sxc.hu, bugdog, 2008

Model railroads continue to be as popular today as they ever have been. One of the first things model railroaders ask each other is what scale they work with. The scale of a model train indicates its size relative to a full-size train. Each scale is associated with a ratio and a fraction, as shown in the table and bar chart below. An HO scale model train has a ratio of 1 to 87, meaning it is $\frac{1}{87}$ as large as an actual train.

Scale	Ratio	As a Fraction
LGB	1 to 22.5	$\frac{1}{22.5}$
#1	1 to 32	$\frac{1}{32}$
O	1 to 43.5	$\frac{1}{43.5}$
S	1 to 64	$\frac{1}{64}$
HO	1 to 87	$\frac{1}{87}$
TT	1 to 120	$\frac{1}{120}$

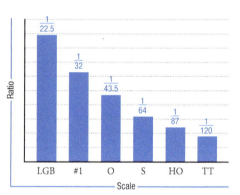

How long is an actual boxcar that has an HO scale model 5 inches long? In this section we will solve this problem using proportions.

A Solving Application Problems Using Proportions

Proportions can be used to solve a variety of word problems. The examples that follow show some of these word problems. In each case we will translate the word problem into a proportion and then solve the proportion using the methods developed in this chapter.

EXAMPLE 1   Recall the problem from the chapter opening. Suppose you drive your car 270 miles in 6 hours using the Pasadena Freeway. If you continue at the same rate, how far will you travel in 10 hours?

Solution   We let x represent the distance you travel in 10 hours. Using x, we translate the problem into the following proportion:

$$\text{Miles} \longrightarrow \frac{x}{10} = \frac{270}{6} \longleftarrow \text{Miles}$$
$$\text{Hours} \longrightarrow \qquad\qquad \longleftarrow \text{Hours}$$

Video Examples
Section 5.5

Practice Problems

1. If you travel 310 miles in 5 hours, and continue at the same rate, how far will you travel in 12 hours?

Answer

1. 744 miles

Notice that the two ratios in the proportion compare the same quantities. That is, both ratios compare miles to hours. In words, this proportion says

x miles is to 10 *hours as* 270 *miles is to* 6 *hours*
$$\downarrow \qquad\qquad \downarrow \qquad\qquad\qquad \downarrow$$
$$\frac{x}{10} \qquad = \qquad \frac{270}{6}$$

Next, we solve the proportion.

$$x \cdot 6 = 10 \cdot 270$$

$$x \cdot 6 = 2{,}700 \qquad\quad 10 \cdot 270 = 2{,}700$$

$$\frac{x \cdot 6}{6} = \frac{2{,}700}{6} \qquad\quad \textit{Divide both sides by 6.}$$

$$x = 450 \text{ miles} \qquad 2{,}700 \div 6 = 450$$

If you continue at the same rate, you will travel 450 miles in 10 hours.

EXAMPLE 2   A baseball player gets 8 hits in the first 18 games of the season. If he continues at the same rate, how many hits will he get in 45 games?

Solution   We let x represent the number of hits he will get in 45 games. Then

x is to 45 *as* 8 *is to* 18.
$$\searrow\quad\downarrow\quad\swarrow$$
$$\text{Hits} \longrightarrow \frac{x}{45} = \frac{8}{18} \longleftarrow \text{Hits}$$
$$\text{Games} \longrightarrow \qquad\qquad \longleftarrow \text{Games}$$

Notice again that the two ratios are comparing the same quantities, hits to games. We solve the proportion as follows:

$$18x = 360 \qquad\quad 45 \cdot 8 = 360$$

$$\frac{18x}{18} = \frac{360}{18} \qquad\quad \textit{Divide both sides by 18.}$$

$$x = 20 \qquad\quad 360 \div 18 = 20$$

If he continues to hit at the rate of 8 hits in 18 games, he will get 20 hits in 45 games.

EXAMPLE 3   A solution contains 4 milliliters of alcohol and 20 milliliters of water. If another solution is to have the same ratio of milliliters of alcohol to milliliters of water and must contain 25 milliliters of water, how much alcohol should it contain?

Solution   We let x represent the number of milliliters of alcohol in the second solution. The problem translates to

x milliliters is to 25 *milliliters as* 4 *milliliters is to* 20 *milliliters.*

$$\text{Alcohol} \longrightarrow \frac{x}{25} = \frac{4}{20} \longleftarrow \text{Alcohol}$$
$$\text{Water} \longrightarrow \qquad\qquad \longleftarrow \text{Water}$$

$$20x = 100 \qquad\qquad 25 \cdot 4 = 100$$

$$\frac{20x}{20} = \frac{100}{20} \qquad\qquad \textit{Divide both sides by 20.}$$

$$x = 5 \text{ milliliters of alcohol} \quad 100 \div 20 = 5$$

2. A baseball pitcher gives up 6 earned runs in 18 innings. If he continues at this rate, how many earned runs will he give up in 81 innings?

3. A solution contains 3.5 g salt in 40 ml of water. If another solution is to have the same ratio of salt to water and it must contain 220 ml of water, how much salt should it contain?

Answers
2. 27 earned runs
3. 19.25 g

4. The scale on a map indicates that $\frac{1}{2}$ inch on the map corresponds to an actual distance of 70 miles. Two cities are $3\frac{1}{4}$ inches apart on the map. What is the actual distance between the two cities?

EXAMPLE 4   The scale on a map indicates that 1 inch on the map corresponds to an actual distance of 85 miles. Two cities are 3.5 inches apart on the map. What is the actual distance between the two cities?

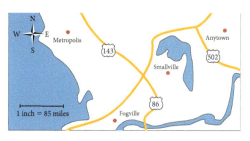

Solution We let x represent the actual distance between the two cities. The proportion is

$$\text{Miles} \longrightarrow \frac{x}{3.5} = \frac{85}{1} \longleftarrow \text{Miles}$$
$$\text{Inches} \longrightarrow \qquad\quad\;\; \longleftarrow \text{Inches}$$

$$x \cdot 1 = 3.5(85)$$

$$x = 297.5 \text{ miles}$$

5. Using the atomic weight information from Example 5, how many pounds of carbon dioxide are produced from burning 8 gallons of gasoline?

EXAMPLE 5   One gallon of gasoline weighs 6.3 pounds, of which 5.5 pounds is carbon. The carbon is combined with hydrogen in gasoline. When gasoline is burned, the carbon and hydrogen separate, and the carbon recombines with two molecules of oxygen from air to form carbon dioxide. The atomic weight of carbon is 12 and the atomic weight of each molecule of oxygen is 16. Show that burning 1 gallon of gasoline produces 20.2 pounds of carbon dioxide.

Solution   First we find the ratio of the weight of carbon to the weight of the whole molecule in carbon dioxide.

Atomic weight of carbon = 12

Atomic weight of carbon dioxide = 12 + 16 + 16 = 44

Ratio of weight of carbon to weight of carbon dioxide = $\dfrac{12}{44} = \dfrac{3}{11}$

Next, since the weight of carbon in one gallon of gasoline is 5.5 pounds, if we let $x =$ the weight of carbon dioxide produced by burning one gallon of gasoline, we have

$$\text{Weight of carbon} \longrightarrow \frac{3}{11} = \frac{5.5}{x} \longleftarrow \text{Weight of carbon}$$
$$\text{Weight of carbon dioxide} \longrightarrow \qquad\quad\;\; \longleftarrow \text{Weight of carbon dioxide}$$

$$3x = 11(5.5) \qquad \text{Fundamental property of proportions}$$

$$\frac{3x}{3} = \frac{11(5.5)}{3} \qquad \text{Divide both sides by 3.}$$

$$x = 20.2 \qquad\quad \text{Round to the nearest tenth.}$$

Each gallon of gasoline burned produces 20.2 pounds of carbon dioxide.

GETTING READY FOR CLASS

After reading through the preceding section, respond in your own words and in complete sentences.

A. Give an example not found in the book of a proportion problem you may encounter.

B. Write a word problem for the proportion $\frac{2}{5} = \frac{4}{x}$.

C. What does it mean to translate a word problem into a proportion?

D. Name some jobs that may frequently require solving proportion problems.

Answers

4. 455 miles

5. 161.3$\overline{3}$ pounds

Vocabulary Review

Choose the correct words to fill in the blanks below.

proportion   quantities   word problems   division

1. Proportions can be used to solve _____ .

2. When translating a word problem into a _____, make sure the two ratios in the proportion compare the same _____ .

3. Once you translate a word problem into a proportion, use _____ to solve for the unknown term.

Problems

A Solve each of the following word problems by translating the statement into a proportion. Be sure to show the proportion used in each case.

1. **Distance** A woman drives her car 235 miles in 5 hours. At this rate how far will she travel in 7 hours?

2. **Distance** An airplane flies 1,260 miles in 3 hours. How far will it fly in 5 hours?

3. **Basketball** A basketball player scores 162 points in 9 games. At this rate how many points will he score in 20 games?

4. **Football** In the first 4 games of the season, a football team scores a total of 68 points. At this rate how many points will the team score in 11 games?

5. **Mixture** A solution contains 8 pints of antifreeze and 5 pints of water. How many pints of water must be added to 24 pints of antifreeze to get a solution with the same concentration?

6. **Nutrition** If 10 ounces of a certain breakfast cereal contains 3 ounces of sugar, how many ounces of sugar does 25 ounces of the same cereal contain?

7. **Map Reading** The scale on a map indicates that 1 inch corresponds to an actual distance of 95 miles. Two cities are 4.5 inches apart on the map. What is the actual distance between the two cities?

8. **Map Reading** A map is drawn so that every 2.5 inches on the map corresponds to an actual distance of 100 miles. If the actual distance between two cities is 350 miles, how far apart are they on the map?

9. **Farming** A farmer knows that of every 50 eggs his chickens lay, only 45 will be marketable. If his chickens lay 1,000 eggs in a week, how many of them will be marketable?

10. **Manufacturing** Of every 17 parts manufactured by a certain machine, 1 will be defective. How many parts were manufactured by the machine if 8 defective parts were found?

Model Trains In the introduction to this section, we indicated that the size of a model train relative to an actual train is referred to as its scale. Each scale is associated with a ratio as shown in the table. For example, an HO model train has a ratio of 1 to 87, meaning it is $\frac{1}{87}$ as large as an actual train.

Scale	Ratio
LGB	1 to 22.5
#1	1 to 32
O	1 to 43.5
S	1 to 64
HO	1 to 87
TT	1 to 120

11. **Boxcar** How long is an actual boxcar that has an HO scale model 5 inches long? Give your answer in inches, then divide by 12 to give the answer in feet.

12. **Length of a Flatcar** How long is an actual flatcar that has an LGB scale model 24 inches long? Give your answer in feet.

13. **Travel Expenses** A traveling salesman figures it costs 21¢ for every mile he drives his car. How much does it cost him a week to drive his car if he travels 570 miles a week?

14. **Travel Expenses** A family plans to drive their car during their annual vacation. The car can go 350 miles on a tank of gas, which is 18 gallons of gas. The vacation they have planned will cover 1,785 miles. How many gallons of gas will that take?

15. **Nutrition** A 6-ounce serving of grapefruit juice contains 159 grams of water. How many grams of water are in 10 ounces of grapefruit juice?

16. **Nutrition** If 100 grams of ice cream contains 13 grams of fat, how much fat is in 250 grams of ice cream?

17. Travel Expenses If a car travels 378.9 miles on 50 liters of gas, how many liters of gas will it take to go 692 miles if the car travels at the same rate? (Round to the nearest tenth.)

18. Nutrition If 125 grams of peas contains 26 grams of carbohydrates, how many grams of carbohydrates does 375 grams of peas contain?

19. Elections During a recent election, 47 of every 100 registered voters in a certain city voted. If there were 127,900 registered voters in that city, how many people voted?

20. Map Reading The scale on a map is drawn so that 4.5 inches corresponds to an actual distance of 250 miles. If two cities are 7.25 inches apart on the map, how many miles apart are they? (Round to the nearest tenth.)

Getting Ready for the Next Section

Simplify.

21. $\dfrac{320}{160}$

22. $21 \cdot 105$

23. $2,205 \div 15$

24. $\dfrac{48}{24}$

Solve each equation.

25. $\dfrac{x}{5} = \dfrac{28}{7}$

26. $\dfrac{x}{4} = \dfrac{6}{3}$

27. $\dfrac{x}{21} = \dfrac{105}{15}$

28. $\dfrac{b}{15} = 2$

Find the Mistake

Each sentence below contains a mistake. Circle the mistake and write the correct word(s) or numbers(s) on the line provided.

1. A basketball player scores 112 points in 8 games. The proportion to find how many points the player will score in 14 games is $\dfrac{112}{8} = \dfrac{14}{x}$. _____

2. The scale of a map indicates that 2 inches corresponds to 250 miles in real life. If two cities on the map are 3.5 inches apart, they are 0.028 miles apart in real life. _____

3. A jellybean company knows that for every 100 jellybeans, 4 will be misshapen. The proportion needed to find how many jelly beans were made if 36 misshapen jelly beans are found is $\dfrac{4}{36} = \dfrac{x}{100}$. _____

4. If burning 1 gallon of gasoline produces 20.2 pounds of carbon dioxide, then burning 12 gallons of gasoline produces approximately 0.59 pounds of carbon dioxide. _____

OBJECTIVES

A Use similar figures to solve for missing sides.

B Draw similar figures.

C Solve applications involving similar figures.

KEY WORDS

similar figures

corresponding sides

proportional

Video Examples
Section 5.6

Note One way to label the important parts of a triangle is to label the vertices with capital letters and the sides with lowercase letters.

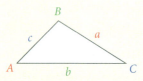

Notice that side *a* is opposite vertex *A*, side *b* is opposite vertex *B*, and side *c* is opposite vertex *C*. Also, because each vertex is the vertex of one of the angles of the triangle, we refer to the three interior angles as *A*, *B*, and *C*.

5.6 Similar Figures

In mathematics, when two or more objects have the same shape, but are different sizes, we say they are similar. If two figures are similar, then their corresponding sides are proportional.

In order to give more details on what we mean by corresponding sides of similar figures, it is helpful to label the parts of a triangle as shown in the margin.

A Similar Triangles

Two triangles that have the same shape are similar when their corresponding sides are proportional, or have the same ratio. The triangles below are similar.

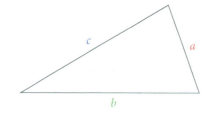
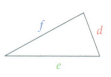

Corresponding Sides	Ratio
side *a* corresponds with side *d*	$\dfrac{a}{d}$
side *b* corresponds with side *e*	$\dfrac{b}{e}$
side *c* corresponds with side *f*	$\dfrac{c}{f}$

Because their corresponding sides are proportional, we write

$$\frac{a}{d} = \frac{b}{e} = \frac{c}{f}$$

EXAMPLE 1   The two triangles below are similar. Find side x.

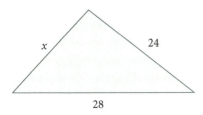

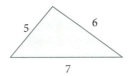

1. The two triangles below are similar. Find side x.

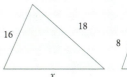

Solution   To find the length x, we set up a proportion of equal ratios. The ratio of x to 5 is equal to the ratio of 24 to 6 and to the ratio of 28 to 7. Algebraically, we have

$$\frac{x}{5} = \frac{24}{6} \quad \text{and} \quad \frac{x}{5} = \frac{28}{7}$$

We can solve either proportion to get our answer. The first gives us

$6x = 5 \cdot 24$ *Fundamental property of proportions*

$6x = 120$ $5 \cdot 24 = 120$

$\dfrac{6x}{6} = \dfrac{120}{6}$ *Divide both sides by 6.*

$x = 20$ $120 \div 6 = 20$

Other Similar Figures

When one shape or figure is either a reduced or enlarged copy of the same shape or figure, we consider them similar. For example, video viewed over the Internet was once confined to a small "postage stamp" size. Now it is common to see larger video over the Internet. Although the width and height has increased, the shape of the video has not changed.

EXAMPLE 2   Suppose the width and height of the two similar drawings are proportional. Find the width, w, of the larger drawing.

2. If the height of the larger drawing from Example 2 is expanded to 80 centimeters, what is the new width of the larger drawing?

16 cm.

32 cm.

12 cm.

w

Solution   We write our proportion as the ratio of the height of the larger drawing to the height of the smaller drawing is equal to the ratio of the width of the larger drawing to the width of the smaller drawing.

$$\frac{32}{16} = \frac{w}{12}$$

$16w = 12 \cdot 32$ *Fundamental property of proportions*

$16w = 384$ $12 \cdot 32 = 384$

$w = 24$ *Divide both sides by 16.*

The width of the larger drawing is 24 centimeters.

3. The base of a triangle is 4 squares of graph paper, and the height is 6 squares. If a similar triangle has a base of 6 squares, what is its height?

B Drawing Similar Figures

EXAMPLE 3   Draw a triangle similar to triangle ABC, if AC is proportional to DF. Make E the third vertex of the new triangle.

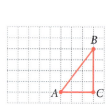
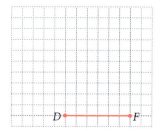

Solution   We see that AC is 3 units in length and BC has a length of 4 units. Since AC is proportional to DF, which has a length of 6 units, we set up a proportion to find the length EF.

$$\frac{EF}{BC} = \frac{DF}{AC}$$

$$\frac{EF}{4} = \frac{6}{3}$$

$3 \cdot EF = 24$ Fundamental property of proportions

$EF = 8$ Divide both sides by 3.

Now we can draw EF with a length of 8 units, then complete the triangle by drawing line DE.

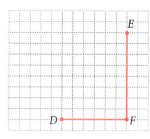
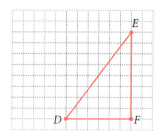

We have drawn triangle DEF similar to triangle ABC.

C Applications

4. If the building from Example 4 is 175 feet high, how long would the shadow that is cast be?

EXAMPLE 4   A building casts a shadow of 105 feet while a 21-foot flagpole casts a shadow that is 15 feet. Find the height of the building.

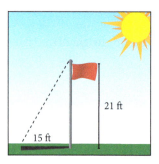
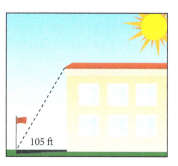

Answers

3. 9 squares

4. 125 feet

Solution   The figure shows both the building and the flagpole, along with their respective shadows. From the figure it is apparent that we have two similar triangles. Letting x = the height of the building, we have

$$\frac{x}{21} = \frac{105}{15}$$

$15x = 2,205$ *Fundamental property of proportions*

$x = 147$ *Divide both sides by 15.*

The height of the building is 147 feet.

EXAMPLE 5   Chessboards are square regardless of their size, which makes them proportional to each other. Suppose you have a chessboard with each side measuring 20 inches. Therefore, the perimeter of that board is 80 inches. Use the measurements of this board to design a life-sized chessboard for your local park. The park can accommodate a board with a side that is 240 inches long. Use a proportion to determine the perimeter P of the new board.

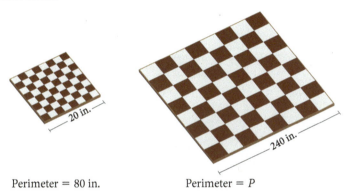

Perimeter = 80 in.   Perimeter = P

Solution   Since the length of a chessboard is proportional to its perimeter, we can set up the following proportion to calculate the perimeter of the life-sized board.

$$\frac{240}{P} = \frac{20}{80}$$

$20P = 19,200$ *Fundamental property of proportions*

$P = 960$ inches   *Divide both sides by 20.*

The perimeter of the new board will be 960 inches.

5. Suppose your chessboard has a side measuring 30 inches with a perimeter of 120 inches. The side of the life-sized board measures 150 inches. Use a proportion to determine the perimeter of the new board.

GETTING READY FOR CLASS

After reading through the preceding section, respond in your own words and in complete sentences.

A. What are similar figures?

B. How do we know if corresponding sides of two triangles are proportional?

C. When labeling a triangle *ABC*, how do we label the sides?

D. How are proportions used when working with similar figures?

Answer

5. 600 inches

Vocabulary Review

Choose the correct words to fill in the blanks below.

shape   size   vertices   similar

1. Similar figures are two or more objects with the same _____ but are of a different _____.

2. If two figures are _____, then their corresponding sides are proportional.

3. When labeling the _____ of a triangle *ABC*, we label the corresponding sides *abc*.

Problems

A In problems 1–4, for each pair of similar triangles, set up a proportion in order to find the unknown.

1.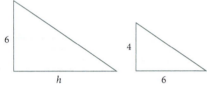

2.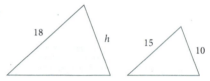

3.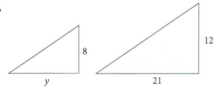

4.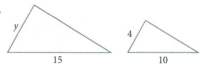

In problems 5–10, for each pair of similar figures, set up a proportion in order to find the unknown.

5.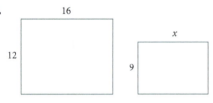

6.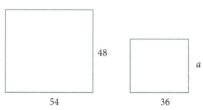

7.

8.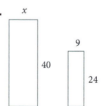

9.

10.

B For each problem, draw a figure on the grid on the right that is similar to the given figure.

11. *AC* is proportional to *DF*.

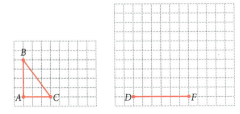

12. *AB* is proportional to *DE*.

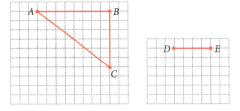

13. *BC* is proportional to *EF*.

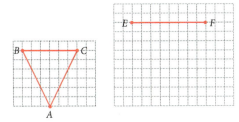

14. *AC* is proportional to *DF*.

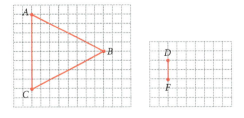

15. *DC* is proportional to *HG*.

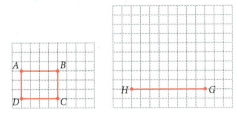

16. *AD* is proportional to *EH*.

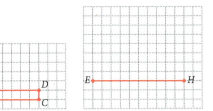

17. *AB* is proportional to *FG*.

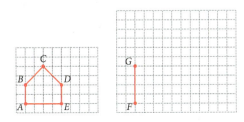

18. *BC* is proportional to *FG*.

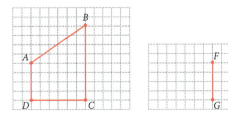

Applying the Concepts

Recall that the perimeters of two chessboards are proportional to the length of each of their sides.

19. **Size of a Chessboard** The perimeter of a chessboard is 50 inches and the length of each side is 12.5 inches. If a life-sized chessboard has a perimeter of 1,000 inches, use proportions to find the length of each side.

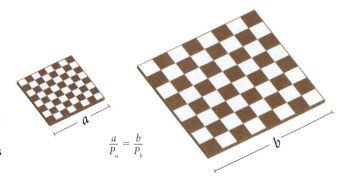

$$\frac{a}{P_a} = \frac{b}{P_b}$$

20. **Size of a Chessboard** The perimeter of a chessboard is 1,280 mm and the length of each side is 320 mm. If a life-sized chessboard has a perimeter of 25,400 mm, use proportions to find the length of each side.

21. **Video Resolution** A new graphics card can increase the resolution of a computer's monitor. Suppose a monitor has a horizontal resolution of 800 pixels and a vertical resolution of 600 pixels. By adding a new graphics card, the resolutions remain in the same proportions, but the horizontal resolution increases to 1,280 pixels. What is the new vertical resolution?

22. **Screen Resolution** The display of a 200 computer monitor is proportional to that of a 230 monitor. A 200 monitor has a horizontal resolution of 1,680 pixels and a vertical resolution of 1,050 pixels. If a 230 monitor has a horizontal resolution of 1,920 pixels, what is its vertical resolution?

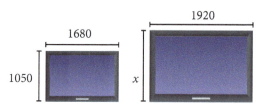

23. **Screen Resolution** The display of a 200 computer monitor is proportional to that of a 170 monitor. A 200 monitor has a horizontal resolution of 1,680 pixels and a vertical resolution of 1,050 pixels. If a 170 monitor has a vertical resolution of 900 pixels, what is its horizontal resolution?

24. **Video Resolution** A new graphics card can increase the resolution of a computer's monitor. Suppose a monitor has a horizontal resolution of 640 pixels and a vertical resolution of 480 pixels. By adding a new graphics card, the resolutions remain in the same proportions, but the vertical resolution increases to 786 pixels. What is the new horizontal resolution?

25. **Height of a Tree** A tree casts a shadow 38 feet long, while a 6-foot man casts a shadow 4 feet long. How tall is the tree?

26. **Height of a Building** A building casts a shadow 128 feet long, while a 24-foot flagpole casts a shadow 32 feet long. How tall is the building?

27. Height of a Child A water tower is 36 feet tall and casts a shadow 54 feet long, while a child casts a shadow 6 feet long. How tall is the child?

28. Height of a Truck A clock tower is 36 feet tall and casts a shadow 30 feet long, while a large truck next to the tower casts a shadow 15 feet long. How tall is the truck?

One Step Further

29. The rectangles shown here are similar.

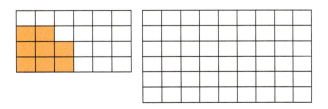

    a. In the smaller figure, what is the ratio of the shaded to nonshaded rectangles?

    b. Shade the larger rectangle so that the ratio of shaded to nonshaded rectangles is the same as in part a.

    c. For each of the figures, what is the ratio of the shaded rectangles to total rectangles?

Find the Mistake

Each sentence below contains a mistake. Circle the mistake and write the correct word(s) or numbers(s) on the line provided.

    1. The two triangles below are similar. The side x is equal to $\frac{4}{3}$. _____

 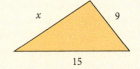

    2. The two triangles below are similar. We can find x by solving the proportion $\frac{12}{x} = \frac{4}{2}$. _____

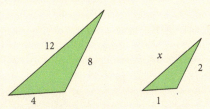

    3. The width of a rectangle on graph paper is 5 squares and the length is 7 squares. If a similar rectangle has a width of 10, then the length would be $\frac{7}{2}$. _____

    4. A pocket dictionary is similar to a regular dictionary. The pocket dictionary is 4 inches wide by 6 inches long. The width of the regular dictionary is 16 inches. You must solve the proportion $\frac{6}{4} = \frac{x}{16}$ to find the remaining side length of the regular dictionary. _____

Olympic Rates

In 2010, the Winter Olympics were held in Vancouver, British Columbia. At the conclusion of these games, Canada became the first host nation to have won the most gold medals since Norway in 1952. The following is a list of some of the sports in which athletes competed during the 2010 Olympic Games:

Alpine skiing

Bobsleigh

Freestyle skiing

Luge

Ski jumping

Biathlon

Cross-country skiing

Ice hockey

Speed skating

Snowboarding

Working in groups, choose a sport from the above list. Research the sport, as well as the 2010 race details and results. Explain how rates and proportions can be used to describe the details of each sport's race and results. Present your findings to the class.

Chapter 5 Summary

Ratio [5.1]

The ratio of a to b is $\frac{a}{b}$. The ratio of two numbers is a way of comparing them using fraction notation.

1. The ratio of 6 to 8 is $\frac{6}{8}$ which can be reduced to $\frac{3}{4}$.

Rates [5.2]

A ratio that compares two different quantities, like miles and hours, gallons and seconds, etc., is called a *rate*.

2. If a car travels 150 miles in 3 hours, then the ratio of miles to hours is considered a rate.

$$\frac{150 \text{ miles}}{3 \text{ hours}} = 50 \frac{\text{miles}}{\text{hour}}$$

$$= 50 \text{ miles per hour}$$

Unit Pricing [5.2]

The *unit price* of an item is the ratio of price to quantity when the quantity is one unit.

3. If a 10-ounce package of frozen peas costs 69¢, then the price per ounce, or unit price, is

$$\frac{69 \text{ cents}}{10 \text{ ounces}} = 6.9 \frac{\text{cents}}{\text{ounce}}$$

$$= 6.9 \text{ cents per ounce}$$

Solving Equations by Division [5.3]

Dividing both sides of an equation by the same number will not change the solution to the equation. For example, the equation $5 \cdot x = 40$ can be solved by dividing both sides by 5.

4. Solve: $5 \cdot x = 40$.

$$5 \cdot x = 40$$

$$\frac{5 \cdot x}{5} = \frac{40}{5} \qquad \text{Divide both sides by 5.}$$

$$x = 8 \qquad 40 \div 5 = 8$$

Proportion [5.4]

A proportion is an equation that indicates that two ratios are equal.
The numbers in a proportion are called *terms* and are numbered as follows:

$$\text{First term} \longrightarrow \frac{a}{b} = \frac{c}{d} \longleftarrow \text{Third term}$$
$$\text{Second term} \longrightarrow \phantom{\frac{a}{b} = } \longleftarrow \text{Fourth term}$$

The first and fourth terms are called the *extremes*. The second and third terms are called the *means*.

$$\text{Means} \longrightarrow \frac{a}{b} = \frac{c}{d} \longleftarrow \text{Extremes}$$

5. The following is a proportion:

$$\frac{6}{8} = \frac{3}{4}$$

Fundamental Property of Proportions [5.4]

In any proportion, the product of the extremes is equal to the product of the means. In symbols,

$$\text{If} \quad \frac{a}{b} = \frac{c}{d} \quad \text{then} \quad ad = bc$$

Finding an Unknown Term in a Proportion [5.4]

6. Find x: $\frac{2}{5} = \frac{8}{x}$.

$2 \cdot x = 5 \cdot 8$

$2 \cdot x = 40$

$\dfrac{2 \cdot x}{2} = \dfrac{40}{2}$

$x = 20$

To find the unknown term in a proportion, we apply the fundamental property of proportions and solve the equation that results by dividing both sides by the number that is multiplied by the unknown. For instance, if we want to find the unknown in the proportion

$$\frac{2}{5} = \frac{8}{x}$$

we use the fundamental property of proportions to set the product of the extremes equal to the product of the means.

Using Proportions to Find Unknown Length in Similar Figures [5.6]

7. Find x.

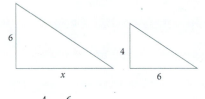

$\dfrac{4}{6} = \dfrac{6}{x}$

$36 = 4x$

$9 = x$

Two triangles that have the same shape are similar when their corresponding sides are proportional, or have the same ratio. The triangles below are similar.

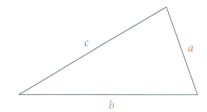 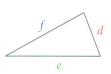

Corresponding Sides	**Ratio**
side a corresponds with side d	$\dfrac{a}{d}$
side b corresponds with side e	$\dfrac{b}{e}$
side c corresponds with side f	$\dfrac{c}{f}$

Because their corresponding sides are proportional, we write

$$\frac{a}{d} = \frac{b}{e} = \frac{c}{f}$$

Write each ratio as a fraction in lowest terms. [5.1]

1. 36 to 16

2. $\frac{4}{9}$ to $\frac{1}{3}$

3. 5 to $3\frac{3}{4}$

4. 0.24 to 0.14

5. $\frac{7}{12}$ to $\frac{5}{12}$

A family of three budgeted the following amounts for some of its monthly bills. Use the pie chart to solve problems 6 and 7.

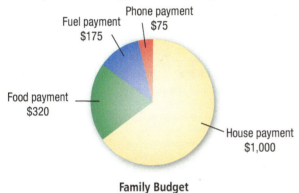

Family Budget

6. Ratio Find the ratio of phone payment to food payment. [5.1]

7. Ratio Find the ratio of house payment to fuel payment. [5.1]

8. Gas Mileage A car travels 348 miles on 12 gallons of gas. What is the gas mileage in miles per gallon? [5.2]

9. Unit Price A restaurant sells different sizes of caffe lattes. The prices are shown below. Give the unit price for each coffee drink and indicate which is the better buy. Round to the nearest cent. [5.2]

16 oz.   20 oz.

$3.45 $3.75

Find the unknown term in each proportion. [5.3, 5.4]

10. $\frac{3}{8} = \frac{21}{x}$

11. $\frac{2.25}{3} = \frac{1.5}{x}$

12. Baseball A baseball player gets 9 hits in his first 18 games of the season. If he continues at the same rate, how many hits will he get in 72 games? [5.5]

13. Map Reading The scale on a map indicates that 1 inch on the map corresponds to an actual distance of 24 miles. Two cities are $3\frac{3}{8}$ inches apart on the map. What is the actual distance between the two cities? [5.5]

Nursing Sometimes body surface area is used to calculate the necessary dosage for a patient. [5.5]

14. The dosage for a drug is 16 mg/m². If an adult has a BSA of 1.8 m², what dosage should he take?

15. Find the dosage an adult should take if her BSA is 1.6 m² and the dosage strength is 26.8 mg/m².

16. The triangles below are similar figures. Find h. [5.6]

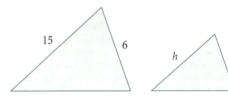

The diagram shows the ratio of the diameters of the planets in our solar system to Earth's diameter. Use the information to answer the following questions if the Earth's diameter is 12,742 km. Round to the nearest kilometer if necessary.

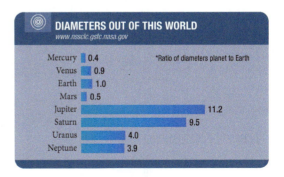

17. What is the diameter of Mars?

18. What is the diameter of Jupiter?

Simplify.

1.   9,341
     296
 + 3,735

2. $2,071 - 1,735$

3. $\dfrac{578}{34}$

4. $(4 \cdot 2) \cdot 6$

5. $24\overline{)12,393}$

6. 4^4

7. $8 \cdot 3^2 - 9$

8. $136 \div 17$

9. $63 + 28$

10. $\dfrac{81}{3}$

11. $(12 - 3) + (509 - 374)$

12. $(4.8)(6.2)$

13. $74.3 - 31.7$

14. $7.3 + 4.27 + 3.09$

15. $29.7 \div 4.5$

16. $\left(\dfrac{1}{2}\right)^4 \left(\dfrac{1}{3}\right)^3$

17. $9 \div \left(\dfrac{3}{4}\right)^2$

18. $\dfrac{3}{8} + \dfrac{5}{12}$

19. $6 \div \left(16 \div 2\dfrac{2}{3}\right)$

20. $16 - 3\dfrac{5}{7}$

Solve.

21. $\dfrac{5}{7} = \dfrac{x}{35}$

22. $\dfrac{3}{8} = \dfrac{9}{x}$

23. Find the perimeter and area of the figure below.

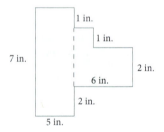

24. Find the perimeter and area of the figure below.

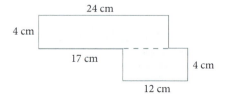

25. Find x if the two rectangles are similar.

26. Ratio If the ratio of men to women in a self-defense class is 2 to 5, and there are 8 men in the class, how many women are in the class?

27. Surfboard Length A surfing company decides that a surfboard would be more efficient if its length were reduced by $1\dfrac{5}{8}$ inches. If the original length was 7 feet $\dfrac{7}{8}$ inches, what will be the new length of the board (in inches)?

28. Teaching A teacher lectures on three sections in two class periods. If she continues at the same rate, on how many sections can the teacher lecture in 46 class periods?

The snapshot shows the amount of memory available in each new generation of iPod. Use the information to answer the following questions.

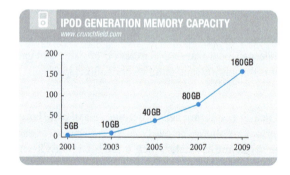

29. Ratio What is the ratio of the 2009 generation to the 2005 generation?

30. Ratio What is the ratio of the 2001 generation to the 2009 generation?

Write each ratio as a fraction in lowest terms. [5.1]

1. 48 to 18

2. $\frac{5}{8}$ to $\frac{3}{4}$

3. 6 to $2\frac{4}{7}$

4. 0.14 to 0.4

5. $\frac{7}{9}$ to $\frac{4}{9}$

A family of three budgeted the following amounts for some of its monthly bills. Use the pie chart to solve problems 6 and 7.

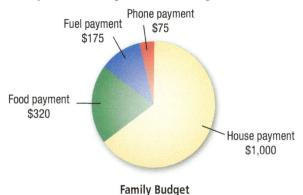

Family Budget

6. Ratio Find the ratio of food payment to phone payment. [5.1]

7. Ratio Find the ratio of house payment to food payment. [5.1]

8. Gas Mileage A car travels 315 miles on 9 gallons of gas. What is the gas mileage in miles per gallon? [5.2]

9. Unit Price A restaurant sells different sizes of Caffe Lattes. The prices are shown below. Give the unit price for each coffee drink and indicate which is a better buy. Round to the nearest cent. [5.2]

Find the unknown term in each proportion. [5.3, 5.4]

10. $\frac{4}{7} = \frac{24}{x}$

11. $\frac{2.5}{5} = \frac{1.5}{x}$

12. Baseball A baseball player gets 13 hits in his first 20 games of the season. If he continues at the same rate, how many hits will he get in 60 games? [5.5]

13. Map Reading The scale on a map indicates that 1 inch on the map corresponds to an actual distance of 15 miles. Two cities are $5\frac{1}{2}$ inches apart on the map. What is the actual distance between the two cities? [5.5]

Nursing Sometimes body surface area is used to calculate the necessary dosage for a patient. [5.5]

14. The dosage for a drug is 27 mg/m². If an adult has a BSA of 1.9 m², what dosage should he take?

15. Find the dosage an adult should take if her BSA is 1.35 m² and the dosage strength is 11.4 mg/m².

16. The triangles below are similar figures. Find h. [5.6]

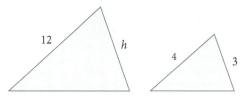

The diagram shows the number of US health club memberships in millions. Use the information to answer the following questions.

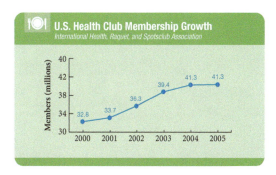

17. What is the ratio of memberships in 2000 to memberships in 2003?

18. What is the ratio of memberships in 2004 to memberships in 2005?

Percent

Everett, Washington
© 2010 Google

The largest building in the world is so big that it could hold all of the Disney Magic Kingdom theme parks inside it and still have room for twelve acres of covered parking! The Boeing facility adjacent to Paine Field in Everett, Washington has over one million light fixtures with a yearly electric bill of $18 million. The build-

©iStockphoto.com/FrankvandenBurgh

ing has the largest floor space in the world and employs 24,000 workers. It was originally opened in 1967 when the company began large-scale manufacturing of commercial aircraft. At that time, Pan American World Airways placed a $525 million order for twenty-five 747 jetliners, requiring the construction of new facilities. Over 300 employees began assembly of the first-ever jumbo jets, and on September 30, 1968, the first 747 rolled out of the factory to worldwide fanfare. Since the inaugural 747, Boeing has expanded its manufacturing programs to include the 767, the 777, and the new 787 Dreamliner.

When the 777 program was launched in 1990, the plant needed to be enlarged to accommodate the increase in production. Suppose you know that the floor space of the Boeing Plant is currently 4.3 million square feet and that the 1990 construction expanded floor space by 50%. What was the floor space of the facility prior to expansion? In order to answer this question, you will need to know how to work problems involving percent, which is one of the topics of this chapter.

OBJECTIVES

A   Understand percents and change percents to decimals.

B   Change decimals to percents.

C   Change percents to fractions.

D   Change fractions to percents.

KEY WORDS

percent

**Video Examples
Section 6.1**

Image © sxc.hu, monmart, 2005

If you manage your own money, you know the importance of a household budget. The following pie chart represents recommended percentages for the various categories to which your money may go. The whole pie chart is represented by 100%. In general, 100% of something is the whole thing.

A Household Budget

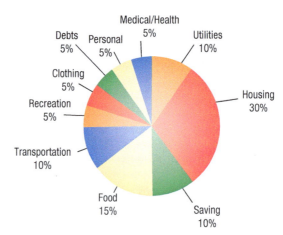

In this section, we will look at the meaning of percent. To begin, we learn to change decimals to percents and percents to decimals.

A The Meaning of Percent

Percent means "per hundred." Writing a number as a percent is a way of comparing the number with the number 100. For example, the number 42% (the % symbol is read "percent") is the same as 42 one-hundredths, that is,

$$42\% = \frac{42}{100}$$

Percents are really fractions (or ratios) with denominator 100.

EXAMPLE 1 Write each percent as a fraction with a denominator of 100.

a. 33%   **b.** 6%   **c.** 160%

Solution

a. $33\% = \dfrac{33}{100}$

b. $6\% = \dfrac{6}{100}$

c. $160\% = \dfrac{160}{100}$

If you are wondering if we could reduce some of these fractions further, the answer is yes. We have not done so because the point of this example is that every percent can be written as a fraction with denominator 100.

Changing Percents to Decimals

To change a percent to a decimal number, we simply use the meaning of percent.

EXAMPLE 2 Change 35.2% to a decimal.

Solution We drop the % symbol and write 35.2 over 100.

$$35.2\% = \frac{35.2}{100} \qquad \text{Use the meaning of percent to convert to a fraction with denominator 100.}$$

$$= 0.352 \qquad \text{Divide 35.2 by 100.}$$

We see from Example 2 that 35.2% is the same as the decimal 0.352. The result is that the % symbol has been dropped and the decimal point has been moved two places to the *left*. Because % always means "per hundred," we will always end up moving the decimal point two places to the left when we change percents to decimals. Because of this, we can write the following rule:

> **RULE** Percent to Decimal
>
> To change a percent to a decimal, drop the % symbol and move the decimal point two places to the *left*.

EXAMPLE 3 Write each percent as a decimal.

a. 37%   **b.** 68%   **c.** 120%   **d.** 0.8%

Solution We drop the % symbol and move the decimal point to the left two places

a. 37% = 0.37

b. 68% = 0.68

Decimal point originally here   **c.** 120% = 1.20   Decimal point moved to here

d. 0.8% = 0.008

EXAMPLE 4 Suppose a cortisone cream is 0.5% hydrocortisone. Writing this number as a decimal, we have

$$0.5\% = 0.005$$

B Changing Decimals to Percents

Now we want to do the opposite of what we just did in Examples 2–4. We want to change decimals to percents. We know that 42% written as a decimal is 0.42, which means that in order to change 0.42 back to a percent, we must move the decimal point two places to the *right* and use the % symbol.

$$0.42 = 42\%$$

Notice that we don't show the new decimal point if it is at the end of the number.

RULE Decimal to Percent

To change a decimal to a percent, we move the decimal point two places to the *right* and use the % symbol.

5. Write each decimal as a percent.
   a. 0.42
   b. 3.86
   c. 0.2
   d. 0.005

EXAMPLE 5 Write each decimal as a percent.

a. 0.27   **b.** 4.89   **c.** 0.5   **d.** 0.09

Solution

a. 0.27 = 27%

b. 4.89 = 489%

c. 0.5 = 0.50 = 50%   *Notice here that we put a 0 after the 5 so we can move the decimal point two places to the right.*

d. 0.09 = 09% = 9%   *Notice that we can drop the 0 at the left without changing the value of the number.*

6. Suppose the player in Example 6 has a batting average of 0.360. Write that number as a percent.

EXAMPLE 6 A softball player has a batting average of 0.650. As a percent, this number is 0.650 = 65.0%.

As you can see from these examples, percent is just a way of comparing numbers to 100. To multiply decimals by 100, we move the decimal point two places to the right. To divide by 100, we move the decimal point two places to the left. Because of this, it is a fairly simple procedure to change percents to decimals and decimals to percents.

C Changing Percents to Fractions

To change a percent to a fraction, drop the % symbol and write the original number over 100.

7. Write each percent as a fraction in lowest terms.
   a. 13%
   b. 20%
   c. 54%

EXAMPLE 7 The pie chart shows who pays for college expenses. Change each percent to a fraction.

Solution   In each case, we drop the percent symbol and write the number over 100. Then we reduce to lowest terms if possible.

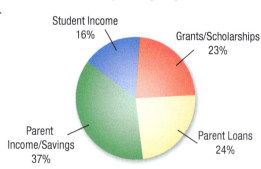

Who Pays College Expenses

Student Income 16%
Grants/Scholarships 23%
Parent Loans 24%
Parent Income/Savings 37%

$$16\% = \frac{16}{100} = \frac{4}{25} \qquad 23\% = \frac{23}{100} \qquad 24\% = \frac{24}{100} = \frac{6}{25} \qquad 37\% = \frac{37}{100}$$

EXAMPLE 8   Change 4.5% to a fraction in lowest terms.

Solution   We begin by writing 4.5 over 100.

$$4.5\% = \frac{4.5}{100}$$

We now multiply the numerator and the denominator by 10 so the numerator will be a whole number.

$$\frac{4.5}{100} = \frac{4.5 \times 10}{100 \times 10} \qquad \textit{Multiply the numerator and the denominator by 10.}$$

$$= \frac{45}{1,000}$$

$$= \frac{9}{200} \qquad \textit{Reduce to lowest terms.}$$

EXAMPLE 9   Change $32\frac{1}{2}\%$ to a fraction in lowest terms.

Solution   Writing $32\frac{1}{2}$ over 100 produces a complex fraction. We change $32\frac{1}{2}$ to an improper fraction and simplify.

$$32\frac{1}{2}\% = \frac{32\frac{1}{2}}{100}$$

$$= \frac{\frac{65}{2}}{100} \qquad \textit{Change } 32\frac{1}{2} \textit{ to the improper fraction } \frac{65}{2}.$$

$$= \frac{65}{2} \times \frac{1}{100} \qquad \textit{Dividing by 100 is the same as multiplying by } \frac{1}{100}.$$

$$= \frac{5 \cdot 13 \cdot 1}{2 \cdot 5 \cdot 20} \qquad \textit{Multiply.}$$

$$= \frac{13}{40} \qquad \textit{Reduce to lowest terms.}$$

Note that we could have changed our original mixed number to a decimal first and then changed to a fraction:

$$32\frac{1}{2}\% = 32.5\% = \frac{32.5}{100} = \frac{32.5 \times 10}{100 \times 10} = \frac{325}{1000} = \frac{5 \cdot 5 \cdot 13}{5 \cdot 5 \cdot 40} = \frac{13}{40}$$

The result is the same in both cases.

D Changing Fractions to Percents

To change a fraction to a percent, we can change the fraction to a decimal and then change the decimal to a percent.

EXAMPLE 10   Suppose the price your bookstore pays for your textbook is $\frac{7}{10}$ of the price you pay for your textbook. Write $\frac{7}{10}$ as a percent.

Solution   We can change $\frac{7}{10}$ to a decimal by dividing 7 by 10.

$$\begin{array}{r} 0.7 \\ 10\overline{)7.0} \\ \underline{7\,0} \\ 0 \end{array}$$

We then change the decimal 0.7 to a percent by moving the decimal point two places to the *right* and using the % symbol.

$$0.7 = 70\%$$

---

8. Change 3.2% to a fraction in lowest terms.

9. Change $53\frac{1}{4}\%$ to a fraction in lowest terms.

10. Write $\frac{3}{5}$ as a percent.

Answers

8. $\frac{4}{125}$

9. $\frac{213}{400}$

10. 60%

You may have noticed that we could have saved some time by simply writing $\frac{7}{10}$ as an equivalent fraction with denominator 100; that is,

$$\frac{7}{10} = \frac{7 \cdot 10}{10 \cdot 10} = \frac{70}{100} = 70\%$$

This is a good way to convert fractions like $\frac{7}{10}$ to percents. It works well for fractions with denominators of 2, 4, 5, 10, 20, 25, and 50, because they are easy to change to fractions with denominators of 100.

11. Change $\frac{5}{8}$ to a percent.

EXAMPLE 11   Change $\frac{3}{8}$ to a percent.

Solution   We begin by dividing 3 by 8.

```
      .375
  8)3.000
    2 4
    ----
      60
      56
    ----
      40
      40
    ----
       0
```

We then change the decimal to a percent by moving the decimal point two places to the right and using the % symbol.

$$\frac{3}{8} = 0.375 = 37.5\%$$

12. Change $\frac{7}{9}$ to a percent.

EXAMPLE 12   Change $\frac{5}{12}$ to a percent.

Solution   We begin by dividing 5 by 12.

```
       .4166
  12)5.0000
     4 8
     ----
      20
      12
     ----
      80
      72
     ----
      80
      72
     ----
       8
```

Note When rounding off, let's agree to round off to the nearest thousandth and then move the decimal point. Our answers in percent form will then be accurate to the nearest tenth of a percent, as in Example 12.

Because the 6s repeat indefinitely, we can use mixed number notation to write

$$\frac{5}{12} = 0.41\overline{6}$$

Or rounding, we can write

$$\frac{5}{12} = 41.7\% \qquad \textit{Round to the nearest tenth of a percent.}$$

13. Change $3\frac{1}{4}$ to a percent.

EXAMPLE 13   Change $2\frac{1}{2}$ to a percent.

Solution   We first change to a decimal and then to a percent.

$$2\frac{1}{2} = 2.5$$

$$= 250\%$$

Answers
11. 62.5%
12. 77.8%
13. 325%

Table 1 lists some of the most commonly used fractions and decimals and their equivalent percents.

Table 1

Fraction	Decimal	Percent
$\frac{1}{2}$	0.5	50%
$\frac{1}{4}$	0.25	25%
$\frac{3}{4}$	0.75	75%
$\frac{1}{3}$	$0.\overline{3}$	$33\frac{1}{3}$%
$\frac{2}{3}$	$0.\overline{6}$	$66\frac{2}{3}$%
$\frac{1}{5}$	0.2	20%
$\frac{2}{5}$	0.4	40%
$\frac{3}{5}$	0.6	60%
$\frac{4}{5}$	0.8	80%

GETTING READY FOR CLASS

After reading through the preceding section, respond in your own words and in complete sentences.

A. What is the relationship between the word *percent* and the number 100?

B. Explain in words how you would change 25% to a decimal.

C. Explain in words how you would change 25% to a fraction.

D. After reading this section you know that $\frac{1}{2}$, 0.5, and 50% are equivalent. Show mathematically why this is true.

Vocabulary Review

Choose the correct words to fill in the blanks below.

ratio   % symbol   hundred   left   decimal   right

1. The word percent means "per _____."
2. A percent is a _____ with a denominator of 100.
3. To change a percent to a decimal, drop the % symbol and move the decimal point two places to the _____.
4. To change a decimal to a percent, move the decimal point two places to the _____ and use the % symbol.
5. To change a percent to a fraction, drop the _____ and write the original number over 100.
6. To change a fraction to a percent, we can change the fraction to a _____ and then change the decimal to a percent.

Problems

A Write each percent as a fraction with denominator 100.

1. 20%   **2.** 40%   **3.** 60%   **4.** 80%

5. 24%   **6.** 48%   **7.** 65%   **8.** 35%

Change each percent to a decimal.

9. 23%   **10.** 34%   **11.** 92%   **12.** 87%

13. 9%   **14.** 7%   **15.** 3.4%   **16.** 5.8%

17. 6.34%   **18.** 7.25%   **19.** 0.9%   **20.** 0.6%

B Change each decimal to a percent.

21. 0.23   **22.** 0.34   **23.** 0.92   **24.** 0.87

25. 0.45   **26.** 0.54   **27.** 0.03   **28.** 0.04

29. 0.6   **30.** 0.9   **31.** 0.8   **32.** 0.5

33. 0.27   **34.** 0.62   **35.** 1.23   **36.** 2.34

C Change each percent to a fraction in lowest terms.

37. 60%

38. 40%

39. 75%

40. 25%

41. 4%

42. 2%

43. 26.5%

44. 34.2%

45. 71.87%

46. 63.6%

47. 0.75%

48. 0.45%

49. $6\frac{1}{4}$%

50. $5\frac{1}{4}$%

51. $33\frac{1}{3}$%

52. $66\frac{2}{3}$%

D Change each fraction or mixed number to a percent.

53. $\frac{1}{2}$

54. $\frac{1}{4}$

55. $\frac{3}{4}$

56. $\frac{2}{3}$

57. $\frac{1}{3}$

58. $\frac{1}{5}$

59. $\frac{4}{5}$

60. $\frac{1}{6}$

61. $\frac{7}{8}$

62. $\frac{1}{8}$

63. $\frac{7}{50}$

64. $\frac{9}{25}$

65. $3\frac{1}{4}$

66. $2\frac{1}{8}$

67. $1\frac{1}{2}$

68. $1\frac{3}{4}$

69. Change $\frac{21}{43}$ to a percent. Round to the nearest tenth of a percent

70. Change $\frac{36}{49}$ to a percent. Round to the nearest tenth of a percent

Applying the Concepts

71. Physiology The human body is between 50% and 75% water. Write each of these percents as a decimal.

72. Alcohol Consumption In the United States, 2.7% of those over 15 years of age drink more than 6.3 ounces of alcohol per day. In France, the same figure is 9%. Write each of these percents as a decimal.

73. iPhone The snapshot below shows what users had before their new iPhone. Use the information to answer the following questions.

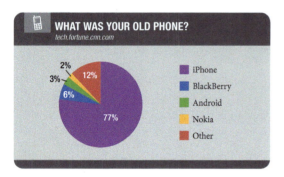

a. Convert each percent to a fraction.

b. Convert each percent to a decimal.

c. About how many times more likely are the respondents to have owned a Blackberry than an Android phone?

74. Foreign Language The chart shows the extent to which Americans say they know a foreign language. Change each percent to a fraction in lowest terms.

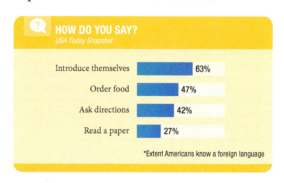

75. Nutrition Although, nutritionally, breakfast is the most important meal of the day, only $\frac{1}{5}$ of the people in the United States consistently eat breakfast. What percent of the population is this?

76. Children in School In Belgium, 96% of all children between 3 and 6 years of age go to school. In Sweden, the number is only 25%. In the United States, it is 60%. Write each of these percents as a fraction in lowest terms.

77. Student Enrollment The pie chart shows Cal Poly enrollment by college. Change each fraction to a percent.

Cal Poly Enrollment

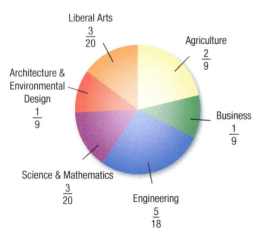

78. iPad The chart shows the percentage of the total iPad sales by state. Use the information to convert the percentage for the following states to a decimal.

a. California   b. Illinois   c. New York

Calculator Problems

Use a calculator to write each fraction as a decimal, and then change the decimal to a percent. Round all answers to the nearest tenth of a percent.

79. $\dfrac{29}{37}$ **80.** $\dfrac{18}{83}$ **81.** $\dfrac{6}{51}$ **82.** $\dfrac{8}{95}$ **83.** $\dfrac{236}{327}$ **84.** $\dfrac{568}{732}$

85. Women in the Military During World War II, $\frac{1}{12}$ of the Soviet armed forces were women. Today only $\frac{1}{450}$ of the Russian armed forces are women. Change both fractions to percents (to the nearest tenth of a percent).

86. Number of Teachers The ratio of the number of teachers to the number of students in secondary schools in Japan is 1 to 17. In the United States, the ratio is 1 to 19. Write each of these ratios as a fraction and then as a percent. Round to the nearest tenth of a percent.

Getting Ready for the Next Section

Multiply.

87. 0.25(74) **88.** 0.15(63) **89.** 0.435(25) **90.** 0.635(45)

Divide. Round the answers to the nearest thousandth, if necessary.

91. $\dfrac{21}{42}$ **92.** $\dfrac{21}{84}$ **93.** $\dfrac{25}{0.4}$ **94.** $\dfrac{31.9}{78}$

Solve for n.

95. $42n = 21$ **96.** $25 = 0.40n$

Find the Mistake

Each sentence below contains a mistake. Circle the mistake and write the correct word(s) or numbers(s) on the line provided.

1. Writing 0.4% as a decimal gives us 0.4. _____

2. To write 3.21 as a percent, divide the number by 100; that is, move the decimal two places to the left.

_____

3. Writing 25% as a fraction in lowest terms gives us $\frac{25}{100}$. _____

4. To change $\frac{5}{8}$ to a percent, we change $\frac{5}{8}$ to 0.625 and then move the decimal two places to the left to get 0.00625%.

_____

Navigation Skills: Prepare, Study, Achieve

Expect to encounter problems you find difficult when taking this course. Also expect to make mistakes. Mistakes highlight possible difficulties you are having and help you learn how to overcome them. We suggest making a list of problems you find difficult. As the course progresses, add new problems to the list, rework the problems on your list, and use the list to study for exams. Be aware of the mistakes you make and what you need to do to ensure you will not make that same mistake twice.

Image © Richard Ling, 2005

KEY WORDS

is

of

Video Examples
Section 6.2

Scientists have discovered a toxin in the spit of a sea snail that works with greater effectiveness but in smaller dosages and without the addictive risk of the painkiller morphine. The marine cone snail dwells on the ocean floor. The snail shoots its harpoon-like teeth coated in toxic saliva into its prey to poison it. Researchers have discovered how to isolate the saliva's toxin and put it into a pill for humans in pain to ingest. A patient will feel the same pain-reducing effects with 1% of a dose of a popular neuropathic painkiller prescribed in hospitals. If the prescription for an adult of the popular painkiller starts at 300 milligrams, how many milligrams of the sea snail drug would be dosed? In this section, we will work some other basic percent problems, similar to this one.

A Solving Percent Problems Using Equations

This section is concerned with three kinds of word problems that are associated with percents. Here is an example of each type:

Type A: What number is 15% of 63?

Type B: What percent of 42 is 21?

Type C: 25 is 40% of what number?

The first method we use to solve all three types of problems involves translating the sentences into equations and then solving the equations. The following translations are used to write the sentences as equations:

English	Mathematics
is	=
of	· (multiply)
a number	n
what number	n
what percent	n

The word *is* always translates to an = sign, the word *of* almost always means multiply, and the number we are looking for can be represented with a variable, such as n or x.

EXAMPLE 1 What number is 15% of 63?

Solution We translate the sentence into an equation as follows:

What number is 15% *of* 63?

$$n = 0.15 \cdot 63$$

Practice Problems

1. What number is 20% of 70?

Answer

1. 14

To do arithmetic with percents, we have to change to decimals. That is why 15% is rewritten as 0.15. Solving the equation, we have

$$n = 0.15 \cdot 63$$

$$n = 9.45$$

Therefore, 15% of 63 is 9.45.

EXAMPLE 2 What percent of 42 is 21?

Solution We translate the sentence as follows:

What percent of 42 is 21?

$$n \cdot 42 = 21$$

We solve for n by dividing both sides by 42.

$$\frac{n \cdot 42}{42} = \frac{21}{42}$$

$$n = \frac{21}{42}$$

$$n = 0.50$$

Because the original problem asked for a percent, we change 0.50 to a percent.

$$n = 50\%$$

Therefore, 21 is 50% of 42.

EXAMPLE 3 25 is 40% of what number?

Solution Following the procedure from the first two examples, we have

25 is 40% of what number?

$$25 = 0.40 \cdot n$$

Again, we changed 40% to 0.40 so we can do the arithmetic involved in the problem. Dividing both sides of the equation by 0.40, we have

$$\frac{25}{0.40} = \frac{0.40 \cdot n}{0.40}$$

$$\frac{25}{0.40} = n$$

$$62.5 = n$$

Therefore, 25 is 40% of 62.5.

As you can see, all three types of percent problems are solved in a similar manner. We write *is* as =, *of* as ·, and *what number* as n. The resulting equation is then solved to obtain the answer to the original question. Here are some more examples:

EXAMPLE 4 What number is 43.5% of 25?

$$n = 0.435 \cdot 25$$

$$n = 10.9 \qquad \text{Round to the nearest tenth.}$$

Therefore, 10.9 is 43.5% of 25.

2. What percent of 148 is 37?

3. 55 is 30% of what number?

4. What number is 38.2% of 45?

5. What percent of 87 is 14.8?

EXAMPLE 5   What percent of 78 is 31.9?

$$n \cdot 78 = 31.9$$

$$\frac{n \cdot 78}{78} = \frac{31.9}{78}$$

$$n = \frac{31.9}{78}$$

$$n = 0.409 \qquad \textit{Round to the nearest thousandth.}$$

$$n = 40.9\%$$

Therefore, 40.9% of 78 is 31.9.

6. 23 is 14% of what number?

EXAMPLE 6   34 is 29% of what number?

$$34 = 0.29 \cdot n$$

$$\frac{34}{0.29} = \frac{0.29 \cdot n}{0.29}$$

$$\frac{34}{0.29} = n$$

$$117.2 = n \qquad\qquad \textit{Round to the nearest tenth.}$$

Therefore, 34 is 29% of 117.2.

7. Suppose the item in Example 7 had 62 calories from fat. What percentage of the total calories would be from fat calories?

EXAMPLE 7   The American Dietetic Association recommends eating foods in which the number of calories from fat is less than 30% of the total number of calories. According to the nutrition label, what percent of the total number of calories are fat calories?

Solution   To solve this problem, we must write the question in the form of one of the three basic percent problems shown in Examples 1–6. Because there are 93 calories from fat and a total of 155 calories, we can write the question this way: 93 is what percent of 155?

Now that we have written the question in the form of one of the basic percent problems, we simply translate it into an equation. Then we solve the equation.

Nutrition Facts		
Serving Size 1 oz		
Servings Per Container About 4		
Amount Per Serving		
Calories 155		Calories from fat 93
		% Daily Value*
Total Fat 11g		**16%**
Saturated Fat 3g		**15%**
Trans Fat 0g		**0%**
Cholesterol 0mg		**0%**
Sodium 148mg		**6%**
Total Carbohydrate 14g		**5%**
Dietary Fiber 1g		**5%**
Sugars 1g		
Protein 2g		
Vitamin A 0%	•	Vitamin C 9%
Calcium 1%	•	Iron 3%
*Percent Daily Values are based on a 2,000 calorie diet		

FIGURE 1

93 *is what percent of* 155?

$$93 = n \cdot 155$$

$$\frac{93}{155} = n$$

$$n = 0.60 = 60\%$$

The number of calories from fat in this food is 60% of the total number of calories. Thus the ADA would not consider this to be a healthy food.

B Solving Percent Problems Using Proportions

We can look at percent problems in terms of proportions also. For example, we know that 24% is the same as $\frac{24}{100}$, which reduces to $\frac{6}{25}$. That is,

$$\frac{24}{100} = \frac{6}{25}$$

$\overbrace{\quad\uparrow\quad}\quad\uparrow\quad\overbrace{\quad\uparrow\quad}$

24 is to 100   as   6 is to 25

We can illustrate this visually with boxes of proportional lengths.

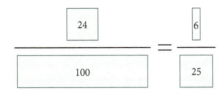

In general, we say

$$\frac{\text{Percent}}{100} = \frac{\text{Amount}}{\text{Base}}$$

$\overbrace{\quad\uparrow\quad}\quad\uparrow\quad\overbrace{\quad\uparrow\quad}$

Percent is to 100   as   amount is to base.

EXAMPLE 8   What number is 15% of 63?

Solution   This is the same problem we worked in Example 1. We let n be the number in question. We reason that n will be smaller than 63 because it is only 15% of 63. The base is 63 and the amount is n. We compare n to 63 as we compare 15 to 100. Our proportion sets up as follows:

15 is to 100   as n is to 63

$$\frac{15}{100} = \frac{n}{63}$$

Solving the proportion, we have

$$15 \cdot 63 = 100n \qquad \textcolor{teal}{\text{Fundamental property of proportions}}$$
$$945 = 100n \qquad \textcolor{teal}{\text{Simplify the left side.}}$$
$$9.45 = n \qquad \textcolor{teal}{\text{Divide each side by 100.}}$$

This gives us the same result we obtained in Example 1.

EXAMPLE 9   What percent of 42 is 21?

Solution   This is the same problem we worked in Example 2. We let n be the percent in question. The amount is 21 and the base is 42. We compare n to 100 as we compare 21 to 42. Here is our reasoning and proportion:

n is to 100   as   21 is to 42

$$\frac{n}{100} = \frac{21}{42}$$

8. What number is 20% of 70?

9. What percent of 148 is 37?

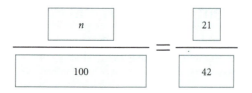

Solving the proportion, we have

$$42n = 21 \cdot 100$$ Fundamental property of proportions

$$42n = 2{,}100$$ Simplify the right side.

$$n = 50$$ Divide each side by 42.

Since n is a percent, our answer is 50%, giving us the same result we obtained in Example 2.

10. 55 is 30% of what number?

EXAMPLE 10   25 is 40% of what number?

Solution   This is the same problem we worked in Example 3. We let n be the number in question. The base is n and the amount is 25. We compare 25 to n as we compare 40 to 100. Our proportion sets up as follows:

$$40 \text{ is to } 100 \quad \text{as} \quad 25 \text{ is to } n$$
$$\downarrow \qquad\qquad \downarrow \qquad\qquad \downarrow$$
$$\frac{40}{100} = \frac{25}{n}$$

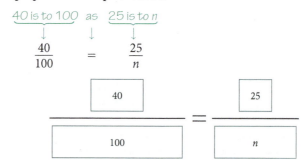

Solving the proportion, we have

$$40 \cdot n = 25 \cdot 100$$ Fundamental property of proportions

$$40 \cdot n = 2{,}500$$ Simplify the right side.

$$n = 62.5$$ Divide each side by 40.

So 25 is 40% of 62.5, which is the same result we obtained in Example 3.

Note   When you work the problems in the problem set, use whichever method you like, unless your instructor indicates that you are to use one method instead of the other.

GETTING READY FOR CLASS

After reading through the preceding section, respond in your own words and in complete sentences.

A. When we translate a sentence such as "What number is 15% of 63?" into symbols, what does each of the following translate to?

   a. is   **b.** of   **c.** what number

B. Using Example 1 in your text as a guide, answer the question below.

   The number 9.45 is what percent of 63?

C. Show that the answer to the question below is the same as the answer to the question in Example 2 of your text.

   The number 21 is what percent of 42?

D. If 21 is 50% of 42, then 21 is what percent of 84?

Answer

10. 183.33

Vocabulary Review

Choose the correct words to fill in the blanks below.

      multiply   fraction   decimal   variable   equals sign

1. In a mathematical sentence, the word *is* always translates to an _____.

2. In a mathematical sentence, the word *of* almost always means _____.

3. When translating a sentence to an equation, the number we are looking for can be represented with

   a _____.

4. To do arithmetic with a percent, change the percent to a _____.

5. Change a percent to a _____ to help solve a percent problem using a proportion.

Problems

A B Solve each of the following problems.

1. What number is 25% of 32?

2. What number is 10% of 80?

3. What number is 20% of 120?

4. What number is 15% of 75?

5. What number is 54% of 38?

6. What number is 72% of 200?

7. What number is 11% of 67?

8. What number is 2% of 49?

9. What percent of 24 is 12?

10. What percent of 80 is 20?

11. What percent of 50 is 5?

12. What percent of 20 is 4?

13. What percent of 36 is 9?

14. What percent of 70 is 14?

15. What percent of 8 is 6?

16. What percent of 16 is 9?

17. 32 is 50% of what number?

18. 16 is 20% of what number?

19. 10 is 20% of what number?

20. 11 is 25% of what number?

21. 37 is 4% of what number?

22. 90 is 80% of what number?

23. 8 is 2% of what number?

24. 6 is 3% of what number?

The following problems can be solved by the same method you used in Problems 1–24.

25. What is 6.4% of 87?

26. What is 10% of 102?

27. 25% of what number is 30?

28. 10% of what number is 22?

29. 28% of 49 is what number?

30. 97% of 28 is what number?

31. 27 is 120% of what number?

32. 24 is 150% of what number?

33. 65 is what percent of 130?

34. 26 is what percent of 78?

35. What is 0.4% of 235,671?

36. What is 0.8% of 721,423?

37. 4.89% of 2,000 is what number?

38. 3.75% of 4,000 is what number?

39. Write a basic percent problem, the solution to which can be found by solving the equation $n = 0.25(350)$.

40. Write a basic percent problem, the solution to which can be found by solving the equation $n = 0.35(250)$.

41. Write a basic percent problem, the solution to which can be found by solving the equation $n \cdot 24 = 16$.

42. Write a basic percent problem, the solution to which can be found by solving the equation $n \cdot 16 = 24$.

43. Write a basic percent problem, the solution to which can be found by solving the equation $46 = 0.75 \cdot n$.

44. Write a basic percent problem, the solution to which can be found by solving the equation $75 = 0.46 \cdot n$.

Applying the Concepts

Nutrition For each nutrition label in Problems 45–48, find what percent of the total number of calories comes from fat calories. Then refer to Example 7 and indicate whether the label is from a food considered healthy by the American Dietetic Association. Round to the nearest tenth of a percent if necessary.

45. Pizza Dough

Nutrition Facts

Serving Size 1/6 of package (65g)
Servings Per Container: 6

Amount Per Serving

Calories 160	Calories from fat 18

	% Daily Value*
Total Fat 2g	3%
Saturated Fat 0.5g	3%
Poly unsaturated Fat 0g	
Monounsaturated Fat 0g	
Cholesterol 0mg	0%
Sodium 470mg	20%
Total Carbohydrate 31g	10%
Dietary Fiber 1g	4%
Sugars 4g	
Protein 5g	

Vitamin A 0%	•	Vitamin C 0%
Calcium 0%	•	Iron 10%

*Percent Daily Values are based on a 2,000 calorie diet

46. Cheez-It Crackers

Nutrition Facts

Serving Size 30 g. (About 27 crackers)
Servings Per Container: 9

Amount Per Serving

Calories 150	Calories from fat 70

	% Daily Value*
Total Fat 8g	12%
Saturated Fat 2g	10%
Trans Fat 0g	
Polysaturated Fat 4g	
Monounsaturated Fat 2g	
Cholesterol 0mg	0%
Sodium 230mg	10%
Total Carbohydrate 17g	6%
Dietary Fiber less than 1g	3%
Sugars 0g	
Protein 3g	

Vitamin A 2%	•	Vitamin C 0%
Calcium 4%	•	Iron 6%

*Percent Daily Values are based on a 2,000 calorie diet

47. Shredded Mozzarella Cheese

Nutrition Facts

Serving Size 1 oz (28.3g)
Servings Per Container: 12

Amount Per Serving

Calories 72	Calories from fat 41

	% Daily Value*
Total Fat 4.5g	7%
Saturated Fat 2.9g	14%
Cholesterol 18mg	6%
Sodium 175mg	7%
Total Carbohydrate 0.8g	0%
Fiber 0g	0%
Sugars 0.3g	
Protein 6.9g	

Vitamin A 3%	•	Vitamin C 0%
Calcium 22%	•	Iron 0%

*Percent Daily Values (DV) are based on a 2,000 calorie diet

48. Canned Corn

Nutrition Facts

Serving Size 1 cup
Servings Per Container About 2 ½

Amount Per Serving

Calories 133	Calories from fat 15

	% Daily Value*
Total Fat 3g	3%
Saturated Fat 1g	1%
Cholesterol 0mg	0%
Sodium 530mg	22%
Total Carbohydrate 30g	10%
Dietary Fiber 3g	13%
Sugars 4g	
Protein 4g	

Vitamin A 0%	•	Vitamin C 23%
Calcium 1%	•	Iron 8%

*Percent Daily Values are based on a 2,000 calorie diet

Getting Ready for the Next Section

Solve each equation.

49. $96 = n \cdot 120$

50. $2{,}400 = 0.48 \cdot n$

51. $114 = 150n$

52. $3{,}360 = 0.42n$

53. What number is 80% of 60?

54. What number is 25% of 300?

Improving Your Quantitative Literacy

55. Survival Rates for Sea Gulls Here is part of a report concerning the survival rates of Western Gulls that appeared on the website of Cornell University:

> *Survival of eggs to hatching is 70%–80%; of hatched chicks to fledgling 50%–70%; of fledglings to age of first breeding <50%.*

Based on this information, give an estimate of the number of gulls of breeding age that would be produced by 1,000 Western Gull eggs.

Find the Mistake

Each sentence below contains a mistake. Circle the mistake and write the correct word(s) or numbers(s) on the line provided.

1. The question, "What number is 28.5% of 30?" translates to $n \cdot 0.285 = 30$. _____

2. Asking "75 is 30% of what number?" gives us 0.004. _____

3. To answer the question, "What number is 45% of 90?", we can solve the proportion $\frac{90}{x} = \frac{40}{100}$. _____

4. Using a proportion to answer the question, "What percent of 65 is 26?" will give us $n = 250\%$. _____

Landmark Review: Checking Your Progress

Write each percent as a fraction with denominator 100.

1. 15%

2. 27%

3. 14%

4. 89%

Change each percent to a decimal.

5. 17%

6. 28%

7. 5%

8. 6.37%

Change each decimal to a percent.

9. 0.38

10. 0.98

11. 0.09

12. 4.87

Change each fraction or mixed number to a percent. Round to the nearest tenth of a percent if necessary

13. $\frac{1}{10}$

14. $\frac{1}{3}$

15. $\frac{1}{7}$

16. $3\frac{1}{5}$

Solve each of the following problems. Round to the nearest hundredth if necessary.

17. What number is 35% of 15?

18. What percent of 85 is 53?

19. 88 is 37% of what number?

6.3 General Applications of Percent

©iStockphoto.com/djgunner

Scientists are blaming a long line of thunderstorms in 2005 for killing millions of trees in the Amazon. The storm winds blew as fast as 90 miles per hour and ripped trees out of the ground or snapped them in half. Some areas of the forest lost 80% of their trees. Scientists now believe that a minimum of 441 million trees were lost. According to an article on livescience.com written by the staff at OurAmazingPlanet, the storm killed upwards of 500,000 trees in Manaus, Brazil, which was 30% of the total number of trees downed by human deforestation in the same year and area. As we progress through this section, we will become more familiar with percent. Then we can work problems similar to one that may ask us to use the information about the trees in Manaus to calculate how many trees deforestation uprooted.

In this section, we continue our study of percent by doing more of the translations that were introduced in the previous section. The better you are at working those problems, the easier it will be for you to get started on the problems in this section.

Video Examples
Section 6.3

A Applications of Percent

EXAMPLE 1   On a 120-question test, a student answered 96 correctly. What percent of the problems did the student work correctly?

Solution   We have 96 correct answers out of a possible 120. The problem can be restated as

$$96 \text{ is what percent of } 120?$$
$$96 = n \cdot 120$$

$$\frac{96}{120} = \frac{n \cdot 120}{120} \qquad \text{Divide both sides by 120.}$$

$$n = \frac{96}{120} \qquad \text{Switch the left and right sides of the equation.}$$

$$n = 0.80 \qquad \text{Divide 96 by 120.}$$

$$= 80\% \qquad \text{Rewrite as a percent.}$$

When we write a test score as a percent, we are comparing the original score to an equivalent score on a 100-question test. That is, 96 correct out of 120 is the same as 80 correct out of 100.

Practice Problems

1. Suppose the test in Example 1 had 130 questions. What percentage of the problems did the student work correctly?

Answer
1. 73.85%

2. How much HCl is in a 60-milliliter bottle that is marked 60% HCl?

EXAMPLE 2   How much HCl (hydrochloric acid) is in a 60-milliliter bottle that is marked 80% HCl?

Solution   If the bottle is marked 80% HCl, that means 80% of the solution is HCl and the rest is water. Because the bottle contains 60 milliliters, we can restate the question as

$$\textit{What is 80\% of 60?}$$
$$n = 0.80 \cdot 60$$
$$n = 48$$

There are 48 milliliters of HCl in 60 milliliters of 80% HCl solution.

3. If the college in Example 3 has 1,500 female students, what is the total number of students in that college?

EXAMPLE 3   If 48% of the students in a certain college are female and there are 2,400 female students, what is the total number of students in the college?

Solution   We restate the problem as

$$\textit{2,400 is 48\% of what number?}$$
$$2{,}400 = 0.48 \cdot n$$
$$\frac{2{,}400}{0.48} = \frac{0.48 \cdot n}{0.48} \qquad \textit{Divide both sides by 0.48.}$$
$$n = \frac{2{,}400}{0.48} \qquad \textit{Switch the left and right sides of the equation.}$$
$$n = 5{,}000$$

There are 5,000 students.

4. If 35% of the students in Example 4 got a B, how many students recieved a B?

EXAMPLE 4   If 25% of the students in elementary algebra courses receive a grade of A, and there are 300 students enrolled in elementary algebra this year, how many students will receive A's?

Solution   After reading the question a few times, we find that it is the same as this question:

$$\textit{What number is 25\% of 300?}$$
$$n = 0.25 \cdot 300$$
$$n = 75$$

Thus, 75 students will receive A's in elementary algebra.

Almost all application problems involving percents can be restated as one of the three basic percent problems we listed in the previous section. It takes some practice before the restating of application problems becomes automatic. You may have to review that section and Examples 1–4 above several times before you can translate word problems into mathematical expressions yourself.

GETTING READY FOR CLASS

After reading through the preceding section, respond in your own words and in complete sentences.

A. On the test mentioned in Example 1, how many questions would the student have answered correctly if she answered 40% of the questions correctly?

B. If the bottle in Example 2 contained 30 milliliters instead of 60, what would the answer be?

C. In Example 3, how many of the students were male?

D. How many of the students mentioned in Example 4 received a grade lower than an A?

Answers

2. 36 ml

3. 3,125 students

4. 105 students

Vocabulary Review

On the lines below, write the three types of problems found in applications that involve percents. (Hint: We first learned of the three types in the previous section, and then put them to use in this section.)

1. Type A: _____

2. Type B: _____

3. Type C: _____

Problems

A Solve each of the following problems by first restating it as one of the three basic percent problems from the previous section. In each case, be sure to show the equation.

1. Test Scores On a 120-question test a student answered 84 correctly. What percent of the problems did the student work correctly?

2. Test Scores An engineering student answered 81 questions correctly on a 90-question trigonometry test. What percent of the questions did she answer correctly? What percent were answered incorrectly?

3. Mixture Problem A solution of alcohol and water is 80% alcohol. The solution is found to contain 32 milliliters of alcohol. How many milliliters total (both alcohol and water) are in the solution?

4. Family Budget A family spends $450 every month on food. If the family's income each month is $1,800, what percent of the family's income is spent on food?

5. Chemistry How much HCl (hydrochloric acid) is in a 60-milliliter bottle that is marked 75% HCl?

6. Chemistry How much acetic acid is in a 5-liter container of acetic acid and water that is marked 80% acetic acid? How much is water?

7. **Farming** A farmer owns 28 acres of land. Of the 28 acres, only 65% can be farmed. How many acres are available for farming? How many are not available for farming?

8. **Number of Students** Of the 420 students enrolled in a basic math class, only 30% are first-year students. How many are first-year students? How many are not?

9. **Number of Students** If 48% of the students in a certain college are female and there are 1,440 female students, what is the total number of students in the college?

10. **Basketball** A basketball player made 63 out of 75 free throws. What percent is this?

11. **Number of Graduates** Suppose 60% of the graduating class in a certain high school goes on to college. If 240 students from this graduating class are going on to college, how many students are there in the graduating class?

12. **Defective Parts** In a shipment of airplane parts, 3% are known to be defective. If 15 parts are found to be defective, how many parts are in the shipment?

13. **Number of Students** Suppose there are 3,200 students at our school. If 52% of them are female, how many female students are there at our school?

14. **Number of Students** In a certain school, 75% of the students in first-year chemistry have had algebra. If there are 300 students in first-year chemistry, how many of them have had algebra?

15. **Population** In a city of 32,000 people, there are 10,000 people under 25 years of age. What percent of the population is under 25 years of age?

16. **Number of Students** If 45 people enrolled in a psychology course but only 35 completed it, what percent of the students completed the course? (Round to the nearest tenth of a percent.)

Calculator Problems

The following problems are similar to Problems 1–16. They should be set up the same way. Then the actual calculations should be done on a calculator.

17. Number of People Of 7,892 people attending an outdoor concert in Los Angeles, 3,972 are over 18 years of age. What percent is this? (Round to the nearest whole number percent.)

18. Manufacturing A car manufacturer estimates that 25% of the new cars sold in one city have defective engine mounts. If 2,136 new cars are sold in that city, how many will have defective engine mounts?

19. Laptops The chart shows the most popular laptops among college students surveyed. If 5,280 students were surveyed, how many preferred a Dell?

20. Video Games The chart shows the most popular video games for home gaming systems. If 12,257 people were surveyed, how many listed Pokemon Heartgold as their favorite?

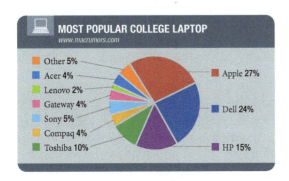

Getting Ready for the Next Section

Multiply.

21. 0.06(550)

22. 0.06(625)

23. 0.03(289,500)

24. 0.03(115,900)

Divide. Write your answers as decimals.

25. 5.44 ÷ 0.04

26. 4.35 ÷ 0.03

27. 19.80 ÷ 396

28. 11.82 ÷ 197

29. $\dfrac{1,836}{0.12}$

30. $\dfrac{115}{0.1}$

31. $\dfrac{90}{600}$

32. $\dfrac{105}{750}$

One Step Further: Batting Averages

Batting averages in baseball are given as decimal numbers, rounded to the nearest thousandth. For example, at the end of the 2011 season, Miguel Cabrera had the highest batting average in the major leagues. He had 197 hits in 572 times at bat, for a batting average of .344. This average is found by dividing the number of hits by the number of times he was at bat and then rounding to the nearest thousandth.

$$\text{Batting average} = \frac{\text{Number of hits}}{\text{Number of times at bat}} = \frac{197}{572} = 0.344$$

Because we can write any decimal number as a percent, we can convert batting averages to percents and use our knowledge of percent to solve problems. Looking at Michael Cabrera's batting average as a percent, we can say that he will get a hit 34.4% of the times he is at bat.

Each of the following problems can be solved by converting batting averages to percents and translating the problem into one of our three basic percent problems. (All numbers are from the end of the 2011 season according to espn.com.)

33. Jose Reyes had the highest batting average in the National League with 181 hits in 537 times at bat. What percent of the time Reyes is at bat can we expect him to get a hit?

34. Matt Kemp had 195 hits in 602 times at bat. What percent of the time can we expect Kemp to get a hit?

35. David Ortiz had a batting average of .309 in 2011. If his batting average remains the same and he has 550 at bats in the 2012 season, how many hits will he have? Round to the nearest hit.

36. Yadier Molina had a batting average of .305 in 2011. If his batting average remains the same and he has 500 at bats in the 2012 season, how many hits will he have? Round to the nearest hit.

37. How many hits must Miguel Cabrera have in his first 50 at bats in 2012 to maintain his average of .344?

38. How many hits must Matt Kemp have in his first 50 at bats in 2012 to maintain his average of .324?

Find the Mistake

Each sentence below contains a mistake. Circle the mistake and write the correct word(s) or numbers(s) on the line provided.

1. On a test with 110 questions, a student answered 98 questions correctly. The percentage of questions the student answered correctly is 112.2%. _____

2. A school track team consists of 12 boys and 10 girls. The total number of girls makes up 54.5% percent of the whole team. _____

3. Suppose 39 students in a college class of 130 students received a B on their tests. To find what percent of students earned a B, solve the proportion $\frac{x}{130} = \frac{39}{100}$. _____

4. Suppose a basketball player made 120 out of 150 free throws attempted. To find what percent of free throws the player made, solve the proportion $\frac{30}{150} = \frac{x}{100}$. _____

6.4 Sales Tax and Commission

Image © Katherine Heistand Shields, 2010

KEY WORDS

sales tax

tax rate

commission

commission rate

Video Examples
Section 6.4

Have an appetite for bugs? A candy company in Pismo Beach, California produces and sells lollipops and other sugary treats with real insects trapped inside! Choose from a wide selection of worms, crickets, scorpions, ants, or butterflies to satisfy that creepy-crawly craving. Suppose you purchase a box of 36 Scorpion Suckers for $81. If sales tax in Pismo Beach at the time of your purchase is 8.75%, how much sales tax will you have to pay in addition to the $81? To solve problems similar to this one, we will first restate them in terms of the problems we have already learned how to solve.

A Sales Tax

EXAMPLE 1 Suppose the sales tax rate in Mississippi is 6% of the purchase price. If the price of a used refrigerator is $550, how much sales tax must be paid?

Solution Because the sales tax is 6% of the purchase price, and the purchase price is $550, the problem can be restated as

What is 6% of $550?

We solve this problem, as we did previously, by translating it into an equation.

What is 6% of $550?

$$n = 0.06 \cdot 550$$

$$n = 33$$

The sales tax is $33. The total price of the refrigerator would be

Purchase price		Sales tax		Total price
$550	+	$33	=	$583

EXAMPLE 2 Suppose the sales tax rate is 4%. If the sales tax on a 10-speed bicycle is $5.44, what is the purchase price, and what is the total price of the bicycle?

Solution We know that 4% of the purchase price is $5.44. We find the purchase price first by restating the problem as

$5.44 is 4% of what number?

$$5.44 = 0.04 \cdot n$$

Practice Problems

1. Suppose the refrigerator in Example 1 cost $700, how much sales tax must be paid?

> *Note* In Example 1, the sales tax rate is 6%, and the sales tax is $33. In most everyday communications, people say "The sales tax is 6%," which is incorrect. The 6% is the tax rate, and the $33 is the actual tax.

2. If the sales tax rate is 6% and the tax on a printer is $73.50, what is the total price of the printer?

Answers

1. $42
2. $1,298.50

We solve the equation by dividing both sides by 0.04.

$$\frac{5.44}{0.04} = \frac{0.04 \cdot n}{0.04} \qquad \text{Divide both sides by 0.04.}$$

$$n = \frac{5.44}{0.04} \qquad \text{Switch the left and right sides of the equation.}$$

$$n = 136 \qquad \text{Divide.}$$

The purchase price is $136. The total price is the sum of the purchase price and the sales tax.

$$
\begin{array}{lll}
\text{Purchase price} & = & \$136.00 \\
\underline{\text{Sales tax}} & = & \underline{+\ 5.44} \\
\text{Total price} & = & \$141.44
\end{array}
$$

3. Suppose the purchase price of a gaming system is $250 and the sales tax is $17.50. What is the sales tax rate?

EXAMPLE 3   Suppose the purchase price of a stereo system is $396 and the sales tax is $19.80. What is the sales tax rate?

Solution   We restate the problem as

19.80 *is what percent of* $396?

$$19.80 = n \cdot 396$$

To solve this equation, we divide both sides by 396.

$$\frac{19.80}{396} = \frac{n \cdot 396}{396} \qquad \text{Divide both sides by 396.}$$

$$n = \frac{19.80}{396} \qquad \text{Switch the left and right sides of the equation.}$$

$$n = 0.05 \qquad \text{Divide.}$$

$$n = 5\% \qquad 0.05 = 5\%$$

The sales tax rate is 5%.

B Commission

Many salespeople work on a *commission* basis. That is, their earnings are a percentage of the amount they sell. The *commission rate* is a percent, and the actual commission they receive is a dollar amount.

4. A car salesman gets 5% of each car he sells as commission. If he sells a car for $30,000, how much money does he earn?

EXAMPLE 4   A real estate agent gets 3% of the price of each house she sells. If she sells a house for $289,500, how much money does she earn?

Solution   The commission is 3% of the price of the house, which is $289,500. We restate the problem as

What is 3% *of* $289,500?

$$n = 0.03 \cdot 289,500$$

$$n = 8,685$$

The commission is $8,685.

Answers

3. 7%

4. $1,500

EXAMPLE 5   Suppose a car salesperson's commission rate is 12%. If the commission on one of the cars is $1,836, what is the purchase price of the car?

Solution   12% of the sales price is $1,836. The problem can be restated as

12% *of what number is* $1,836?

$$0.12 \cdot n = 1,836$$

$$\frac{0.12 \cdot n}{0.12} = \frac{1,836}{0.12}$$ *Divide both sides by 0.12.*

$$n = 15,300$$

The car sells for $15,300.

EXAMPLE 6   If the commission on a $600 dining room set is $90, what is the commission rate?

Solution   The commission rate is a percentage of the selling price. What we want to know is

$90 is what percent of $600?

$$90 = n \cdot 600$$

$$\frac{90}{600} = \frac{n \cdot 600}{600}$$ *Divide both sides by 600.*

$$n = \frac{90}{600}$$ *Switch the left and right sides of the equation.*

$$n = 0.15$$ *Divide.*

$$n = 15\%$$ *Change to a percent.*

The commission rate is 15%.

> **GETTING READY FOR CLASS**
>
> *After reading through the preceding section, respond in your own words and in complete sentences.*
>
> **A.** Explain the difference between the sales tax and the sales tax rate.
>
> **B.** Rework Example 1 using a sales tax rate of 7% instead of 6%.
>
> **C.** Suppose the bicycle in Example 2 was purchased in California, where the sales tax rate in 2010 was 8.25%. How much more would the bicycle have cost?
>
> **D.** Explain the difference between commission and the commission rate.

5. If a real estate agent's comission rate is 6% and the commission on a property is $15,000, what is the purchase price of the property?

6. If the commission on a $500 sofa is $75, what is the commission rate?

Answers

5. $250,000
6. 15%

EXERCISE SET 6.4

Problems

A These problems should be solved by the methods shown in this section. In each case, show the equation needed to solve the problem. Write neatly, and show your work.

1. **Sales Tax** Suppose the sales tax rate in Mississippi is 7% of the purchase price. If a new food processor sells for $750, how much is the sales tax?

2. **Sales Tax** If the sales tax rate is 5% of the purchase price, how much sales tax is paid on a television that sells for $980?

3. **Sales Tax and Purchase Price** Suppose the sales tax rate in Michigan is 6%. How much is the sales tax on a $45 concert ticket? What is the total price?

4. **Sales Tax and Purchase Price** Suppose the sales tax rate in Hawaii is 4%. How much sales tax is charged on a new car if the purchase price is $16,400? What is the total price?

5. **Total Price** The sales tax rate is 4%. If the sales tax on a 10-speed bicycle is $6, what is the purchase price? What is the total price?

6. **Total Price** The sales tax on a new microwave oven is $30. If the sales tax rate is 5%, what is the purchase price? What is the total price?

7. **Tax Rate** Suppose the purchase price of a dining room set is $450. If the sales tax is $22.50, what is the sales tax rate?

8. **Tax Rate** If the purchase price of a bottle of California wine is $24 and the sales tax is $1.50, what is the sales tax rate?

B

9. **Commission** A real estate agent has a commission rate of 3%. If a piece of property sells for $94,000, what is her commission?

10. **Commission** A tire salesperson has a 12% commission rate. If he sells a set of radial tires for $400, what is his commission?

11. **Commission and Purchase Price** Suppose a salesperson gets a commission rate of 12% on the lawnmowers she sells. If the commission on one of the mowers is $24, what is the purchase price of the lawnmower?

12. **Commission and Purchase Price** If an appliance salesperson gets 9% commission on all the appliances she sells, what is the price of a refrigerator if her commission is $67.50?

13. **Commission Rate** If the commission on an $800 washer is $112, what is the commission rate?

14. **Commission Rate** A realtor makes a commission of $3,600 on a $90,000 house he sells. What is his commission rate?

Calculator Problems

The following problems are similar to Problems 1–14. Set them up in the same way, but use a calculator for the calculations.

15. **Sales Tax** The sales tax rate on a certain item is 5.5%. If the purchase price is $216.95, how much is the sales tax? (Round to the nearest cent.)

16. **Purchase Price** If the sales tax rate is 4.75% and the sales tax is $18.95, what is the purchase price? What is the total price? (Both answers should be rounded to the nearest cent.)

17. Tax Rate The purchase price for a new suit is $229.50. If the sales tax is $10.33, what is the sales tax rate? (Round to the nearest tenth of a percent.)

18. Commission If the commission rate for a mobile home salesperson is 11%, what is the commission on the sale of a $15,794 mobile home?

19. Selling Price Suppose the commission rate on the sale of used cars is 13%. If the commission on one of the cars is $519.35, what did the car sell for?

20. Commission Rate If the commission on the sale of $79.40 worth of clothes is $14.29, what is the commission rate? (Round to the nearest percent.)

Getting Ready for the Next Section

Multiply.

21. 0.05(22,000)

22. 0.176(1,793,000)

23. 0.25(300)

24. 0.12(450)

Divide. Write your answers as decimals.

25. $4 \div 25$

26. $7 \div 35$

Subtract.

27. $25 - 21$

28. $1,793,000 - 315,568$

29. $450 - 54$

30. $300 - 75$

Add.

31. $396 + 19.8$

32. $22,000 + 1,100$

One Step Further: Luxury Taxes

Suppose a luxury tax requires an additional tax of 10% on a portion of the purchase price of certain luxury items. For expensive cars, it must be paid on the part of the purchase price that exceeded $30,000. For example, if you purchased a Jaguar XJ-S for $53,000, you would pay a luxury tax of 10% of $23,000, because the purchase price, $53,000, is $23,000 above $30,000.

33. If you purchased a Jaguar XJ-S for $53,000 in California, where the sales tax rate was 6%, how much would you pay in luxury tax and how much would you pay in sales tax?

34. If you purchased a Mercedes 300E for $43,500 in California, where the sales tax rate was 6%, how much more would you pay in sales tax than luxury tax?

35. How much would you have saved if you had purchased the Jaguar mentioned in Problem 33 in Alaska which has no sales tax?

36. How much would you have saved if you bought a car with a purchase price of $45,000 without the luxury tax?

37. How much would you save on a car with a sticker price of $31,500, if you persuaded the car dealer to reduce the price to $29,900?

38. Suppose one of the cars you were interested in had a sticker price of $35,500, while another had a sticker price of $28,500. If you expected to pay full price for either car, how much did you save if you bought the less expensive car?

Find the Mistake

Each sentence below contains a mistake. Circle the mistake and write the correct word(s) or numbers(s) on the line provided.

1. Suppose the sale tax rate on a new computer is 8%. If the computer cost $650, then the total price of purchase would be $52. _____

2. If a new shirt that costs $32 has sales tax equal to $1.92, then the sales tax rate is 8%. _____

3. A car salesman's commission rate is 7%. To find his commission on a $15,000 sale of a 2005 Ford truck, we must divide the product of 100 and 15,000 by 7. _____

4. A saleswoman makes a commission of $6.80 on a sale of $85 worth of clothing. To find the woman's commission rate, solve the proportion $\frac{85}{6.8} = \frac{x}{100}$. _____

6.5 Percent Increase or Decrease, and Discount

Image © sxc.hu, atroszko, 2006

Insurance companies gather statistics about stopping distances of cars. The following table and bar chart show some statistics for cars with different tires traveling 20 miles per hour on ice. The percent decrease column on the table shows how stopping distances of a car with special tires relate to the stopping distance of a car with regular tires.

KEY WORDS

percent increase

percent decrease

discount

Video Examples
Section 6.5

	Stopping Distance	Percent Decrease
Regular tires	150 ft	0
Snow tires	151 ft	−1%
Studded snow tires	120 ft	20%
Reinforced tire chains	75 ft	50%

Table courtesy of The Casualty Adjuster's Guide

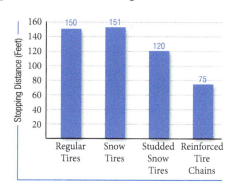

A Percent Increases and Decreases

Many times it is more effective to state increases or decreases in terms of percents, rather than the actual number, because with percent we are comparing everything to 100.

EXAMPLE 1   If a person earns $22,000 a year and gets a 5% increase in salary, what is the new salary?

Solution   We can find the dollar amount of the salary increase by finding 5% of $22,000.

$$0.05 \times 22,000 = 1,100$$

The increase in salary is $1,100. The new salary is the old salary plus the raise.

$22,000	Old salary
+ 1,100	Raise (5% of $22,000)
$23,100	New salary

Practice Problems

1. If a person earns $30,000 per year and gets and 8% raise, what is the new salary?

EXAMPLE 2 In 1997, there were approximately 1,477,000 arrests for driving under the influence of alcohol or drugs (DUI) in the United States. By 2007, the number of arrests for DUI had decreased 3.4% from the 1997 number. How many people were arrested for DUI in 2007? Round the answer to the nearest thousand.

Solution The decrease in the number of arrests is 3.4% of 1,477,000, or

$$0.034 \times 1,477,000 = 50,218$$

Subtracting this number from 1,477,000, we have the number of DUI arrests in 2007.

1,477,000	*Number of arrests in 1997*
− 50,218	*Decrease of 3.4%*
1,426,782	*Number of arrests in 2007*

To the nearest thousand, there were approximately 1,427,000 arrests for DUI in 2007.

EXAMPLE 3 Shoes that usually sell for $25 are on sale for $21. What is the percent decrease in price?

Solution We must first find the decrease in price. Subtracting the sale price from the original price, we have

$$\$25 - \$21 = \$4$$

The decrease is $4. To find the percent decrease (from the original price), we have

4 *is what percent of* 25?

$$4 = n \cdot 25$$

$$\frac{4}{25} = \frac{n \cdot 25}{25} \qquad \text{Divide both sides by 25.}$$

$$n = \frac{4}{25} \qquad \text{Switch the left and right sides of the equation.}$$

$$n = 0.16 \qquad \text{Divide.}$$

$$n = 16\% \qquad \text{Change to a percent.}$$

The shoes that sold for $25 have been reduced by 16% to $21. In a problem like this, $25 is the *original* (or *marked*) price, $21 is the *sale price*, $4 is the *discount*, and 16% is the *rate of discount*.

B Discount

In Example 3, $4 was the discount amount for the shoes on sale. Now we will work some examples that deal directly with *discount*.

EXAMPLE 4 During a clearance sale, a suit that usually sells for $300 is marked "25% off." What is the discount? What is the sale price?

Solution To find the discount, we restate the problem as

What is 25% *of* 300?

$$n = 0.25 \cdot 300$$

$$n = 75$$

The discount is $75. The sale price is the original price less the discount.

$300	*Original price*
− 75	*Less the discount (25% of $300)*
$225	*Sale price*

2. If a $30,000 car decreased in value by 12.4%, what is it worth?

3. A jacket that usually sells for $120 is on sale for $108. What is the percent decrease in price?

4. A pair of shoes that usually sells for $150 is on clearance for 30% off. What is the discount? What is the sale price?

5. A refrigerator that normally sells for $800 is on sale at 15% off. If the sales tax rate is 7%, what is the total bill for the refrigerator?

EXAMPLE 5   A man buys a washing machine on sale. The machine usually sells for $450, but it is on sale at 12% off. If the sales tax rate is 5%, how much is the total bill for the washer?

Solution   First, we have to find the sale price of the washing machine, and we begin by finding the discount.

$$What\ is\ 12\%\ of\ \$450?$$
$$n = 0.12 \cdot 450$$

$$n = 54$$

The washing machine is marked down $54. The sale price is

$450	Original price
− 54	Discount (12% of $450)
$396	Sale price

Because the sales tax rate is 5%, we find the sales tax as follows:

$$What\ is\ 5\%\ of\ 396?$$
$$n = 0.05 \cdot 396$$

$$n = 19.80$$

The sales tax is $19.80. The total price the man pays for the washing machine is

$396.00	Sale price
+ 19.80	Sales tax
$415.80	Total price

GETTING READY FOR CLASS

After reading through the preceding section, respond in your own words and in complete sentences.

A. Suppose the person mentioned in Example 1 was earning $32,000 per year and received the same percent increase in salary. How much more would the raise have been?

B. Suppose the shoes mentioned in Example 3 were on sale for $20, instead of $21. Calculate the new percent decrease in price.

C. Suppose a store owner pays $225 for a suit, and then marks it up $75, to $300. Find the percent increase in price.

D. What is discount?

Vocabulary Review

Read the following description of a television on sale.

An LCD HD television that usually sells for $400 is on sale for $340. The television's price has been reduced $60, which is a 15% percent decrease.

Now match the following quantities mentioned in the above description with their correct labels.

1. $400 _____

2. $340 _____

3. $60 _____

4. 15% _____

a. Discount

b. Original price

c. Rate of discount

d. Sale price

Problems

A B Solve each of these problems using the method developed in this section.

1. Salary Increase If a person earns $23,000 a year and gets a 7% increase in salary, what is the new salary?

2. Salary Increase A computer programmer's yearly income of $57,000 is increased by 8%. What is the dollar amount of the increase, and what is her new salary?

3. Tuition Increase The yearly tuition at a college is presently $3,000. Next year it is expected to increase by 17%. What will the tuition at this school be next year?

4. Price Increase A supermarket increased the price of cheese that sold for $1.98 per pound by 3%. What is the new price for a pound of this cheese? (Round to the nearest cent.)

5. Car Value In one year, a new car decreased in value by 20%. If it sold for $16,500 when it was new, what was it worth after 1 year?

6. Calorie Content A certain light beer has 20% fewer calories than the regular beer. If the regular beer has 120 calories per bottle, how many calories are in the same-sized bottle of the light beer?

7. Salary Increase A person earning $3,500 a month gets a raise of $350 per month. What is the percent increase in salary?

8. Rate Increase A student reader is making $6.50 per hour and gets a $0.70 raise. What is the percent increase? (Round to the nearest tenth of a percent.)

9. Shoe Sale Shoes that usually sell for $25 are on sale for $20. What is the percent decrease in price?

10. Enrollment Decrease The enrollment in a certain elementary school was 410. The next year, the enrollment in the same school was 328. Find the percent decrease in enrollment from one year to the next.

11. Soda Consumption The chart shows the consumption of soda in gallons per person per year in different countries. What is the increase in percent of consumption in The United States as compared to Norway? Round to the nearest percent.

POP A TOP OF POP
Euromonitor (2003)

United States	51.7
Mexico	33.3
Norway	32.2
Ireland	32.0
Canada	30.9

12. Farmers' Markets The chart shows the rise in farmers' markets throughout the country. What is the percent increase in farmers' markets from 1994 to 2010? Round to the nearest tenth of a percent.

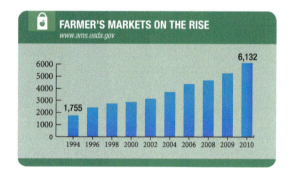

FARMER'S MARKETS ON THE RISE
www.ams.usda.gov

1,755 ... 6,132
1994 1996 1998 2000 2002 2004 2006 2008 2009 2010

13. Discount During a clearance sale, a three-piece suit that usually sells for $300 is marked "15% off." What is the discount? What is the sale price?

14. Sale Price On opening day, a new music store offers a 12% discount on all electric guitars. If the regular price on a guitar is $550, what is the sale price?

15. Total Price A man buys a washing machine that is on sale. The washing machine usually sells for $450 but is on sale at 20% off. If the sales tax rate in his state is 6%, how much is the total bill for the washer?

16. Total Price A bedroom set that normally sells for $1,450 is on sale for 10% off. If the sales tax rate is 5%, what is the total price of the bedroom set if it is bought while on sale?

Calculator Problems

Set up the following problems the same way you set up Problems 1–16. Then use a calculator to do the calculations.

17. Salary Increase A teacher making $43,752 per year gets a 6.5% raise. What is the new salary?

18. Utility Increase A homeowner had a $15.90 electric bill in December. In January, the bill was $17.81. Find the percent increase in the electric bill from December to January. (Round to the nearest whole number.)

19. Soccer The rules for soccer state that the playing field must be from 100 to 120 yards long and 55 to 75 yards wide. The 1999 Women's World Cup was played at the Rose Bowl on a playing field 116 yards long and 72 yards wide. The diagram below shows the smallest possible soccer field, the largest possible soccer field, and the soccer field at the Rose Bowl.

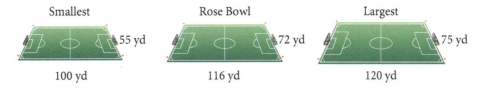

Smallest — 55 yd — 100 yd

Rose Bowl — 72 yd — 116 yd

Largest — 75 yd — 120 yd

  a. Percent Increase A team plays on the smallest field, then plays in the Rose Bowl. What is the percent increase in the area of the playing field from the smallest field to the Rose Bowl? Round to the nearest tenth of a percent.

  b. Percent Increase A team plays a soccer game in the Rose Bowl. The next game is on a field with the largest dimensions. What is the percent increase in the area of the playing field from the Rose Bowl to the largest field? Round to the nearest tenth of a percent.

20. Football The diagrams below show the dimensions of playing fields for the National Football League (NFL), the Canadian Football League (CFL), and Arena Football.

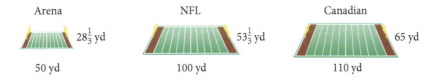

Arena — $28\frac{1}{3}$ yd — 50 yd

NFL — $53\frac{1}{3}$ yd — 100 yd

Canadian — 65 yd — 110 yd

  a. Percent Increase Kurt Warner made a successful transition from Arena Football to the NFL, winning the Most Valuable Player award. What was the percent increase in the area of the fields he played on in moving from Arena Football to the NFL? Round to the nearest percent.

  b. Percent Decrease Doug Flutie played in the Canadian Football League before moving to the NFL. What was the percent decrease in the area of the fields he played on in moving from the CFL to the NFL? Round to the nearest tenth of a percent.

Getting Ready for the Next Section

Multiply. Round to nearest hundredth if necessary.

21. $0.07(2,000)$

22. $0.12(8,000)$

23. $600(0.04)\left(\dfrac{1}{6}\right)$

24. $900(0.06)\left(\dfrac{1}{4}\right)$

25. $10,150(0.06)\left(\dfrac{1}{4}\right)$

26. $10,302.25(0.06)\left(\dfrac{1}{4}\right)$

Add.

27. $3,210 + 224.7$

28. $900 + 13.50$

29. $10,000 + 150$

30. $10,150 + 152.25$

31. $10,302.25 + 154.53$

32. $10,456.78 + 156.85$

Simplify.

33. $2,000 + 0.07(2,000)$ **34.** $8,000 + 0.12(8,000)$ **35.** $3,000 + 0.07(3,000)$ **36.** $9,000 + 0.12(9,000)$

Find the Mistake

Each sentence below contains a mistake. Circle the mistake and write the correct word(s) or numbers(s) on the line provided.

1. If a new model of a car increases 12% from and old model's price of $24,000, then the new selling price is $2,880.

_____

2. A lawnmower goes on sale from $98 to $63.70. The percent decrease of the lawnmower's price is 65%.

_____

3. A backpack that normally sells for $75 is on sale. The new price of $45 shows a percent increase of 40%

_____

4. A designer pair of sunglasses is on sale from $125 for 20% off. If the sales tax is 6% of the sale price, then the total bill for the glasses would be $107.50.   _____

6.6 Interest

Image © sxc.hu, LeoSynapse, 2006

OBJECTIVES

A Solve application problems involving annual interest.

B Solve application problems involving simple interest.

C Solve application problems involving compound interest.

KEY WORDS

interest

principal

interest rate

simple interest

compound interest

A Interest

Anyone who has borrowed money from a bank or other lending institution, or who has invested money in a savings account, is aware of interest. Interest is the amount of money paid for the use of money. If we put $500 in a savings account that pays 6% annually, the interest will be 6% of $500, or 0.06(500) = $30. The amount we invest ($500) is called the *principal*, the percent (6%) is the *interest rate*, and the money earned ($30) is the *interest*. Let's begin by working some examples that involve annual interest.

EXAMPLE 1 A man invests $2,000 in a savings plan that pays 7% per year. How much money will be in the account at the end of 1 year?

Solution We first find the interest by taking 7% of the principal, $2,000.

$$\text{Interest} = 0.07(\$2,000)$$
$$= \$140$$

The interest earned in 1 year is $140. The total amount of money in the account at the end of a year is the original amount plus the $140 interest.

$2,000	*Original investment (principal)*
+ 140	*Interest (7% of $2,000)*
$2,140	*Amount after 1 year*

The amount in the account after 1 year is $2,140.

EXAMPLE 2 A farmer borrows $8,000 from his local bank at 12%. How much does he pay back to the bank at the end of the year to pay off the loan?

Solution The interest he pays on the $8,000 is

$$\text{Interest} = 0.12(\$8,000)$$
$$= \$960$$

At the end of the year, he must pay back the original amount he borrowed ($8,000) plus the interest at 12%.

$8,000	*Amount borrowed (principal)*
+ 960	*Interest at 12%*
$8,960	*Total amount to pay back*

The total amount that the farmer pays back is $8,960.

**Video Examples
Section 6.6**

Practice Problems

1. A woman invests $3,000 in an account that pays 8% per year in interest. How much is in the account after 1 year?

2. Ryan finances a $23,000 car at 8% interest for 1 year. What is the total amount he he will pay back at the end of the loan?

Answers

1. $3,240
2. $24,840

B Simple Interest

There are many situations in which interest on a loan is figured on other than a yearly basis. Many short-term loans are for only 30 or 60 days. In these cases, we can use a formula to calculate the interest that has accumulated. This type of interest is called *simple interest*.

The formula is $I = P \cdot R \cdot T$ where

$$I = \text{Interest}$$

$$P = \text{Principal}$$

$$R = \text{Interest rate (the percent)}$$

$$T = \text{Time (in years, 1 year} = 360 \text{ days)}$$

We could have used this formula to find the interest in Examples 1 and 2. In those two cases, T is 1. When the length of time is in days rather than years, it is common practice to use 360 days for 1 year, and we write T as a fraction. Examples 3 and 4 illustrate this procedure.

3. Suppose the loan in Example 3 is at an interest rate of 6%. How much interest will be paid if the loan is paid back in 90 days?

EXAMPLE 3 A student takes out an emergency loan for tuition, books, and supplies. The loan is for $600 at an interest rate of 4%. How much interest does the student pay if the loan is paid back in 60 days?

Solution The principal P is $600, the rate R is 4% = 0.04, and the time T is $\frac{60}{360}$. Notice that T must be given in years, and 60 days $= \frac{60}{360}$ year. Applying the formula, we have

$$I = P \cdot R \cdot T$$

$$I = 600 \times 0.04 \times \frac{60}{360}$$

$$I = 600 \times 0.04 \times \frac{1}{6} \qquad \frac{60}{360} = \frac{1}{6}$$

$$I = 4 \qquad \text{Multiply.}$$

The interest is $4.

4. If the woman in Example 4 withdrew the money after 180 days, how much did she withdraw?

EXAMPLE 4 A woman deposits $900 in an account that pays 6% annually. If she withdraws all the money in the account after 90 days, how much does she withdraw?

Solution We have $P = \$900$, $R = 0.06$, and $T = 90$ days $= \frac{90}{360}$ year. Using these numbers in the formula, we have

$$I = P \cdot R \cdot T$$

$$I = 900 \times 0.06 \times \frac{90}{360}$$

$$I = 900 \times 0.06 \times \frac{1}{4} \qquad \frac{90}{360} = \frac{1}{4}$$

$$I = 13.5 \qquad \text{Multiply.}$$

The interest earned in 90 days is $13.50. If the woman withdraws all the money in her account, she will withdraw

$900.00	Original amount (principal)
+ 13.50	Interest for 90 days
$913.50	Total amount withdrawn

The woman will withdraw $913.50.

C Compound Interest

A second common kind of interest is *compound interest*. Compound interest includes interest paid on interest. We can use what we know about simple interest to help us solve problems involving compound interest.

EXAMPLE 5   A homemaker puts $3,000 into a savings account that pays 7% compounded annually. How much money is in the account at the end of 2 years?

Solution   Because the account pays 7% annually, the simple interest at the end of 1 year is 7% of $3,000.

$$\text{Interest after 1 year} = 0.07(\$3,000)$$
$$= \$210$$

Because the interest is paid annually, at the end of 1 year the total amount of money in the account is

$3,000	Original amount
+ 210	Interest for 1 year
$3,210	Total in account after 1 year

The interest paid for the second year is 7% of this new total, or

$$\text{Interest paid the second year} = 0.07(\$3,210)$$
$$= \$224.70$$

At the end of 2 years, the total in the account is

$3,210.00	Amount at the beginning of second year
+ 224.70	Interest paid for second year
$3,434.70	Account after 2 years

At the end of 2 years, the account totals $3,434.70. The total interest earned during this 2-year period is $210 (first year) + $224.70 (second year) = $434.70.

You may have heard of savings and loan companies that offer interest rates that are compounded quarterly. If the interest rate is, say, 6% and it is compounded quarterly, then after every 90 days ($\frac{1}{4}$ of a year) the interest is added to the account. If it is compounded semiannually, then the interest is added to the account every 6 months. Most accounts have interest rates that are compounded daily, which means the simple interest is computed daily and added to the account.

EXAMPLE 6   If $10,000 is invested in a savings account that pays 6% compounded quarterly, how much is in the account at the end of a year?

Solution   The interest for the first quarter ($\frac{1}{4}$ of a year) is calculated using the formula for simple interest.

$$I = P \cdot R \cdot T$$
$$I = \$10,000 \times 0.06 \times \frac{1}{4} \quad \text{First quarter}$$
$$I = \$150$$

At the end of the first quarter, this interest is added to the original principal. The new principal is $10,000 + $150 = $10,150. Again, we apply the formula to calculate the interest for the second quarter.

$$I = \$10,150 \times 0.06 \times \frac{1}{4} \quad \text{Second quarter}$$
$$I = \$152.25$$

5. Suppose the woman in Example 5 put $5,000 in the savings account. How much is there at the end of 4 years?

Note   If the interest earned in Example 5 were calculated using the formula for simple interest, $I = P \cdot R \cdot T$, the amount of money in the account at the end of two years would be $3,420.00.

6. If $12,000 is invested in an account that pays 5% compounded quarterly, how much is in the account at the end of the year?

Answers
5. $6,553.98
6. $12,611.34

The principal at the end of the second quarter is $10,150 + $152.25 = $10,302.25. The interest earned during the third quarter is

$$I = \$10,302.25 \times 0.06 \times \frac{1}{4} \quad \text{Third quarter}$$

$$I = \$154.53 \qquad \qquad \text{Round to the nearest cent.}$$

The new principal is $10,302.25 + $154.53 = $10,456.78. Interest for the fourth quarter is

$$I = \$10,456.78 \times 0.06 \times \frac{1}{4} \quad \text{Fourth quarter}$$

$$I = \$156.85 \qquad \qquad \text{Round to the nearest cent.}$$

The total amount of money in this account at the end of 1 year is

$$\$10,456.78 + \$156.85 = \$10,613.63$$

USING TECHNOLOGY   Compound Interest from a Formula

We can summarize the previous work above with a formula that allows us to calculate compound interest for any interest rate and any number of compounding periods. If we invest P dollars at an annual interest rate r, compounded n times a year, then the amount of money in the account after t years is given by the formula

$$A = P\left(1 + \frac{r}{n}\right)^{nt}$$

Using numbers from Example 6 to illustrate, we have

P = Principal = $10,000
r = annual interest rate = 0.06
n = number of compounding periods = 4 (interest is compounded quarterly)
t = number of years = 1

Substituting these numbers into the formula above, we have

$$A = 10,000\left(1 + \frac{0.06}{4}\right)^{4 \cdot 1}$$

$$= 10,000(1 + 0.015)^4$$
$$= 10,000(1.015)^4$$

To simplify this last expression on a calculator, we have

Scientific calculator: 10,000 $\boxed{\times}$ 1.015 $\boxed{y^x}$ 4 $\boxed{=}$
Graphing calculator: 10,000 $\boxed{\times}$ 1.015 $\boxed{\wedge}$ 4 $\boxed{\text{ENTER}}$

In either case, the answer is $10,613.63551, which rounds to $10,613.64.

Note The reason that this answer is different than the result we obtained in Example 6 is that, in Example 6, we rounded each calculation as we did it. The calculator will keep all the digits in all of the intermediate calculations.

GETTING READY FOR CLASS

After reading through the preceding section, respond in your own words and in complete sentences.

A. What is the difference between the interest rate and the interest?

B. What is simple interest and how is it different than compound interest?

C. How much does the student in Example 3 pay back if the loan is paid off after a year, instead of after 60 days?

D. In Example 6, how much money would the account contain at the end of 1 year if it were compounded annually, instead of quarterly?

Vocabulary Review

Choose the correct words to fill in the blanks below.

interest rate   simple   principal   compound

1. For an investment, the amount we invest is called the _____ , the percent is called the interest rate, and the money earned is called the interest.

2. In the formula $I = P \cdot R \cdot T$, I = interest, P = principal, R = _____ , and T = time.

3. _____ interest is a percent of money earned on the principal investment only.

4. _____ interest is interest earned on interest added to the principal.

Problems

A B These problems are similar to the examples found in this section. They should be set up and solved in the same way. (Problems 1–12 involve simple interest.)

1. **Savings Account** A man invests $2,000 in a savings plan that pays 8% per year. How much money will be in the account at the end of 1 year?

2. **Savings Account** How much simple interest is earned on $5,000 if it is invested for 1 year at 5%?

3. **Savings Account** A savings account pays 7% per year. How much interest will $9,500 invested in such an account earn in a year?

4. **Savings Account** A local bank pays 5.5% annual interest on all savings accounts. If $600 is invested in this type of account, how much will be in the account at the end of a year?

5. **Bank Loan** A farmer borrows $8,000 from his local bank at 7%. How much does he pay back to the bank at the end of the year when he pays off the loan?

6. **Bank Loan** If $400 is borrowed at a rate of 12% for 1 year, how much is the interest?

7. **Bank Loan** A bank lends one of its customers $2,000 at 8% for 1 year. If the customer pays the loan back at the end of the year, how much does he pay the bank?

8. **Bank Loan** If a loan of $2,000 at 20% for 1 year is to be paid back in one payment at the end of the year, how much does the borrower pay the bank?

9. **Student Loan** A student takes out an emergency loan for tuition, books, and supplies. The loan is for $600 with an interest rate of 5%. How much interest does the student pay if the loan is paid back in 60 days?

10. **Short-Term Loan** If a loan of $1,200 at 9% is paid off in 90 days, what is the interest?

11. **Savings Account** A woman deposits $800 in a savings account that pays 5%. If she withdraws all the money in the account after 120 days, how much does she withdraw?

12. **Savings Account** $1,800 is deposited in a savings account that pays 6%. If the money is withdrawn at the end of 30 days, how much interest is earned?

C The problems that follow involve compound interest.

13. **Compound Interest** A woman puts $5,000 into a savings account that pays 6% compounded annually. How much money is in the account at the end of 2 years?

14. **Compound Interest** A savings account pays 5% compounded annually. If $10,000 is deposited in the account, how much is in the account at the end of 2 years?

15. **Compound Interest** If $8,000 is invested in a savings account that pays 5% compounded quarterly, how much is in the account at the end of a year?

16. **Compound Interest** Suppose $1,200 is invested in a savings account that pays 6% compounded semiannually. How much is in the account at the end of $1\frac{1}{2}$ years?

Calculator Problems

The following problems should be set up in the same way in which Problems 1–16 have been set up. Then the calculations should be done on a calculator.

17. **Savings Account** A woman invests $917.26 in a savings account that pays 6.25% annually. How much is in the account at the end of a year?

18. **Business Loan** The owner of a clothing store borrows $6,210 for 1 year at 11.5% interest. If he pays the loan back at the end of the year, how much does he pay back?

19. Compound Interest Suppose $10,000 is invested in each account below. In each case, find the amount of money in the account at the end of 5 years.

   a. Annual interest rate = 6%, compounded quarterly

   b. Annual interest rate = 6%, compounded monthly

   c. Annual interest rate = 5%, compounded quarterly

   d. Annual interest rate = 5%, compounded monthly

20. Compound Interest Suppose $5,000 is invested in each account below. In each case, find the amount of money in the account at the end of 10 years.

   a. Annual interest rate = 5%, compounded quarterly

   b. Annual interest rate = 6%, compounded quarterly

   c. Annual interest rate = 7%, compounded quarterly

   d. Annual interest rate = 8%, compounded quarterly

Extending the Concepts

The following problems are percent problems. Use any of the methods developed in this chapter to solve them.

21. Movie Making The bar chart shows the production costs for some of the most expensive movies to win an Academy Award for Best Picture. Find the percent increase from each Best Picture to the next. Round to the nearest percent.

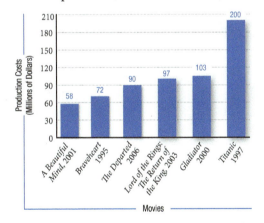

22. Movie Making The table below shows how much money each of the Academy Award winners shown at the left has grossed worldwide. Using the information from the previous problem, find the ratio of gross revenue to cost. Write your answer rounded to the nearest percent.

Grossed Revenue	
A Beautiful Mind (2001)	$314,000,000
Braveheart (1995)	$211,000,000
The Departed (2006)	$290,000,000
Lord of the Rings: The Return of the King (2003)	$1,120,000,000
Gladiator (2000)	$458,000,000
Titanic (1997)	$1,843,000,000

Find the Mistake

Each sentence below contains a mistake. Circle the mistake and write the correct word(s) or numbers(s) on the line provided.

  1. A woman invests $1,500 into an account with a 6% annual interest rate. She will have $90 in her account by the end of one year. _____

  2. A business man invests $2,750 into an account that has an 8% interest rate per year. To find out how much money will be in the man's account after 72 days, you must multiply the product of 2,750 and 0.08 by 72. _____

  3. If a person invests $10,000 into an account that is compounded annually at 6%, then after two years, the account will contain $11,200. _____

  4. A woman deposits $4,000 into a savings account that pays 7% compounded quarterly. At the end of the year, the account contains $4,280. _____

Supplies Needed

A Access to research information, such as a library or the internet

Everyday Math

Most people don't realize how often we use math skills in our daily lives. The obvious skills are used doing things such as paying for groceries, calculating supplies for a home project, or cooking from a recipe. This project will help shed light on some lesser known ways of using math. Choose an occupation from the list below. Research what and how math skills are used in the chosen occupation. Present your findings to the class.

Accountant

Agriculturist

Biologist

Carpenter

Chemist

Computer Programmer

Engineer

Geologist

Graphic Designer

Lawyer

Manager

Marketer

Nurse and Doctor

Meteorologist

Politician

Repair Technician (plumber, electrician, mechanic, etc.)

Teacher

Chapter 6 Summary

The Meaning of Percent [6.1]

Percent means "per hundred." It is a way of comparing numbers to the number 100.

EXAMPLES

1. 42% means 42 per hundred or $\frac{42}{100}$.

Changing Percents to Decimals [6.1]

To change a percent to a decimal, drop the % symbol and move the decimal point two places to the *left*.

2. 75% = 0.75

Changing Decimals to Percents [6.1]

To change a decimal to a percent, move the decimal point two places to the *right*, and use the % symbol.

3. 0.25 = 25%

Changing Percents to Fractions [6.1]

To change a percent to a fraction, drop the % symbol, and use a denominator of 100. Reduce the resulting fraction to lowest terms if necessary.

4. $6\% = \frac{6}{100} = \frac{3}{50}$

Changing Fractions to Percents [6.1]

To change a fraction to a percent, either write the fraction as a decimal and then change the decimal to a percent, or write the fraction as an equivalent fraction with denominator 100, drop the 100, and use the % symbol.

5. $\frac{3}{4} = 0.75 = 75\%$

or

$\frac{9}{10} = \frac{90}{100} = 90\%$

Basic Word Problems Involving Percents [6.2]

There are three basic types of word problems:

Type A: What number is 14% of 68?

Type B: What percent of 75 is 25?

Type C: 25 is 40% of what number?

6. Translating to equations, we have:

Type A: $n = 0.14(68)$; $n = 9.52$
Type B: $75n = 25$; $n = 0.33$
Type C: $25 = 0.40n$; $n = 62.5$

Applications of Percent [6.3, 6.4, 6.5, 6.6]

To solve application problems, we write *is* as =, *of* as · (multiply), and *what number* or *what percent* as *n*. We then solve the resulting equation to find the answer to the original question.

There are many different kinds of application problems involving percent. They include problems on income tax, sales tax, commission, discount, percent increase and decrease, and interest. Generally, to solve these problems, we restate them as an equivalent problem of Type A, B, or C above. Problems involving simple interest can be solved using the formula

$$I = P \cdot R \cdot T$$

where I = interest, P = principal, R = interest rate, and T = time (in years). It is standard procedure with simple interest problems to use 360 days = 1 year.

COMMON MISTAKE

1. A common mistake is forgetting to change a percent to a decimal when working problems that involve percents in the calculations. We always change percents to decimals before doing any calculations.

2. Moving the decimal point in the wrong direction when converting percents to decimals or decimals to percents is another common mistake. Remember, *percent* means "per hundred." Rewriting a number expressed as a percent as a decimal will make the numerical part smaller.

    25% = 0.25

Write each percent as a decimal. [6.1]

1. 56%   **2.** 3%   **3.** 0.4%

Write each decimal as a percent. [6.1]

4. 0.32   **5.** 0.7   **6.** 1.64

Write each percent as a fraction or a mixed number in lowest terms. [6.1]

7. 85%   **8.** 128%   **9.** 8.4%

Write each number as a percent. [6.1]

10. $\dfrac{13}{25}$ **11.** $\dfrac{5}{8}$ **12.** $1\dfrac{9}{20}$

13. What number is 20% of 64? [6.2]

14. What percent of 50 is 30? [6.2]

15. 64 is 80% of what number? [6.2]

16. Driver's Test On a 25-question driver's test, a student answered 21 questions correctly. What percent of the questions did the student answer correctly? [6.3]

17. Commission A salesperson gets a 4% commission rate on all computers she sells. If she sells $25,000 in computers in one day, what is her commission? [6.4]

18. Discount A washing machine that usually sells for $830 is marked down to $539.50. What is the discount? What is the discount rate? [6.5]

19. Total Price A tennis racket that normally sells for $95 is on sale for 25% off. If the sales tax rate is 8.5%, what is the total price of the tennis racket if it is purchased during the sale? Round to the nearest cent. [6.5]

20. Percent Increase A driver gets into a car accident and his insurance increases by 17%. If he paid $860 before the accident, how much is he paying now? [6.5]

21. Simple Interest If $9,600 is invested at 5% simple interest for 8 months, how much interest is earned? [6.6]

22. Compound Interest How much interest will be earned on a savings account that pays 9% compounded annually, if $18,000 is invested for 3 years? [6.6]

The chart shows the distribution of iPhones throughout the world. If 190,000,000 iPhones have been sold worldwide, use the information to answer the following questions..

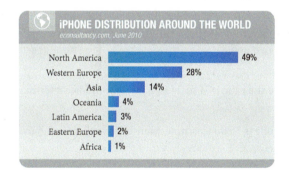

23. How many iPhones have been sold in North America?

24. How many iPhones have been sold in Eastern Europe?

Simplify.

1. 4,381
     623
   + 407

2.   3,062
     − 1,971

3. 13(613)

4. $1,335 \div 25$

5. $5.07\overline{)162.24}$

6. $\dfrac{6}{5} - \dfrac{2}{5}$

7. $\left(\dfrac{3}{5}\right)^3$

8. $7.462 + 2.04$

9. $3 - 2.714$

10. $3.5(0.34)$

11. $\dfrac{8}{21} \cdot \dfrac{7}{24}$

12. $\dfrac{16}{3} \div \dfrac{4}{9}$

13. $\dfrac{5}{6} + \dfrac{3}{8}$

14. $7\dfrac{2}{3} - 3\dfrac{5}{6}$

15. $4 \cdot 3\dfrac{5}{8}$

16. Subtract $2\dfrac{3}{5}$ from 8.6.

17. Find the quotient of $2\dfrac{1}{3}$ and $\dfrac{1}{6}$.

18. Translate into symbols, and then simplify: Three times the sum of 6 and 8.

19. Write the ratio of 4 to 24 as a fraction in lowest terms.

20. If 1 mile is 5,280 feet, how many feet are there in 3.6 miles?

21. If 1 square yard is 1,296 square inches, how many square inches are in $\dfrac{7}{8}$ square yard?

22. Write $\dfrac{5}{8}$ as a percent.

23. Convert 32% to a fraction.

24. Solve the equation $\dfrac{3}{x} = \dfrac{6}{7}$.

25. $2 \cdot 4^2 + 2 \cdot 3^3 - 6 \cdot 2^3$

26. What number is 13% of 30?

27. 217 is what percent of 620?

28. 26.6 is 28% of what number?

29. Unit Pricing If a six-pack of Coke costs $7.95, what is the price per can to the nearest cent?

30. Unit Pricing A quart of 2% reduced-fat milk contains four 1-cup servings. If the quart costs $3.65, find the price per serving to the nearest cent.

31. Temperature Use the formula $C = \dfrac{5(F - 32)}{9}$ to find the temperature in degrees Celsius when the Fahrenheit temperature is 95°F.

32. Percent Increase Kendra is earning $2,800 a month when she receives a raise to $3,066 a month. What is the percent increase in her monthly salary?

33. Driving Distance If Ethan drives his car 225 miles in 3 hours, how far will he drive in 5 hours if he drives at the same rate?

34. Movie Tickets A movie theater has a total of 375 seats. If they have a sellout crowd for a matinee and each ticket costs $8.50, how much money will ticket sales bring in that afternoon?

35. Geometry Find the perimeter and area of a square with side 6.2 inches.

36. Hourly Pay Jean tutors in the math lab and earns $76.50 in one week. If she works 9 hours that week, what is her hourly pay?

The chart shows the minimum amount of space required for each of the sports fields or courts. Use the information to answer the following questions. Round to the nearest hundredth if necessary.

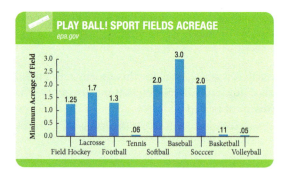

37. If a school will be building 4 baseball diamonds and 7 basketball courts, what is the minimum acreage it will need?

38. If a city will be adding a field hockey facility, two soccer fields, and five volleyball courts, what is the minimum acreage necessary?

Write each percent as a decimal. [6.1]

1. 27%   **2.** 6%   **3.** 0.9%

Write each decimal as a percent. [6.1]

4. 0.64   **5.** 0.3   **6.** 1.49

Write each percent as a fraction or a mixed number in lowest terms. [6.1]

7. 45%   **8.** 136%   **9.** 7.2%

Write each number as a percent. [6.1]

10. $\dfrac{13}{20}$ **11.** $\dfrac{7}{8}$ **12.** $2\dfrac{1}{4}$

13. What number is 25% of 48? [6.2]

14. What percent of 80 is 28? [6.2]

15. 30 is 40% of what number? [6.2]

16. Driver's Test On a 25-question driver's test, a student answered 24 questions correctly. What percent of the questions did the student answer correctly? [6.3]

17. Commission A salesperson gets a 6% commission rate on all computers she sells. If she sells $15,000 in computers in one day, what is her commission? [6.4]

18. Discount A washing machine that usually sells for $725 is marked down to $580. What is the discount? What is the discount rate? [6.5]

19. Total Price A tennis racket that normally sells for $179 is on sale for 30% off. If the sales tax rate is 8%, what is the total price of the tennis racket if it is purchased during the sale? Round to the nearest cent. [6.5]

20. Percent Increase A driver gets into a car accident and his insurance increases by 14%. If he paid $760 before the accident, how much is he paying now? [6.5]

21. Simple Interest If $6,000 is invested at 7% simple interest for 4 months, how much interest is earned? [6.6]

22. Compound Interest How much interest will be earned on a savings account that pays 8% compounded annually, if $16,000 is invested for 2 years? [6.6]

The diagram shows the number of finishers for the Nike Women's Marathon in San Francisco. Use the information to answer the following questions.

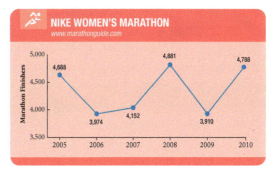

23. What is the percent increase in finishers from 2009 to 2010? Round to the nearest hundredth.

24. If the trend from 2009 to 2010 holds, how many finishers will there be in 2011? Round to the nearest person.

Measurement

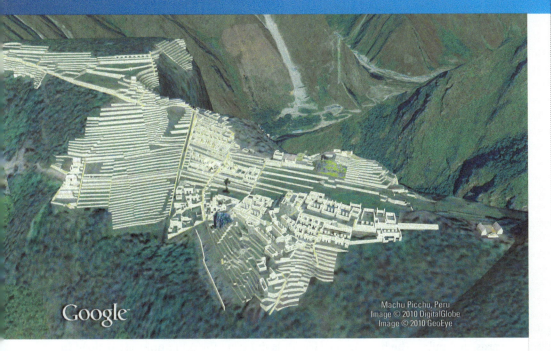
Machu Picchu, Peru
Image © 2010 DigitalGlobe
Image © 2010 GeoEye

The ruins of Machu Picchu stand high in the Andes mountains of Peru and are among the most beautiful and mysterious ancient building sites in the world. Most scholars believe that the Incas began construction on the site around 1400 AD. Archaeological excavations have found temples, baths, palaces, storage rooms, and 150 houses. Builders used blocks weighing up to 50 tons to create the structures. Despite the primitive technology of the time, the blocks form tight mortar-less joints that the blade of a knife cannot fit between. Alignment of some of the structures also shows remarkable knowledge of astronomy still being studied today. And the surrounding agricultural terraces and local springs are believed to have produced enough food and water to make the small city self-supporting.

The ruins are also very well preserved, adding to their archaeological value. At 2,350 meters above sea level, they sit on the edge of a sheer cliff and are invisible from the river that flows through the Urubamba Valley 600 meters below. Though the Spanish conquerors reportedly searched for the location a century after its construction, they never found it. As a result, it is one of the only Incan settlements that was not plundered by the conquering armies.

Suppose you want to convert the above heights to the more familiar units of feet and miles. You could use a table like the one shown below and a process called unit analysis, which is one of the topics of this chapter. Unit analysis can be used to compare any units of measure, such as those for length, weight, temperature, and volume. Once you have worked through this chapter, come back and convert these distances so you can begin planning your trip to Machu Picchu!

Conversion Factors Between the Metric and U.S. Systems of Measurement	
Length	
inches and centimeters	2.54 cm = 1 in.
feet and meters	1 m = 3.28 ft
miles and kilometers	1.61 km = 1 mi

OBJECTIVES

A Convert units of length using the U.S. system of measurement.

B Convert units of length using the metric system.

C Work problems using unit analysis.

KEY WORDS

U.S. system of measurement

length

conversion factor

metric system of measurement

unit analysis

Video Examples
Section 7.1

Image © sxc.hu, verzerk, 2005

In this section, we will become more familiar with the units used to measure length. We will look at the U.S. system of measurement and the metric system of measurement.

A U.S. System of Measurement

Measuring the length of an object is done by assigning a number to its length. To let other people know what that number represents, we include with it a unit of measure. The most common units used to represent *length* in the U.S. system are inches, feet, yards, and miles. The basic unit of length is the foot. The other units are defined in terms of feet, as Table 1 shows.

Table 1

12 inches (in.)	=	1 foot (ft)
1 yard (yd)	=	3 feet (ft)
1 mile (mi)	=	5,280 feet ft)

As you can see from the table, the abbreviations for inches, feet, yards, and miles are in., ft, yd, and mi, respectively. What we haven't indicated, even though you may not have realized it, is what 1 foot represents. We have defined all our units associated with length in terms of feet, but we haven't said what a foot is.

There is a long history of the evolution of what is now called a foot. At different times in the past, a foot has represented different arbitrary lengths. Currently, a foot is defined to be exactly 0.3048 meter (the basic measure of length in the metric system), where a meter is 1,650,763.73 wavelengths of the orange-red line in the spectrum of krypton-86 in a vacuum (this doesn't mean much to me either). The reason a foot and a meter are defined this way is that we always want them to measure the same length. Because the wavelength of the orange-red line in the spectrum of krypton-86 will always remain the same, so will the length that a foot represents.

Now that we have said what we mean by 1 foot (even though we may not understand the technical definition), we can go on and look at some examples that involve converting from one kind of unit to another.

EXAMPLE 1   Convert 5 feet to inches.

Solution   Because 1 foot = 12 inches, we can multiply 5 by 12 inches to get

$$5 \text{ feet} = 5 \times 12 \text{ inches}$$

$$= 60 \text{ inches}$$

Practice Problems

1. Convert 3 feet to inches.

Answer

1. 36 inches

This method of converting from feet to inches probably seems fairly simple. But as we go further in this chapter, the conversions from one kind of unit to another will become more complicated. For these more complicated problems, we need another way to show conversions so that we can be certain to end them with the correct unit of measure. For example, since 1 ft = 12 in., we can say that there are 12 in. per 1 ft or 1 ft per 12 in.; that is,

$$\frac{12 \text{ in.}}{1 \text{ ft}} \longleftarrow \textit{Per} \qquad \text{or} \qquad \frac{1 \text{ ft}}{12 \text{ in.}} \longleftarrow \textit{Per}$$

We call the expressions $\frac{12 \text{ in.}}{1 \text{ ft}}$ and $\frac{1 \text{ ft}}{12 \text{ in.}}$ *conversion factors*. The fraction bar is read as "per." Both these conversion factors are really just the number 1. That is,

$$\frac{12 \text{ in.}}{1 \text{ ft}} = \frac{12 \text{ in.}}{12 \text{ in.}} = 1$$

We already know that multiplying a number by 1 leaves the number unchanged. So, to convert from one unit to the other, we can multiply by one of the conversion factors without changing value. Both the conversion factors above say the same thing about the units feet and inches. They both indicate that there are 12 inches in every foot. The one we choose to multiply by depends on what units we are starting with and what units we want to end up with. If we start with feet and we want to end up with inches, we multiply by the conversion factor

$$\frac{12 \text{ in.}}{1 \text{ ft}}$$

The units of feet will divide out and leave us with inches.

$$5 \text{ feet} = 5 \text{ ft} \times \frac{12 \text{ in.}}{1 \text{ ft}}$$
$$= 5 \times 12 \text{ in.}$$
$$= 60 \text{ in.}$$

The key to this method of conversion lies in setting the problem up so that the correct units divide out to simplify the expression. We are treating units such as feet in the same way we treated factors when reducing fractions. If a factor is common to the numerator and the denominator, we can divide it out and simplify the fraction. The same idea holds for units such as feet.

We can rewrite Table 1 so that it shows the conversion factors associated with units of length, as shown in Table 2.

Table 2

Units of Length in the U.S. System		
The relationship between	is	To convert one to the other, multiply by
feet and inches	12 in. = 1 ft	$\frac{12 \text{ in.}}{1 \text{ ft}}$ or $\frac{1 \text{ ft}}{12 \text{ in.}}$
feet and yards	1 yd = 3 ft	$\frac{3 \text{ ft}}{1 \text{ yd}}$ or $\frac{1 \text{ yd}}{3 \text{ ft}}$
feet and miles	1 mi = 5,280 ft	$\frac{5,280 \text{ ft}}{1 \text{ mi}}$ or $\frac{1 \text{ mi}}{5,280 \text{ ft}}$

EXAMPLE 2 The most common ceiling height in houses is 8 feet. How many yards is this?

Solution To convert 8 feet to yards, we multiply by the conversion factor $\frac{1 \text{ yd}}{3 \text{ ft}}$ so that feet will divide out and we will be left with yards.

$$8 \text{ ft} = 8 \text{ ft} \times \frac{1 \text{ yd}}{3 \text{ ft}} \qquad \textit{Multiply by correct conversion factor.}$$
$$= \frac{8}{3} \text{ yd} \qquad \textit{8} \times \frac{1}{3} = \frac{8}{3}$$
$$= 2\frac{2}{3} \text{ yd} \qquad \textit{Or 2.67 yd rounded to the nearest hundredth.}$$

2. Suppose a house has high ceilings. Convert 12 feet to yards.

3. A football field is 53.33 yards wide. How many inches wide is a football field? Round your answer to the nearest inch.

EXAMPLE 3   A football field is 100 yards long. How many inches long is a football field?

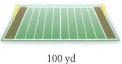

100 yd

Solution   In this example, we must convert yards to feet and then feet to inches. We choose the conversion factors that will allow all the units except inches to divide out.

$$100 \text{ yd} = 100 \text{ yd} \times \frac{3 \text{ ft}}{1 \text{ yd}} \times \frac{12 \text{ in.}}{1 \text{ ft}}$$

$$= 100 \times 3 \times 12 \text{ in.}$$

$$= 3{,}600 \text{ in.}$$

B Metric Units of Length

In the metric system, the standard unit of length is a meter. A meter is a little longer than a yard (about 3.4 inches longer). The other units of length in the metric system are written in terms of a meter. The metric system uses prefixes to indicate what part of the basic unit of measure is being used. For example, in *milli*meter the prefix *milli* means "one thousandth" of a meter. Table 3 gives the meanings of the most common metric prefixes.

Table 3

The Meaning of Metric Prefixes

Prefix	Meaning
milli	0.001
centi	0.01
deci	0.1
deka	10
hecto	100
kilo	1,000

Table 4

Metric Units of Length

The relationship between	is	To convert from one to the other, multiply by	
millimeters (mm) and meters (m)	1,000 mm = 1 m	$\frac{1{,}000 \text{ mm}}{1 \text{ m}}$ or	$\frac{1 \text{ m}}{1{,}000 \text{ mm}}$
centimeters (cm) and meters	100 cm = 1 m	$\frac{100 \text{ cm}}{1 \text{ m}}$ or	$\frac{1 \text{ m}}{100 \text{ cm}}$
decimeters (dm) and meters	10 dm = 1 m	$\frac{10 \text{ dm}}{1 \text{ m}}$ or	$\frac{1 \text{ m}}{10 \text{ dm}}$
dekameters (dam) and meters	1 dam = 10 m	$\frac{10 \text{ m}}{1 \text{ dam}}$ or	$\frac{1 \text{ dam}}{10 \text{ m}}$
hectometers (hm) and meters	1 hm = 100 m	$\frac{100 \text{ m}}{1 \text{ hm}}$ or	$\frac{1 \text{ hm}}{100 \text{ m}}$
kilometers (km) and meters	1 km = 1,000 m	$\frac{1{,}000 \text{ m}}{1 \text{ km}}$ or	$\frac{1 \text{ km}}{1{,}000 \text{ m}}$

We can use these prefixes to write the other units of length and conversion factors for the metric system, as given in Table 4.

We use the same method to convert between units in the metric system as we did with the U.S. system. We choose the conversion factor that will allow the units we start with to divide out, leaving the units we want to end up with.

Answer

3. 1,920 inches

EXAMPLE 4   Convert 25 millimeters to meters.

Solution   To convert from millimeters to meters, we multiply by the conversion factor $\frac{1\,m}{1{,}000\,mm}$.

$$25\text{ mm} = 25\text{ mm} \times \frac{1\text{ m}}{1{,}000\text{ mm}}$$

$$= \frac{25\text{ m}}{1{,}000}$$

$$= 0.025\text{ m}$$

EXAMPLE 5   Convert 36.5 centimeters to decimeters.

Solution   We convert centimeters to meters and then meters to decimeters.

$$36.5\text{ cm} = 36.5\text{ cm} \times \frac{1\text{ m}}{100\text{ cm}} \times \frac{10\text{ dm}}{1\text{ m}}$$

$$= \frac{36.5 \times 10}{100}\text{ dm}$$

$$= 3.65\text{ dm}$$

C Unit Analysis

The most common units of length in the metric system are millimeters, centimeters, meters, and kilometers. The other units of length we have listed in our table of metric lengths are not as widely used. The method we have used to convert from one unit of length to another in Examples 1–5 is called *unit analysis*. If you take a chemistry class, you will see it used many times. The same is true of many other science classes as well.

We can summarize the procedure used in unit analysis with the following steps:

HOW TO   Steps Used in Unit Analysis

Step 1: Identify the units you are starting with.

Step 2: Identify the units you want to end with.

Step 3: Find conversion factors that will bridge the starting units and the ending units.

Step 4: Set up the multiplication problem so that all units except the units you want to end with will divide out.

EXAMPLE 6   A sheep rancher is making new lambing pens for the upcoming lambing season. Each pen is a rectangle 6 feet wide and 8 feet long. The fencing material he wants to use sells for $1.36 per foot. If he is planning to build five separate lambing pens (they are separate because he wants a walkway between them), how much will he have to spend for fencing material?

Solution   To find the amount of fencing material he needs for one pen, we find the perimeter of a pen.

Perimeter = 6 + 6 + 8 + 8 = 28 feet

4. Convert 35 millimeters to meters.

5. Convert 50.5 centimeters to decimeters.

6. Suppose the rancher in Example 6 is building 7 pens. How much will he spend?

We set up the solution to the problem using unit analysis. Our starting unit is *pens* and our ending unit is *dollars*. Here are the conversion factors that will form a bridge between pens and dollars:

$$1 \text{ pen} = 28 \text{ feet of fencing}$$

$$1 \text{ foot of fencing} = 1.36 \text{ dollars}$$

Next we write the multiplication problem, using the conversion factors, that will allow all the units except dollars to divide out.

$$5 \text{ pens} = 5 \text{ pens} \times \frac{28 \text{ feet of fencing}}{1 \text{ pen}} \times \frac{1.36 \text{ dollars}}{1 \text{ foot of fencing}}$$

$$= 5 \times 28 \times 1.36 \text{ dollars}$$

$$= \$190.40$$

7. Convert the advertised speed in Example 7 to yards per minute.

EXAMPLE 7   In 1993, a ski resort in Vermont advertised their new high-speed chair lift as "the world's fastest chair lift, with a speed of 1,100 feet per second." Show why the speed cannot be correct.

Solution   To solve this problem, we can convert feet per second into miles per hour, a unit of measure we are more familiar with on an intuitive level. Here are the conversion factors we will use:

$$1 \text{ mile} = 5,280 \text{ feet}$$

$$1 \text{ hour} = 60 \text{ minutes}$$

$$1 \text{ minute} = 60 \text{ seconds}$$

$$1,100 \text{ ft/second} = \frac{1,100 \text{ feet}}{1 \text{ second}} \times \frac{1 \text{ mile}}{5,280 \text{ feet}} \times \frac{60 \text{ seconds}}{1 \text{ minute}} \times \frac{60 \text{ minutes}}{1 \text{ hour}}$$

$$= \frac{1,100 \times 60 \times 60 \text{ miles}}{5,280 \text{ hours}}$$

$$= 750 \text{ miles/hour}$$

GETTING READY FOR CLASS

After reading through the preceding section, respond in your own words and in complete sentences.

A. What is a conversion factor?

B. Write the relationship between feet and miles. That is, write an equality that shows how many feet are in every mile.

C. Explain how the metric system uses prefixes to indicate units of measure.

D. Briefly explain how you would use unit analysis to solve a problem.

Answer

7. 22,000 yd/min

Vocabulary Review

Choose the correct words to fill in the blanks below.

    conversion factor   divide   foot   unit analysis   fraction bar   meter

1. In the U.S. system of measurement, the basic unit of length is the _____.

2. In the metric system of measurement, the basic unit of length is the _____.

3. A _____ is a way to convert one unit of measurement to another without changing the value.

4. The _____ in a conversion factor is read as "per."

5. The process of converting one unit of measurement to another is called _____.

6. For the last step of unit analysis, set up a multiplication problem so that all units except the units you want to end with will _____ out.

Problems

A Make the following conversions in the U.S. system by multiplying by the appropriate conversion factor. Write your answers as whole numbers or mixed numbers.

1. 5 ft to inches

2. 9 ft to inches

3. 10 ft to inches

4. 20 ft to inches

5. 2 yd to feet

6. 8 yd to feet

7. 4.5 yd to inches

8. 9.5 yd to inches

9. 27 in. to feet

10. 36 in. to feet

11. 2.5 mi to feet

12. 6.75 mi to feet

13. 48 in. to yards

14. 56 in. to yards

B Make the following conversions in the metric system by multiplying by the appropriate conversion factor. Write your answers as whole numbers or decimals.

15. 18 m to centimeters

16. 18 m to millimeters

17. 4.8 km to meters

18. 8.9 km to meters

19. 5 dm to centimeters

20. 12 dm to millimeters

21. 248 m to kilometers

22. 969 m to kilometers

23. 67 cm to millimeters

24. 67 mm to centimeters

25. 3,498 cm to meters

26. 4,388 dm to meters

27. 63.4 cm to decimeters

28. 89.5 cm to decimeters

Applying the Concepts

29. Softball If the distance between first and second base in softball is 60 feet, how many yards is it from first to second base?

30. Tower Height A transmitting tower is 100 feet tall. How many inches is that?

31. High Jump If a person high jumps 6 feet 8 inches, how many inches is the jump?

6 ft 8 in

32. Desk Width A desk is 48 inches wide. What is the width in yards?

33. Ceiling Height Suppose the ceiling of a home is 2.44 meters above the floor. Express the height of the ceiling in centimeters.

34. Notebook Width Standard-sized notebook paper is 21.6 centimeters wide. Express this width in millimeters.

35. Dollar Width A dollar bill is about 6.5 centimeters wide. Express this width in millimeters.

36. Pencil Length Most new pencils are 19 centimeters long. Express this length in meters.

37. Surveying A unit of measure sometimes used in surveying is the *chain*. There are 80 chains in 1 mile. How many chains are in 37 miles?

38. Surveying Another unit of measure used in surveying is a *link*; 1 link is about 8 inches. About how many links are there in 5 feet?

39. Metric System A very small unit of measure in the metric system is the *micron* (abbreviated μm). There are 1,000 μm in 1 millimeter. How many microns are in 12 centimeters?

40. Metric System Another very small unit of measure in the metric system is the *angstrom* (abbreviated Å). There are 10,000,000 Å in 1 millimeter. How many angstroms are in 15 decimeters?

41. Horse Racing In horse racing, 1 *furlong* is 220 yards. How many feet are in 12 furlongs?

42. Sailing A *fathom* is a measurement of the depth of water and is equivalent to 6 feet. How many yards are in 19 fathoms?

43. Cell Phones The typical wavelength used by most cell phones today is between 12 and 35 centimeters. Convert these numbers to meters.

44. Credit Card A typical credit card is 85.6 millimeters long. Convert this to meters.

45. Human Brain The power used by the human brain is 16 watts. Convert this to kilowatts.

46. DNA The length of a DNA strand in the human genome is about 70.9 inches long. Convert this number to feet. Round to the nearest tenth.

47. Speed Limit The maximum speed limit on part of Highway 101 in California is 55 miles/hour. Convert 55 miles/hour to feet/second. (Round to the nearest tenth.)

48. Speed Limit The maximum speed limit on part of Highway 5 in California is 65 miles/hour. Convert 65 miles/hour to feet/second. (Round to the nearest tenth.)

49. Track and Field A person who runs the 100-yard dash in 10.5 seconds has an average speed of 9.52 yards/second. Convert 9.52 yards/second to miles/hour. (Round to the nearest tenth.)

50. Track and Field A person who runs a mile in 8 minutes has an average speed of 0.125 miles/minute. Convert 0.125 miles/minute to miles/hour.

51. Speed of a Bullet The bullet from a rifle leaves the barrel traveling 1,500 feet/second. Convert 1,500 feet/second to miles/hour. (Round to the nearest whole number.)

52. Speed of a Bullet A bullet from a machine gun on a B-17 Flying Fortress in World War II had a muzzle speed of 1,750 feet/second. Convert 1,750 feet/second to miles/hour. (Round to the nearest whole number.)

53. Farming A farmer is fencing a pasture that is $\frac{1}{2}$ mile wide and 1 mile long. If the fencing material sells for $1.15 per foot, how much will it cost him to fence all four sides of the pasture?

54. Cost of Fencing A family with a swimming pool puts up a chain-link fence around the pool. The fence forms a rectangle 12 yards wide and 24 yards long. If the chain-link fence sells for $2.50 per foot, how much will it cost to fence all four sides of the pool?

55. Farming A 4-H Club group is raising lambs to show at the County Fair. Each lamb eats $\frac{1}{8}$ of a bale of alfalfa a day. If the alfalfa costs $10.50 per bale, how much will it cost to feed one lamb for 120 days?

56. Farming A 4-H Club group is raising pigs to show at the County Fair. Each pig eats 2.4 pounds of grain a day. If the grain costs $5.25 per pound, how much will it cost to feed one pig for 60 days?

Calculator Problems

Set up the following conversions as you have been doing. Then perform the calculations on a calculator.

57. Change 751 miles to feet.

58. Change 639.87 centimeters to meters.

59. Change 4,982 yards to inches.

60. Change 379 millimeters to kilometers.

61. Mount Whitney is the highest point in California. It is 14,494 feet above sea level. Give its height in miles to the nearest tenth.

62. The tallest mountain in the United States is Mount McKinley in Alaska. It is 20,320 feet tall. Give its height in miles to the nearest tenth.

63. California has 3,427 miles of shoreline. How many feet is this?

64. The tip of the television tower at the top of the Empire State Building in New York City is 1,472 feet above the ground. Express this height in miles to the nearest hundredth.

Getting Ready for the Next Section

Perform the indicated operations.

65. 12×12

66. 36×24

67. $1 \times 4 \times 2$

68. $5 \times 4 \times 2$

69. $10 \times 10 \times 10$

70. $100 \times 100 \times 100$

71. $75 \times 43{,}560$

72. $55 \times 43{,}560$

73. $864 \div 144$

74. $1,728 \div 144$

75. $256 \div 640$

76. $960 \div 240$

77. $45 \times \dfrac{9}{1}$

78. $36 \times \dfrac{9}{1}$

79. $1,800 \times \dfrac{1}{4}$

80. $2,000 \times \dfrac{1}{4} \times \dfrac{1}{10}$

81. 1.5×30

82. 1.5×45

83. $2.2 \times 1,000$

84. $3.5 \times 1,000$

85. 67.5×9

86. 43.5×9

Find the Mistake

Each sentence below contains a mistake. Circle the mistake and write the correct word(s) or numbers(s) on the line provided.

1. The average length for a house cat is roughly 18 inches. To show how many feet this is, multiply by the conversion factor $\dfrac{12 \text{ inches}}{1 \text{ foot}}$. _____

2. The length of a parking space is based on the average car length, which measures approximately 15 feet. This length converted to yards is 45 yards. _____

3. A cell phone measures 100 mm in length. To show how many meters this is, multiply 100 mm by the conversion factor $\dfrac{1 \text{ mm}}{1,000 \text{ m}}$. _____

4. To complete a unit analysis problem, make sure the units you want to end with will divide out.

_____

Navigation Skills: Prepare, Study, Achieve

Think about your current study routine. Has it been successful? There are many things that you must consider when creating a routine. One important aspect is the environment in which you choose to study. Think about the location you typically study. Are you able to focus there without distraction? Consider what things may distract you and find a place to study where these things are absent. Other important aspects of a productive study routine are time of day you choose to study and sights and sounds around you during your study time.

7.2 Unit Analysis II: Area and Volume

A Area

Currents in the Pacific Ocean have helped to form the Great Pacific Garbage Patch. The patch is an area in the ocean at least the size of Texas where a high concentration of trash and pollutants have accumulated. Dutch scientists and engineers have come up with a plan to turn the garbage patch into a self-sustaining island that would be recognized as an individual nation. The scientists' goal for the island is to recycle plastic waste on the spot, therefore removing the pollutants from the water.

Known as Recycled Island, the proposed size for the island is 4,000 square miles. This measurement is one we've worked with previously to represent the area of an object. Let's review some other formulas for finding the area of a square, a rectangle, and a triangle. Then we'll use unit analysis to solve problems involving such areas.

KEY WORDS

area

volume

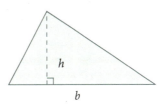

Area = (side)(side)
= (side)^2
$A = s^2$

Area = (length)(width)
$A = lw$

Area = $\frac{1}{2}$(base)(height)
$A = \frac{1}{2}bh$

FIGURE 1

**Video Examples
Section 7.2**

EXAMPLE 1 Find the number of square inches in 1 square foot.

Solution We can think of 1 square foot as 1 ft^2 = 1 ft × ft. To convert from feet to inches, we use the conversion factor 1 foot = 12 inches. Because the unit foot appears twice in 1 ft^2, we multiply by our conversion factor twice.

$$1 \text{ ft}^2 = 1 \text{ ft} \times \text{ft} \times \frac{12 \text{ in.}}{1 \text{ ft}} \times \frac{12 \text{ in.}}{1 \text{ ft}} = 12 \times 12 \text{ in.} \times \text{in.} = 144 \text{ in}^2$$

Now that we know that 1 ft^2 is the same as 144 in^2, we can use this fact as a conversion factor to convert between square feet and square inches. Depending on which units we are converting from, we would use either

$$\frac{144 \text{ in}^2}{1 \text{ ft}^2} \qquad \text{or} \qquad \frac{1 \text{ ft}^2}{144 \text{ in}^2}$$

Practice Problems

1. Find the number of square feet in 1 square yard.

Answer
1. 9 ft^2

2. Suppose there is a 2 inch frame around the poster in Example 2. How many square feet of wall space will be covered?

EXAMPLE 2   A rectangular poster measures 36 inches by 24 inches. How many square feet of wall space will the poster cover?

Solution   One way to work this problem is to find the number of square inches the poster covers, and then convert square inches to square feet.

$$\text{Area of poster} = \text{length} \times \text{width} = 36 \text{ in.} \times 24 \text{ in.} = 864 \text{ in}^2$$

To finish the problem, we convert square inches to square feet.

$$864 \text{ in}^2 = 864 \text{ in}^2 \times \frac{1 \text{ ft}^2}{144 \text{ in}^2}$$

$$= \frac{864}{144} \text{ ft}^2$$

$$= 6 \text{ ft}^2$$

Table 1 gives the most common units of area in the U.S. system of measurement, along with the corresponding conversion factors.

Table 1

U.S. Units of Area

The relationship between	is	To convert from one to the other, multiply by
square inches and square feet	$144 \text{ in}^2 = 1 \text{ ft}^2$	$\frac{144 \text{ in}^2}{1 \text{ ft}^2}$ or $\frac{1 \text{ ft}^2}{144 \text{ in}^2}$
square yards and square feet	$9 \text{ ft}^2 = 1 \text{ yd}^2$	$\frac{9 \text{ ft}^2}{1 \text{ yd}^2}$ or $\frac{1 \text{ yd}^2}{9 \text{ ft}^2}$
acres and square feet	$1 \text{ acre} = 43{,}560 \text{ ft}^2$	$\frac{43{,}560 \text{ ft}^2}{1 \text{ acre}}$ or $\frac{1 \text{ acre}}{43{,}560 \text{ ft}^2}$
acres and square miles	$640 \text{ acres} = 1 \text{ mi}^2$	$\frac{640 \text{ acres}}{1 \text{ mi}^2}$ or $\frac{1 \text{ mi}^2}{640 \text{ acres}}$

3. Suppose the dressmaker in Example 3 ordered 50 yards of fabric. How many square feet would he have ordered?

EXAMPLE 3   A dressmaker orders a bolt of material that is 1.5 yards wide and 30 yards long. How many square feet of material were ordered?

Solution   The area of the material in square yards is

$$A = 1.5 \times 30$$
$$= 45 \text{ yd}^2$$

Converting this to square feet, we have

$$45 \text{ yd}^2 = 45 \text{ yd}^2 \times \frac{9 \text{ ft}^2}{1 \text{ yd}^2}$$

$$= 405 \text{ ft}^2$$

4. Suppose the farmer in Example 4 has 125 acres. How many square feet does he have?

EXAMPLE 4   A farmer has 75 acres of land. How many square feet of land does the farmer have?

Solution   Changing acres to square feet, we have

$$75 \text{ acres} = 75 \text{ acres} \times \frac{43{,}560 \text{ ft}^2}{1 \text{ acre}}$$

$$= 75 \times 43{,}560 \text{ ft}^2$$

$$= 3{,}267{,}000 \text{ ft}^2$$

Answers
2. 7.78 ft²
3. 675 ft²
4. 5,445,000 ft²

EXAMPLE 5   A new shopping center is to be constructed on 256 acres of land. How many square miles is this?

Solution   Multiplying by the conversion factor that will allow acres to divide out, we have

$$256 \text{ acres} = 256 \text{ acres} \times \frac{1 \text{ mi}^2}{640 \text{ acres}}$$

$$= \frac{256}{640} \text{ mi}^2$$

$$= 0.4 \text{ mi}^2$$

5. How many square miles is 348 acres?

Units of area in the metric system are considerably simpler than those in the U.S. system because metric units are given in terms of powers of 10. Table 2 lists the conversion factors that are most commonly used.

Table 2

Metric Units of Area

The relationship between	is	To convert from one to the other, multiply by	
square millimeters and square centimeters	$1 \text{ cm}^2 = 100 \text{ mm}^2$	$\frac{100 \text{ mm}^2}{1 \text{ cm}^2}$ or	$\frac{1 \text{ cm}^2}{100 \text{ mm}^2}$
square centimeters and square decimeters	$1 \text{ dm}^2 = 100 \text{ cm}^2$	$\frac{100 \text{ cm}^2}{1 \text{ dm}^2}$ or	$\frac{1 \text{ dm}^2}{100 \text{ cm}^2}$
square decimeters and square meters	$1 \text{ m}^2 = 100 \text{ dm}^2$	$\frac{100 \text{ dm}^2}{1 \text{ m}^2}$ or	$\frac{1 \text{ m}^2}{100 \text{ dm}^2}$
square meters and ares (a)	$1 \text{ a} = 100 \text{ m}^2$	$\frac{100 \text{ m}^2}{1 \text{ a}}$ or	$\frac{1 \text{ a}}{100 \text{ m}^2}$
ares and hectares (ha)	$1 \text{ ha} = 100 \text{ a}$	$\frac{100 \text{ a}}{1 \text{ ha}}$ or	$\frac{1 \text{ ha}}{100 \text{ a}}$

EXAMPLE 6   How many square millimeters are in 1 square meter?

Solution   We start with 1 m² and end up with square millimeters.

$$1 \text{ m}^2 = 1 \text{ m}^2 \times \frac{100 \text{ dm}^2}{1 \text{ m}^2} \times \frac{100 \text{ cm}^2}{1 \text{ dm}^2} \times \frac{100 \text{ mm}^2}{1 \text{ cm}^2}$$

$$= 100 \times 100 \times 100 \text{ mm}^2$$

$$= 1{,}000{,}000 \text{ mm}^2$$

6. How many square centimeters are in 1 square meter?

B Volume

Table 3 lists the units of volume in the U.S. system and their conversion factors.

Table 3

Units of Volume in the U.S. System

The relationship between	is	To convert from one to the other, multiply by	
cubic inches (in³) and cubic feet (ft³)	$1 \text{ ft}^3 = 1{,}728 \text{ in}^3$	$\frac{1{,}728 \text{ in}^3}{1 \text{ ft}^3}$ or	$\frac{1 \text{ ft}^3}{1{,}728 \text{ in}^3}$
cubic feet and cubic yards (yd³)	$1 \text{ yd}^3 = 27 \text{ ft}^3$	$\frac{27 \text{ ft}^3}{1 \text{ yd}^3}$ or	$\frac{1 \text{ yd}^3}{27 \text{ ft}^3}$
fluid ounces (fl oz) and pints (pt)	$1 \text{ pt} = 16 \text{ fl oz}$	$\frac{16 \text{ fl oz}}{1 \text{ pt}}$ or	$\frac{1 \text{ pt}}{16 \text{ fl oz}}$
pints and quarts (qt)	$1 \text{ qt} = 2 \text{ pt}$	$\frac{2 \text{ pt}}{1 \text{ qt}}$ or	$\frac{1 \text{ qt}}{2 \text{ pt}}$
quarts and gallons (gal)	$1 \text{ gal} = 4 \text{ qt}$	$\frac{4 \text{ qt}}{1 \text{ gal}}$ or	$\frac{1 \text{ gal}}{4 \text{ qt}}$

Answers
5. 0.54 mi²
6. 10,000 cm²

7. How many fluid ounces are in a 1 gallon container?

EXAMPLE 7   What is the capacity (volume) in pints of a 1-gallon container of milk?

Solution   We change from gallons to quarts and then quarts to pints by multiplying by the appropriate conversion factors as given in Table 3.

$$1 \text{ gal} = 1 \text{ gal} \times \frac{4 \text{ qt}}{1 \text{ gal}} \times \frac{2 \text{ pt}}{1 \text{ qt}}$$

$$= 1 \times 4 \times 2 \text{ pt}$$

$$= 8 \text{ pt}$$

A 1-gallon container has the same capacity as 8 one-pint containers.

8. How many pints are produced by the dairy herd in Example 8 each day?

EXAMPLE 8   A dairy herd produces 1,800 quarts of milk each day. How many gallons is this equivalent to?

Solution   Converting 1,800 quarts to gallons, we have

$$1,800 \text{ qt} = 1,800 \text{ qt} \times \frac{1 \text{ gal}}{4 \text{ qt}}$$

$$= \frac{1,800}{4} \text{ gal}$$

$$= 450 \text{ gal}$$

We see that 1,800 quarts is equivalent to 450 gallons.

> *Note*   As you can see from the table and the discussion above, a cubic centimeter (cm³) and a milliliter (mL) are equal. Both are one thousandth of a liter. It is also common in some fields (like medicine) to abbreviate the term cubic centimeter as cc. Although we will use the notation mL when discussing volume in the metric system, you should be aware that 1 mL = 1 cm³ = 1 cc.

In the metric system, the basic unit of measure for volume is the liter. A liter is the volume enclosed by a cube that is 10 cm on each edge, as shown in Figure 2. We can see that a liter is equivalent to 1,000 cm³.

The other units of volume in the metric system use the same prefixes we encountered previously. The units with prefixes centi, deci, and deka are not as common as the others, so in Table 4 we include only liters, milliliters, hectoliters, and kiloliters.

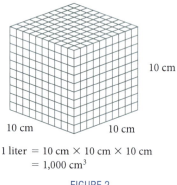

1 liter = 10 cm × 10 cm × 10 cm
      = 1,000 cm³

FIGURE 2

Table 4

Metric Units of Volume		
The relationship between	is	To convert from one to the other, multiply by
milliliters (mL) and liters	1 liter (L) = 1,000 mL	$\frac{1,000 \text{ mL}}{1 \text{ liter}}$ or $\frac{1 \text{ liter}}{1,000 \text{ mL}}$
hectoliters (hL) and liters	100 liters = 1 hL	$\frac{100 \text{ liters}}{1 \text{ hL}}$ or $\frac{1 \text{ hL}}{100 \text{ liters}}$
kiloliters (kL) and liters	1,000 liters (L) = 1 kL	$\frac{1,000 \text{ liters}}{1 \text{ kL}}$ or $\frac{1 \text{ kL}}{1,000 \text{ liters}}$

Answers

7. 128 fl oz

8. 3,600 pints

Here is an example of conversion from one unit of volume to another in the metric system.

EXAMPLE 9   A sports car has a 2.2-liter engine. What is the displacement (volume) of the engine in milliliters?

Solution   Using the appropriate conversion factor from Table 4, we have

$$2.2 \text{ liters} = 2.2 \text{ liters} \times \frac{1{,}000 \text{ mL}}{1 \text{ liter}}$$

$$= 2.2 \times 1{,}000 \text{ mL}$$

$$= 2{,}200 \text{ mL}$$

9. A truck has a 4.9 liter engine. What is the displacement of the engine in mL?

GETTING READY FOR CLASS

After reading through the preceding section, respond in your own words and in complete sentences.

A. Write the formula for the area of each of the following:

      a. A square of side s

      b. A rectangle with length l and width w

B. What is the relationship between square feet and square inches?

C. Fill in the numerators below so that each conversion factor is equal to 1.

a.   **b.**   **c.**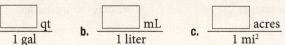

D. List two examples of units of volume in the U.S. system and two examples of units of volume in the metric system.

EXERCISE SET 7.2

Vocabulary Review

Choose the correct words to fill in the blanks below.

> multiply   U.S. system   liter   metric system   square inches   square foot

1. In the U.S. system of measurement, the basic unit of measure for area is a _____.

2. The relationship between _____ and square feet is 144 in^2 = 1 ft^2.

3. To convert from acres to square feet, _____ by the conversion factor $\frac{43{,}560 \text{ ft}^2}{1 \text{ acre}}$.

4. Cubic inches, pints, and gallons are units of volume in the _____.

5. The units of area in the _____ are given in terms of powers of 10.

6. In the metric system, the basic unit of measure for volume is the _____.

Problems

A B Use the tables given in this section to make the following conversions. Be sure to show the conversion factor used in each case.

1. 3 ft^2 to square inches

2. 5 ft^2 to square inches

3. 288 in^2 to square feet

4. 720 in^2 to square feet

5. 30 acres to square feet

6. 92 acres to square feet

7. 2 mi^2 to acres

8. 7 mi^2 to acres

9. 1,920 acres to square miles

10. 3,200 acres to square miles

11. 12 yd^2 to square feet

12. 20 yd^2 to square feet

13. 17 cm^2 to square millimeters

14. 150 mm^2 to square centimeters

15. 2.8 m^2 to square centimeters

16. 10 dm^2 to square millimeters

17. 1,200 mm^2 to square meters

18. 19.79 cm^2 to square meters

19. 5 ares to square meters

20. 12 ares to square centimeters

21. 7 ha to ares

22. 3.6 ha to ares

23. 342 ares to hectares

24. 986 ares to hectares

25. 5 yd^2 to square feet

Make the following conversions using the conversion factors given in Tables 3 and 4.

26. 3.8 yd^3 to cubic feet

27. 3 pt to fluid ounces

28. 8 pt to fluid ounces

29. 2 gal to quarts

30. 12 gal to quarts

31. 2.5 gal to pints

32. 7 gal to pints

33. 15 qt to fluid ounces

34. 5.9 qt to fluid ounces

35. 64 pt to gallons

36. 256 pt to gallons

37. 12 pt to quarts

38. 18 pt to quarts

39. 243 ft^3 to cubic yards

40. 864 ft^3 to cubic yards

41. 5 L to milliliters

42. 9.6 L to milliliters

43. 127 mL to liters

44. 93.8 mL to liters

45. 4 kL to milliliters

46. 3 kL to milliliters

47. 14.92 kL to liters

48. 4.71 kL to liters

Applying the Concepts

49. Sports The diagrams below show the dimensions of playing fields for the National Football League (NFL), the Canadian Football League, and Arena Football. Find the area of each field and then convert each area to acres. Round answers to the nearest hundredth.

Arena   NFL   Canadian

$28\frac{1}{3}$ yd $53\frac{1}{3}$ yd   65 yd

50 yd   100 yd   110 yd

50. Soccer The rules for soccer state that the playing field must be from 100 to 120 yards long and 55 to 75 yards wide. The 1999 Women's World Cup was played at the Rose Bowl on a playing field 116 yards long and 72 yards wide. The diagram below shows the smallest possible soccer field, the largest possible soccer field, and the soccer field at the Rose Bowl. Find the area of each one and then convert the area of each to acres. Round to the nearest tenth, if necessary.

Smallest   Rose Bowl   Largest

55 yd   72 yd   75 yd

100 yd   116 yd   120 yd

51. Swimming Pool A public swimming pool measures 100 meters by 30 meters and is rectangular. What is the area of the pool in ares?

52. Construction A family decides to put tiles in the entryway of their home. The entryway has an area of 6 square meters. If each tile is 5 centimeters by 5 centimeters, how many tiles will it take to cover the entryway?

53. Landscaping A landscaper is putting in a brick patio. The area of the patio is 110 square meters. If the bricks measure 10 centimeters by 20 centimeters, how many bricks will it take to make the patio? Assume no space between bricks.

54. Sewing A dressmaker is using a pattern that requires 2 square yards of material. If the material is on a bolt that is 54 inches wide, how long must a piece of material from the bolt be to ensure sure there is enough material for the pattern?

55. **Passport** A typical passport has an area of 17.1 square inches. Convert this to square feet. Round to the nearest thousandth.

56. **Volleyball** A volleyball court has an area of about 162 square meters. Convert this to square decimeters.

57. **Pentagon** The total floor area of the Pentagon is about 620,000 square meters. Convert this to square kilometers.

58. **Manhattan** The area of Manhattan Island is about 33.8 square miles. Covert this to acres.

59. **Walt Disney World** Walt Disney World covers an area of 122 square kilometers. Covert this to hectares.

60. **Blood Plasma** A 6-foot tall man weighing about 175 pounds has about 7.1 pints of blood plasma in his body. Convert this volume to quarts.

61. **Lung Capacity** A 6-foot tall man weighing about 150 pounds has a lung capacity of 4.2 liters. Covert this volume to milliliters.

62. **Heart Volume** A 6-foot tall man weighing about 181 pounds has a heart with a volume of 0.2 gallons. Convert this volume to pints.

63. **Red Blood Cells** A 5.5-foot tall woman weighing about 130 pounds has about 1.5 liters of red blood cells. Convert this volume to milliliters.

64. **Red Blood Cells** A 5'3" woman weighing 124 pounds has about 214 cubic inches of blood. Convert this volume to cubic feet. Round to the nearest thousandth.

65. **Filling Coffee Cups** If a regular-size coffee cup holds about $\frac{1}{2}$ pint, about how many cups can be filled from a 1-gallon coffee maker?

66. **Filling Glasses** If a regular-size drinking glass holds about 0.25 liter of liquid, how many glasses can be filled from a 750-milliliter container?

67. Capacity of a Refrigerator A refrigerator has a capacity of 20 cubic feet. What is the capacity of the refrigerator in cubic inches?

68. Volume of a Tank The gasoline tank on a car holds 18 gallons of gas. What is the volume of the tank in quarts?

69. Filling Glasses How many 8-fluid-ounce glasses of water will it take to fill a 3-gallon aquarium?

70. Filling a Container How many 5-milliliter test tubes filled with water will it take to fill a 1-liter container?

Calculator Problems

Set up the following problems as you have been doing. Then use a calculator to perform the actual calculations. Round all answers to two decimal places where appropriate.

71. Geography Lake Superior is the largest of the Great Lakes. It covers 31,700 square miles of area. What is the area of Lake Superior in acres?

72. Geography The state of California consists of 156,360 square miles of land and 2,330 square miles of water. Write the total area (both land and water) in acres.

73. Geography Death Valley National Monument contains 2,067,795 acres of land. How many square miles is this?

74. Geography The Badlands National Monument in South Dakota was established in 1929. It covers 243,302 acres of land. What is the area in square miles?

75. Convert 93.4 qt to gallons.

76. Convert 7,362 fl oz to gallons.

77. How many cubic feet are contained in 796 cubic yards?

78. The engine of a car has a displacement of 440 cubic inches. What is the displacement in cubic feet?

79. Volume of Water The Grand Coulee Dam holds 10,585,000 cubic yards of water. What is the volume of water in cubic feet?

80. Volume of Water Hoover Dam was built in 1936 on the Colorado River in Nevada. It holds a volume of 4,400,000 cubic yards of water. What is this volume in cubic feet?

Getting Ready for the Next Section

Perform the indicated operations.

81. 12×16

82. 15×16

83. $3 \times 2,000$

84. $5 \times 2,000$

85. $3 \times 1,000 \times 100$

86. $5 \times 1,000 \times 100$

87. 50×250

88. 75×200

89. $12,500 \times \dfrac{1}{1,000}$

90. $15,000 \times \dfrac{1}{1,000}$

Find the Mistake

Each sentence below contains a mistake. Circle the mistake and write the correct word(s) or numbers(s) on the line provided.

1. A rectangular computer monitor is 24 inches by 30 inches. To find the area of the computer monitor in square feet, divide the product of 24 and 30 by 12. _____

2. A family bought 320 acres of land. The area of this land in square miles is 204,800 mi^2. _____

3. A 3-quart container of ice cream can hold 1.5 pints of ice cream. _____

4. To find how many milliliters four 2-liter soda bottles can hold, you must find the product of 8 L and $\dfrac{1\,\text{mL}}{1,000\,\text{L}}$.

_____

Video Examples
Section 7.3

Image © sxc.hu, gnmills, 2007

The indigenous Polynesian people of New Zealand, called the Māori, revered the longfin eel. The Māori believed the eel was a guardian of sacred places, and representations of the eel appeared frequently in the culture's mythology. They often maintained special ponds in a river where they kept the eels and fed them daily. The eels live for decades, growing to 6 feet in length and 80 pounds in weight!

A Weight

Pounds are a common unit of measure for weight in the U.S. system. In this section, we'll use unit analysis to work problems involving weight. The most common units of weight in the U.S. system are ounces, pounds, and tons. The relationships among these units are given in Table 1.

Table 1

Units of Weight in the U.S. System

The relationship between	is	To convert from one to the other, multiply by
ounces (oz) and pounds (lb)	1 lb = 16 oz	$\dfrac{16 \text{ oz}}{1 \text{ lb}}$ or $\dfrac{1 \text{ lb}}{16 \text{ oz}}$
pounds and tons (T)	1 T = 2,000 lb	$\dfrac{2{,}000 \text{ lb}}{1 \text{ T}}$ or $\dfrac{1 \text{T}}{2{,}000 \text{ lb}}$

EXAMPLE 1 Convert 12 pounds to ounces.

Solution Using the conversion factor from the table, and applying the method we have been using, we have

$$12 \text{ lb} = 12 \text{ lb} \times \frac{16 \text{ oz}}{1 \text{ lb}}$$
$$= 12 \times 16 \text{ oz}$$
$$= 192 \text{ oz}$$

12 pounds is equivalent to 192 ounces.

EXAMPLE 2 Convert 3 tons to pounds.

Solution We use the conversion factor from the table. We have

$$3 \text{ T} = 3 \text{ T} \times \frac{2{,}000 \text{ lb}}{1 \text{ T}}$$
$$= 6{,}000 \text{ lb}$$

6,000 pounds is the equivalent of 3 tons.

In the metric system, the basic unit of weight is a gram. We use the same prefixes we have already used to write the other units of weight in terms of grams. Table 2 lists the most common metric units of weight and their conversion factors.

Table 2

Metric Units of Weight

The relationship between	is	To convert from one to the other, multiply by
milligrams (mg) and grams (g)	1 g = 1,000 mg	$\frac{1{,}000 \text{ mg}}{1 \text{ g}}$ or $\frac{1 \text{ g}}{1{,}000 \text{ mg}}$
centigrams (cg) and grams	1 g = 100 cg	$\frac{100 \text{ cg}}{1 \text{ g}}$ or $\frac{1 \text{ g}}{100 \text{ cg}}$
kilograms (kg) and grams	1,000 g = 1 kg	$\frac{1{,}000 \text{ g}}{1 \text{ kg}}$ or $\frac{1 \text{ kg}}{1{,}000 \text{ g}}$
metric tons (t) and kilograms	1,000 kg = 1 t	$\frac{1{,}000 \text{ kg}}{4 \text{ t}}$ or $\frac{1 \text{ t}}{1{,}000 \text{ kg}}$

EXAMPLE 3   Convert 3 kilograms to centigrams.

Solution   We convert kilograms to grams and then grams to centigrams.

$$3 \text{ kg} = 3 \text{ kg} \times \frac{1{,}000 \text{ g}}{1 \text{ kg}} \times \frac{100 \text{ cg}}{1 \text{ g}}$$

$$= 3 \times 1{,}000 \times 100 \text{ cg}$$

$$= 300{,}000 \text{ cg}$$

3. Convert 4 metric tons to kilograms.

EXAMPLE 4   A bottle of vitamin C contains 50 tablets. Each tablet contains 250 milligrams of vitamin C. What is the total number of grams of vitamin C in the bottle?

Solution   We begin by finding the total number of milligrams of vitamin C in the bottle. Since there are 50 tablets, and each contains 250 mg of vitamin C, we can multiply 50 by 250 to get the total number of milligrams of vitamin C.

$$\text{Milligrams of vitamin C} = 50 \times 250 \text{ mg}$$

$$= 12{,}500 \text{ mg}$$

Next, we convert 12,500 mg to grams.

$$12{,}500 \text{ mg} = 12{,}500 \text{ mg} \times \frac{1 \text{ g}}{1{,}000 \text{ mg}}$$

$$= \frac{12{,}500}{1{,}000} \text{ g}$$

$$= 12.5 \text{ g}$$

The bottle contains 12.5 g of vitamin C.

4. Suppose the bottle in Example 4 contained 75 tablets. How many grams of vitamin C are in the bottle?

Answers
3. 4,000 kg
4. 18.75 g

EXERCISE SET 7.3

Problems

A Use the conversion factors in Tables 1 and 2 to make the following conversions.

1. 8 lb to ounces

2. 5 lb to ounces

3. 2 T to pounds

4. 5 T to pounds

5. 192 oz to pounds

6. 176 oz to pounds

7. 1,800 lb to tons

8. 10,200 lb to tons

9. 1 T to ounces

10. 3 T to ounces

11. $3\frac{1}{2}$ lb to ounces

12. $5\frac{1}{4}$ lb to ounces

13. $6\frac{1}{2}$ T to pounds

14. $4\frac{1}{5}$ T to pounds

15. 2 kg to grams

16. 5 kg to grams

17. 4 cg to milligrams

18. 3 cg to milligrams

19. 2 kg to centigrams

20. 5 kg to centigrams

21. 5.08 g to centigrams

22. 7.14 g to centigrams

23. 450 cg to grams

24. 979 cg to grams

25. 478.95 mg to centigrams

26. 659.43 mg to centigrams

27. 1,578 mg to grams

28. 1,979 mg to grams

29. 42,000 cg to kilograms

30. 97,000 cg to kilograms

Applying the Concepts

31. Fish Oil A bottle of fish oil contains 60 soft gels, each containing 800 mg of the omega-3 fatty acid. How many total grams of the omega-3 fatty acid are in this bottle?

32. Fish Oil A bottle of fish oil contains 50 soft gels, each containing 300 mg of the omega-6 fatty acid. How many total grams of the omega-6 fatty acid are in this bottle?

33. B-Complex A certain B-complex vitamin supplement contains 50 mg of riboflavin, or vitamin B_2. A bottle contains 80 vitamins. How many total grams of riboflavin are in this bottle?

34. B-Complex A certain B-complex vitamin supplement contains 30 mg of thiamine, or vitamin B_1. A bottle contains 80 vitamins. How many total grams of thiamine are in this bottle?

35. Aspirin A bottle of low-strength aspirin contains 120 tablets. Each tablet contains 81 mg of aspirin. How many total grams of aspirin are in this bottle?

36. Aspirin A bottle of maximum-strength aspirin contains 90 tablets. Each tablet contains 500 mg of aspirin. How many total grams of aspirin are in this bottle?

37. Dairy Cow A typical dairy cow has a mass of 0.77 tons. Convert this to pounds.

38. Snowflake The mass of a typical snowflake is 0.000003 kilograms. Convert this to milligrams.

39. Brain Mass A 6'2" man has a brain mass of about 1.4 kilograms. Convert this to grams.

40. Brain Mass A 4'2" child has a brain mass of about 48 ounces. Convert this to pounds.

41. Penny A United States penny has a mass of 2.5 grams. Convert this to centigrams.

42. Quarter A United States quarter has a mass of 0.0125 pounds. Convert this to ounces.

Coca Cola Bottles The soft drink Coke is sold throughout the world. Although the size of the bottle varies between different countries, a "six-pack" is sold everywhere. For each of the problems, find the number of liters in a "six-pack" from the given bottle size. Write each fraction or mixed number as a decimal.

Country	Bottle size	Liters in a 6-pack
43. Estonia	500 mL	
44. Israel	350 mL	
45. Jordan	250 mL	
46. Kenya	300 mL	

Getting Ready for the Next Section

Perform the indicated operations. Round to the nearest hundredth if necessary.

47. 8×2.54

48. 9×3.28

49. $3 \times 1.06 \times 2$

50. $3 \times 5 \times 3.79$

51. $80.5 \div 1.61$

52. $96.6 \div 1.61$

53. $125 \div 2.20$

54. $165 \div 2.20$

55. $2,000 \div 16.39$ (Round your answer to the nearest whole number.)

56. $2,200 \div 16.39$ (Round your answer to the nearest whole number.)

57. $\frac{9}{5}(120) + 32$

58. $\frac{9}{5}(40) + 32$

59. $\frac{5(102 - 32)}{9}$

60. $\frac{5(101.6 - 32)}{9}$

One Step Further

61. Ethanol Ethanol has a density of 0.789 grams per cubic centimeter. Convert this to kilograms per cubic meter.

62. Gold Gold has a density of 0.697 pounds per cubic inch. Convert this to ounces per cubic inch.

Find the Mistake

Each of the following problems contains a conversion factor. Note whether the conversion factor is correct or incorrect for the corresponding problem. If incorrect, write the correct conversion factor on the line provided.

1. To convert 48 ounces to pounds, multiply by $\frac{1\text{ lb}}{16\text{ oz}}$. _____

2. To convert a car's weight of 2.5 tons to pounds, multiply by $\frac{2.5\text{ T}}{2{,}000\text{ lbs}}$. _____

3. Converting 6,000 g to kg gives us 6 kg after multiplying by $\frac{1{,}000\text{ kg}}{1\text{ g}}$. _____

4. A piece of chocolate candy weighs 150 cg. To find the weight in grams of a package that contains 32 candies, multiply the product of 150 and 32 by $\frac{1\text{ g}}{100\text{ cg}}$. _____

Landmark Review: Checking Your Progress

Make the following conversions.

1. 7 ft to inches

2. 27 in to yards

3. 7 mi to inches

4. 15 ft to miles

5. 38 m to centimeters

6. 14.3 centimeters to decimeters

7. 43 decimeters to meters

8. 115 cm to meters

9. 8 ft^2 to square inches

10. 2,487 acres to square miles

11. 14 mi^2 to acres

12. 5 yd^2 to square feet

13. 7 ares to square meters

14. 14.3 cm^2 to square meters

15. 350 mm^2 to square centimeters

16. 4.2 ha to ares

17. 7 lb to ounces

18. 15 oz to pounds

19. 2 T to ounces

20. 1,500 lb to tons

21. 6.5 kg to grams

22. 5 cg to milligrams

23. 1,759 mg to grams

24. 859 cg to grams

OBJECTIVES

A Convert units of measurement between the metric system and the U.S. system.

B Solve problems involving temperature on both the Fahrenheit and the Celsius scales.

KEY WORDS

temperature

degree

Video Examples
Section 7.4

A Convert Between Systems

Because most of us have always used the U.S. system of measurement in our everyday lives, we are much more familiar with it on an intuitive level than we are with the metric system. We have an intuitive idea of how long feet and inches are, how much a pound weighs, and what a square yard of material looks like. The metric system is actually much easier to use than the U.S. system. The reason some of us have such a hard time with the metric system is that we don't have the feel for it that we do for the U.S. system. We have trouble visualizing how long a meter is or how much a gram weighs. The following list is intended to give you something to associate with each basic unit of measurement in the metric system.

1. A meter is just a little longer than a yard.
2. The length of the edge of a sugar cube is about 1 centimeter.
3. A liter is just a little larger than a quart.
4. A sugar cube has a volume of approximately 1 milliliter.
5. A paper clip weighs about 1 gram.
6. A 2-pound can of coffee weighs about 1 kilogram.

Table 1

Conversion Factors Between the Metric and U.S. Systems of Measurement		
The relationship between	is	To convert from one to the other, multiply by
Length		
inches and centimeters	2.54 cm = 1 in.	$\frac{2.54 \text{ cm}}{1 \text{ in.}}$ or $\frac{1 \text{ in.}}{2.54 \text{ cm}}$
feet and meters	1 m = 3.28 ft	$\frac{3.28 \text{ ft}}{1 \text{ m}}$ or $\frac{1 \text{ m}}{3.28 \text{ ft}}$
miles and kilometers	1.61 km = 1 mi	$\frac{1.61 \text{ km}}{1 \text{ mi}}$ or $\frac{1 \text{ mi}}{1.61 \text{ km}}$
Area		
square inches and square centimeters	6.45 cm² = 1 in²	$\frac{6.45 \text{ cm}^2}{1 \text{ in}^2}$ or $\frac{1 \text{ in}^2}{6.45 \text{ cm}^2}$
square meters and square yards	1.196 yd² = 1 m²	$\frac{1.196 \text{ yd}^2}{1 \text{ m}^2}$ or $\frac{1 \text{ m}^2}{1.196 \text{ yd}^2}$
acres and hectares	1 ha = 2.47 acres	$\frac{2.47 \text{ acres}}{1 \text{ ha}}$ or $\frac{1 \text{ ha}}{2.47 \text{ acres}}$

Volume

cubic inches and milliliters	16.39 mL = 1 in³	$\dfrac{16.39\ \text{mL}}{1\ \text{in}^3}$ or $\dfrac{1\ \text{in}^3}{16.39\ \text{mL}}$
liters and quarts	1.06 qt = 1 liter	$\dfrac{1.06\ \text{qt}}{1\ \text{liter}}$ or $\dfrac{1\ \text{liter}}{1.06\ \text{qt}}$
gallons and liters	3.79 liters = 1 gal	$\dfrac{3.79\ \text{liters}}{1\ \text{gal}}$ or $\dfrac{1\ \text{gal}}{3.79\ \text{liters}}$

Weight

ounces and grams	28.3 g = 1 oz	$\dfrac{28.3\ \text{g}}{1\ \text{oz}}$ or $\dfrac{1\ \text{oz}}{28.3\ \text{g}}$
kilograms and pounds	2.20 lb = 1 kg	$\dfrac{2.20\ \text{lb}}{1\ \text{kg}}$ or $\dfrac{1\ \text{kg}}{2.20\ \text{lb}}$

There are many other conversion factors that we could have included in Table 1. We have listed only the most common ones. Almost all of them are approximations. That is, most of the conversion factors are decimals that have been rounded to the nearest hundredth. If we want more accuracy, we obtain a table that has more digits in the conversion factors.

EXAMPLE 1 Convert 8 inches to centimeters.

Solution Choosing the appropriate conversion factor from Table 1, we have

$$8\ \text{in.} = 8\ \text{in.} \times \frac{2.54\ \text{cm}}{1\ \text{in.}}$$
$$= 8 \times 2.54\ \text{cm}$$
$$= 20.32\ \text{cm}$$

EXAMPLE 2 Convert 80.5 kilometers to miles.

Solution Using the conversion factor that takes us from kilometers to miles, we have

$$80.5\ \text{km} = 80.5\ \text{km} \times \frac{1\ \text{mi}}{1.61\ \text{km}}$$
$$= \frac{80.5}{1.61}\ \text{mi}$$
$$= 50\ \text{mi}$$

So 50 miles is equivalent to 80.5 kilometers. If we travel at 50 miles per hour in a car, we are moving at the rate of 80.5 kilometers per hour.

EXAMPLE 3 Convert 3 liters to pints.

Solution Because Table 1 doesn't list a conversion factor that will take us directly from liters to pints, we first convert liters to quarts, and then convert quarts to pints.

$$3\ \text{liters} = 3\ \text{liters} \times \frac{1.06\ \text{qt}}{1\ \text{liter}} \times \frac{2\ \text{pt}}{1\ \text{qt}}$$
$$= 3 \times 1.06 \times 2\ \text{pt}$$
$$= 6.36\ \text{pt}$$

Practice Problems

1. Convert 4 feet to meters.

2. Convert 400 inches to kilometers.

3. Convert 5 liters to gallons.

Answers

1. 1.22 meters
2. 0.01 km
3. 1.32 gallons

4. A truck engine has a 4 liter displacement. What is the displacement in quarts?

EXAMPLE 4   The engine in a car has a 2-liter displacement. What is the displacement in cubic inches?

Solution   We convert liters to milliliters and then milliliters to cubic inches.

$$2 \text{ liters} = 2 \text{ liters} \times \frac{1,000 \text{ mL}}{1 \text{ liter}} \times \frac{1 \text{ in}^3}{16.39 \text{ mL}}$$

$$= \frac{2 \times 1,000}{16.39} \text{ in}^3 \qquad \textit{This calculation should be done on a calculator.}$$

$$= 122 \text{ in}^3 \qquad \textit{Round to the nearest cubic inch.}$$

5. What is the person's weight from Example 5 in grams?

EXAMPLE 5   If a person weighs 125 pounds, what is her weight in kilograms?

Solution   Converting from pounds to kilograms, we have

$$125 \text{ lb} = 125 \text{ lb} \times \frac{1 \text{ kg}}{2.20 \text{ lb}}$$

$$= \frac{125}{2.20} \text{ kg}$$

$$= 56.8 \text{ kg} \qquad \textit{Round to the nearest tenth.}$$

B Temperature

We end this section with a discussion of temperature in both systems of measurement.

In the U.S. system, we measure temperature on the Fahrenheit scale. On this scale, water boils at 212 degrees and freezes at 32 degrees. When we write 32 degrees measured on the Fahrenheit scale, we use the notation

32°F (read, "32 degrees Fahrenheit")

In the metric system, the scale we use to measure temperature is the Celsius scale (formerly called the centigrade scale). On this scale, water boils at 100 degrees and freezes at 0 degrees. When we write 100 degrees measured on the Celsius scale, we use the notation

100°C (read, "100 degrees Celsius")

Table 2 is intended to give you a feel for the relationship between the two temperature scales. Table 3 gives the formulas, in both symbols and words, that are used to convert between the two scales.

Table 2

Situation	Temperature Fahrenheit	Temperature Celsius
Water freezes	32°F	0°C
Room temperature	68°F	20°C
Normal body temperature	98.6°F	37°C
Water boils	212°F	100°C
Bake cookies	350°F	176.7°C
Broil meat	554°F	290°C

Table 3

To convert from	Formula In symbols	Formula In words
Fahrenheit to Celsius	$C = \dfrac{5(F - 32)}{9}$	Subtract 32, multiply by 5, and then divide by 9.
Celsius to Fahrenheit	$F = \dfrac{9}{5}C + 32$	Multiply by $\dfrac{9}{5}$, and then add 32.

The following examples show how we use the formulas given in Table 3.

EXAMPLE 6   Convert 120°C to degrees Fahrenheit.

Solution   We use the formula

$$F = \frac{9}{5}C + 32$$

and replace C with 120.

When → $\quad C = 120$

the formula → $F = \dfrac{9}{5}C + 32$

becomes → $\quad F = \dfrac{9}{5}(120) + 32$

$\qquad\qquad F = 216 + 32$

$\qquad\qquad F = 248$

We see that 120°C is equivalent to 248°F; they both mean the same temperature.

EXAMPLE 7   A man with the flu has a temperature of 102°F. What is his temperature on the Celsius scale?

Solution

When → $\quad F = 102$

the formula → $C = \dfrac{5(F - 32)}{9}$

becomes → $\quad C = \dfrac{5(102 - 32)}{9}$

$\qquad\qquad C = \dfrac{5(70)}{9}$

$\qquad\qquad C = 38.9 \qquad$ *Round to the nearest tenth.*

The man's temperature, rounded to the nearest tenth, is 38.9°C on the Celsius scale.

GETTING READY FOR CLASS

After reading through the preceding section, respond in your own words and in complete sentences.

A. Write the equality that gives the relationship between centimeters and inches.

B. Write the equality that gives the relationship between grams and ounces.

C. Fill in the numerators below so that each conversion factor is equal to 1.

  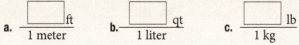

a. $\dfrac{\boxed{}\,\text{ft}}{1 \text{ meter}}$ b. $\dfrac{\boxed{}\,\text{qt}}{1 \text{ liter}}$ c. $\dfrac{\boxed{}\,\text{lb}}{1 \text{ kg}}$

D. Is it a hot day if the temperature outside is 37°C?

6. Convert 120°F to Celsius.

7. Convert 102°C to Fahrenheit.

EXERCISE SET 7.4

Problems

A B Use Tables 1 and 3 to make the following conversions.

1. 6 in. to centimeters

2. 1 ft to centimeters

3. 4 m to feet

4. 2 km to feet

5. 6 m to yards

6. 15 mi to kilometers

7. 20 mi to meters (round to the nearest hundred meters)

8. 600 m to yards

9. 5 m^2 to square yards

10. 2 in^2 to square centimeters

11. 10 ha to acres

12. 50 ares to acres

13. 500 in^3 to milliliters

14. 400 in^3 to liters

15. 2 L to quarts

16. 15 L to quarts

17. 20 gal to liters

18. 15 gal to liters

19. 12 oz to grams

20. 1 lb to grams (round to the nearest 10 grams)

21. 15 kg to pounds

22. 10 kg to ounces

23. 185°C to degrees Fahrenheit

24. 20°C to degrees Fahrenheit

25. 86°F to degrees Celsius

26. 122°F to degrees Celsius

Calculator Problems

Set up the following problems as we have set up the examples in this section. Then use a calculator for the calculations and round your answers to the nearest hundredth.

27. 10 cm to inches

28. 100 mi to kilometers

29. 25 ft to meters

30. 400 mL to cubic inches

31. 49 qt to liters

32. 65 L to gallons

33. 500 g to ounces

34. 100 lb to kilograms

35. Weight Give your weight in kilograms.

36. Height Give your height in meters and centimeters.

37. Sports The 100-yard dash is a popular race in track. How far is 100 yards in meters?

38. Engine Displacement A 351-cubic-inch engine has a displacement of how many liters?

39. Sewing 25 square yards of material is how many square meters?

40. Weight How many grams does a 5 lb 4 oz roast weigh?

41. Speed 55 miles per hour is equivalent to how many kilometers per hour?

42. Capacity A 1-quart container holds how many liters?

43. Sports A high jumper jumps 6 ft 8 in. How many meters is this?

44. Farming A farmer owns 57 acres of land. How many hectares is that?

45. Body Temperature A person has a temperature of 101°F. What is the person's temperature, to the nearest tenth, on the Celsius scale?

46. Air Temperature If the temperature outside is 30°C, is it a better day for water skiing or for snow skiing?

Getting Ready for the Next Section

Perform the indicated operations.

47. $15 + 60$

48. $25 + 60$

49.
$$\begin{array}{r} 37 \\ + 45 \\ \hline \end{array}$$

50.
$$\begin{array}{r} 37 \\ + 46 \\ \hline \end{array}$$

51. $3 + 0.25$

52. $2 + 0.75$

53. $82 - 60$

54. $73 - 60$

55.
$$\begin{array}{r} 75 \\ - 34 \\ \hline \end{array}$$

56.
$$\begin{array}{r} 85 \\ - 42 \\ \hline \end{array}$$

57. 12×4

58. 8×4

59. $3 \times 60 + 15$

60. $2 \times 65 + 45$

61. $3 + 16 \times \dfrac{1}{60}$

62. $2 + 45 \times \dfrac{1}{60}$

63. If fish costs $6.00 per pound, find the cost of 15 pounds of fish.

64. If fish costs $5.00 per pound, find the cost of 14 pounds of fish.

One Step Further

65. Caffeine Caffeine has a density of 1.23 grams per cubic centimeter. Convert this to ounces per cubic inch.

66. Caffeine A lethal dosage of caffeine is 192 milligrams per kilogram. This means a person would have to consume 192 milligrams of caffeine for every kilogram they weigh for them to get a lethal dose of caffeine. If a person weighs 150 pounds, how many grams of caffeine would they have to take for it to be lethal.

67. Ibuprofen Ibuprofen is considered toxic if a person exceeds 1,255 milligrams per kilogram. If a person weights 120 pounds, how many grams is considered a toxic dose?

68. Molasses Molasses has a density of 1.42 grams per cubic centimeter. Convert this to pounds per cubic foot. Round to the nearest hundredth.

Find the Mistake

Each problem below contains a mistake. Circle the mistake and write the correct number(s) or word(s) on the line provided.

1. To convert 6 miles to km, divide 6 miles by $\dfrac{1.61 \text{ km}}{1 \text{ mi}}$. _____

2. Suppose a pasta recipe requires a half gallon of water, but your measuring cup only measures milliliters. To find how many milliliters are equal to a half gallon of water, multiply $\frac{1}{2}$ gal by $\dfrac{3.79 \text{ ml}}{1 \text{ gal}}$. _____

3. To convert 25° Celsius to Fahrenheit, use the formula $C = \dfrac{5(F-32)}{9}$ to get 77° F. _____

4. A cookie recipe requires you to bake them at 325° F. This temperature in Celsius is 33° C (rounded to the nearest degree).

_____

7.5 Operations with Time, and Mixed Units

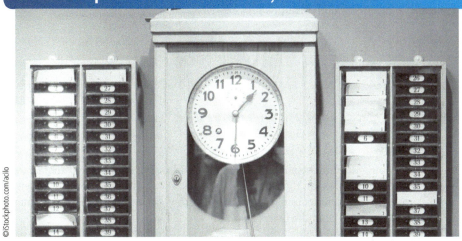
©iStockphoto.com/acilo

Many occupations require the use of a time card. A time card records the number of hours and minutes an employee spends at work. At the end of a work week, the hours and minutes are totaled separately, and then the minutes are converted to hours.

A Time and Mixed Units

In this section, we will perform operations with mixed units of measure. For instance, mixed units are used when we use 2 hours 30 minutes, rather than 2 and a half hours, or 5 feet 9 inches, rather than five and three-quarter feet. As you will see, many of these types of problems arise in everyday life.

The relationship between	is	To convert from one to the other, multiply by
minutes and seconds	1 min = 60 sec	$\dfrac{1 \text{ min}}{60 \text{ sec}}$ or $\dfrac{60 \text{ sec}}{1 \text{ min}}$
hours and minutes	1 hr = 60 min	$\dfrac{1 \text{ hr}}{60 \text{ min}}$ or $\dfrac{60 \text{ min}}{1 \text{ hr}}$

Video Examples
Section 7.5

EXAMPLE 1 Convert 3 hours 15 minutes to
a. minutes. **b.** hours.

Solution

a. To convert to minutes, we multiply the hours by the conversion factor then add minutes.

$$3 \text{ hr } 15 \text{ min} = 3 \text{ hr} \times \frac{60 \text{ min}}{1 \text{ hr}} + 15 \text{ min}$$

$$= 180 \text{ min} + 15 \text{ min}$$

$$= 195 \text{ min}$$

b. To convert to hours, we multiply the minutes by the conversion factor then add hours.

$$3 \text{ hr } 15 \text{ min} = 3 \text{ hr} + 15 \text{ min} \times \frac{1 \text{ hr}}{60 \text{ min}}$$

$$= 3 \text{ hr} + 0.25 \text{ hr}$$

$$= 3.25 \text{ hr}$$

2. Add 3 hours 45 minutes and 4 hours 20 minutes.

EXAMPLE 2   Add 5 minutes 37 seconds and 7 minutes 45 seconds.
Solution   First, we align the units properly.

$$
\begin{array}{r}
5 \text{ min} \quad 37 \text{ sec} \\
+ \ 7 \text{ min} \quad 45 \text{ sec} \\
\hline
12 \text{ min} \quad 82 \text{ sec}
\end{array}
$$

Since there are 60 seconds in every minute, we write 82 seconds as 1 minute 22 seconds. We have

$$12 \text{ min } 82 \text{ sec} = 12 \text{ min} + 1 \text{ min } 22 \text{ sec}$$
$$= 13 \text{ min } 22 \text{ sec}$$

The idea of adding the units separately is similar to adding mixed numbers. That is, we align the whole numbers with the whole numbers and the fractions with the fractions.

Similarly, when we subtract units of time, we "borrow" 60 seconds from the minutes column, or 60 minutes from the hours column.

3. Subtract 43 minutes from 4 hours 10 minutes.

EXAMPLE 3   Subtract 34 minutes from 8 hours 15 minutes.
Solution   Again, we first line up the numbers in the hours column, and then the numbers in the minutes column.

$$
\begin{array}{r}
8 \text{ hr} \quad 15 \text{ min} \\
- \qquad\quad 34 \text{ min} \\
\hline
\end{array}
$$

Since there are 60 minutes in an hour, we borrow 1 hour from the hours column and add 60 minutes to the minutes column. Then we subtract.

$$
\begin{array}{r}
7 \text{ hr} \quad 75 \text{ min} \\
- \qquad\quad 34 \text{ min} \\
\hline
7 \text{ hr} \quad 41 \text{ min}
\end{array}
$$

Next we see how to multiply and divide using units of measure.

4. Suppose Jake purchased 6 of the halibut in Example 4. What is the cost?

EXAMPLE 4   Jake purchases 4 halibut. The fish cost $6.00 per pound, and each weighs 3 lb 12 oz. What is the cost of the fish?
Solution   First, we multiply each unit by 4.

$$
\begin{array}{r}
3 \text{ lb} \quad 12 \text{ oz} \\
\times \qquad\quad 4 \\
\hline
12 \text{ lb} \quad 48 \text{ oz}
\end{array}
$$

To convert the 48 ounces to pounds, we multiply the ounces by the conversion factor

$$12 \text{ lb } 48 \text{ oz} = 12 \text{ lb} + 48 \text{ oz} \times \frac{1 \text{ lb}}{16 \text{ oz}}$$

$$= 12 \text{ lb} + 3 \text{ lb}$$

$$= 15 \text{ lb}$$

Finally, we multiply the 15 lb and $6.00/lb for a total price of $90.00.

GETTING READY FOR CLASS

After reading through the preceding section, respond in your own words and in complete sentences.

A. Explain the difference between saying *2 and a half hours* and saying *2 hours and 30 minutes.*

B. How are operations with mixed units of measure similar to operations with mixed numbers?

C. Why do we borrow 1 from the hours column and add 60 to the minutes column when subtracting in Example 3?

D. Give an example of when you may have to use multiplication with mixed units of measure.

Vocabulary Review

Choose the correct words to fill in the blanks below.

columns   seconds   borrow   minutes

1. To convert from _____ to _____, multiply by the conversion factor $\frac{60 \text{ sec}}{1 \text{ min}}$.

2. Adding mixed units is similar to adding mixed numbers, such that we align the units separately in _____ and add.

3. To subtract units of time, we may need to _____ 60 seconds from the minutes column, or 60 minutes from the hours column.

Problems

A Use the tables of conversion factors given in this section and other sections in this chapter to make the following conversions. (Round your answers to the nearest hundredth if necessary.)

1. Convert 4 hours 30 minutes to
   a. minutes.

   b. hours.

2. Convert 2 hours 45 minutes to
   a. minutes.

   b. hours.

3. Convert 5 hours 20 minutes to
   a. minutes.

   b. hours.

4. Convert 4 hours 40 minutes to
   a. minutes.

   b. hours.

5. Convert 6 minutes 30 seconds to
   a. seconds.

   b. minutes.

6. Convert 8 minutes 45 seconds to
   a. seconds.

   b. minutes.

7. Convert 5 minutes 20 seconds to
   a. seconds.

   b. minutes.

8. Convert 4 minutes 40 seconds to
   a. seconds.

   b. minutes.

9. Convert 2 pounds 8 ounces to
   a. ounces.

   b. pounds.

10. Convert 3 pounds 4 ounces to
    a. ounces.

    b. pounds.

11. Convert 4 pounds 12 ounces to
    a. ounces.

    b. pounds.

12. Convert 5 pounds 16 ounces to
    a. ounces.

    b. pounds.

13. Convert 4 feet 6 inches to
    a. inches.

    b. feet.

14. Convert 3 feet 3 inches to
    a. inches.

    b. feet.

15. Convert 5 feet 9 inches to
    a. inches.

    b. feet.

16. Convert 3 feet 4 inches to
    a. inches.

    b. feet.

17. Convert 2 gallons 1 quart to
    a. quarts.

    b. gallons.

18. Convert 3 gallons 2 quarts to
    a. quarts.

    b. gallons.

Perform the indicated operation. Again, remember to use the appropriate conversion factor.

19. Add 4 hours 47 minutes and 6 hours 13 minutes.

20. Add 5 hours 39 minutes and 2 hours 21 minutes.

21. Add 8 feet 10 inches and 13 feet 6 inches.

22. Add 16 feet 7 inches and 7 feet 9 inches.

23. Add 4 pounds 12 ounces and 6 pounds 4 ounces.

24. Add 11 pounds 9 ounces and 3 pounds 7 ounces.

25. Subtract 2 hours 35 minutes from 8 hours 15 minutes.

26. Subtract 3 hours 47 minutes from 5 hours 33 minutes.

27. Subtract 3 hours 43 minutes from 7 hours 30 minutes.

28. Subtract 1 hour 44 minutes from 6 hours 22 minutes.

29. Subtract 4 hours 17 minutes from 5 hours 9 minutes.

30. Subtract 2 hours 54 minutes from 3 hours 7 minutes.

Applying the Concepts

Triathlon The Ironman Triathlon World Championship, held each October in Kona on the island of Hawaii, consists of three parts: a 2.4-mile ocean swim, a 112-mile bike race, and a 26.2-mile marathon. The table shows the results from the 2010 event. Use the table to answer Problems 31–34.

Triathlete	Swim Time (Hr:Min:Sec)	Bike Time (Hr:Min:Sec)	Run Time (Hr:Min:Sec)	Total Time (Hr:Min:Sec)
Chris McCormack	0:51:36	4:31:51	2:43:31	
Andreas Raelert	0:51:27	4:32:27	2:44:25	

31. Fill in the total time column.

32. How much faster was Chris's total time than Andreas's?

33. How much faster was Andreas's swim time than Chris's?

34. How much faster was Chris than Andreas in the run?

35. Cost of Fish Fredrick is purchasing four whole salmon. The fish cost $4.00 per pound, and each weighs 6 lb 8 oz. What is the cost of the fish?

36. Cost of Steak Mike is purchasing eight top sirloin steaks. The meat costs $4.00 per pound, and each steak weighs 1 lb 4 oz. What is the total cost of the steaks?

37. Stationary Bike Maggie rides a stationary bike for 1 hour and 15 minutes, 4 days a week. After 2 weeks, how many hours has she spent riding the stationary bike?

38. Gardening Scott works in his garden for 1 hour and 5 minutes, 3 days a week. After 4 weeks, how many hours has Scott spent gardening?

39. Cost of Fabric Allison is making a quilt. She buys 3 yards and 1 foot each of six different fabrics. The fabrics cost $7.50 a yard. How much will Allison spend?

40. Cost of Lumber Trish is building a fence. She buys six fence posts at the lumberyard, each measuring 5 ft 4 in. The lumber costs $3 per foot. How much will Trish spend?

41. Molecular Weight Silver nitrate has a molecular weight of 169.9 grams per mole. If you have a solution containing 2.1 moles, how many grams of silver nitrate do you have?

42. Molecular Weight Potassium chloride has a molecular weight of 74.6 grams per mole. How many moles do you have if you have 52.3 grams of potassium chloride? Round to the nearest tenth.

43. Cost of Wheat Wheat is being sold for 560 cents per bushel. If a farmer sells 5,231 bushels, how many dollars will he make?

44. Cost of Corn Corn is being sold for 403 cents per bushel. If a farmer sells 3,503 bushels, how many dollars will he make?

One Step Further

45. In 2010, the horse Animal Kingdom won the Kentucky Derby with a time of 2:02.04, or two minutes and 2.04 seconds. The record time for the Kentucky Derby is still held by Secretariat, who won the race with a time of 1:59.40 in 1973. How much faster did Secretariat run in 1973 than Animal Kingdom?

46. In 2010, the horse Drosselmeyer won the Belmont Stakes with a time of 2:31.57, or two minutes and 31.57 seconds. The record time for the Belmont Stakes is still held by Secretariat, who won the race with a time of 2:24.00 in 1973. How much faster did Secretariat run in 1973 than Drosselmeyer?

Find the Mistake

Each problem below contains a mistake. Circle the mistake and write the correct number(s) or word(s) on the line provided.

1. Converting 6 feet 6 inches to inches gives us 72 inches. _____

2. The correct way to write the sum of 2 hours, 55 min and 4 hours, 10 min is 6 hours and 65 minutes.

_____

3. The correct way to subtract 27 minutes from 6 hours and 12 minutes is to add 60 minutes to the 6 hours.

_____

4. Jane is buying two cups of frozen yogurt. One cup contains 3 ounces and the other contains 11.9 grams. If each ounce cost $1.50, then the total purchase price will be $4.50. _____

Hunting for Treasure

Geocaching is a modern-day treasure hunt. For this popular outdoor activity, people hunt for hidden containers called geocaches using a GPS receiver or mobile device. According to the official geocaching website, there are over 1.5 million geocaches hidden around the world and over 5 million participants. These participants use GPS coordinates to find a geocache. Each geocache includes a logbook that the participant signs upon discovery. Then the geocacher returns the container to its original location.

For this project, you are going to create your own treasure hunt. However, instead of GPS, you will rely on the tried and true treasure map. Break into an even amount of groups. Each group should decide what to hide as its hidden treasure. It can be as simple as a pencil or a folded piece of paper. Then draw a map that leads your treasure hunter from your classroom's door to the hidden item. Your map should be drawn using centimeters for distance. Provide a key that states 1 cm on the map equals a unit of measurement from the U.S. System (e.g., inch, foot, or yard), which will be needed to find the treasure. For instance, 1 cm on the map could equal 3 feet on land. Once your map is complete and your treasure is hidden, exchange maps with another group. Convert your metric measurements to the indicated U.S. System unit of measure and use a ruler to find the treasure. Happy hunting!

Supplies Needed

A An object to hide

B A piece of paper

C A pen or pencil

D A metric ruler

E A large measuring device such as a measuring tape or a yard stick

Conversion Factors [7.1, 7.2, 7.3, 7.4, 7.5]

EXAMPLES

1. Convert 5 feet to inches.

$5 \text{ ft} = 5 \text{ ft} \times \dfrac{12 \text{ in.}}{1 \text{ ft}}$

$= 5 \times 12 \text{ in.}$

$= 60 \text{ in.}$

To convert from one kind of unit to another, we choose an appropriate conversion factor from one of the tables given in this chapter. For example, if we want to convert 5 feet to inches, we look for conversion factors that give the relationship between feet and inches. There are two conversion factors for feet and inches:

$$\frac{12 \text{ in.}}{1 \text{ ft}} \quad \text{and} \quad \frac{1 \text{ ft}}{12 \text{ in.}}$$

Length [7.1]

2. Convert 8 feet to yards.

$8 \text{ ft} = 8 \text{ ft} \times \dfrac{1 \text{ yd}}{3 \text{ ft}}$

$= \dfrac{8}{3} \text{ yd}$

$= 2\dfrac{2}{3} \text{ yd}$

U.S. System

The relationship between	is	To convert from one to the other, multiply by
feet and inches	12 in. = 1 ft	$\dfrac{12 \text{ in.}}{1 \text{ ft}}$ or $\dfrac{1 \text{ ft}}{12 \text{ in.}}$
feet and yards	1 yd = 3 ft	$\dfrac{3 \text{ ft}}{1 \text{ yd}}$ or $\dfrac{1 \text{ yd}}{3 \text{ ft}}$
feet and miles	1 mi = 5,280 ft	$\dfrac{5,280 \text{ ft}}{1 \text{ mi}}$ or $\dfrac{1 \text{ mi}}{5,280 \text{ ft}}$

Metric System

The relationship between	is	To convert from one to the other, multiply by
millimeters (mm) and meters (m)	1,000 mm = 1 m	$\dfrac{1,000 \text{ mm}}{1 \text{ m}}$ or $\dfrac{1 \text{ m}}{1,000 \text{ mm}}$
centimeters (cm) and meters	100 cm = 1 m	$\dfrac{100 \text{ cm}}{1 \text{ m}}$ or $\dfrac{1 \text{ m}}{100 \text{ cm}}$
decimeters (dm) and meters	10 dm = 1 m	$\dfrac{10 \text{ dm}}{1 \text{ m}}$ or $\dfrac{1 \text{ m}}{10 \text{ dm}}$
dekameters (dam) and meters	1 dam = 10 m	$\dfrac{10 \text{ m}}{1 \text{ dam}}$ or $\dfrac{1 \text{ dam}}{10 \text{ m}}$
hectometers (hm) and meters	1 hm = 100 m	$\dfrac{100 \text{ m}}{1 \text{ hm}}$ or $\dfrac{1 \text{ hm}}{100 \text{ m}}$
kilometers (km) and meters	1 km = 1,000 m	$\dfrac{1,000 \text{ m}}{1 \text{ km}}$ or $\dfrac{1 \text{ km}}{1,000 \text{ m}}$

Area [7.2]

4. Convert 256 acres to square miles.

$256 \text{ acres} = 256 \text{ acres} \times \dfrac{1 \text{ mi}^2}{640 \text{ acres}}$

$= \dfrac{256}{640} \text{ mi}^2$

$= 0.4 \text{ mi}^2$

U.S. System

The relationship between	is	To convert from one to the other, multiply by
square inches and square feet	144 in² = 1 ft²	$\dfrac{144 \text{ in}^2}{1 \text{ ft}^2}$ or $\dfrac{1 \text{ ft}^2}{144 \text{ in}^2}$
square yards and square feet	9 ft² = 1 yd²	$\dfrac{9 \text{ ft}^2}{1 \text{ yd}^2}$ or $\dfrac{1 \text{ yd}^2}{9 \text{ ft}^2}$
acres and square feet	1 acre = 43,560 ft²	$\dfrac{43,560 \text{ ft}^2}{1 \text{ acre}}$ or $\dfrac{1 \text{ acre}}{43,560 \text{ ft}^2}$
acres and square miles	640 acres = 1 mi²	$\dfrac{640 \text{ acres}}{1 \text{ mi}^2}$ or $\dfrac{1 \text{ mi}^2}{640 \text{ acres}}$

Metric System

The relationship between	is	To convert from one to the other, multiply by	
square millimeters and square centimeters	$1 \text{ cm}^2 = 100 \text{ mm}^2$	$\dfrac{100 \text{ mm}^2}{1 \text{ cm}^2}$ or	$\dfrac{1 \text{ cm}^2}{100 \text{ mm}^2}$
square centimeters and square decimeters	$1 \text{ dm}^2 = 100 \text{ cm}^2$	$\dfrac{100 \text{ cm}^2}{1 \text{ dm}^2}$ or	$\dfrac{1 \text{ dm}^2}{100 \text{ cm}^2}$
square decimeters and square meters	$1 \text{ m}^2 = 100 \text{ dm}^2$	$\dfrac{100 \text{ dm}^2}{1 \text{ m}^2}$ or	$\dfrac{1 \text{ m}^2}{100 \text{ dm}^2}$
square meters and ares (a)	$1 \text{ a} = 100 \text{ m}^2$	$\dfrac{100 \text{ m}^2}{1 \text{ a}}$ or	$\dfrac{1 \text{ a}}{100 \text{ m}^2}$
ares and hectares (ha)	$1 \text{ ha} = 100 \text{ a}$	$\dfrac{100 \text{ a}}{1 \text{ ha}}$ or	$\dfrac{1 \text{ ha}}{100 \text{ a}}$

Volume [7.2]

U.S. System

The relationship between	is	To convert from one to the other, multiply by	
cubic inches (in^3) and cubic feet (ft^3)	$1 \text{ ft}^3 = 1{,}728 \text{ in}^3$	$\dfrac{1{,}728 \text{ in}^3}{1 \text{ ft}^3}$ or	$\dfrac{1 \text{ ft}^3}{1{,}728 \text{ in}^3}$
cubic feet and cubic yards (yd^3)	$1 \text{ yd}^3 = 27 \text{ ft}^3$	$\dfrac{27 \text{ ft}^3}{1 \text{ yd}^3}$ or	$\dfrac{1 \text{ yd}^3}{27 \text{ ft}^3}$
fluid ounces (fl oz) and pints (pt)	$1 \text{ pt} = 16 \text{ fl oz}$	$\dfrac{16 \text{ fl oz}}{1 \text{ pt}}$ or	$\dfrac{1 \text{ pt}}{16 \text{ fl oz}}$
pints and quarts (qt)	$1 \text{ qt} = 2 \text{ pt}$	$\dfrac{2 \text{ pt}}{1 \text{ qt}}$ or	$\dfrac{1 \text{ qt}}{2 \text{ pt}}$
quarts and gallons (gal)	$1 \text{ gal} = 4 \text{ qt}$	$\dfrac{4 \text{ qt}}{1 \text{ gal}}$ or	$\dfrac{1 \text{ gal}}{4 \text{ qt}}$

5. Convert 452 hectoliters to liters.

$$452 \text{ hL} = 452 \text{ hL} \times \frac{100 \text{ L}}{1 \text{ hL}}$$

$$= 45{,}200 \text{ hL}$$

Metric System

The relationship between	is	To convert from one to the other, multiply by	
milliliters (mL) and liters	$1 \text{ liter (L)} = 1{,}000 \text{ mL}$	$\dfrac{1{,}000 \text{ mL}}{1 \text{ liter}}$ or	$\dfrac{1 \text{ liter}}{1{,}000 \text{ mL}}$
hectoliters (hL) and liters	$100 \text{ liters} = 1 \text{ hL}$	$\dfrac{100 \text{ liters}}{1 \text{ hL}}$ or	$\dfrac{1 \text{ hL}}{100 \text{ liters}}$
kiloliters (kL) and liters	$1{,}000 \text{ liters (L)} = 1 \text{ kL}$	$\dfrac{1{,}000 \text{ liters}}{1 \text{ kL}}$ or	$\dfrac{1 \text{ kL}}{1{,}000 \text{ liters}}$

Weight [7.3]

U.S. System

The relationship between	is	To convert from one to the other, multiply by	
ounces (oz) and pounds (lb)	$1 \text{ lb} = 16 \text{ oz}$	$\dfrac{16 \text{ oz}}{1 \text{ lb}}$ or	$\dfrac{1 \text{ lb}}{16 \text{ oz}}$
pounds and tons (T)	$1 \text{ T} = 2{,}000 \text{ lb}$	$\dfrac{2{,}000 \text{ lb}}{1 \text{ T}}$ or	$\dfrac{1 \text{ T}}{2{,}000 \text{ lb}}$

6. Convert 12 pounds to ounces.

$$12 \text{ lb} = 12 \text{ lb} \times \frac{16 \text{ oz}}{1 \text{ lb}}$$

$$= 12 \times 16 \text{ oz}$$

$$= 192 \text{ oz}$$

Metric System

The relationship between	is	To convert from one to the other, multiply by
milligrams (mg) and grams (g)	1 g = 1,000 mg	$\dfrac{1,000 \text{ mg}}{1 \text{ g}}$ or $\dfrac{1 \text{ g}}{1,000 \text{ mg}}$
centigrams (cg) and grams	1 g = 100 cg	$\dfrac{100 \text{ cg}}{1 \text{ g}}$ or $\dfrac{1 \text{ g}}{100 \text{ cg}}$
kilograms (kg) and grams	1,000 g = 1 kg	$\dfrac{1,000 \text{ g}}{1 \text{ kg}}$ or $\dfrac{1 \text{ kg}}{1,000 \text{ g}}$
metric tons (t) and kilograms	1,000 kg = 1 t	$\dfrac{1,000 \text{ kg}}{4 \text{ t}}$ or $\dfrac{1 \text{ t}}{1,000 \text{ kg}}$

Converting Between the Systems [7.4]

7. Convert 8 inches to centimeters.

$8 \text{ in.} = 8 \text{ in.} \times \dfrac{2.54 \text{ cm}}{1 \text{ in.}}$

$\phantom{8 \text{ in.}} = 8 \times 2.54 \text{ cm}$

$\phantom{8 \text{ in.}} = 20.32 \text{ cm}$

Conversion Factors

The relationship between	is	To convert from one to the other, multiply by
Length		
inches and centimeters	2.54 cm = 1 in.	$\dfrac{2.54 \text{ cm}}{1 \text{ in.}}$ or $\dfrac{1 \text{ in.}}{2.54 \text{ cm}}$
feet and meters	1 m = 3.28 ft	$\dfrac{3.28 \text{ ft}}{1 \text{ m}}$ or $\dfrac{1 \text{ m}}{3.28 \text{ ft}}$
miles and kilometers	1.61 km = 1 mi	$\dfrac{1.61 \text{ km}}{1 \text{ mi.}}$ or $\dfrac{1 \text{ mi}}{1.61 \text{ km}}$
Area		
square inches and square centimeters	$6.45 \text{ cm}^2 = 1 \text{ in}^2$	$\dfrac{6.45 \text{ cm}^2}{1 \text{ in}^2}$ or $\dfrac{1 \text{ in}^2}{6.45 \text{ cm}^2}$
square meters and square yards	$1.196 \text{ yd}^2 = 1 \text{ m}^2$	$\dfrac{1.196 \text{ yd}^2}{1 \text{ m}^2}$ or $\dfrac{1 \text{ m}^2}{1.196 \text{ yd}^2}$
acres and hectares	1 ha = 2.47 acres	$\dfrac{2.47 \text{ acres}}{1 \text{ ha}}$ or $\dfrac{1 \text{ ha}}{2.47 \text{ acres}}$
Volume		
cubic inches and milliliters	$16.39 \text{ mL} = 1 \text{ in}^3$	$\dfrac{16.39 \text{ mL}}{1 \text{ in}^3}$ or $\dfrac{1 \text{ in}^3}{16.39 \text{ mL}}$
liters and quarts	1.06 qt = 1 liter	$\dfrac{1.06 \text{ qt}}{1 \text{ liter}}$ or $\dfrac{1 \text{ liter}}{1.06 \text{ qt}}$
gallons and liters	3.79 liters = 1 gal	$\dfrac{3.79 \text{ liters}}{1 \text{ gal}}$ or $\dfrac{1 \text{ gal}}{3.79 \text{ liters}}$
Weight		
ounces and grams	28.3 g = 1 oz	$\dfrac{28.3 \text{ g}}{1 \text{ oz}}$ or $\dfrac{1 \text{ oz}}{28.3 \text{ g}}$
kilograms and pounds	2.20 lb = 1 kg	$\dfrac{2.20 \text{ lb}}{1 \text{ kg}}$ or $\dfrac{1 \text{ kg}}{2.20 \text{ lb}}$

Temperature [7.4]

8. Convert 120°C to degrees Fahrenheit.

$F = \dfrac{9}{5} C + 32$

$F = \dfrac{9}{5} (120) + 32$

$F = 216 + 32$
$F = 248$

To convert from	Formula in symbols	Formula in words
Fahrenheit to Celsius	$C = \dfrac{5(F - 32)}{9}$	Subtract 32, multiply by 5, and then divide by 9.
Celsius to Fahrenheit	$F = \dfrac{9}{5}C + 32$	Multiply by $\dfrac{9}{5}$, and then add 32.

Time [7.5]

The relationship between	is	To convert from one to the other, multiply by
minutes and seconds	1 min = 60 sec	$\dfrac{1 \text{ min}}{60 \text{ sec}}$ or $\dfrac{60 \text{ sec}}{1 \text{ min}}$
hours and minutes	1 hr = 60 min	$\dfrac{1 \text{ hr}}{60 \text{ min}}$ or $\dfrac{60 \text{ min}}{1 \text{ hr}}$

Use the tables in the chapter to make the following conversions.[7.1, 7.2, 7.3, 7.4]

1. 9 yd to feet

2. 570 m to kilometers

3. 2.5 acres to square feet

4. 792 in² to square feet

5. 16.4 L to milliliters

6. 5 mi to kilometers

7. 13 L to quarts

8. 77°F to degrees Celsius (round to the nearest tenth.)

9. 46 T to pounds

10. 136 oz to pounds

11. 990 in² to square feet

12. 7.3 a to square meters

13. 167.4 ft³ to cubic yards

14. 65°C to degrees Fahrenheit

Work the following problems. Round answers to the nearest hundredth. [7.1, 7.2, 7.3, 7.4]

15. How many gallons are there in a 1.5-liter bottle of cola?

16. Change 627 yd to inches.

17. A motorcycle engine has a displacement of 650 mL. What is the displacement in cubic feet?

18. Change 94 qt to liters.

19. Change 498 ft to meters.

20. How many liters are contained in an 15-quart container?

21. 37 cm to inches

22. 42 mi to kilometers

23. 47 in² to square centimeters

24. 23 ha to acres

25. 57 qt to liters

26. 47 lb to kilograms

27. Construction A 6-square-foot pantry floor is to be tiled using tiles that measure 4 inches by 4 inches. How many tiles will be needed to cover the pantry floor?

28. Filling an Aquarium How many 16-fluid ounce glasses of water will it take to fill a 9-gallon aquarium?

29. Change 7 hours 15 minutes to [7.5]
   a. minutes.
   b. hours.

30. Add 5 pounds 11 ounces and 10 pounds 7 ounces. [7.5]

The chart shows the annual sales for the top frozen pizza retailers in the United States. Use the information to answer the following questions.

TOP FROZEN PIZZA SELLERS
www.aibonline.org, 2009

DiGiorno	$591,262,700
Tombstone	$270,412,700
Red Baron	$256,308,000
California Pizza Kitchen	$175,750,800
Totino's Party Pizza	$152,630,700

31. If there are 12.07 Mexican pesos in 1 US dollar, convert the sales of DiGiorno and Red Baron to pesos.

32. If there are 19,487.01 Vietnamese dong to 1 US dollar, convert the sales of Tombstone and Totino's to dong. You will need to use an online conversion calculator.

Simplify.

1.   3,420
    679
  + 7,524

2.   7,000
  − 5,999

3. $378 \div 14$

4. $6(3 \cdot 9)$

5. $24\overline{)8,565}$

6. 3^5

7. $16 + 72 \div 2^2$

8. $\dfrac{468}{52}$

9. $17 + 39$

10. $(12 + 6) + (84 - 36)$

11. $\dfrac{60}{4}$

12. $11.5(3.9)$

13. $6.2 + 11.36 + 4.09$

14. $52.6 - 3.82$

15. $3.2\overline{)43.2}$

16. $\left(\dfrac{1}{3}\right)^2\left(\dfrac{1}{4}\right)^3$

17. $12 \div \left(\dfrac{2}{3}\right)^2$

18. $\dfrac{7}{48} + \dfrac{5}{12}$

19. $\left(15 \div 1\dfrac{2}{3}\right) \div 4$

20. $13 - 4\dfrac{3}{4}$

21. $\dfrac{5}{8}(3.6) - \dfrac{1}{2}(0.3)$

22. $\dfrac{3}{8}(4.8) - \dfrac{1}{4}(2.9)$

23. $14 + \dfrac{7}{13} \div \dfrac{21}{26}$

Solve.

24. $4 \cdot x = 17$

25. $36 = 8 \cdot y$

26. $\dfrac{6}{7} = \dfrac{18}{x}$

27. Find the perimeter and area of the figure below.

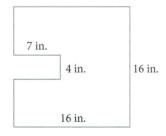

7 in.

4 in.   16 in.

16 in.

28. Find the perimeter of the figure below.

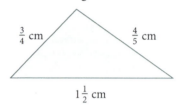

$\frac{3}{4}$ cm $\frac{4}{5}$ cm

$1\frac{1}{2}$ cm

29. Find the difference between 35 and 17.

30. If a car travels 288 miles in $4\frac{1}{2}$ hours, what is its rate in miles per hour?

31. What number is 32% of 6,450?

32. Factor 630 into a product of prime factors.

33. Find $\frac{3}{5}$ of the product of 15 and 6.

34. If 5,280 feet = 1 mile, convert 8,484 feet to miles. Round to the nearest tenth.

The diagram shows the number of viewers who watched the top five shows on NBC during one week. Use the information to answer the following questions. Round to the nearest hundredth if necessary.

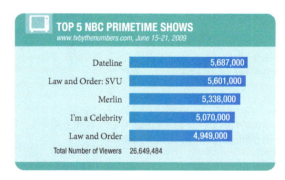

TOP 5 NBC PRIMETIME SHOWS
www.tvbythenumbers.com, June 15-21, 2009

Dateline	5,687,000
Law and Order: SVU	5,601,000
Merlin	5,338,000
I'm a Celebrity	5,070,000
Law and Order	4,949,000
Total Number of Viewers	26,649,484

35. Of the viewers who watched these shows, what percentage watched Dateline?

36. Of the viewers who watched these shows, what percentage watched the two Law and Order shows?

Use the tables in the chapter to make the following conversions. [7.1, 7.2, 7.3, 7.4]

1. 3 yd to feet

2. 640 m to kilometers

3. 4 acres to square feet

4. 864 in² to square feet

5. 12 L to milliliters

6. 3 mi to kilometers

7. 8 L to quarts

8. 90°F to degrees Celsius (round to the nearest tenth.)

9. 2.6 T to pounds

10. 112 oz to pounds

11. 648 in² to square feet

12. 6.3 a to square meters

13. 116.1 ft³ to cubic yards

14. 20°C to degrees Fahrenheit

Work the following problems. Round answers to the nearest hundredth. [7.1, 7.2, 7.3, 7.4]

15. How many gallons are there in a 2-liter bottle of cola?

16. Change 362 yd to inches.

17. A car engine has a displacement of 376 in³. What is the displacement in cubic feet?

18. Change 65 qt to liters.

19. Change 375 ft to meters.

20. How many liters are contained in an 11-quart container?

21. 27 cm to inches

22. 9 mi to kilometers

23. 29 in² to square centimeters

24. 17 ha to acres

25. 36 qt to liters

26. 23 lb to kilograms

27. Construction A 30-square-foot pantry floor is to be tiled using tiles that measure 6 inches by 6 inches. How many tiles will be needed to cover the pantry floor?

28. Filling an Aquarium How many 8-fluid ounce glasses of water will it take to fill a 12-gallon aquarium?

29. Change 2 hours 45 minutes to [7.5]
   a. minutes.
   b. hours.

30. Add 4 pounds 9 ounces and 2 pounds 7 ounces. [7.5]

The chart shows the amount of caffeine in different kinds of soda, measured in milligrams. Use the information to make the following conversions.

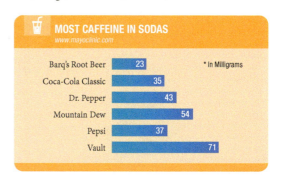

MOST CAFFEINE IN SODAS
www.mayoclinic.com

* In Milligrams

Barq's Root Beer	23
Coca-Cola Classic	35
Dr. Pepper	43
Mountain Dew	54
Pepsi	37
Vault	71

31. Convert the amount of caffeine found in all the sodas to grams.

32. Convert the amount of caffeine found in all the sodas to ounces.

Geometry

Krausnick, Germany
Image © 2010 GeoContent

Google

When you think of a tropical vacation, you picture sandy beaches and warm tropical waters. Now suppose you could take a tropical vacation just south of Berlin, Germany, and you could stay indoors the entire trip. You can do so at the Tropical Islands Resort in Krausnick, Germany, inside one of the largest buildings in the world. The structure was originally built in 2000 as an airplane manufacturing hanger for the CL 160 Airship, a craft that was never built. The hanger went unused until it was purchased in 2003 and transformed into an indoor vacation destination. With constant temperatures set at 79°F, the resort offers many attractions to visitors. Some vacationers relax at the spa or swim in the artificial lagoon. Others take advantage of the helium balloon rides or the tallest water slide tower in Germany. There is even an opportunity to get a suntan: one side of the building is made of special transparent glass, allowing UV rays to pass through!

Suppose you wanted to get some exercise while on your vacation. If you knew that the resort has a rectangular floor that measures 1181 feet long by 689 feet wide, how far would you walk if you walked the perimeter? How would you figure out how much surface the floor of the resort covers? To answer either of these questions, you will need to understand some basic geometric concepts. You will also need to know how to calculate perimeter and area, two of the topics we will work more with in this chapter.

8.1 Perimeter and Circumference

Video Examples
Section 8.1

Are you a fan of volleyball? What about soccer? Martial arts? Dance? Or would you just prefer to listen to music? What if there was a sport that included all of the above. That's right, it's called Bossaball. Teams of 3 to 5 players play on a court similar to the shape and size a volleyball court, only it's inflatable. And to top it off, each side of the court has a trampoline near the net from which a designated attacker does tricks and jumps to hit the ball! A typical attacker will hit the ball with his or her hand, but a select few prefer to execute a bicycle kick (like those seen on the soccer field) to slam the ball down on the defensive team. Each team can contact the ball a maximum of eight times to return it to the other team. The offense scores a point when the ball falls in bounds on the defensive side. If the ball lands on the defensive trampoline, the offense earns 3 points. The first team to reach 30 points wins the set.

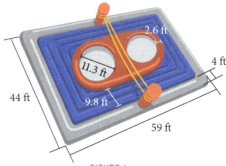

FIGURE 1

Imagine you are organizing a bossaball tournament and you need to reserve a venue large enough to fit the inflatable court. Examine the dimensions of a bossaball court in Figure 1. How would we use the measurements of the court to calculate the perimeter? How would you find the circumference of the trampoline? We will explore more about *perimeter* and *circumference* in this section. Then return to this introduction and try to answer these questions.

A Perimeter

We begin this section by reviewing the definition of a polygon, and the definition of perimeter.

DEFINITION polygon

A *polygon* is a closed geometric figure, with at least three sides, in which each side is a straight line segment.

The most common polygons are squares, rectangles, and triangles.

DEFINITION perimeter

The *perimeter* of any polygon is the sum of the lengths of the sides, and it is denoted with the letter *P*.

To find the perimeter of a polygon we add all the lengths of the sides together. Here are the most common polygons, along with the formula for the perimeter of each.

Square

s

$P = 4s$

Rectangle

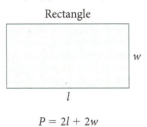

w

l

$P = 2l + 2w$

Triangle

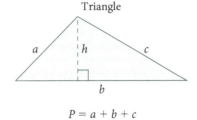

a h c

b

$P = a + b + c$

We can justify our formulas as follows. If each side of a square is s units long, then the perimeter is found by adding all four sides together.

$$\text{Perimeter} = P = s + s + s + s = 4s$$

Likewise, if a rectangle has a length of l and a width of w, then to find the perimeter we add all four sides together.

$$\text{Perimeter} = P = l + l + w + w = 2l + 2w$$

EXAMPLE 1 Find the perimeter of the given rectangle.

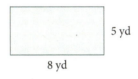

5 yd

8 yd

Solution The rectangle has a width of 5 yards and a length of 8 yards. We can use the formula for $P = 2l + 2w$ to find the perimeter.

$$P = 2(8) + 2(5)$$
$$= 16 + 10$$
$$= 26 \text{ yards}$$

EXAMPLE 2 Find the perimeter of each of the following stamps. Write your answer as a decimal, rounded to the nearest tenth, if necessary.

a. Each side is 35.0 millimeters.

b. Base $= 2\frac{5}{8}$ inches, Other two sides $= 1\frac{7}{8}$ inches

c. Length $= 1.56$ inches, Width $= 0.99$ inches

1. Find the perimeter.

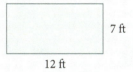

7 ft

12 ft

2. Find the perimeter of the stamps in Example 2 if they had the following dimensions. Round to the nearest tenth if necessary.

a. Each side is 42 millimeters.

b. Base $= 3\frac{4}{5}$ inches
Other two sides $= 1\frac{2}{5}$ inches

c. Length $= 3.86$ inches
Width $= 1.34$ inches

Answers

1. 38 ft

2. a. 168 mm **b.** $6\frac{3}{5}$ inches $= 6.6$ in.

c. 10.4 in.

Solution   We can add all the sides together, or we can apply our formulas. Let's apply the formulas.

a. $P = 4s = 4 \cdot 35 = 140$ mm

b. $P = a + b + c = 2\frac{5}{8} + 1\frac{7}{8} + 1\frac{7}{8} = 6\frac{3}{8} \approx 6.4$ in.

c. $P = 2l + 2w = 2(1.56) + 2(0.99) = 5.1$ in.

B Circumference

The *circumference* of a circle is the distance around the outside, just as the perimeter of a polygon is the distance around the outside. The circumference of a circle can be found by measuring its radius or diameter and then using the appropriate formula. The *radius* of a circle is the distance from the center of the circle to the circle itself. The radius is denoted by the letter r. The *diameter* of a circle is the distance from one side to the other, through the center. The diameter is denoted by the letter d. In the figure below, we can see that the diameter is twice the radius, or $d = 2r$.

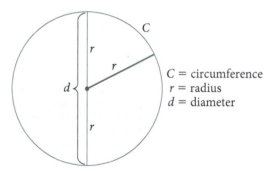

C = circumference
r = radius
d = diameter

The relationship between the circumference and the diameter or radius is not as obvious. As a matter of fact, it takes some fairly complicated mathematics to show just what the relationship between the circumference and the diameter is.

If you took a string and actually measured the circumference of a circle by wrapping the string around the circle and then measured the diameter of the same circle, you would find that the ratio of the circumference to the diameter, $\frac{C}{d}$, would be approximately equal to 3.14. The actual ratio of C to d in any circle is an irrational number. It can't be written in decimal form. We use the symbol π (Greek pi) to represent this ratio. In symbols, the relationship between the circumference and the diameter in any circle is

$$\frac{C}{d} = \pi$$

Knowing what we do about the relationship between division and multiplication, we can rewrite this formula as

$$C = \pi d$$

This is the formula for the circumference of a circle. When we do the actual calculations, we will use the approximation 3.14 for π.

Because $d = 2r$, the same formula written in terms of the radius is

$$C = 2\pi r$$

EXAMPLE 3   Find the circumference of each coin.

a. 1 Euro coin (Round to the nearest whole number.)
   Diameter = 23.25 millimeters

b. Susan B. Anthony dollar (Round to the nearest hundredth.)
   Radius = 0.52 inch

Solution   Applying our formulas for circumference, we have

a. $C = \pi d \approx (3.14)(23.25) \approx 73$ mm

b. $C = 2\pi r \approx 2(3.14)(0.52) \approx 3.27$ in.

3. Find the circumference of the following circles.
   a. Diameter = 14.2 mm
   b. Radius = 5.1 mm

Circles and the Earth

There are many circles found on the surface of the earth. The most familiar are the latitude and longitude lines. Of these circles, the ones with the largest circumference are called *great circles*. All of the longitude lines are great circles. Of the latitude lines, only the equator is a great circle.

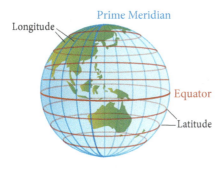

EXAMPLE 4   If the circumference of the earth is approximately 24,900 miles at the equator, what is the diameter of the earth to the nearest 10 miles?

Solution   We substitute 24,900 for C in the formula $C = \pi d$, and then we solve for d.

$$24,900 = \pi d$$

$$24,900 \approx 3.14d \qquad \text{Substitute 3.14 for } \pi.$$

$$d \approx \frac{24,900}{3.14} \qquad \text{Divide each side by 3.14.}$$

$$d \approx 7,930 \text{ miles}$$

4. If the circumference of a desk globe is 90 inches, what is its diameter to the nearest tenth of an inch?

GETTING READY FOR CLASS

After reading through the preceding section, respond in your own words and in complete sentences.

A. What is the perimeter of a polygon?

B. How are perimeter and circumference related?

C. What does π represent?

D. Using symbols, how does the radius of a circle relate to the circle's circumference?

Answers

3. a. 44.6 mm   **b.** 32.0 mm

4. 28.7 in

Vocabulary Review

Choose the correct words to fill in the blanks below.

perimeter   radius   diameter   polygon   circumference

1. A _____ is a closed geometric figure, with at least three sides, in which each side is a straight line segment.

2. The _____ of any polygon is the sum of the lengths of the sides.

3. The _____ of a circle is the distance around the outside, just as the perimeter of a polygon is the distance around the outside.

4. The _____ of a circle is the distance from the center of the circle to the circle itself.

5. The _____ of a circle is the distance from one side to the other, through the center.

Problems

A Find the perimeter of each figure. The first two figures are squares.

1.

8 in.

2.

9 cm

3.

30 yd

100 yd

4.

1 m

0.5 m

5.

10 in.   12 in.

14 in.

6.

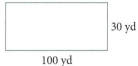

1.2 cm   3.0 cm

3.6 cm

7.

$2\frac{3}{8}$ in.

$2\frac{3}{8}$ in.   2 in.   4 in.

7 in.

8.

4 in.

$7\frac{3}{5}$ in.   4 in. $5\frac{1}{5}$ in.

13 in.

9.

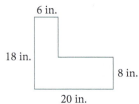

6 in.

18 in.

8 in.

20 in.

10.

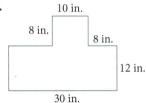

10 in.

8 in.   8 in.

12 in.

30 in.

11.

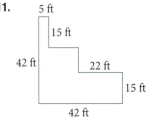

5 ft

15 ft

42 ft   22 ft

15 ft

42 ft

12.

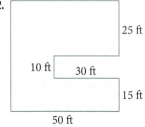

25 ft

10 ft   30 ft

15 ft

50 ft

A B Find the perimeter of each figure. Use 3.14 for π.

13.

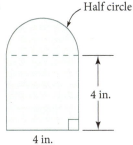

Half circle

4 in.

4 in.

14.

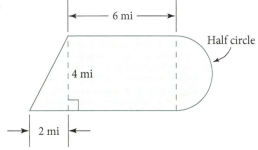

6 mi

Half circle

4 mi

2 mi

B Find the circumference of each circle. Use 3.14 for π.

15.

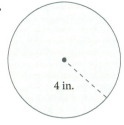

4 in.

16.

2 in.

Applying the Concepts

17. Painting The painting below has a length of 32 inches and a width of 24 inches. Find the perimeter.

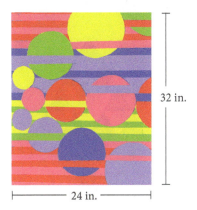

32 in.

24 in.

18. Painting The painting below has a length of 76 centimeters and a width of 102 centimeters. Find the perimeter.

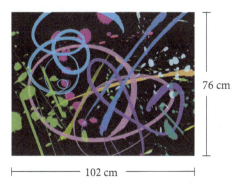

76 cm

102 cm

19. Circumference A dinner plate has a radius of 6 inches. Find the circumference.

20. Circumference A salad plate has a radius of 3 inches. Find the circumference.

21. Circumference The radius of the earth is approximately 3,900 miles. Find the circumference of the earth at the equator. (The equator is a circle around the earth that divides the earth into two equal halves.)

22. Circumference The radius of the moon is approximately 1,100 miles. Find the circumference of the moon around its equator.

23. Perimeter of a Banknote A 10-euro banknote has a width of 67 millimeters and a length of 127 millimeters. Find the perimeter.

24. Perimeter of a Dollar A $10 bill has a width of 2.56 inches and a length of 6.14 inches. Find the perimeter.

25. Circumference of a Game Piece A checkers game piece is 0.75 inches in diameter. Find the circumference.

26. Circumference of a Bown A salad bowl has a diameter of 29.21 centimeters. Find the circumference of the bowl.

27. Perimeter of an iPad An iPad has a length of 9.5 inches and a width of 7.31 inches. Find the perimeter of an iPad.

28. Perimeter of an iPod Nano An iPod Nano has a length of 37.5 mm and a width of 40.9 mm. Find the perimeter of an iPod Nano.

29. Perimeter of the Sierpinski Triangle The diagram shows one stage of what is known as the Sierpinski triangle. Each triangle in the diagram has three equal sides. The large triangle is made up of 4 smaller triangles. If each side of the large triangle is 2 inches, and each side of the smaller triangles is 1 inch, what is the perimeter of the shaded region?

30. Perimeter of the Sierpinski Triangle The diagram here shows another stage of the Sierpinski triangle. Each triangle in the diagram has three equal sides. The largest triangle is made up of a number of smaller triangles. If each side of the large triangle is 2 inches, and each side of the smallest triangles is 0.5 inches, what is the perimeter of the shaded region?

31. Geometry Suppose a rectangle has a perimeter of 12 inches. If the length and the width are whole numbers, give all the possible values for the width. Assume the width is shorter side and the length is the longer side.

32. Geometry Suppose a rectangle has a perimeter of 10 inches. If the length and the width are whole numbers, give all the possible values for the width. Assume the width is the shorter side and the length is the longer side.

33. Geometry If a rectangle has a perimeter of 20 feet, is it possible for the rectangle to be a square? Explain your answer.

34. Geometry If a rectangle has a perimeter of 10 feet, is it possible for the rectangle to be a square? Explain your answer.

35. Geometry A rectangle has a perimeter of 9.5 inches. If the length is 2.75 inches, find the width.

36. Geometry A rectangle has a perimeter of 11 inches. If the width is 2.5 inches, find the length.

Getting Ready for the Next Section

Simplify each expression. Round your answers to the nearest hundredth.

37. $\frac{1}{2} \cdot 6 \cdot 3$

38. $\frac{1}{2} \cdot 4 \cdot 2$

39. $\frac{1}{2} + 2\frac{5}{8} + 1\frac{1}{4}$

40. $\frac{1}{2}(6.6)(3.3)$

41. $3.14(14.5)^2$

42. $3.14(5)^2$

43. $144 - 36(3.14)$

44. $100 - 25(3.14)$

Find the Mistake

Each problem below contains a mistake. Circle the mistake and write the correct number(s) or word(s) on the line provided.

1. To find the perimeter of a triangle with side lengths of 4 inches, 3 inches, and 12 inches, find the product of the side lengths. _____

2. A standard piece of paper measures 8.5 inches by 11 inches and has a perimeter of 93.5 inches. _____

3. To find the circumference of a circle with a radius of 7.2 feet, multiply the diameter by 2 to get 45.22 feet.
_____

4. The circumference of a coin with a radius of 11.25 mm is 22.5 mm. _____

Navigation Skills: Prepare, Study, Achieve

Your academic self-image is how you see yourself as a student and the level of success you see yourself achieving. Do you believe you are capable of learning any subject and succeeding in any class you take? If you believe in yourself and work hard by applying the appropriate study methods, you will succeed. If you have a poor outlook for a class, most likely your performance in that class will match that outlook. Self-doubt or questioning the purpose of this course will negatively affect your focus. Furthermore, an inner dialogue of negative statements, such as "I'll never be able to learn this material" or "I'm never going to use this stuff," distract you from achieving success. Consider replacing those thoughts with three positive statements you can say when you notice your mind participating in a negative inner dialogue. Make a commitment to change your attitude for the better. Begin by thinking positively, having confidence in your abilities, and utilizing your resources if you are having difficulty. Asking for help is a sign of a successful student.

8.2 Area

©iStockphoto.com/Fitzer

Imagine you are an artist who has just been asked to showcase your work in a local gallery. You have framed a number of pieces of varying sizes to display. To ensure they will all fit, you need to know the square footage of the wall where your artwork will hang. If you know that the wall is 30 feet long by 10 feet tall, how will you calculate its square footage? Square footage (ft^2) is one way to measure area.

Recall that the area of a flat object is a measure of the amount of surface the object has. The area of the rectangle below is 8 square centimeters, because it takes 8 square centimeters to cover it.

**Video Examples
Section 8.2**

1 cm²	1 cm²	1 cm²	1 cm²
1 cm²	1 cm²	1 cm²	1 cm²

⎫ 2 cm

⎣___ 4 cm ___⎦

A Area

As we have noted previously, the area of this rectangle can also be found by multiplying the length and the width.

$$\text{Area} = (\text{length}) \cdot (\text{width})$$
$$= (4 \text{ centimeters}) \cdot (2 \text{ centimeters})$$
$$= (4 \cdot 2) \cdot (\text{centimeters} \cdot \text{centimeters})$$
$$= 8 \text{ square centimeters}$$

From this example, and others, we conclude that the area of any rectangle is the product of the length and width.

Here are the most common geometric figures along with the formula for the area of each one. The only formulas that are new to us are the ones that accompany the parallelogram and the circle.

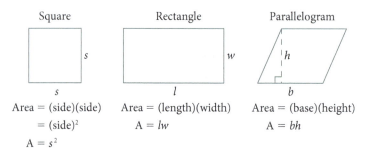

Square	Rectangle	Parallelogram
Area = (side)(side)	Area = (length)(width)	Area = (base)(height)
= (side)²	A = lw	A = bh
A = s^2		

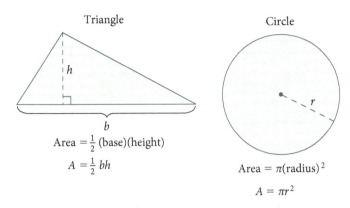

EXAMPLE 1 The parallelogram below has a base of 5 centimeters and a height of 2 centimeters. Find the area.

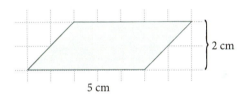

Solution   If we apply our formula, we have

$$\text{Area} = (\text{base})(\text{height})$$

$$A = bh$$

$$= 5 \cdot 2$$

$$= 10 \text{ cm}^2$$

Or, we could simply count the number of square centimeters it takes to cover the object. There are 8 complete squares and 4 half-squares, giving a total of 10 squares for an area of 10 square centimeters. Counting the squares in this manner helps us see why the formula for the area of a parallelogram is the product of the base and the height.

To justify our formula in general, we simply rearrange the parts to form a rectangle.

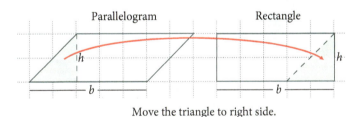

Move the triangle to right side.

EXAMPLE 2 The triangle below has a base of 6 centimeters and a height of 3 cm. Find the area.

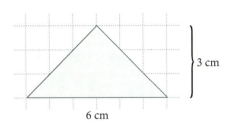

Practice Problems

1. Find the area of the parallelogram in Example 1 if the base is 6 centimeters and the height is 2 centimeters?

2. Find the area of the triangle in Example 2 if the height is 4 centimeters and the base is 8 centimeters.

Answers
1. 12 cm²
2. 16 cm²

Solution   If we apply our formula, we have

$$\text{Area} = \frac{1}{2}(\text{base})(\text{height})$$

$$A = \frac{1}{2}bh$$

$$= \frac{1}{2} \cdot 6 \cdot 3$$

$$= 9 \text{ cm}^2$$

As was the case in Example 1, we can also count the number of square centimeters it takes to cover the triangle. There are 6 complete squares and 6 half-squares, giving a total of 9 squares for an area of 9 square centimeters.

3. Find the area of the stamps in Example 3 if they have the following dimensions. Round to the nearest tenth if necessary.

a. Each side is 42 millimeters.

b. Base $= 3\frac{4}{5}$ inches

Height $= 2\frac{1}{4}$ inches

c. Length $= 3.86$ inches

Width $= 1.34$ inches

EXAMPLE 3   Find the area of each of the following stamps.

a. Each side is 35.0 millimeters.

b. Write your answer as a decimal, rounded to the nearest hundredth.

Base $= 2\frac{5}{8}$ inches

Height $= 1\frac{1}{4}$ inches

c. Round to the nearest hundredth.

Length $= 1.56$ inches

Width $= 0.99$ inches

Solution   Applying our formulas for area, we have

a. $A = s^2 = (35 \text{ mm})^2 = 1{,}225 \text{ mm}^2$

b. $A = \frac{1}{2}bh = \frac{1}{2}\left(2\frac{5}{8}\text{ in.}\right)\left(1\frac{1}{4}\text{ in.}\right) = \frac{1}{2} \cdot \frac{21}{8} \cdot \frac{5}{4}\text{ in}^2 \approx 1.64 \text{ in}^2$

c. $A = lw = (1.56 \text{ in.})(0.99 \text{ in.}) \approx 1.54 \text{ in}^2$

4. Find the area of a circle if the radius is 16 millimeters. Round to the nearest whole number.

EXAMPLE 4   Suppose a circle has a radius of 14.5 millimeters. Find the area of the circle to the nearest whole number.

Solution   Using our formula for the area of a circle, and using 3.14 for π, we have

$$A = \pi r^2 \qquad \qquad \text{Formula for area}$$

$$\approx 3.14(14.5)^2 \qquad \text{Substitute in values.}$$

$$\approx 3.14(210.25) \qquad \text{Square 14.5.}$$

$$\approx 660.185 \text{ mm}^2 \qquad \text{Multiply.}$$

$$\approx 660 \text{ mm}^2 \qquad \quad \text{Round to the nearest whole number.}$$

Answers

3. a. $1{,}764 \text{ mm}^2$ **b.** $8\frac{11}{20} \text{ in}^2$

c. 5.2 in^2

4. 804 mm^2

EXAMPLE 5   Find the area of the shaded portion of this figure.

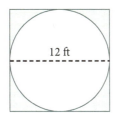

Solution   We have a circle inscribed in a square. We notice the diameter of the circle is the same length as one side of the square. To find the area of the shaded region, we subtract the area of the circle from the area of the square as follows:

$$A = 12^2 - 6^2\pi$$

$$= 144 - 36\pi$$

$$\approx 30.96 \text{ ft}^2$$

5. Find the area of the shaded portion of the figure in Example 5 if the diameter is 20 feet.

GETTING READY FOR CLASS

After reading through the preceding section, respond in your own words and in complete sentences.

A. Suppose a rectangle is 8 inches long and 3 inches wide. How many square inches will it take to cover the rectangle?

B. A rectangle measures 4 feet by 6 feet. What units will you assign to the perimeter and to the area?

C. Briefly explain how the formula for the area of a rectangle is similar to the formula for the area of a parallelogram.

D. How do you find the area of a circle?

Answer

5. 86 ft^2

Vocabulary Review

Choose the correct words to fill in the blanks below.

| parallelogram | triangle | square | area | circle |

1. The _____ of a flat object is a measure of the amount of surface the object has.

2. The formula for the area of a _____ is $A = s \cdot s = s^2$.

3. The formula for the area of a _____, $A = bh$, is similar to the formula for the area of a rectangle.

4. The formula for the area of a _____ is $A = \frac{1}{2}bh$.

5. The formula for the area of a _____ is $A = \pi r^2$.

Problems

A Find the area enclosed by each figure.

1.

5 cm

5 cm

2.

10 ft

10 ft

3.

14 m

24 m

4.

0.3 in.

1.2 in.

5.

6 ft

10 ft

6.

6 ft

8 ft

7.
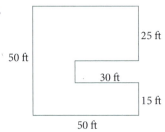
25 ft

50 ft

30 ft

15 ft

50 ft

8.
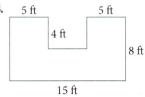
5 ft   5 ft

4 ft

8 ft

15 ft

9.
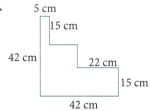
5 cm

15 cm

42 cm

22 cm

15 cm

42 cm

10.
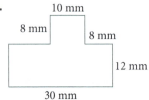
10 mm

8 mm

8 mm

12 mm

30 mm

11.
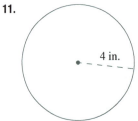
4 in.

12.

2 in.

13.

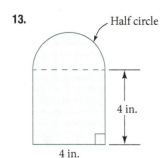

Half circle

4 in.

4 in.

14.

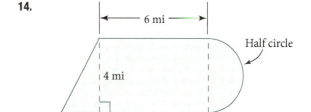

6 mi

Half circle

4 mi

2 mi

15. Find the area of each object.

a. A square with side 10 inches

b. A circle with radius 10 inches

16. Find the area of each object.

a. A square with side 6 centimeters

b. A triangle with a base and height of 6 centimeters

c. A circle with radius 6 centimeters (Round to the nearest whole number.)

17. Find the area of the triangle with base 19 inches and height 14 inches.

18. Find the area of the triangle with base 13 inches and height 8 inches.

19. The base of a triangle is $\frac{4}{3}$ feet and the height is $\frac{2}{3}$ feet. Find the area.

20. The base of a triangle is $\frac{8}{7}$ feet and the height is $\frac{14}{5}$ feet. Find the area.

Applying the Concepts

21. Area A swimming pool is 20 feet wide and 40 feet long. If it is surrounded by square tiles, each of which is 1 foot by 1 foot, how many tiles are there surrounding the pool?

22. Area A garden is rectangular with a width of 8 feet and a length of 12 feet. If it is surrounded by a walkway 2 feet wide, how many square feet of area does the walkway cover?

23. Area of a Stamp Imagine the image area of a stamp has a width of 0.84 inches and a length of 1.41 inches. Find the area of the image. Round to the nearest hundredth.

24. Area of a Stamp Assume the image area of a stamp has a width of 21.4 millimeters and a length of 35.8 millimeters. Find the area of the image. Round to the nearest whole number.

25. Area of a Euro A 10-euro banknote has a width of 67 millimeters and a length of 127 millimeters. Find the area.

26. Area of a Dollar A $10 bill has a width of 2.56 inches and a length of 6.14 inches. Find the area of the bill. Round to the nearest hundredth.

27. Comparing Areas The side of a square is 5 feet long. If all four sides are increased by 2 feet, by how much is the area increased?

28. Comparing Areas The length of a side in a square is 20 inches. If all four sides are decreased by 4 inches, by how much is the area decreased?

29. Area of a Coin The diameter of this $1 coin is 26.5 mm. Find the area of one side of the coin. Round to the nearest hundredth.

30. Area of a Coin The Susan B. Anthony dollar has a radius of 0.52 inches. Find the area of one side of the coin to the nearest hundredth.

31. a. Each side of the red square in the corner is 1 centimeter, and all squares are the same size. On the grid below, draw three more squares. Each side of the first one will be 2 centimeters, each side of the second square will be 3 centimeters, and each side of the third square will be 4 centimeters.

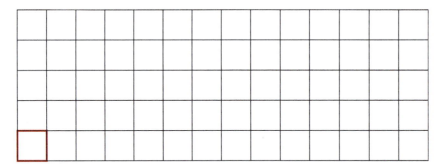

b. Use the squares you have drawn above to complete each of the following tables.

Perimeters of Squares

Length of each side (in centimeters)	Perimeter (in centimeters)
1	
2	
3	
4	

Areas of Squares

Length of each side (in centimeters)	Area (in square centimeters)
1	
2	
3	
4	

32. a. The lengths of the sides of the squares in the grid below are all 1 centimeter. The red square has a perimeter of 12 centimeters. On the grid below, draw two different rectangles, each with a perimeter of 12 centimeters.

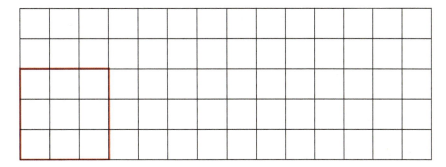

   b. Find the area of each of the three figures in part a.

33. The circle here is said to be *inscribed* in the square. If the area of the circle is 64π square centimeters, find the length of one of the diagonals of the square (the distance from *A* to *D*). Round to the nearest tenth.

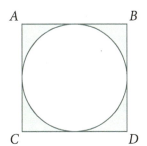

34. Suppose you have a pizza box with each side measuring 12 inches long. If your pizza just fits in the box, what is the diameter, area and circumference of your pizza?

Getting Ready for the Next Section

Simplify.

35. $2 \cdot 3 \cdot 4$

36. $2(12)(15)$

37. $12 + 20 + 40$

38. $54 + 40 + 46$

39. $2(3.14)(0.125)(6)$

40. $314 \div 2$

41. $78.5 + 311.5 + 157$

42. $\frac{1}{2}[4(3.14)(100)]$

Find the Mistake

Each problem below contains a mistake. Circle the mistake and write the correct number(s) or word(s) on the line provided.

1. To find the area of a parallelogram with a base of 6 inches and a height of 4 inches, you must add the base and height measurements to get an area of 24 square inches. _____

2. To find the area of a triangle with a base of 6 cm and a height of 3 cm, use the formula $A = bh$ to get an area of 9 cm². _____

3. To find the area of a circle with a diameter of 12 feet, use the formula $A = \pi r^2$ to get an area of 452.16 ft². _____

4. To find the area of the shaded region below, add the area of the circle to the area of the square. _____

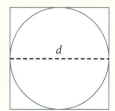

Landmark Review: Checking Your Progress

Find the perimeter or circumference and area of each figure Use 3.14 for π if necessary.

1.

7 in.
7 in.

2.

27 cm
27 cm

3.
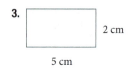
2 cm
5 cm

4.

3 mi
8 mi

5.
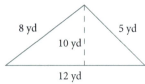
8 yd   5 yd
10 yd
12 yd

6.
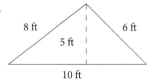
8 ft   6 ft
5 ft
10 ft

7.
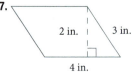
2 in.   3 in.
4 in.

8.
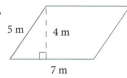
5 m   4 m
7 m

9.

5 in.

10.

28 mm

8.3 Surface Area

The Moeraki boulders are large spherical boulders scattered on a beach in New Zealand. The boulders are made up of a hard mineral that once filled the pores in a sedimentary rock layer. Over time, the sedimentary rock eroded, leaving behind the boulders in the sand. The native Maori people believe the boulders are fossilized wreckage of an ancient canoe than has washed ashore.

The largest Moeraki boulder measures just under 7 feet in diameter. In this section, we will learn how to compute the surface area of any sphere, such as a Moeraki boulder, given its radius (which is half the diameter). We will also find the surface area of other three dimensional shapes, such as cubes an cylinders. Let's first examine the surface area of a rectangular solid.

A Surface Area of a Rectangular Solid

The figure below shows a closed box with length l, width w, and height h. The surfaces of the box are labeled as sides, top, bottom, front, and back. A box like this is called a *rectangular solid*. In general, a rectangular solid is a closed figure in which all sides are rectangular that meet at right angles.

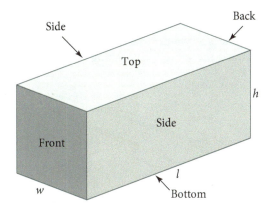

**Video Examples
Section 8.3**

To find the *surface area* S of the box, we add the areas of each of the six surfaces.

$$\text{Surface area} = \text{side} + \text{side} + \text{front} + \text{back} + \text{top} + \text{bottom}$$

$$S = l \cdot h + l \cdot h + h \cdot w + h \cdot w + l \cdot w + l \cdot w$$

Therefore, the formula for the surface area of the box is

$$S = 2lh + 2hw + 2lw$$

Practice Problems

1. Find the surface area of the box shown here:

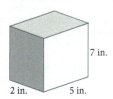

2 in.   7 in.   5 in.

EXAMPLE 1 Find the surface area of the box shown here:

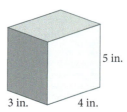

5 in.

3 in.   4 in.

Solution To find the surface area, we find the area of each surface individually, and then we add them together.

$$\text{Surface area} = 2(3 \text{ in.})(4 \text{ in.}) + 2(3 \text{ in.})(5 \text{ in.}) + 2(4 \text{ in.})(5 \text{ in.})$$

$$= 24 \text{ in}^2 + 30 \text{ in}^2 + 40 \text{ in}^2$$

$$= 94 \text{ in}^2$$

The total surface area is 94 square inches.

B Surface Area of a Cylinder

Here are the formulas for the surface area of some right circular cylinders. A right cylinder is a cylinder whose base is a circle and whose sides are perpendicular to the base.

Open at both ends

h

r

$S = 2\pi rh$

Closed at one end

h

r

$S = \pi r^2 + 2\pi rh$

Closed at both ends

h

r

$S = 2\pi r^2 + 2\pi rh$

2. Find the amount of material used in the straw in Example 2 if the length is 10 inches and the radius is 0.24 inches. Round to the nearest tenth.

EXAMPLE 2 The drinking straw shown below has a radius of 0.125 inch and a length of 6 inches. How much material was used to make the straw?

Solution Since a straw is a cylinder that is open at both ends, we find the amount of material needed to make the straw by calculating the surface area.

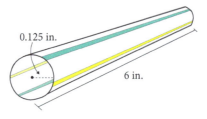

0.125 in.

6 in.

$$S = 2\pi rh$$

$$\approx 2(3.14)(0.125)(6)$$

$$= 4.71 \text{ in}^2$$

It takes 4.71 square inches of material to make the straw.

Answers

1. 118 in²
2. 15.1 in²

C Surface Area of a Sphere

The figure below shows a sphere and the formula for its surface area.

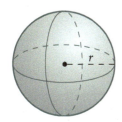

Surface Area = $4\pi(\text{radius})^2$

$S = 4\pi r^2$

EXAMPLE 3   The figure below is composed of a right circular cylinder with half a sphere on top. (A half-sphere is called a hemisphere.) Find the surface area of the figure assuming it is closed on the bottom.

3. Find the surface area of the figure in Example 3 if the radius is 7 inches and the height is 12 inches. Round to the nearest tenth.

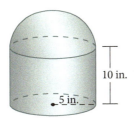

10 in.

5 in.

Solution   The total surface area is found by adding the surface area of the cylinder to the surface area of the hemisphere.

S = surface area of bottom of cylinder + surface area of side of cylinder + surface area of hemisphere

$$= \pi r^2 + 2\pi rh + \frac{1}{2}(4\pi r^2)$$

$$\approx (3.14)(5)^2 + 2(3.14)(5)(10) + \frac{1}{2}[4(3.14)(5^2)]$$

$$= 78.5 + 314 + 157$$

$$= 549.5 \text{ in}^2$$

The total surface area is 549.5 square inches.

GETTING READY FOR CLASS

After reading through the preceding section, respond in your own words and in complete sentences.

A. What is a rectangular solid?

B. How do the formulas for a cylinder open at both ends and a cylinder closed at both ends differ?

C. What is a hemisphere?

D. How are a circle and a sphere related?

Vocabulary Review

Choose the correct words to fill in the blanks below.

surface area   closed   rectangular solid   sphere   open   box

1. A _____ is a closed figure in which all sides are rectangular that meet at right angles.

2. To find the surface area of a _____, add the areas of each of the six surfaces.

3. The formula for the _____ of a box is given by $S = 2lh + 2hw + 2lw$, where l represents length, h represents height, and w represents width.

4. The surface area of a cylinder _____ at both ends is given by $S = 2\pi rh$.

5. The surface area of a cylinder _____ at both ends is given by $S = 2\pi r^2 + 2\pi rh$.

6. The surface area of a _____ is given by $S = 4\pi r^2$.

Problems

A Find the surface area of each figure.

1.

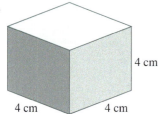

4 cm
4 cm   4 cm

2.

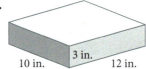

3 in.
10 in.   12 in.

3.

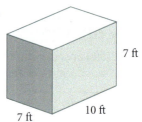

7 ft
7 ft   10 ft

4.

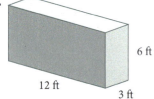

6 ft
12 ft   3 ft

5.

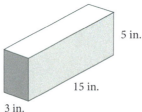

5 in.
15 in.
3 in.

6.

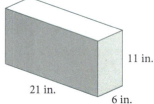

11 in.
21 in.   6 in.

B C Assume all cylinders are closed at both ends. Round to the nearest hundredth, if necessary.

7.

8 ft
2 ft

8.

8 ft
4 ft

9.

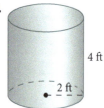

4 ft
2 ft

10.

4 ft
4 ft

11.

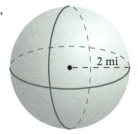

2 mi

12.

3.9 in.

13.

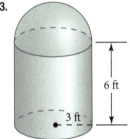

6 ft
3 ft

14.

3 ft
6 ft

Applying the Concepts

15. Surface Area of a Coin The Sacagawea dollar coin has a diameter of 26.5 mm, and a thickness of 2.00 mm. Find the surface area of the coin to the nearest hundredth.

16. Surface Area of a Coin A Susan B. Anthony dollar has a radius of 0.52 inch and a thickness of 0.0079 inch. Find the surface area of the coin. Round to the nearest ten thousandth.

17. Travertine at the Getty Over 1.2 million square feet of travertine stone from Italy was used to construct the Getty Museum. It was mined in large slabs that measured 6 meters high, 12 meters wide, and 2 meters deep. Find the surface area of one of these slabs.

18. Travertine at the Getty After large slabs of travertine were mined in Italy, they were cut into smaller blocks that were used to construct the Getty Museum. According to the website *www.getty.edu*, "On average, each block at the Getty Center is 76 × 76 centimeters and weighs 115 kilograms, with a typical thickness of 8 centimeters." Find the surface area of one of these stones.

Making a Cylinder Make an $8\frac{1}{2}$ by 11 inch piece of graph paper into a cylinder as shown. Use the diagrams to help with Problems 19 and 20.

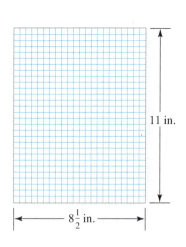

11 in.

$8\frac{1}{2}$ in.

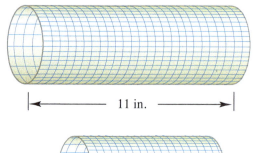

11 in.

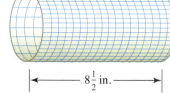

$8\frac{1}{2}$ in.

19. a. If the length of the cylinder is 11 inches, what is the largest possible radius? Round to the nearest thousandth.

b. Using the radius from part a, and the formula for the surface area of a cylinder, find the surface area of the rolled piece of graph paper. Round to the nearest tenth.

c. Find the area of the original piece of graph paper by multiplying length by width. Does this area match the area you found in part b?

20. a. If the length of the cylinder is $8\frac{1}{2}$ inches, what is the largest possible radius? Round to the nearest thousandth.

b. Using the radius from part a, and the formula for the surface area of a cylinder, find the surface area of the rolled piece of graph paper. Round to the nearest tenth.

c. Find the area of the original piece of graph paper by multiplying length by width. Does this area match the area you found in part b?

21. A living room is 10 feet long and 8 feet wide. If the ceiling is 8 feet high, what is the total surface area of the four walls?

22. A family room is 12 feet wide and 14 feet long. If the ceiling is 8 feet high, what is the total surface area of the four walls?

One Step Further

23. The surface of the earth is 70% water. If the diameter of the earth is 8,000 miles, how many square miles of the earth are covered by land.

24. The surface of the earth is 70% water. How many square kilometers of the earth are covered by land if the diameter is 12,874 kilometers?

Surface Area of a Cone The surface area of a cone is given by the formula $S = \pi r^2 + \pi r l$ where l is called the slant height. We can use the Pythagorean theorem to find the slant height (see right figure). Using the formula, find the slant height and surface area of the following cones:

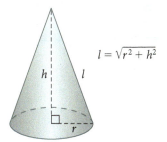

$l = \sqrt{r^2 + h^2}$

25.

26.

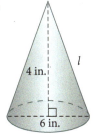

Getting Ready for the Next Section

Simplify each expression.

27. 3^3

28. $15 \cdot 3 \cdot 5$

29. $\dfrac{1}{2} \cdot \dfrac{4}{3}$

30. 0.125^2

Simplify each expression. Round to the nearest thousandth.

31. $3.14(0.125)^2(6)$

32. $\dfrac{2}{3}(392.6)$

33. $3.14(25)(10)$

34. $785 \div 3$

Find the Mistake

Each problem below contains a mistake. Circle the mistake and write the correct number(s) or word(s) on the line provided.

1. To find the surface area of a rectangular solid, we add the areas of the top, the bottom, and one side.

_____

2. The surface area of this box is 124 in². _____

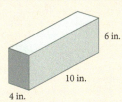

6 in.

10 in.

4 in.

3. Suppose a soup company needs new labels for their cans. Each label will wrap around the entire side of the can. To find the surface area of the label, use the formula $S = 2\pi r^2 + 2\pi r h$. _____

4. The surface area of a sphere with radius 4.2 in is 55.4 in². _____

KEY WORDS

volume

cubic foot

cone

parallelogram

Video Examples
Section 8.4

8.4 Volume

©iStockphoto.com/CvE

The Cube Houses in Rotterdam, Netherlands are apartments shaped like cubes tilted onto one of their points. Each apartment has three triangular floors of living space within the cube. Let's suppose each exterior side of one apartment is 7.5 meters long. How might we use this measurement to calculate the total *volume* of the apartment? In this section, we will further explore volume and work problems similar to this one.

Volume is the measure of the space enclosed by a solid. For instance, if each edge of a cube is 3 feet long, then we can think of the cube as being made up of a number of smaller cubes, each of which is 1 foot long, 1 foot wide, and 1 foot high. Each of these smaller cubes is called a *cubic foot*. To count the number of them in the larger cube, think of the large cube as having three layers. You can see that the top layer contains 9 cubic feet. Because there are three layers, the total number of cubic feet in the large cube is $9 \cdot 3 = 27$.

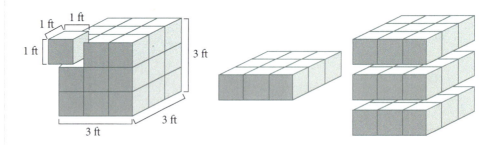

A Volume of a Rectangular Solid

On the other hand, if we multiply the length, the width, and the height of the cube, we have the same result.

$$\text{Volume} = (3 \text{ feet})(3 \text{ feet})(3 \text{ feet})$$

$$= (3 \cdot 3 \cdot 3)(\text{feet} \cdot \text{feet} \cdot \text{feet})$$

$$= 27 \text{ cubic feet}$$

For our first example, we will confine our discussion of volume to the volume of a *rectangular solid*. Recall that rectangular solids are the three-dimensional equivalents of rectangles: Opposite sides are parallel, and any two sides that meet, join at right angles. A rectangular solid is shown, along with the formula used to calculate its volume.

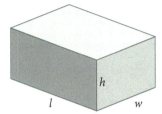

Volume = (length)(width)(height)
$$V = lwh$$

EXAMPLE 1   Find the volume of a rectangular solid with a length of 15 inches, a width of 3 inches, and a height of 5 inches.

Solution   To find the volume, we apply the formula shown above.

$$V = l \cdot w \cdot h$$
$$= (15 \text{ in.})(3 \text{ in.})(5 \text{ in.})$$
$$= 225 \text{ in}^3$$

Practice Problems

1. Find the volume of a rectangular solid with a length of 12 inches, a width of 4 inches, and a height of 6 inches.

B Volume of a Cylinder

Here is the formula for the volume of a right circular cylinder—a cylinder whose base is a circle and whose sides are perpendicular to the base.

Volume = π(radius)^2(height)
$$V = \pi r^2 h$$

EXAMPLE 2   The drinking straw shown below has a radius of 0.125 inch and a length of 6 inches. To the nearest thousandth, find the volume of liquid that it will hold.

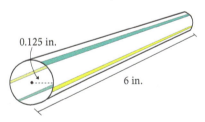

0.125 in.

6 in.

2. Find the volume of liquid the straw in Example 2 will hold if it has a radius of 0.24 inches and a length of 10 inches. Round to the nearest thousandth.

Solution   The total volume is found from the formula for the volume of a right circular cylinder. In this case, the radius is $r = 0.125$, and the height is $h = 6$. We approximate π with 3.14.

$$V = \pi r^2 h$$
$$\approx (3.14)(0.125)^2(6)$$
$$\approx (3.14)(0.015625)(6)$$
$$\approx 0.294 \text{ in}^3 \text{ to the nearest thousandth}$$

Answers

1. 288 in^3
2. 1.809 in^3

C Volume of a Cone

Next, we have the formula for the volume of a cone. As you can see, the relevant dimensions of a cone are the radius of its circular base and its height. You will also notice that this formula involves both a fraction, the number $\frac{1}{3}$, and a decimal, the number π, for which we have been using 3.14.

$\text{Volume} = \frac{1}{3}\pi(\text{radius})^2(\text{height})$

$V = \frac{1}{3}\pi r^2 h$

3. Find the volume of the cone in Example 3 if the height is 7 centimeters and the radius is 4 centimeters.

EXAMPLE 3   Find the volume of the given cone.

5 cm

3 cm

Solution   The volume is found by using the formula for the volume of a cone. In this case, the radius = 3 cm, and the height = 5 cm. Again, we use 3.14 for π.

$$V = \frac{1}{3}(3.14)(3^2)(5)$$

$$\approx \frac{1}{3}(3.14)(9)(5)$$

$$\approx (3.14)(3)(5)$$

$$\approx (3.14)(15)$$

$$\approx 47.1 \text{ cm}^3$$

D Volume of a Sphere

Next, we have a sphere and the formula for its volume. Once again, the formula contains both the fraction $\frac{4}{3}$ and the number π.

$\text{Volume} = \frac{4}{3}\pi(\text{radius})^3$

$= \frac{4}{3}\pi r^3$

Answer

3. 117.2 cm³

EXAMPLE 4   The figure here is composed of a right circular cylinder with half a sphere on top. (A half-sphere is called a hemisphere.) To the nearest tenth, find the total volume enclosed by the figure.

10 in.

.5 in.

Solution   The total volume is found by adding the volume of the cylinder to the volume of the hemisphere.

$V = $ volume of cylinder $+$ volume of hemisphere

$$= \pi r^2 h + \frac{1}{2} \cdot \frac{4}{3} \pi r^3$$

$$\approx (3.14)(5)^2(10) + \frac{1}{2} \cdot \frac{4}{3}(3.14)(5)^3$$

$$\approx (3.14)(25)(10) + \frac{1}{2} \cdot \frac{4}{3}(3.14)(125)$$

$$\approx 785 + \frac{2}{3}(392.5) \qquad \text{Multiply: } \frac{1}{2} \cdot \frac{4}{3} = \frac{4}{6} = \frac{2}{3}.$$

$$\approx 785 + \frac{785}{3} \qquad \text{Multiply: } 2(392.5) = 785.$$

$$\approx 785 + 261.7 \qquad \text{Divide } 785 \text{ by } 3, \text{ and round to the nearest tenth.}$$

$$\approx 1{,}046.7 \text{ in}^3$$

GETTING READY FOR CLASS

After reading through the preceding section, respond in your own words and in complete sentences.

A. If the dimensions of a rectangular solid are given in inches, what units will be associated with the volume?

B. What is the relationship between area and volume?

C. What formulas from this section involve both a fraction and a decimal?

D. State the volume formula for a cube, a cylinder, a cone, and a sphere.

4. Find the volume of the figure in Example 4 if the height is 12 inches and the radius is 4 inches.

EXERCISE SET 8.4

Problems

A–D Find the volume of each figure. Round to the nearest hundredth, if necessary.

1.

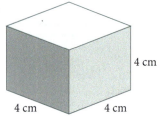

4 cm

4 cm   4 cm

2.

3 in.

10 in.   12 in.

3.

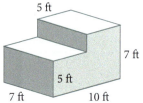

5 ft

7 ft

5 ft

7 ft   10 ft

4.

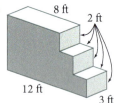

8 ft

2 ft

12 ft   3 ft

5.

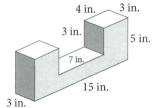

4 in.   3 in.

3 in.   5 in.

7 in.

15 in.

3 in.

6.

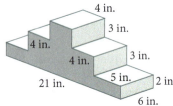

4 in.

3 in.

4 in.

4 in.   3 in.

21 in.   5 in.   2 in.

6 in.

7.

8 ft

2 ft

8.

8 ft

4 ft

9.

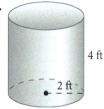

4 ft

2 ft

10.

4 ft

4 ft

11.

2 mi

12.

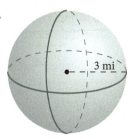

3 mi

13.

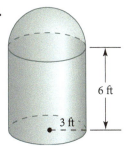

6 ft

3 ft

14.

3 ft

6 ft

15.

7.1 in.

3.9 in.

16.

3.4 in.

1.1 in.

Applying the Concepts

17. Volume of a Coin The Sacagawea dollar coin has a diameter of 26.5 mm, and a thickness of 2.00 mm. Find the volume of the coin to the nearest hundredth.

18. Volume of a Coin The Susan B. Anthony dollar coin has a radius of 0.52 inch and a thickness of 0.0079 inch. Find the volume of the coin. Round to the nearest ten thousandth.

19. Ice Cream An ice cream cone has a radius of 2.3 cm and a height of 6.2 cm. The cone is filled with ice cream and one additional "scoop" in the shape of a sphere with the same radius as the cone is placed on top. What is the amount of ice cream in cubic centimeters? Round to the nearest hundredth.

20. Ice Cream An ice cream cone has a radius of 1.7 inches and a height of 4.6 inches. The cone is filled with ice cream and one and a half additional "scoops" in the shape of a sphere with the same radius as the cone are placed on top. What is the amount of ice cream in cubic inches? Round to the nearest hundredth.

Engine Size The size of an engine is a measure of volume. To calculate the size of an engine is to find the volume of a cylinder and multiply by the number of cylinders. The size of a cylinder is given as a bore size, or diameter, and a stroke size, or height, of the cylinder.

21. Find the size of an 8-cylinder Chevy big-block engine with a bore of 4.25 inches and a stroke of 3.76 inches. Round to the nearest hundredth.

22. Find the size of an 8-cylinder Chevy big-block engine with a bore of 4.47 inches and a stroke of 4.00 inches. Round to the nearest hundredth.

Triangular Pyramid The formula for finding the volume of a triangular pyramid is one-third the area of the base times the height, or $V = \frac{1}{3}Bh$, where B = the area of the base.

Find the volume of each pyramid. Round to the nearest hundredth.

23.

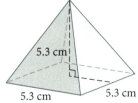

5.3 cm

5.3 cm   5.3 cm

24.

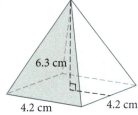

6.3 cm

4.2 cm   4.2 cm

25.

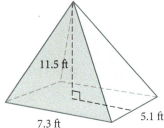

11.5 ft

7.3 ft   5.1 ft

26.

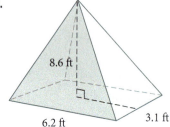

8.6 ft

6.2 ft   3.1 ft

27. **Travertine at the Getty** As mentioned in the previous section, construction material for the Getty Museum included large slabs of travertine. After the slabs were mined in Italy, they were cut into smaller blocks that were used to construct the museum. According to the website *www.getty.edu*, "on the average each block at the Getty Center is 76 × 76 centimeters and weighs 115 kilograms, with a typical thickness of 8 centimeters." Find the volume of an average block of travertine.

28. **Travertine at the Getty** Find the volume of a large slab of travertine if the slab is 6 meters high, 12 meters wide, and 2 meters deep.

Find the Mistake

Each problem below contains a mistake. Circle the mistake and write the correct number(s) or word(s) on the line provided.

1. The volume of a rectangular solid with a length of 4 mm, a height of 9 mm, and a width of 4 mm is found by adding $4 + 9 + 4$ to get $17 mm^3$. _____

2. The volume of a right circular cylinder is found by using the formula $V = r^2h$. _____

3. Finding the volume of a cone with a radius of 7 cm and a height of 10 cm would give $V = 1{,}539\ cm^3$.

   _____

4. To find the volume of a tennis ball, use the formula for the volume of a sphere $V = \frac{4}{3}\pi r^2$.

   _____

Supplies Needed

A A paper squares, 15 cm × 15 cm

B A ruler

Origami Pinwheel

Origami is the art of folding paper that originated in Japan. Elaborate shapes can be created from a simple square of paper.

1. Cut out a square with a side length of 15 cm. Calculate the perimeter and surface area of the square. Using the steps below, fold the square into a pinwheel.

1. Fold your square in half diagonally both ways, and unfold.

2. Fold all four corners to the center, and unfold.

3. Fold the square in half, and unfold.

4. Fold the right and left sides into center. Do not unfold.

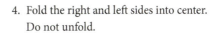

5. Pull the center of the top edge down to the center of square. The sides should point out and extend past the right and left folded sides.

6. Repeat step 5 with the bottom edge.

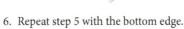

7. Fold the left top flap up.

8. Fold the right bottom flap down.

2. Find the outer perimeter (cm) and surface area (cm²) of the pinwheel.

3. Repeat steps 1 and 2 for a 10 cm square and a 20 cm square.

4. What is the relationship between the squares' original measurements and the measurements of each pinwheel.

Extra Challenge Using another square, fold any shape that has a perimeter larger than that of the original square.

Chapter **8** Summary

EXAMPLES

1. a. Find the perimeter and the area.

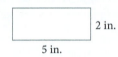

5 in.   2 in.

$P = 2 \cdot 5 + 2 \cdot 2$
$\quad = 14 \text{ in.}$
$A = 5 \cdot 2$
$\quad = 10 \text{ in}^2$

b. Find the area.

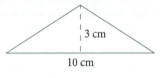

3 cm

10 cm

$A = \frac{1}{2}(10)(3)$
$\quad = 15 \text{ cm}^2$

Formulas for Perimeter and Area of Polygons [8.1, 8.2]

Square

s

s

$P = 4s$
$A = s^2$

Rectangle

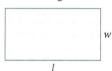

w

l

$P = 2l + 2w$
$A = lw$

Triangle

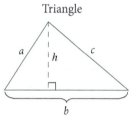

a h c

b

$P = a + b + c$
$A = \frac{1}{2}bh$

Parallelogram

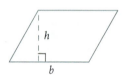

h

b

$A = bh$

Formulas for Diameter and Radius of a Circle [8.1, 8.2]

2. If the radius of a circle is 5.7 feet, find the diameter.
$d = 2(5.7)$
$\quad = 11.4 \text{ ft}$

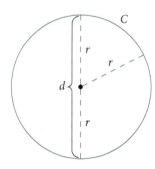

C

r

r

d

r

$C = \text{circumference}$
$r = \text{radius}$
$d = \text{diameter}$
$d = 2r$
$r = \frac{d}{2}$

Formulas for Circumference and Area of a Circle [8.1, 8.2]

3. Find the circumference and the area.

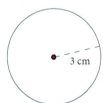

3 cm

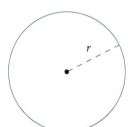

r

$C = 2\pi r$
$A = \pi r^2$

$C = 2(3.14)(3)$
$\quad = 18.84 \text{ cm}$
$A = 3.14(3)^2$
$\quad = 28.26 \text{ cm}^2$

Formulas for Surface Area and Volume [8.3, 8.4]

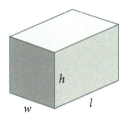

Volume = lwh

Surface Area = $2lh + 2hw + 2lw$

Volume = $\pi r^2 h$

Surface Area = $2\pi r^2 + 2\pi rh$

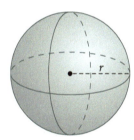

Volume = $\frac{4}{3}\pi r^3$

Surface Area = $4\pi r^2$

Volume = $\frac{1}{3}\pi r^2 h$

4. Find the volume and the surface area.

a.

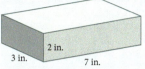

$V = 7 \cdot 3 \cdot 2$

$\quad = 42$ in³

$S = 2(3)(2) + 2(3)(7) + 2(2)(7)$

$\quad = 12 + 42 + 28$

$\quad = 82$ in²

b.

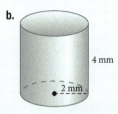

$V = (3.14)(2)^2 \cdot 4$

$\quad = 50.24$ mm³

$S = 2(3.14)(2)^2 + 2(3.14)(2)(4)$

$\quad = 75.4$ mm²

c.

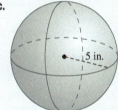

$V = \frac{4}{3}(3.14)5^3$

$\quad = 523$ in³, to the nearest whole number

$S = 4(3.14)5^2$

$\quad = 314$ in²

d. Find volume only.

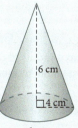

$V = \frac{1}{3}(3.14)(4)^2(6)$

$\quad = 100.5$ cm²

Find the perimeter of each figure. [8.1]

1.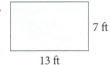

7 ft

13 ft

2.

5.3 m   8.9 m

11.6 m

3.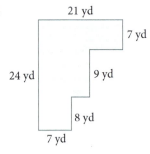

21 yd

7 yd

24 yd   9 yd

8 yd

7 yd

4. Find the circumference of the circle.
Use 3.14 for π. [8.1]

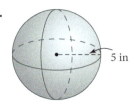

8 in

5. Find the perimeter of a square with a side of 9 cm. [8.1]

Find the area enclosed by each figure. [8.2]

6.

11 cm

11 cm

7.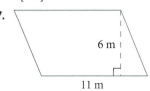

6 m

11 m

8.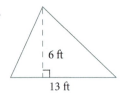

6 ft

13 ft

9.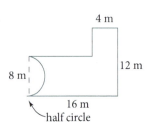

4 m

8 m   12 m

16 m
half circle

10. A circle has a radius of 6 mm. Find its area. Round to the nearest whole number. [8.2]

Find the surface area of each figure. [8.3]

11.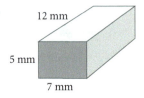

12 mm

5 mm

7 mm

12.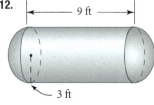

9 ft

3 ft

13.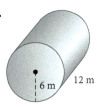

6 m   12 m

14.

5 in

Find the volume of each figure. [8.4]

15.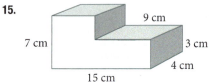

9 cm

7 cm   3 cm

4 cm

15 cm

16.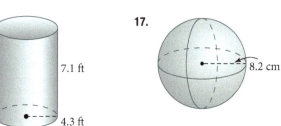

7.1 ft

4.3 ft

17.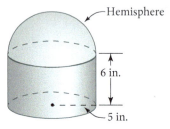

8.2 cm

18. Hemisphere

6 in.

5 in.

19. Find the volume and surface area of the rectangular solid shown here. [8.3, 8.4]

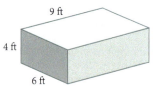

9 ft

4 ft

6 ft

20. Perimeter of a Garden A a rectangular garden is $17\frac{3}{4}$ feet long and $8\frac{1}{2}$ feet wide. What length of fence is needed to enclose the garden?

21. Volume of a Coin A 5-yen coin has a diameter of 22 mm, a thickness of 1.5 mm and a hole in the center with a diameter of 5 mm. Find the volume of the coin to the nearest whole number.

Simplify.

1. 9,371
  − 3,426

2. 64(204)

3. 861 ÷ 35

4. $0.5\overline{)7.625}$

5. $\dfrac{19}{24} - \dfrac{11}{24}$

6. $\left(\dfrac{5}{9}\right)^2$

7. 7.503 + 6.31

8. 13 − 4.714

9. 6.3(0.34)

10. $\dfrac{9}{14} \cdot \dfrac{7}{12}$

11. $\dfrac{8}{15} \div \dfrac{2}{5}$

12. $\dfrac{3}{16} + \dfrac{7}{12}$

13. $3\dfrac{3}{8} - 1\dfrac{3}{16}$

14. Add $3\dfrac{3}{8}$ to 5.625.

15. $6 + 2[(3 + 5) \cdot 3]$

16. $12 \div 4[(5 - 3) \cdot 2]$

17. $6 - 2^3 + 16 \div 8$

18. $12(4) - 4^2 + 4(6 + 10)$

19. $9 - 3(2) + 3^2$

20. $4(3 + 5) - 6 \div 3$

21. $\dfrac{9(4) + 6(8)}{2(6 - 2)}$

22. $\dfrac{3}{4}(12 - 8) - 6 + 3^3$

23. Find the product of $\dfrac{2}{5}$ and $3\dfrac{3}{4}$.

24. Translate and simplify: The difference of three times 4 and 7.

25. Write the ratio of 5 feet (converted to inches) to 48 inches as a fraction in lowest terms.

26. Convert 15 feet per second to miles per hour. Round to the nearest tenth.

27. Convert 5.9 square miles to hectares.

28. Write $\dfrac{1}{5}$ as a percent.

29. Convert 65% to a fraction.

30. Solve the equation $\dfrac{2}{x} = \dfrac{10}{13}$.

31. Write 65,209 in expanded form.

32. What number is 24% of 950?

33. 24 is what percent of 64?

34. $\dfrac{9}{2}$ is 25% of what number?

35. If 1 mile is 1.61 kilometers, convert 24 feet per second to kilometers per hour.

36. Textbook Prices A used copy of a textbook sells for $75. If the full price of a new book is $185, what is the percent increase to buy a new book?

37. Sales Tax A blu-ray player costs $175. If the sales tax rate is 8.5%, what is the total cost of the blu-ray player?

Find the perimeter and area of the following figures.

38.

7.5 cm   11 cm   4.5 cm   16 cm

39.

9 ft   17 ft

Find the perimeter of each figure. [8.1]

1.

9 in
17 in

2.
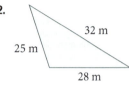
32 m
25 m
28 m

3.
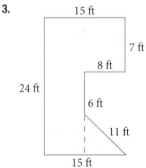
15 ft
7 ft
8 ft
24 ft
6 ft
11 ft
15 ft

4. Find the circumference of the circle. Use 3.14 for π. [8.1]

12 yd

5. Find the perimeter of a square with side 27 millimeters. [8.1]

Find the area enclosed by each figure. [8.2]

6.

16 in
16 in

7.
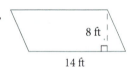
8 ft
14 ft

8.
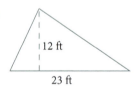
12 ft
23 ft

9.

9 yd
5 yd
4 yd
Half Circle

10. A circle has radius of 13 meters. Find its area. Round to the nearest whole number. [8.2]

Find the surface area of each figure. [8.3]

11.
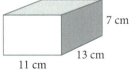
7 cm
13 cm
11 cm

12.
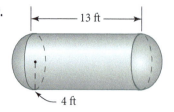
13 ft
4 ft

13.
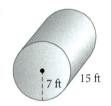
7 ft
15 ft

14.

2 cm

Find the volume of each figure. [8.4]

15.
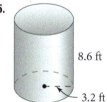
8.6 ft
3.2 ft

16.
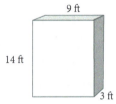
9 ft
14 ft
3 ft

17.
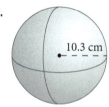
10.3 cm

18.
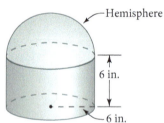
Hemisphere
6 in.
6 in.

19. Find the volume and surface area of the rectangular solid shown below. [8.3, 8.4]
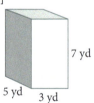
7 yd
5 yd
3 yd

20. Perimeter of a Pool A rectangular pool is $21\frac{3}{4}$ feet long and $11\frac{1}{3}$ feet wide. What is the perimeter of the pool?

21. Volume of a Coin The British one-pound coin has a diameter of 22.5 mm and a thickness of 3.15 mm. Find the volume of the coin, to the nearest whole number.

Linear Equations and Inequalities in One Variable

Estádio Jornalista Mário Filho Stadium
Rio de Janiero, Brazil
© 2010 Google

The honor of hosting the 2014 FIFA World Cup for soccer has fallen to Brazil, making it just the fifth country to host two World Cup tournaments. Expected to be one of the main venues of the competition, Estádio Jornalista Mário Filho in Rio de Janiero is the sixth largest stadium in the world. The stadium was originally constructed for the 1950 World Cup, where it acquired the nickname "Maracana" after a nearby river. The Maracana hosted the 1950 tournament final where Uruguay upset the home team Brazil in a stunning 2-1 win. At that time, the crowd was reported to have been 200,000 soccer fans, including those who illegally stormed the gates and entered after the start of the match. That crowd is still the largest ever to watch a soccer match. Due to an accident in 1992 resulting in two deaths, renovations greatly reduced Maracana's capacity, which is currently listed at 103,045.

Suppose the 2014 FIFA World Cup lists their ticket prices in the same manner as the 2010 World Cup in South Africa. The following chart shows the 2010 prices by seating category in US dollars.

	Category 1	Category 2	Category 3	Wheelchair
Opening Match	$450.00	$300.00	$200.00	$70.00
Group Stage	$160.00	$120.00	$80.00	$20.00
Round of 16	$200.00	$150.00	$100.00	$50.00
Quarter Finals	$300.00	$200.00	$150.00	$75.00
Semi-Finals	$600.00	$400.00	$250.00	$100.00
3rd/4th Place	$300.00	$200.00	$150.00	$75.00
The Final	$900.00	$600.00	$400.00	$150.00

Suppose the Stadium Maracana sells out for the final game with 100,000 paid tickets. If the seating prices hold, and there are 50,000 Category 2 tickets sold, 15,000 Category 3 tickets sold, and 5,000 Wheelchair tickets sold, how many Category 1 tickets will be sold? An equation to answer this question looks like this:

$$100,000 = 50,000 + 15,000 + 5,000 + x$$

Solving the equation above requires the use of the addition property of equality, one of the topics we will cover in this chapter.

9.1 Addition Property of Equality

A Check the solution to an equation by substitution.

B Use the addition property of equality to solve an equation.

KEY WORDS

solution set

equivalent equation

addition property of equality

©iStockphoto.com/IvicaNS

Did you know you can roughly calculate the outside temperature by counting the number of times a cricket chirps? Researchers at the Library of Congress suggest that you focus on one cricket and count the number of chirps in 15 seconds. Then add 37 to that number and you'll get the approximate outside temperature in degrees Fahrenheit. We can also use this information to write the following linear equation:

$$t = n + 37$$

where t is the outside temperature in degrees Fahrenheit and n is the number of cricket chirps in 15 seconds. Suppose the temperature outside is 70 degrees Fahrenheit. After plugging this value into the equation, we can solve for n to find the number of chirps; that is, we must find all replacements for the variable that make the equation a true statement. To do so, we must first understand how to use the addition property of equality, which we will cover in this section.

**Video Examples
Section 9.1**

Note The symbol { } is used in roster notation to represent a list of numbers in a set.

Practice Problems

1. Is 4 a solution to $2x + 3 = 7$?

Note We use a question mark over the equal signs to show that we don't know yet whether the two sides of the equation are equal.

Answer

1. No

A Solutions to Equations

DEFINITION solution set

The *solution set* for an equation is the set of all numbers that when used in place of the variable make the equation a true statement

For example, the equation $x + 2 = 5$ has the solution set {3} because when x is 3 the equation becomes the true statement $3 + 2 = 5$, or $5 = 5$.

EXAMPLE 1 Is 5 a solution to $2x - 3 = 7$?

Solution We substitute 5 for x in the equation, and then simplify to see if a true statement results. A true statement means we have a solution; a false statement indicates the number we are using is not a solution.

$$\text{When} \rightarrow \qquad x = 5$$
$$\text{the equation} \rightarrow \quad 2x - 3 = 7$$
$$\text{becomes} \rightarrow \quad 2(5) - 3 \overset{?}{=} 7$$
$$10 - 3 \overset{?}{=} 7$$
$$7 = 7 \qquad \textit{A true statement}$$

Because $x = 5$ turns the equation into the true statement $7 = 7$, we know 5 is a solution to the equation.

EXAMPLE 2   Is -2 a solution to $8 = 3x + 4$?

Solution   Substituting -2 for x in the equation, we have

$$8 \overset{?}{=} 3(-2) + 4$$

$$8 \overset{?}{=} -6 + 4$$

$$8 = -2 \qquad \textcolor{green}{\textit{A false statement}}$$

$$x \neq -2$$

Substituting -2 for x in the equation produces a false statement. Therefore, $x = -2$ is not a solution to the equation.

2. Is $\dfrac{4}{3}$ a solution to $8 = 3x + 4$?

> *Note*   The symbol $\neq$ means "does not equal".

The important thing about an equation is its solution set. Therefore, we make the following definition to classify together all equations with the same solution set.

> **DEFINITION**   equivalent equation
>
> Two or more equations with the same solution set are said to be *equivalent equations*.

Equivalent equations may look different but must have the same solution set.

EXAMPLE 3

a. $x + 2 = 5$ and $x = 3$ are equivalent equations because both have solution set $\{3\}$.

b. $a - 4 = 3$, $a - 2 = 5$, and $a = 7$ are equivalent equations because they all have solution set $\{7\}$.

c. $y + 3 = 4$, $y - 8 = -7$, and $y = 1$ are equivalent equations because they all have solution set $\{1\}$.

3. Answer the following.
   a. Are the equations $x + 3 = 9$ and $x = 6$ equivalent?
   b. Are the equations $a - 5 = 3$ and $a = 6$ equivalent?
   c. Are $y + 2 = 5$, $y - 1 = 2$, and $y = 3$ equivalent equations?

B Addition Property of Equality

If two numbers are equal and we increase (or decrease) both of them by the same amount, the resulting quantities are also equal. We can apply this concept to equations. Adding the same amount to both sides of an equation always produces an equivalent equation—one with the same solution set. This fact about equations is called the *addition property of equality* and can be stated more formally as follows:

> **PROPERTY**   Addition Property of Equality
>
> For any three algebraic expressions A, B, and C,
>
> *In symbols*: if $A = B$
>
> then $A + C = B + C$
>
> *In words*: Adding the same quantity to both sides of an equation will not change the solution set.

> *Note*   We will use this property many times in the future. Be sure you understand it completely by the time you finish this section.

This property is just as simple as it seems. We can add any amount to both sides of an equation and always be sure we have not changed the solution set.

Consider the equation $x + 6 = 5$. We want to solve this equation for the value of x. To solve for x means we want to end up with x on one side of the equal sign and a number on the other side. Because we want x by itself, we will add -6 to both sides.

$$x + 6 + (-6) = 5 + (-6) \qquad \textcolor{green}{\text{Addition property of equality}}$$

$$x + 0 = -1$$

$$x = -1$$

All three equations say the same thing about x. They all say that x is -1. All three equations are equivalent. The last one is just easier to read.

Answers
2. Yes
3. a. Yes **b.** No **c.** Yes

Here are some further examples of how the addition property of equality can be used to solve equations.

4. Solve $x - 3 = 10$ for x.

EXAMPLE 4   Solve the equation $x - 5 = 12$ for x.

Solution   Because we want x alone on the left side, we choose to add 5 to both sides.

$$x - 5 + 5 = 12 + 5 \qquad \textit{Addition property of equality}$$
$$x + 0 = 17$$
$$x = 17$$

To check our solution to Example 4, we substitute 17 for x in the original equation:

$$\text{When} \rightarrow \qquad x = 17$$
$$\text{the equation} \rightarrow \qquad x - 5 = 12$$
$$\text{becomes} \rightarrow \qquad 17 - 5 \overset{?}{=} 12$$
$$12 = 12 \qquad \textit{A true statement}$$

As you can see, our solution checks. The purpose for checking a solution to an equation is to catch any mistakes we may have made in the process of solving the equation.

5. Solve $a + \dfrac{2}{3} = -\dfrac{1}{6}$ for a.

EXAMPLE 5   Solve $a + \dfrac{3}{4} = -\dfrac{1}{2}$ for a.

Solution   Because we want a by itself on the left side of the equal sign, we add the opposite of $\dfrac{3}{4}$ to each side of the equation.

$$a + \frac{3}{4} + \left(-\frac{3}{4}\right) = -\frac{1}{2} + \left(-\frac{3}{4}\right) \qquad \textit{Addition property of equality}$$
$$a + 0 = -\frac{1}{2} \cdot \frac{2}{2} + \left(-\frac{3}{4}\right) \qquad \textit{LCD on the right side is 4.}$$
$$a = -\frac{2}{4} + \left(-\frac{3}{4}\right) \qquad \textit{$\frac{2}{4}$ is equivalent to $\frac{1}{2}$.}$$
$$a = -\frac{5}{4} \qquad \textit{Add fractions.}$$

The solution is $a = -\dfrac{5}{4}$. To check our result, we replace a with $-\dfrac{5}{4}$ in the original equation. The left side then becomes $-\dfrac{5}{4} + \dfrac{3}{4}$, which reduces to $-\dfrac{1}{2}$, so our solution checks.

6. Solve $2.7 + x = 8.1$ for x.

EXAMPLE 6   Solve $7.3 + x = -2.4$ for x.

Solution   Again, we want to isolate x, so we add the opposite of 7.3 to both sides.

$$7.3 + (-7.3) + x = -2.4 + (-7.3) \qquad \textit{Addition property of equality}$$
$$0 + x = -9.7$$
$$x = -9.7$$

Sometimes it is necessary to simplify each side of an equation before using the addition property of equality. The reason we simplify both sides first is that we want as few terms as possible on each side of the equation before we use the addition property of equality. The following examples illustrate this procedure.

7. Solve $-2x + 3 + 3x = 6 + 8$ for x.

EXAMPLE 7   Solve $-x + 2 + 2x = 7 + 5$ for x.

Solution   We begin by combining similar terms on each side of the equation. Then we use the addition property to solve the simplified equation.

$$x + 2 = 12 \qquad \textit{Simplify both sides first.}$$
$$x + 2 + (-2) = 12 + (-2) \qquad \textit{Addition property of equality}$$
$$x + 0 = 10$$
$$x = 10$$

Answers

4. 13

5. $-\dfrac{5}{6}$

6. 5.4

7. 11

EXAMPLE 8   Solve $4(2a - 3) - 7a = 2 - 5$.

Solution   We must begin by applying the distributive property to separate terms on the left side of the equation. Following that, we combine similar terms and then apply the addition property of equality.

$$4(2a - 3) - 7a = 2 - 5 \qquad \textit{Original equation}$$
$$8a - 12 - 7a = 2 - 5 \qquad \textit{Distributive property}$$
$$a - 12 = -3 \qquad \textit{Simplify each side.}$$
$$a - 12 + 12 = -3 + 12 \qquad \textit{Add 12 to each side.}$$
$$a = 9$$

To check our solution, we replace a with 9 in the original equation.

$$4(2 \cdot 9 - 3) - 7 \cdot 9 \stackrel{?}{=} 2 - 5$$
$$4(15) - 63 \stackrel{?}{=} -3$$
$$60 - 63 \stackrel{?}{=} -3$$
$$-3 = -3 \qquad \textit{A true statement}$$

We can also add a term involving a variable to both sides of an equation.

EXAMPLE 9   Solve $3x - 5 = 2x + 7$.

Solution   We can solve this equation in two steps. First, we add $-2x$ to both sides of the equation. When this has been done, x appears on the left side only. Second, we add 5 to both sides.

$$3x + (-2x) - 5 = 2x + (-2x) + 7 \qquad \textit{Add } -2x \textit{ to both sides.}$$
$$x - 5 = 7 \qquad \textit{Simplify each side.}$$
$$x - 5 + 5 = 7 + 5 \qquad \textit{Add 5 to both sides.}$$
$$x = 12 \qquad \textit{Simplify each side.}$$

A Note on Subtraction

Although the addition property of equality is stated for addition only, we can subtract the same number from both sides of an equation as well. Because subtraction is defined as addition of the opposite, subtracting the same quantity from both sides of an equation does not change the solution.

$$x + 2 = 12 \qquad \textit{Original equation}$$
$$x + 2 - 2 = 12 - 2 \qquad \textit{Subtract 2 from each side.}$$
$$x = 10$$

GETTING READY FOR CLASS

After reading through the preceding section, respond in your own words and in complete sentences.

A. What is a solution set for an equation?

B. What are equivalent equations?

C. Explain in words the addition property of equality.

D. How do you check a solution to an equation?

8. Solve $5(3a - 4) - 14a = 25$.

Note   Remember that terms cannot be combined until the parentheses are removed.

Note   Again, we place a question mark over the equal sign because we don't know yet whether the expressions on the left and right side of the equal sign will be equal.

9. Solve $5x - 4 = 4x + 6$.

Note   In my experience teaching, I find that students make fewer mistakes if they think in terms of addition rather than subtraction. So, you are probably better off if you continue to use the addition property just the way we have used it in the examples in this section. But, if you are curious as to whether you can subtract the same number from both sides of an equation, the answer is yes.

Answers
8. 45
9. 10

EXERCISE SET 9.1

Vocabulary Review

Choose the correct words to fill in the blanks below.

equivalent   addition   subtraction   solution set

1. The set of numbers used in place of the variable in an equation are called the _____ for that equation.

2. Two or more equations with the same solution set are said to be _____ equations.

3. The _____ property of equality states that adding the same quantity to both sides of an equation will not change the solution set.

4. Since subtraction is defined as addition of the opposite, we can apply the addition property of equality to both sides of a _____ problem and the solution will not change.

Problems

A Check to see if the number to the right of each of the following equations is the solution to the equation.

1. $2x + 1 = 5; 2$ **2.** $4x + 3 = 7; 1$ **3.** $3x + 4 = 19; 5$ **4.** $3x + 8 = 14; 2$

5. $2x - 4 = 2; 4$ **6.** $5x - 6 = 9; 3$ **7.** $2x + 1 = 3x + 3; -2$ **8.** $4x + 5 = 2x - 1; -6$

9. $x - 4 = 2x + 1; -4$ **10.** $x - 8 = 3x + 2; -5$ **11.** $x - 2 = 3x + 1; \dfrac{-3}{2}$ **12.** $x + 4 = 4x - 1; \dfrac{5}{3}$

B Solve the following equations.

13. $x - 3 = 8$ **14.** $x - 2 = 7$ **15.** $x + 2 = 6$ **16.** $x + 5 = 4$

17. $a + \dfrac{1}{2} = -\dfrac{1}{4}$ **18.** $a + \dfrac{1}{3} = -\dfrac{5}{6}$ **19.** $x + 2.3 = -3.5$ **20.** $x + 7.9 = 23.4$

21. $y + 11 = -6$ **22.** $y - 3 = -1$ **23.** $x - \dfrac{5}{8} = -\dfrac{3}{4}$ **24.** $x - \dfrac{2}{5} = -\dfrac{1}{10}$

25. $m - 6 = -10$ **26.** $m - 10 = -6$ **27.** $6.9 + x = 3.3$ **28.** $7.5 + x = 2.2$

29. $5 = a + 4$ **30.** $12 = a - 3$ **31.** $-\dfrac{5}{9} = x - \dfrac{2}{5}$ **32.** $-\dfrac{7}{8} = x - \dfrac{4}{5}$

Simplify both sides of the following equations as much as possible, and then solve.

33. $4x + 2 - 3x = 4 + 1$

34. $5x + 2 - 4x = 7 - 3$

35. $8a - \dfrac{1}{2} - 7a = \dfrac{3}{4} + \dfrac{1}{8}$

36. $9a - \dfrac{4}{5} - 8a = \dfrac{3}{10} - \dfrac{1}{5}$

37. $-3 - 4x + 5x = 18$

38. $10 - 3x + 4x = 20$

39. $-11x + 2 + 10x + 2x = 9$

40. $-10x + 5 - 4x + 15x = 0$

41. $-2.5 + 4.8 = 8x - 1.2 - 7x$

42. $-4.8 + 6.3 = 7x - 2.7 - 6x$

43. $2y - 10 + 3y - 4y = 18 - 6$

44. $15 - 21 = 8x + 3x - 10x$

The following equations contain parentheses. Apply the distributive property to remove the parentheses, then simplify each side before using the addition property of equality.

45. $2(x + 3) - x = 4$

46. $5(x + 1) - 4x = 2$

47. $-3(x - 4) + 4x = 3 - 7$

48. $-2(x - 5) + 3x = 4 - 9$

49. $5(2a + 1) - 9a = 8 - 6$

50. $4(2a - 1) - 7a = 9 - 5$

51. $-(x + 3) + 2x - 1 = 6$

52. $-(x - 7) + 2x - 8 = 4$

53. $4y - 3(y - 6) + 2 = 8$

54. $7y - 6(y - 1) + 3 = 9$

55. $-3(2m - 9) + 7(m - 4) = 12 - 9$

56. $-5(m - 3) + 2(3m + 1) = 15 - 8$

Solve the following equations by the method used in Example 9 in this section. Check each solution in the original equation.

57. $4x = 3x + 2$

58. $6x = 5x - 4$

59. $8a = 7a - 5$

60. $9a = 8a - 3$

61. $2x = 3x + 1$

62. $4x = 3x + 5$

63. $3y + 4 = 2y + 1$

64. $5y + 6 = 4y + 2$

65. $2m - 3 = m + 5$

66. $8m - 1 = 7m - 3$

67. $4x - 7 = 5x + 1$

68. $3x - 7 = 4x - 6$

69. $5x - \dfrac{2}{3} = 4x + \dfrac{4}{3}$

70. $3x - \dfrac{5}{4} = 2x + \dfrac{1}{4}$

71. $8a - 7.1 = 7a + 3.9$

72. $10a - 4.3 = 9a + 4.7$

Applying the Concepts

73. Movie Tickets A movie theater has a total of 300 seats. For a special Saturday night preview, they reserve 20 seats to give away to their VIP guests at no charge. If x represents the number of tickets they can sell for the preview, then x can be found by solving the equation $x + 20 = 300$.

   a. Solve the equation for x.

   b. If tickets for the preview are $7.50 each, what is the maximum amount of money they can make from ticket sales?

74. Movie Tickets A movie theater has a total of 250 seats. For a special Saturday night preview, they reserve 45 seats to give away to their VIP guests at no charge. If x represents the number of tickets they can sell for the preview, then x can be found by solving the equation $x + 45 = 250$.

   a. Solve the equation for x.

   b. If tickets for the preview are $8.00 each, what is the maximum amount of money they can make from ticket sales?

75. Geometry Two angles are complementary angles. If one of the angles is 23°, then solving the equation $x + 23° = 90°$ will give you the other angle. Solve the equation.

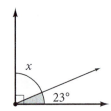

76. Geometry Two angles are supplementary angles. If one of the angles is 23°, then solving the equation $x + 23° = 180°$ will give you the other angle. Solve the equation.

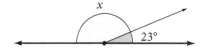

77. Theater Tickets The El Portal Center for the Arts in North Hollywood, California, holds a maximum of 400 people. The two balconies hold 86 and 89 people each; the rest of the seats are at the stage level. Solving the equation $x + 86 + 89 = 400$ will give you the number of seats on the stage level.

   a. Solve the equation for x.

   b. If tickets on the stage level are $30 each, and tickets in either balcony are $25 each, what is the maximum amount of money the theater can bring in for a show?

78. Geometry The sum of the angles in the triangle on the swing set is 180°. Use this fact to write an equation containing x. Then solve the equation.

79. Using the equation $t = n + 40$ given in the introduction of this section, complete the following table.

n	t
32	
	90
44	
	136

80. Using the equation $T = 40 + O$ which relates overtime hours to total hours, complete the following table.

O	T
10	
	51
17	
	62

81. Geometry The three angles shown in the triangle at the front of the tent in the following figure add up to 180°. Use this fact to write an equation containing x, and then solve the equation to find the number of degrees in the angle at the top of the triangle.

82. Suppose Ava and Kendra are bridesmaids in a friend's wedding. Answer the following questions about the money they spend.

a. Suppose Kendra spent $210 for a bridesmaid dress, then bought new shoes to go with the dress. If the total bill was $289, does the equation $210 + x = 289$ describe this situation? If so, what does x represent?

b. Suppose Ava buys a bridesmaid dress on sale and then spends $120 on her makeup, nails, and hair. If the total bill is $219, and x represents a positive number, does the equation $219 + x = 120$ describe the situation?

Getting Ready for the Next Section

To understand all of the explanations and examples in the next section you must be able to work the problems below.

Simplify.

83. $\dfrac{3}{2}\left(\dfrac{2}{3}y\right)$

84. $\dfrac{5}{2}\left(-\dfrac{2}{5}y\right)$

85. $\dfrac{1}{5}(5x)$

86. $-\dfrac{1}{4}(-4a)$

87. $\dfrac{1}{5}(30)$

88. $-\dfrac{1}{4}(24)$

89. $\dfrac{3}{2}(4)$

90. $\dfrac{1}{26}(13)$

91. $12\left(-\dfrac{3}{4}\right)$

92. $12\left(\dfrac{1}{2}\right)$

93. $\dfrac{3}{2}\left(-\dfrac{5}{4}\right)$

94. $\dfrac{5}{3}\left(-\dfrac{6}{5}\right)$

95. $13 + (-5)$

96. $-13 + (-5)$

97. $-\dfrac{3}{4} + \left(-\dfrac{1}{2}\right)$

98. $-\dfrac{7}{10} + \left(-\dfrac{1}{2}\right)$

99. $7x + (-4x)$

100. $5x + (-2x)$

101. $3x + (-12x)$

102. $5x - 9x$

Find the Mistake

Each sentence below contains a mistake. Circle the mistake and write the correct word or expression on the line provided.

1. The number 2 is a solution to the equation $3x - 9 = 8x + 1$. _____

2. Saying that the number 12 is a solution to the equation $\frac{1}{6}y + 3 = 5$ is a false statement. _____

3. The addition property of equality states that $A(C) = B(C)$ if $A = B$. _____

4. To solve the equation $\frac{5}{6}b + 6 = 11$ for b, begin by adding -11 to both sides. _____

Navigation Skills: Prepare, Study, Achieve

It is very common for students to feel burned out or feel a decrease in motivation to continue putting in the hard work it takes to succeed in this course. Many factors may contribute to this burn out and vary depending on the student. It is important to recognize these factors and implement ways to combat them. If you notice yourself feeling burned out, the following suggestions may help:

- Join a study group to help provide emotional support during tough times in this class.
- Take breaks during your study sessions.
- Begin each study session with the most difficult topics, then work through the easier topics toward the end of your session.
- Revisit your study calendar to see if you have overbooked your time.
- Change up your time and location for studying.
- Examine your diet and introduce healthier and more balanced meals.
- Get adequate restful sleep between studying and going to class.
- Schedule regular physical activity during your day.
- Set short-term and long-term goals.

Image © Vandenwyngaert Stephane

OBJECTIVES

A Use the multiplication property of equality to solve an equation.

B Use the addition and multiplication properties of equality together to solve an equation.

KEY WORDS

multiplication property of equality

Henry Ford made famous the mass production assembly line in an effort to streamline his automobile manufacturing. He realized that an assembler was taking 8.5 hours to complete his entire task because the assembler had to retrieve parts on his own as needed. Ford asked that parts be delivered to an assembler, which reduced the time it took to complete the task to 2.5 minutes! Later, Ford installed the moving conveyor belt, reducing the completion time even further.

Imagine yourself working on an assembly line. If it takes you 2.5 minutes to assemble your item, how many items can you assemble in an hour's time? To answer this question, we must use the given information to set up the following equation:

$$60 = 2.5x$$

Notice we converted an hour into 60 minutes to match the units for your assembly time. In this section, we will introduce you to the multiplication property of equality. You will use this property to solve linear equations, such as the one for the assembly line.

Video Examples
Section 9.2

A Multiplication Property of Equality

In the previous section, we found that adding the same number to both sides of an equation never changed the solution set. The same idea holds for multiplication by numbers other than zero. We can multiply both sides of an equation by the same nonzero number and always be sure we have not changed the solution set. (The reason we cannot multiply both sides by zero will become apparent later.) This fact about equations is called the *multiplication property of equality*, which can be stated formally as follows:

PROPERTY Multiplication Property of Equality

For any three algebraic expressions A, B, and C, where $C \neq 0$,

In symbols: if $A = B$

then $AC = BC$

In words: Multiplying both sides of an equation by the same nonzero number will not change the solution set.

Suppose we want to solve the equation $5x = 30$. We have $5x$ on the left side but would like to have just x. We choose to multiply both sides by $\frac{1}{5}$ which allows us to divide out the fives and isolate the variable. The solution follows.

$$5x = 30$$

$$\frac{1}{5}(5x) = \frac{1}{5}(30) \qquad \text{Multiplication property of equality}$$

$$\left(\frac{1}{5} \cdot 5\right)x = \frac{1}{5}(30) \qquad \text{Associative property}$$

$$1x = 6 \qquad\qquad \text{Note: } 1x \text{ is the same as } x.$$

$$x = 6$$

We chose to multiply by $\frac{1}{5}$ because it is the reciprocal of 5 and $\left(\frac{1}{5}\right)(5) = 1$. We can see that multiplication by any number except zero will not change the solution set. If, however, we were to multiply both sides by zero, the result would always be $0 = 0$. Although the statement $0 = 0$ is true, we have lost our variable and cannot solve the equation. This is the only restriction of the multiplication property of equality. We are free to multiply both sides of an equation by any number except zero in an effort to isolate the variable.

Here are some more examples that use the multiplication property of equality.

EXAMPLE 1   Solve $-4a = 24$ for a.

Solution   Because we want the variable a alone on the left side, we choose to multiply both sides by $-\frac{1}{4}$.

$$-\frac{1}{4}(-4a) = -\frac{1}{4}(24) \qquad \text{Multiplication property of equality}$$

$$\left[-\frac{1}{4}(-4)\right]a = -\frac{1}{4}(24) \qquad \text{Associative property}$$

$$a = -6$$

EXAMPLE 2   Solve $-\frac{t}{3} = 5$ for t.

Solution   Because division by 3 is the same as multiplication by $\frac{1}{3}$, we can write $-\frac{t}{3}$ as $-\frac{1}{3}t$. To solve the equation, we multiply each side by the reciprocal of $-\frac{1}{3}$, which is -3.

$$-\frac{t}{3} = 5 \qquad\qquad \text{Original equation}$$

$$-\frac{1}{3}t = 5 \qquad\qquad \text{Dividing by 3 is equivalent to multiplying by } \frac{1}{3}.$$

$$-3\left(-\frac{1}{3}t\right) = -3(5) \qquad \text{Multiply each side by } -3.$$

$$t = -15$$

EXAMPLE 3   Solve $\frac{2}{3}y = \frac{4}{9}$.

Solution   We can multiply both sides by $\frac{3}{2}$ and have $1y$ on the left side.

$$\frac{3}{2}\left(\frac{2}{3}y\right) = \frac{3}{2}\left(\frac{4}{9}\right) \qquad \text{Multiplication property of equality}$$

$$\left(\frac{3}{2} \cdot \frac{2}{3}\right)y = \frac{3}{2}\left(\frac{4}{9}\right) \qquad \text{Associative property}$$

$$y = \frac{2}{3} \qquad\qquad \text{Simplify } \frac{3}{2}\left(\frac{4}{9}\right) = \frac{12}{18} = \frac{2}{3}.$$

EXAMPLE 4   Solve $5 + 8 = 10x + 20x - 4x$.

Solution   Our first step will be to simplify each side of the equation.

$$13 = 26x \qquad\qquad \text{Simplify by combining similar terms.}$$

$$\frac{1}{26}(13) = \frac{1}{26}(26x) \qquad \text{Multiplication property of equality}$$

$$\frac{13}{26} = x$$

$$\frac{1}{2} = x \qquad\qquad \text{Reduce to lowest terms.}$$

$$x = \frac{1}{2}$$

Practice Problems

1. Solve $3a = -27$ for a.

2. Solve $\frac{t}{4} = 6$ for t.

3. Solve $\frac{2}{5}y = \frac{3}{4}$.

Note   Notice in Examples 1 through 3 that if the variable is being multiplied by a number like -4 or $\frac{2}{3}$, we always multiply by the number's reciprocal, $-\frac{1}{4}$ or $\frac{3}{2}$, to end up with just the variable on one side of the equation.

4. Solve $3 + 7 = 6x + 8x - 9x$.

Answers

1. -9
2. 24
3. $\frac{15}{8}$
4. 2

B Solving Equations Using Addition and Multiplication

In the next three examples, we will use both the addition property of equality and the multiplication property of equality.

EXAMPLE 5   Solve $6x + 5 = -13$ for x.

Solution   We begin by adding -5 to both sides of the equation.

$$6x + 5 + (-5) = -13 + (-5) \qquad \text{Add } -5 \text{ to both sides.}$$

$$6x = -18 \qquad \text{Simplify.}$$

$$\frac{1}{6}(6x) = \frac{1}{6}(-18) \qquad \text{Multiply both sides by } \frac{1}{6}.$$

$$x = -3$$

5. Solve $4x + 3 = -13$ for x.

6. Solve $7x = 4x + 15$ for x.

EXAMPLE 6   Solve $5x = 2x + 12$ for x.

Solution   We begin by adding $-2x$ to both sides of the equation.

$$5x + (-2x) = 2x + (-2x) + 12 \qquad \text{Add } -2x \text{ to both sides.}$$

$$3x = 12 \qquad \text{Simplify.}$$

$$\frac{1}{3}(3x) = \frac{1}{3}(12) \qquad \text{Multiply both sides by } \frac{1}{3}.$$

$$x = 4 \qquad \text{Simplify.}$$

Note   Notice that in Example 6 we used the addition property of equality first to combine all the terms containing x on the left side of the equation. Once this had been done, we used the multiplication property to isolate x on the left side.

EXAMPLE 7   Solve $3x - 4 = -2x + 6$ for x.

Solution   We begin by adding $2x$ to both sides.

$$3x + 2x - 4 = -2x + 2x + 6 \qquad \text{Add } 2x \text{ to both sides.}$$

$$5x - 4 = 6 \qquad \text{Simplify.}$$

Now we add 4 to both sides.

$$5x - 4 + 4 = 6 + 4 \qquad \text{Add 4 to both sides.}$$

$$5x = 10 \qquad \text{Simplify.}$$

$$\frac{1}{5}(5x) = \frac{1}{5}(10) \qquad \text{Multiply by } \frac{1}{5}.$$

$$x = 2 \qquad \text{Simplify.}$$

7. Solve $2x - 5 = -3x + 10$ for x.

The next example involves fractions. You will see that the properties we use to solve equations containing fractions are the same as the properties we used to solve the previous equations. Also, the LCD that we used previously to add fractions can be used with the multiplication property of equality to simplify equations containing fractions.

EXAMPLE 8   Solve $\frac{2}{3}x + \frac{1}{2} = -\frac{3}{4}$.

Solution   We can solve this equation by applying our properties and working with the fractions, or we can begin by eliminating the fractions.

Method 1   Working with the fractions.

$$\frac{2}{3}x + \frac{1}{2} + \left(-\frac{1}{2}\right) = -\frac{3}{4} + \left(-\frac{1}{2}\right) \qquad \text{Add } -\frac{1}{2} \text{ to each side.}$$

$$\frac{2}{3}x = -\frac{5}{4} \qquad \text{Note that } -\frac{3}{4} + \left(-\frac{1}{2}\right) = -\frac{3}{4} + \left(-\frac{2}{4}\right).$$

$$\frac{3}{2}\left(\frac{2}{3}x\right) = \frac{3}{2}\left(-\frac{5}{4}\right) \qquad \text{Multiply each side by } \frac{3}{2}.$$

$$x = -\frac{15}{8}$$

8. Solve $\frac{3}{5}x + \frac{1}{2} = -\frac{7}{10}$.

Answers

5. -4

6. 5

7. 3

8. -2

Note   Our original equation
has denominators of 3, 2, and
4. The LCD for these three
denominators is 12, and it
has the property that all three
denominators will divide it
evenly. Therefore, if we multiply
both sides of our equation by 12,
each denominator will divide
into 12 and we will be left with
an equation that does not contain
any denominators other than 1.

Method 2   Eliminating the fractions in the beginning.

$$12\left(\frac{2}{3}x + \frac{1}{2}\right) = 12\left(-\frac{3}{4}\right)$$ Multiply each side by the LCD 12.

$$12\left(\frac{2}{3}x\right) + 12\left(\frac{1}{2}\right) = 12\left(-\frac{3}{4}\right)$$ Distributive property on the left side

$$8x + 6 = -9$$ Multiply.

$$8x = -15$$ Add -6 to each side.

$$x = -\frac{15}{8}$$ Multiply each side by $\frac{1}{8}$.

As the third line in Method 2 indicates, multiplying each side of the equation by the LCD eliminates all the fractions from the equation.

As you can see, both methods yield the same solution.

A Note on Division

Because *division* is defined as multiplication by the reciprocal, multiplying both sides of an equation by the same number is equivalent to dividing both sides of the equation by the reciprocal of that number; that is, multiplying each side of an equation by $\frac{1}{3}$ and dividing each side of the equation by 3 are equivalent operations. If we were to solve the equation $3x = 18$ using division instead of multiplication, the steps would look like this:

$$3x = 18$$ Original equation

$$\frac{3x}{3} = \frac{18}{3}$$ Divide each side by 3.

$$x = 6$$

Using division instead of multiplication on a problem like this may save you some writing. However, with multiplication, it is easier to explain "why" we end up with just one x on the left side of the equation. (The "why" has to do with the associative property of multiplication.) Continue to use multiplication to solve equations like this one until you understand the process completely. Then, if you find it more convenient, you can use division instead of multiplication.

GETTING READY FOR CLASS

After reading through the preceding section, respond in your own words and in complete sentences.

A. Explain in words the multiplication property of equality.

B. If an equation contains fractions, how do you use the multiplication property of equality to clear the equation of fractions?

C. Why is it okay to divide both sides of an equation by the same nonzero number?

D. Give an example of using an LCD with the multiplication property of equality to simplify an equation.

Vocabulary Review

Choose the correct words to fill in the blanks below.

    multiplication   division   nonzero   reciprocal

1. Multiplying both sides of an equation by the same _____ number will not change the solution set.

2. $AC = BC$, where $C \neq 0$ is the _____ property of equality written in symbols.

3. _____ is defined as multiplication by the reciprocal.

4. Dividing both sides of the equation by the same number is equivalent to multiplying both sides of the equation by the _____ of that number.

Problems

A Solve the following equations using the multiplication property of equality. Be sure to show your work.

1. $5x = 10$ **2.** $6x = 12$ **3.** $7a = 28$ **4.** $4a = 36$

5. $-8x = 4$ **6.** $-6x = 2$ **7.** $8m = -16$ **8.** $5m = -25$

9. $-3x = -9$ **10.** $-9x = -36$ **11.** $-7y = -28$ **12.** $-15y = -30$

13. $2x = 0$ **14.** $7x = 0$ **15.** $-5x = 0$ **16.** $-3x = 0$

17. $\dfrac{x}{3} = 2$ **18.** $\dfrac{x}{4} = 3$ **19.** $-\dfrac{m}{5} = 10$ **20.** $-\dfrac{m}{7} = 1$

21. $-\dfrac{x}{2} = -\dfrac{3}{4}$ **22.** $-\dfrac{x}{3} = \dfrac{5}{6}$ **23.** $\dfrac{2}{3}a = 8$ **24.** $\dfrac{3}{4}a = 6$

25. $-\dfrac{3}{5}x = \dfrac{9}{5}$ **26.** $-\dfrac{2}{5}x = \dfrac{6}{15}$ **27.** $-\dfrac{5}{8}y = -20$ **28.** $-\dfrac{7}{2}y = -14$

Simplify both sides as much as possible, and then solve.

29. $-4x - 2x + 3x = 24$

30. $7x - 5x + 8x = 20$

31. $4x + 8x - 2x = 15 - 10$

32. $5x + 4x + 3x = 4 + 8$

33. $-3 - 5 = 3x + 5x - 10x$

34. $10 - 16 = 12x - 6x - 3x$

35. $18 - 13 = \frac{1}{2}a + \frac{3}{4}a - \frac{5}{8}a$

36. $20 - 14 = \frac{1}{3}a + \frac{5}{6}a - \frac{2}{3}a$

37. $\frac{1}{2}x - \frac{1}{3}x = 2 - 4$

38. $\frac{1}{3}x - \frac{1}{4}x = \frac{1}{2} + 2$

39. $\frac{1}{2}x + \frac{1}{3}x = \frac{1}{2} - 3$

40. $\frac{1}{4}y + \frac{1}{3}y = 2 - 16$

Solve the following equations by multiplying both sides by -1.

41. $-x = 4$

42. $-x = -3$

43. $-x = -4$

44. $-x = 3$

45. $15 = -a$

46. $-15 = -a$

47. $-y = \frac{1}{2}$

48. $-y = -\frac{3}{4}$

B Solve each of the following equations using the addition and multiplication properties where necessary.

49. $3x - 2 = 7$

50. $2x - 3 = 9$

51. $2a + 1 = 3$

52. $5a - 3 = 7$

53. $\frac{1}{8} + \frac{1}{2}x = \frac{1}{4}$

54. $\frac{1}{3} + \frac{1}{7}x = -\frac{8}{21}$

55. $6x = 2x - 12$

56. $8x = 3x - 10$

57. $2y = -4y + 18$

58. $3y = -2y - 15$

59. $-7x = -3x - 8$

60. $-5x = -2x - 12$

61. $8x + 4 = 2x - 5$ **62.** $5x + 6 = 3x - 6$ **63.** $x + \dfrac{1}{2} = \dfrac{1}{4}x - \dfrac{5}{8}$ **64.** $\dfrac{1}{3}x + \dfrac{2}{5} = \dfrac{1}{5}x - \dfrac{2}{5}$

65. $6m - 3 = m + 2$ **66.** $6m - 5 = m + 5$ **67.** $\dfrac{1}{2}m - \dfrac{1}{4} = \dfrac{1}{12}m + \dfrac{1}{6}$ **68.** $\dfrac{1}{2}m - \dfrac{5}{12} = \dfrac{1}{12}m + \dfrac{5}{12}$

69. $9y + 2 = 6y - 4$ **70.** $6y + 14 = 2y - 2$ **71.** $\dfrac{3}{2}y + \dfrac{1}{3} = y - \dfrac{2}{3}$ **72.** $\dfrac{3}{2}y + \dfrac{7}{2} = \dfrac{1}{2}y - \dfrac{1}{2}$

Applying the Concepts

73. Break-Even Point Movie theaters pay a certain price for the movies that you and I see. Suppose a theater pays $1,500 for each showing of a popular movie. If they charge $7.50 for each ticket they sell, then the equation

$$7.5x = 1,500$$

gives the number of tickets they must sell to equal the $1,500 cost of showing the movie. This number is called the break-even point. Solve the equation for x to find the break-even point.

74. Basketball Laura plays basketball for her community college. In one game she scored 13 points total, with a combination of free throws, field goals, and three-pointers. Each free throw is worth 1 point, each field goal is 2 points, and each three-pointer is worth 3 points. If she made 1 free throw and 3 field goals, then solving the equation

$$1 + 3(2) + 3x = 13$$

will give us the number of three-pointers she made. Solve the equation to find the number of three-point shots Laura made.

75. Taxes Suppose 21% of your monthly pay is withheld for federal income taxes and another 8% is withheld for Social Security, state income tax, and other miscellaneous items. If you are left with $987.50 a month in take-home pay, then the amount you earned before the deductions were removed from your check is given by the equation

$$G - 0.21G - 0.08G = 987.5$$

Solve this equation to find your gross income.

76. Rhind Papyrus The *Rhind Papyrus* is an ancient document that contains mathematical riddles. One problem asks the reader to find a quantity such that when it is added to one-fourth of itself the sum is 15. The equation that describes this situation is

$$x + \dfrac{1}{4}x = 15$$

Solve this equation.

Angles The sum of the measures of the interior angles in a triangle is 180°. Given the following angle measures, set up an equation and solve for x.

77.

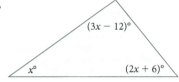

78.

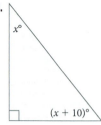

Angles Solve for x.

79.

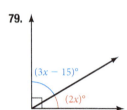

80.

81. Music Downloads Suppose you wish to purchase music through the internet for a personal playlist. It costs $10 to join and $1.50 per song. You have $40 in credit to use toward joining and making your playlist. How many songs can you purchase without going over budget?

82. Music Downloads Ellis and Celia wish to purchase music through the internet. Ellis finds a site that costs $20 to join and $1.25 per song thereafter. Celia finds one that costs $10 to join and $1.50 per song thereafter. How many songs must they purchase so that they have spent the same amount?

83. Recall from section 1.6 the formula $P = 14.7 + 0.4x$, which gives the pressure (in lb/in²) at a diving depth of x feet. Use this formula to find the diving depths which would result in the following pressures:

  a. 22.7 lb/in²

  b. 28.7 lb/in²

  c. 44.7 lb/in²

84. Suppose the total bill from a water company is given by $B = 7.5 + 0.0058w$, where B is the total bill and w is the water usage (in cubic feet). Use this formula to find the water usage resulting in the following monthly water bills.

  a. $56.80

  b. $94.50

  c. $167.00

85. The formula for Franco's pay is $P = 320 + 12(h - 40)$, where P is the pay (in $) and h is the hours worked. How many hours must Franco work to earn $500 in one week?

86. The formula for Esmeralda's weekly pay is $P = 400 + 15(h - 40)$, where P is the pay (in $) and h is the hours worked. How many hours must Esmeralda work to earn $520 in one week?

Getting Ready for the Next Section

To understand all of the explanations and examples in the next section you must be able to work the problems below.

Solve each equation.

87. $2x = 4$

88. $3x = 24$

89. $30 = 5x$

90. $0 = 5x$

91. $0.17x = 510$

92. $0.1x = 400$

93. $0.12x = 78$

94. $0.35x = 420$

Apply the distributive property and then simplify if possible.

95. $3(x - 5) + 4$

96. $5(x - 3) + 2$

97. $0.09(x + 2,000)$

98. $0.04(x + 7,000)$

99. $7 - 3(2y + 1)$

100. $4 - 2(3y + 1)$

101. $3(2x - 5) - (2x - 4)$

102. $4(3x - 2) - (6x - 5)$

Find the Mistake

Each sentence below contains a mistake. Circle the mistake and write the correct word or number on the line provided.

1. The multiplication property of equality states that multiplying one side of an equation by the same nonzero number will not change the solution set. _____

2. To solve $-\frac{2b}{9} = 3$ for b, we multiply both sides of the equation by the opposite of $-\frac{2}{9}$. _____

3. To begin solving the equation $\frac{5}{6}x + \frac{1}{2} = -\frac{1}{4}$, we can either add $\frac{5}{6}$ to each side of the equation or multiply both sides by the LCD 12. _____

4. To solve the equation $-\frac{8}{9}y = -3$, we begin by multiplying both sides of the equation by the LCD of 9 and 3. _____

A Solve a linear equation in one variable.

linear equation in one variable

Video Examples
Section 9.3

Note You may have some previous experience solving equations. Even so, you should solve the equations in this section using the method developed here. Your work should look like the examples in the text. If you have learned shortcuts or a different method of solving equations somewhere else, you can always go back to them later. What is important now is that you are able to solve equations by the methods shown here.

9.3 Solving Linear Equations in One Variable

©iStockphoto.com/mevans

In the spring of 2010, over 500 robots took over a California town for the seventh annual RoboGames, held at the San Mateo County Expo Center. Robots of all shapes and sizes competed in events, such as stair climbing, obstacle running, kung-fu, sumo wrestling, basketball, and even combat battles. Advance tickets to the games were $20, and entry fees for most events were $35 each. Suppose you purchased an advance ticket and entered a robot into x number of events, spending $125 total. We could write the following equation based on the above information, and then solve for x to determine how many events you entered.

$$35x + 20 = 125$$

In this section, we will use the material we have developed in the first two sections of this chapter to build a method for solving any linear equation such as the one above.

A Solving Linear Equations

The equations we have worked with in the first two sections of this chapter are called *linear equations in one variable*. Let's begin with a formal definition.

> **DEFINITION** linear equation in one variable
>
> A *linear equation in one variable* is any equation that can be put in the form $Ax + B = C$, where A, B, and C are real numbers and a is not zero.

Each of the equations we will solve in this section is a linear equation in one variable. The steps we use to solve a linear equation in one variable are listed here:

> **HOW TO** Strategy for Solving Linear Equations in One Variable
>
> **Step 1a:** Use the distributive property to separate terms, if necessary.
>
> **1b:** If fractions are present, consider multiplying both sides by the LCD to eliminate the fractions. If decimals are present, consider multiplying both sides by a power of 10 to clear the equation of decimals.
>
> **1c:** Combine similar terms on each side of the equation.
>
> **Step 2:** Use the addition property of equality to get all variable terms on one side of the equation and all constant terms on the other side. A variable term is a term that contains the variable (for example, $5x$). A constant term is a term that does not contain the variable (the number 3, for example).
>
> **Step 3:** Use the multiplication property of equality to get x (that is, $1x$) by itself on one side of the equation.
>
> **Step 4:** Check your solution in the original equation to be sure that you have not made a mistake in the solution process.

As you work through the examples in this section, it is not always necessary to use all four steps when solving equations. The number of steps used depends on the equation. In Example 1, there are no fractions or decimals in the original equation, so step 1b will not be used. Likewise, after applying the distributive property to the left side of the equation in Example 1, there are no similar terms to combine on either side of the equation, making step 1c also unnecessary.

Practice Problems

EXAMPLE 1   Solve $2(x + 3) = 10$.

1. Solve $3(x + 4) = 6$.

Solution   To begin, we apply the distributive property to the left side of the equation to separate terms.

Step 1a:	$2x + 6 = 10$	Distributive property
Step 2:	$2x + 6 + (-6) = 10 + (-6)$	Addition property of equality
	$2x = 4$	
Step 3:	$\frac{1}{2}(2x) = \frac{1}{2}(4)$	Multiply each side by $\frac{1}{2}$.
	$x = 2$	The solution is 2.

The solution to our equation is 2. We check our work (to be sure we have not made either a mistake in applying the properties or an arithmetic mistake) by substituting 2 into our original equation and simplifying each side of the result separately.

Step 4:
When → $x = 2$

the equation → $2(x + 3) = 10$

becomes → $2(2 + 3) \overset{?}{=} 10$

                 $2(5) \overset{?}{=} 10$

             $10 = 10$ A true statement

Our solution checks.

The general method of solving linear equations is actually very simple. We can add any number to both sides of the equation and multiply both sides by any nonzero number. The equation may change in form, but the solution set will not.

The examples that follow show a variety of equations and their solutions. When you have finished this section and worked the problems in the problem set, the steps in the solution process should be a description of how you operate when solving equations.

EXAMPLE 2   Solve $3(x - 5) + 4 = 13$ for x.

2. Solve $2(x - 3) + 3 = 9$ for x.

Solution   Our first step will be to apply the distributive property to the left side of the equation.

Step 1a:	$3x - 15 + 4 = 13$	Distributive property
Step 1c:	$3x - 11 = 13$	Simplify the left side.
Step 2:	$3x - 11 + 11 = 13 + 11$	Add 11 to both sides.
	$3x = 24$	
Step 3:	$\frac{1}{3}(3x) = \frac{1}{3}(24)$	Multiply both sides by $\frac{1}{3}$.
	$x = 8$	The solution is 8.

Step 4:
When → $x = 8$

the equation → $3(x - 5) + 4 = 13$

becomes → $3(8 - 5) + 4 \overset{?}{=} 13$

              $3(3) + 4 \overset{?}{=} 13$

             $9 + 4 \overset{?}{=} 13$

             $13 = 13$ A true statement

Answers

1. -2

2. 6

3. Solve.

$$7(x - 3) + 5 = 4(3x - 2) - 8$$

EXAMPLE 3   Solve $5(x - 3) + 2 = 5(2x - 8) - 3$.

Solution   In this case, we first apply the distributive property to each side of the equation.

Step 1a:	$5x - 15 + 2 = 10x - 40 - 3$	*Distributive property*
Step 1c:	$5x - 13 = 10x - 43$	*Simplify each side.*
Step 2:	$5x + (-5x) - 13 = 10x + (-5x) - 43$	*Add $-5x$ to both sides.*
	$-13 = 5x - 43$	
	$-13 + 43 = 5x - 43 + 43$	*Add 43 to both sides.*
	$30 = 5x$	
Step 3:	$\dfrac{1}{5}(30) = \dfrac{1}{5}(5x)$	*Multiply both sides by $\frac{1}{5}$.*
	$6 = x$	*The solution is 6.*

Step 4:   Replacing x with 6 in the original equation, we have

$$5(6 - 3) + 2 \overset{?}{=} 5(2 \cdot 6 - 8) - 3$$

$$5(3) + 2 \overset{?}{=} 5(12 - 8) - 3$$

$$5(3) + 2 \overset{?}{=} 5(4) - 3$$

$$15 + 2 \overset{?}{=} 20 - 3$$

$$17 = 17 \qquad \textit{A true statement}$$

> *Note*   It makes no difference on which side of the equal sign x ends up. Most people prefer to have x on the left side because we read from left to right, and it seems to sound better to say x is 6 rather than 6 is x. Both expressions, however, have exactly the same meaning.

4. Solve the equation.

$$0.06x + 0.04(x + 7,000) = 680$$

EXAMPLE 4   Solve the equation $0.08x + 0.09(x + 2,000) = 690$.

Solution   We can solve the equation in its original form by working with the decimals, or we can eliminate the decimals first by using the multiplication property of equality and solving the resulting equation. Both methods follow.

Method 1

Working with the decimals.

	$0.08x + 0.09(x + 2,000) = 690$	*Original equation*
Step 1a:	$0.08x + 0.09x + 0.09(2,000) = 690$	*Distributive property*
Step 1c:	$0.17x + 180 = 690$	*Simplify the left side.*
Step 2:	$0.17x + 180 + (-180) = 690 + (-180)$	*Add -180 to each side.*
	$0.17x = 510$	
Step 3:	$\dfrac{0.17x}{0.17} = \dfrac{510}{0.17}$	*Divide each side by 0.17.*
	$x = 3,000$	

Note that we divided each side of the equation by 0.17 to obtain the solution. This is still an application of the multiplication property of equality because dividing by 0.17 is equivalent to multiplying by $\frac{1}{0.17}$.

Method 2

Eliminating the decimals in the beginning.

	$0.08x + 0.09(x + 2,000) = 690$	*Original equation*
Step 1a:	$0.08x + 0.09x + 180 = 690$	*Distributive property*
Step 1b:	$100(0.08x + 0.09x + 180) = 100(690)$	*Multiply both sides by 100.*
	$8x + 9x + 18,000 = 69,000$	

Step 1c:　　　　$17x + 18,000 = 69,000$　　　Simplify the left side.

Step 2:　　　　　　　　$17x = 51,000$　　　Add $-18,000$ to each side.

Step 3:　　　　　$\dfrac{17x}{17} = \dfrac{51,000}{17}$　　　Divide each side by 17.

　　　　　　　　　　　$x = 3,000$

Substituting 3,000 for x in the original equation, we have

Step 4:　　$0.08(3,000) + 0.09(3,000 + 2,000) \overset{?}{=} 690$

　　　　　　　$0.08(3,000) + 0.09(5,000) \overset{?}{=} 690$

　　　　　　　　　　$240 + 450 \overset{?}{=} 690$

　　　　　　　　　　　　$690 = 690$　　　A true statement

EXAMPLE 5　　Solve $7 - 3(2y + 1) = 16$.

Solution　We begin by multiplying -3 times the sum of $2y$ and 1.

Step 1a:　　　$7 - 6y - 3 = 16$　　　　Distributive property

Step 1c:　　　　$-6y + 4 = 16$　　　　Simplify the left side.

Step 2:　　$-6y + 4 + (-4) = 16 + (-4)$　　Add -4 to both sides.

　　　　　　　　　$-6y = 12$

Step 3:　　$-\dfrac{1}{6}(-6y) = -\dfrac{1}{6}(12)$　　Multiply both sides by $-\dfrac{1}{6}$.

　　　　　　　　　　$y = -2$

There are two things to notice about the example that follows. First, the distributive property is used to remove parentheses that are preceded by a negative sign. Second, the addition property and the multiplication property are not shown in as much detail as in the previous examples.

EXAMPLE 6　　Solve $3(2x - 5) - (2x - 4) = 6 - (4x + 5)$.

Solution　When we apply the distributive property to remove the grouping symbols and separate terms, we have to be careful with the signs. Remember, we can think of $-(2x - 4)$ as $-1(2x - 4)$, so that

　　　　$-(2x - 4) = -1(2x - 4) = -2x + 4$

It is not uncommon for students to make a mistake with this type of simplification and write the result as $-2x - 4$, which is incorrect. Here is the complete solution to our equation:

$$3(2x - 5) - (2x - 4) = 6 - (4x + 5) \quad \text{Original equation}$$

$$6x - 15 - 2x + 4 = 6 - 4x - 5 \quad \text{Distributive property}$$

$$4x - 11 = -4x + 1 \quad \text{Simplify each side.}$$

$$8x - 11 = 1 \quad \text{Add 4x to each side.}$$

$$8x = 12 \quad \text{Add 11 to each side.}$$

$$x = \frac{12}{8} \quad \text{Multiply each side by } \frac{1}{8}.$$

$$x = \frac{3}{2} \quad \text{Reduce to lowest terms.}$$

The solution, $\frac{3}{2}$, checks when replacing x in the original equation.

5. Solve $4 - 2(3y + 1) = -16$.

6. Solve.
　$4(3x - 2) - (6x - 5) = 6 - (3x + 1)$

Answers

5. 3

6. $\dfrac{8}{9}$

7. Solve $2(t + 2) - 2 = 2(t + 3) - 4$.

EXAMPLE 7   Solve $-2(3t - 2) = 2(4 - 3t) - 4$.

Solution   We apply the distributive property to simplify both sides of the equation.

$$-2(3t - 2) = 2(4 - 3t) - 4 \qquad \textit{Original equation}$$
$$-6t + 4 = 8 - 6t - 4 \qquad \textit{Distributive property}$$
$$-6t + 4 = -6t + 4 \qquad \textit{Simplify each side.}$$
$$-6t = -6t \qquad \textit{Add } -4 \textit{ to each side.}$$
$$0 = 0 \qquad \textit{Add } 6t \textit{ to each side.}$$

We have eliminated the variable and are left with a true statement. This means that any real number that we use in place of the variable will be a solution to the equation. The solution is all real numbers.

8. Solve $-2(3y - 2) = 3(3 - 2y)$.

EXAMPLE 8   Solve $4(y - 5) + 3 = 2(2y - 3)$.

Solution   Again, we apply the distributive property.

$$4(y - 5) + 3 = 2(2y - 3) \qquad \textit{Original equation}$$
$$4y - 20 + 3 = 4y - 6 \qquad \textit{Distributive property}$$
$$4y - 17 = 4y - 6 \qquad \textit{Simplify each side.}$$
$$-17 = -6 \qquad \textit{Add } -4y \textit{ to each side.}$$

We have eliminated the variable and are left with a false statement. This means that no real numbers used in place of the variable will be a solution to the equation. There is no solution.

GETTING READY FOR CLASS

After reading through the preceding section, respond in your own words and in complete sentences.

A. What is the first step in solving a linear equation containing parentheses?

B. When solving a linear equation, why should you get all variable terms on one side and all constant terms on the other before using the multiplication property of equality?

C. What is the last step in solving a linear equation?

D. If an equation contains decimals, what can you do to eliminate the decimals?

Answers

7. All real numbers

8. No solution

Vocabulary Review

The following is a list of steps for solving a linear equation in one variable. Choose the correct words to fill in the blanks below.

solution	addition	multiplication	distributive
LCD	decimals	similar	constant

Step 1a: Use the _____ property to separate terms.

   1b: To clear any fractions, multiply both sides of the equation by the _____. To clear _____, multiply both sides by a power of 10.

   1c: Combine _____ terms.

Step 2: Use the _____ property of equality to get all variable terms on one side of the equation and all _____ terms on the other side.

Step 3: Use the _____ property of equality to get x by itself on one side of the equation.

Step 4: Check your _____ in the original equation.

Problems

A Solve each of the following equations using the four steps shown in this section.

1. $2(x + 3) = 12$ **2.** $3(x - 2) = 6$ **3.** $6(x - 1) = -18$ **4.** $4(x + 5) = 16$

5. $2(4a + 1) = -6$ **6.** $3(2a - 4) = 12$ **7.** $14 = 2(5x - 3)$ **8.** $-25 = 5(3x + 4)$

9. $-2(3y + 5) = 14$ **10.** $-3(2y - 4) = -6$ **11.** $-5(2a + 4) = 0$ **12.** $-3(3a - 6) = 0$

13. $1 = \frac{1}{2}(4x + 2)$ **14.** $1 = \frac{1}{3}(6x + 3)$ **15.** $3(t - 4) + 5 = -4$ **16.** $5(t - 1) + 6 = -9$

Solve each equation.

17. $4(2x + 1) - 7 = 1$ **18.** $6(3y + 2) - 8 = -2$

19. $\frac{1}{2}(x - 3) = \frac{1}{4}(x + 1)$ **20.** $\frac{1}{3}(x - 4) = \frac{1}{2}(x - 6)$

21. $-0.7(2x - 7) = 0.3(11 - 4x)$

22. $-0.3(2x - 5) = 0.7(3 - x)$

23. $-2(3y + 3) = 3(1 - 2y) - 9$

24. $-5(4y - 3) = 2(2 - 10y) + 11$

25. $\frac{3}{4}(8x - 4) + 3 = \frac{2}{5}(5x + 10) - 1$

26. $\frac{5}{6}(6x + 12) + 1 = \frac{2}{3}(9x - 3) + 5$

27. $0.06x + 0.08(100 - x) = 6.5$

28. $0.05x + 0.07(100 - x) = 6.2$

29. $6 - 5(2a - 3) = 1$

30. $-8 - 2(3 - a) = 0$

31. $0.2x - 0.5 = 0.5 - 0.2(2x - 13)$

32. $0.4x - 0.1 = 0.7 - 0.3(6 - 2x)$

33. $2(t - 3) + 3(t - 2) = 28$

34. $-3(t - 5) - 2(2t + 1) = -8$

35. $5(x - 2) - (3x + 4) = 3(6x - 8) + 10$

36. $3(x - 1) - (4x - 5) = 2(5x - 1) - 7$

37. $2(5x - 3) - (2x - 4) = 5 - (6x + 1)$

38. $3(4x - 2) - (5x - 8) = 8 - (2x + 3)$

39. $-(3x + 1) + 7 = 4 - (3x + 2)$

40. $-(6x + 2) = 8 - 2(3x + 1)$

41. $x + (2x - 1) = 2$

42. $x + (5x + 2) = 20$

43. $x - (3x + 5) = -3$

44. $x - (4x - 1) = 7$

45. $15 = 3(x - 1)$

46. $12 = 4(x - 5)$

47. $4x - (-4x + 1) = 5$

48. $-2x - (4x - 8) = -1$

49. $5x - 8(2x - 5) = 7$

50. $3x + 4(8x - 15) = 10$

51. $2(4x - 1) = 4(2x + 3)$

52. $2(3y - 3) = 3(2y + 4)$

53. $0.2x + 0.5(12 - x) = 3.6$

54. $0.3x + 0.6(25 - x) = 12$

55. $0.5x + 0.2(18 - x) = 5.4$

56. $0.1x + 0.5(40 - x) = 32$

57. $x + (x + 3)(-3) = x - 3$

58. $x - 2(x + 2) = x - 2$

59. $5(x + 2) + 3(x - 1) = -9$

60. $4(x + 1) + 3(x - 3) = 2$

61. $3(x - 3) + 2(2x) = 5$

62. $2(x - 2) + 3(5x) = 30$

63. $-5(y + 3) = -5y - 15$

64. $3(4y - 1) - 5 = 4(3y - 2)$

65. $3x + 2(x - 2) = 6$

66. $5x - (x - 5) = 25$

67. $50(x - 5) = 30(x + 5)$

68. $34(x - 2) = 26(x + 2)$

69. $0.08x + 0.09(x + 2{,}000) = 860$

70. $0.11x + 0.12(x + 4{,}000) = 940$

71. $0.10x + 0.12(x + 500) = 214$

72. $0.08x + 0.06(x + 800) = 104$

73. $5x + 10(x + 8) = 245$

74. $5x + 10(x + 7) = 175$

75. $5x + 10(x + 3) + 25(x + 5) = 435$

76. $5(x + 3) + 10x + 25(x + 7) = 390$

The next two problems are intended to give you practice reading, and paying attention to, the instructions that accompany the problems you are working. Working these problems is an excellent way to get ready for a test or a quiz.

77. Work each problem according to the instructions.

   a. Solve $4x - 5 = 0$.

   b. Solve $4x - 5 = 25$.

   c. Add $(4x - 5) + (2x + 25)$.

   d. Solve $4x - 5 = 2x + 25$.

   e. Multiply $4(x - 5)$.

   f. Solve $4(x - 5) = 2x + 25$.

78. Work each problem according to the instructions.

   a. Solve $3x + 6 = 0$.

   b. Solve $3x + 6 = 4$.

   c. Add $(3x + 6) + (7x + 4)$.

   d. Solve $3x + 6 = 7x + 4$.

   e. Multiply $3(x + 6)$.

   f. Solve $3(x + 6) = 7x + 4$.

79. Solve the equation $35x + 20 = 125$, given at the introduction of this section, to find the number of robot events that were entered.

80. A traveling circus with thrill rides charges $6 for admission plus $1.50 for each ride you take. If you pay $16.50 total, how many rides did you take?

81. A stock trading company charges a commission of $15 per trade plus $0.025 per dollar value of stock traded. If your commission is $120, what was the value of the stock you traded?

82. Brad earns a base monthly salary of $2,400 plus $0.018 commission on his total sales. If his monthly pay is $3,246, what were his sales that month?

Getting Ready for the Next Section

To understand all of the explanations and examples in the next section you must be able to work the problems below.

Solve each equation.

83. $40 = 2x + 12$

84. $80 = 2x + 12$

85. $12 + 2y = 6$

86. $3x + 18 = 6$

87. $24x = 6$

88. $45 = 0.75x$

89. $70 = x \cdot 210$

90. $15 = x \cdot 80$

Apply the distributive property.

91. $\dfrac{1}{2}(-3x + 6)$

92. $\dfrac{2}{3}(-6x + 9)$

93. $-\dfrac{1}{4}(-5x + 20)$

94. $-\dfrac{1}{3}(-4x + 12)$

95. $12\left(\dfrac{1}{3}x - \dfrac{1}{4}\right)$

96. $-8\left(-\dfrac{1}{4}x + \dfrac{1}{8}\right)$

97. $-24\left(-\dfrac{1}{3}x - \dfrac{1}{8}\right)$

98. $-15\left(\dfrac{3}{5}x - \dfrac{2}{3}\right)$

Find the Mistake

Each sentence below contains a mistake. Circle the mistake and write the correct word(s) or equation on the line provided.

1. A linear equation in one variable is any equation that can be put in the form of $y = Ax + B$. _____

2. $12 + 6y - 4 = 3y$ is not a linear equation in one variable because it cannot be simplified to the form $Ax + B = C$.

_____

3. To solve the linear equation $3(2x + 7) = 2(x - 1)$, begin by dividing both sides of the equation by $(x - 1)$.

_____

4. To solve a linear equation, make sure to keep similar terms on opposite sides of the equation.

_____

9.4 Formulas

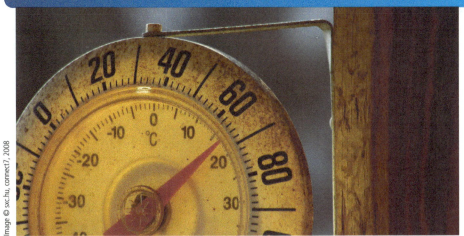

OBJECTIVES

A Find the value of a variable in a formula given replacements for the other variables.

B Solve a formula for one of its variables.

C Continue to find the complement and supplement of an angle.

D Solve basic percent problems.

KEY WORDS

formula

solving for a variable

Recall our discuss from Section 1 about finding the outside temperature by listening to a cricket chirp. What if we needed to find the number of chirps based on a Celsius temperature? The formula to find the outside temperature in Celsius is

$$t = \left(\frac{n}{3}\right) + 4$$

where t is the outside temperature in degrees Celsius and n is the number of cricket chirps in 25 seconds. In this section, we will learn about formulas and how to solve a formula for one of its variables. For instance, we will learn how the Celsius temperature formula could be solved for n and become

$$n = 3(t - 4)$$

Then we could provide a numerical replacement for t and easily calculate a value for n.

In this section, we continue solving equations by working with formulas. Here is the definition of a formula.

> **DEFINITION** formula
>
> In mathematics, a **formula** is an equation that contains more than one variable.

The equation $P = 2l + 2w$, which tells us how to find the perimeter of a rectangle, is an example of a formula.

A Solving Formulas

To begin our work with formulas, we will consider some examples in which we are given numerical replacements for all but one of the variables.

Video Examples
Section 9.4

1. Suppose the livestock pen in Example 1 has a perimeter of 80 feet. If the width is still 6 feet, what is the new length?

EXAMPLE 1   The perimeter P of a rectangular livestock pen is 40 feet. If the width w is 6 feet, find the length.

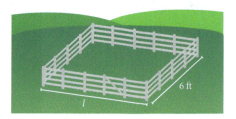

Solution   First, we substitute 40 for P and 6 for w in the formula $P = 2l + 2w$. Then we solve for l.

When → $P = 40$ and $w = 6$

the formula → $P = 2l + 2w$

becomes → $40 = 2l + 2(6)$

$40 = 2l + 12$ Multiply 2 and 6.

$28 = 2l$ Add -12 to each side.

$14 = l$ Multiply each side by $\frac{1}{2}$.

To summarize our results, if a rectangular pen has a perimeter of 40 feet and a width of 6 feet, then the length must be 14 feet.

2. Find x when y is 9 in the formula $3x + 2y = 6$.

EXAMPLE 2   Find y when $x = 4$ in the formula $3x + 2y = 6$.

Solution   We substitute 4 for x in the formula and then solve for y.

When → $x = 4$

the formula → $3x + 2y = 6$

becomes → $3(4) + 2y = 6$

$12 + 2y = 6$ Multiply 3 and 4.

$2y = -6$ Add -12 to each side.

$y = -3$ Multiply each side by $\frac{1}{2}$

B Solving for a Variable

In the next examples we will solve a formula for one of its variables without being given numerical replacements for the other variables.

Consider the formula for the area of a triangle:

$A = \frac{1}{2}bh$

where A = area, b = length of the base, and h = height of the triangle.

Suppose we want to solve this formula for h. What we must do is isolate the variable h on one side of the equal sign. We begin by multiplying both sides by 2, because it is the reciprocal of $\frac{1}{2}$.

$$2 \cdot A = 2 \cdot \frac{1}{2}bh$$

$$2A = bh$$

Then we divide both sides by b.

$$\frac{2A}{b} = \frac{bh}{b}$$

$$h = \frac{2A}{b}$$

The original formula $A = \frac{1}{2}bh$ and the final formula $h = \frac{2A}{b}$ both give the same relationship among A, b, and h. The first one has been solved for A and the second one has been solved for h.

RULE Solving for a Variable

To solve a formula for one of its *variables*, we must isolate that variable on either side of the equal sign. All other variables and constants will appear on the other side.

EXAMPLE 3   Solve $3x + 2y = 6$ for y.

Solution   To solve for y, we must isolate y on the left side of the equation. To begin, we use the addition property of equality to add $-3x$ to each side.

$$3x + 2y = 6 \qquad \text{Original formula}$$
$$3x + (-3x) + 2y = (-3x) + 6 \qquad \text{Add } -3x \text{ to each side.}$$
$$2y = -3x + 6 \qquad \text{Simplify the left side.}$$
$$\frac{1}{2}(2y) = \frac{1}{2}(-3x + 6) \qquad \text{Multiply each side by } \frac{1}{2}.$$
$$y = -\frac{3}{2}x + 3$$

3. Solve $5x - 4y = 20$ for y.

EXAMPLE 4   Solve $h = vt - 16t^2$ for v.

Solution   Let's begin by interchanging the left and right sides of the equation. That way, the variable we are solving for, v, will be on the left side.

$$vt - 16t^2 = h \qquad \text{Symmetric property}$$
$$vt - 16t^2 + 16t^2 = h + 16t^2 \qquad \text{Add } 16t^2 \text{ to each side.}$$
$$vt = h + 16t^2$$
$$\frac{vt}{t} = \frac{h + 16t^2}{t} \qquad \text{Divide each side by } t.$$
$$v = \frac{h + 16t^2}{t}$$

4. Solve $P = 2w + 2l$ for l.

We know we are finished because we have isolated the variable we are solving for on the left side of the equation and it does not appear on the other side.

EXAMPLE 5   Solve $\frac{y - 1}{x} = \frac{3}{2}$ for y.

Solution   Although we will do more extensive work with formulas of this form later in the book, we need to know how to solve this particular formula for y in order to understand some things in the next chapter. We begin by multiplying each side of the formula by x. Doing so will simplify the left side of the equation, and make the rest of the solution process simple.

$$\frac{y - 1}{x} = \frac{3}{2} \qquad \text{Original formula}$$
$$x \cdot \frac{y - 1}{x} = \frac{3}{2} \cdot x \qquad \text{Multiply each side by } x.$$
$$y - 1 = \frac{3}{2}x \qquad \text{Simplify each side.}$$
$$y = \frac{3}{2}x + 1 \qquad \text{Add 1 to each side.}$$

5. Solve $\frac{y - 2}{x} = \frac{4}{3}$ for y.

This is our solution. If we look back to the first step, we can justify our result on the left side of the equation this way: Dividing by x is equivalent to multiplying by its reciprocal $\frac{1}{x}$. Here is what it looks like when written out completely:

$$x \cdot \frac{y - 1}{x} = x \cdot \frac{1}{x}(y - 1) = 1(y - 1) = (y - 1)$$

Answers

3. $y = \frac{5}{4}x - 5$

4. $l = \frac{P - 2w}{2}$

5. $y = \frac{4}{3}x + 2$

C Complementary and Supplementary Angles

FACTS FROM GEOMETRY Complementary and Supplementary Angles

Complementary angles are defined as angles that add to 90°; that is, if x and y are complementary angles, then

$$x + y = 90°$$

If we solve this formula for y, we obtain a formula equivalent to our original formula:

$$y = 90° - x$$

Because y is the complement of x, we can generalize by saying that the complement of angle x is the angle $90° - x$. By a similar reasoning process, since supplementary angles are angles that add to 180° we can say that the supplement of angle x is the angle $180° - x$. To summarize, if x is an angle, then

The complement of x is $90° - x$, and

The supplement of x is $180° - x$

If you go on to take a trigonometry class, you will see this formula again.

6. Find the complement and the supplement of 35°.

EXAMPLE 6   Find the complement and the supplement of 25°.

Solution   We can use the formulas above with $90° - x$ and $180° - x$.

The complement of 25° is $90° - 25° = 65°$.

The supplement of 25° is $180° - 25° = 155°$.

D Basic Percent Problems

The last examples in this section show how basic percent problems can be translated directly into equations. To understand these examples, you must recall that *percent* means "per hundred" that is, 75% is the same as $\frac{75}{100}$, 0.75, and, in reduced fraction form, $\frac{3}{4}$. Likewise, the decimal 0.25 is equivalent to 25%. To change a decimal to a percent, we move the decimal point two places to the right and write the % symbol. To change from a percent to a decimal, we drop the % symbol and move the decimal point two places to the left. The table that follows gives some of the most commonly used fractions and decimals and their equivalent percents.

Fraction	Decimal	Percent
$\frac{1}{2}$	0.5	50%
$\frac{1}{4}$	0.25	25%
$\frac{3}{4}$	0.75	75%
$\frac{1}{3}$	$0.33\overline{3}$	$33\frac{1}{3}$%
$\frac{2}{3}$	0.667	$66\frac{2}{3}$%
$\frac{1}{5}$	0.2	20%
$\frac{2}{5}$	0.4	40%

Answer

6. 55°; 145°

EXAMPLE 7   What number is 25% of 60?

Solution   To solve a problem like this, we let $x =$ the number in question (that is, the number we are looking for). Then, we translate the sentence directly into an equation by using an equal sign for the word "is" and multiplication for the word "of." Here is how it is done:

$$\underbrace{\text{What number}}\ \ \underset{\downarrow}{\text{is}}\ \underset{\downarrow}{25\%}\ \underset{\downarrow}{\text{of}}\ \underset{\downarrow}{60?}$$

$$x \quad\quad = 0.25 \cdot 60$$
$$= 15$$

Notice that we must write 25% as a decimal in order to do the arithmetic in the problem. The number 15 is 25% of 60.

7. What number is 25% of 74?

EXAMPLE 8   What percent of 24 is 6?

Solution   Translating this sentence into an equation, as we did in Example 7, we have

$$\underbrace{\text{What percent}}\ \underset{\downarrow}{\text{of}}\ \underset{\downarrow}{24}\ \underset{\downarrow}{\text{is}}\ \underset{\downarrow}{6?}$$

$$x \quad \cdot\ 24 = 6$$
$$24x = 6$$

Next, we multiply each side by $\frac{1}{24}$. (This is the same as dividing each side by 24.)

$$\frac{1}{24}(24x) = \frac{1}{24}(6)$$

$$x = \frac{6}{24}$$

$$= \frac{1}{4}$$

$$= 0.25, \text{ or } 25\%$$

25% of 24 is 6, or in other words, the number 6 is 25% of 24.

8. What percent of 84 is 21?

EXAMPLE 9   45 is 75% of what number?

Solution   Again, we translate the sentence directly.

$$\underset{\downarrow}{45}\ \underset{\downarrow}{\text{is}}\ \underset{\downarrow}{75\%}\ \underset{\downarrow}{\text{of}}\ \underbrace{\text{what number?}}$$

$$45 = 0.75 \cdot \quad\quad x$$

Next, we multiply each side by $\frac{1}{0.75}$ (which is the same as dividing each side by 0.75).

$$\frac{1}{0.75}(45) = \frac{1}{0.75}(0.75x)$$

$$\frac{45}{0.75} = x$$

$$60 = x$$

The number 45 is 75% of 60.

9. 35 is 40% of what number?

Answers
7. 18.5
8. 25%
9. 87.5

10. In another pizza ingredient, pepperoni, 75% of the calories are from fat. If one serving of pepperoni contains 132 calories, how many of those calories are from fat?

EXAMPLE 10   The American Dietetic Association (ADA) recommends eating foods in which the calories from fat are less than 30% of the total calories. The nutrition labels from two ingredients you may use for a pizza are shown in Figure 1. For each ingredient, what percent of the total calories come from fat?

PIZZA DOUGH

Nutrition Facts

Serving Size 1/6 of package (65g)
Servings Per Container: 6

Amount Per Serving

Calories 160   Calories from fat 18

% Daily Value*

Total Fat 2g	3%
Saturated Fat 0.5g	3%
Poly unsaturated Fat 0g	
Monounsaturated Fat 0g	
Cholesterol 0mg	0%
Sodium 470mg	20%
Total Carbohydrate 31g	10%
Dietary Fiber 1g	4%
Sugars 4g	
Protein 5g	

Vitamin A 0%   •   Vitamin C 0%

Calcium 0%   •   Iron 10%

*Percent Daily Values are based on a 2,000 calorie diet

MOZZARELLA CHEESE

Nutrition Facts

Serving Size 1 oz (28.3g)
Servings Per Container: 12

Amount Per Serving

Calories 72   Calories from fat 41

% Daily Value*

Total Fat 4.5g	7%
Saturated Fat 2.9g	14%
Cholesterol 18mg	6%
Sodium 175mg	7%
Total Carbohydrate 0.8g	0%
Fiber 0g	0%
Sugars 0.3g	
Protein 6.9g	

Vitamin A 3%   •   Vitamin C 0%

Calcium 22%   •   Iron 0%

*Percent Daily Values (DV) are based on a 2,000 calorie diet

FIGURE 1

Solution   The information needed to solve this problem is located towards the top of each label. Each serving of the pizza dough contains 160 calories, of which 18 calories come from fat. To find the percent of total calories that come from fat, we must answer this question:

18 is what percent of 160?

For the mozzarella cheese, one serving contains 72 calories, of which 41 calories come from fat. To find the percent of total calories that come from fat, we must answer this question:

41 is what percent of 72?

Translating each equation into symbols, we have

18 is what percent of 160	41 is what percent of 72
$18 = x \cdot 160$	$41 = x \cdot 72$
$x = \dfrac{18}{160}$	$x = \dfrac{41}{72}$
$x = 0.11$ to the nearest hundredth	$x = 0.57$ to the nearest hundredth
$x = 11\%$	$x = 57\%$

Comparing the two ingredients, 11% of the calories in the pizza dough are fat calories, whereas 57% of the calories in the mozzarella cheese are fat calories. According to the ADA, pizza dough is the healthier ingredient.

GETTING READY FOR CLASS

After reading through the preceding section, respond in your own words and in complete sentences.

A. What is a formula?

B. How do you solve a formula for one of its variables?

C. What are complementary angles?

D. Translate the following question into an equation: what number is 40% of 20?

Vocabulary Review

Choose the correct words to fill in the blanks below.

isolate   variable   complementary   formula   supplement

1. A formula is an equation with more than one _____.

2. $P = 2l + 2w$ is the _____ for finding the perimeter of a rectangle.

3. To solve for a variable in a formula, _____ that variable on either side of the equals sign.

4. If x and y are _____ angles, then $x + y = 90°$.

5. The _____ of angle x is the angle $180° - x$.

Problems

A Use the formula $P = 2l + 2w$ to find the length l of a rectangular lot if

1. The width w is 50 feet and the perimeter P is 300 feet.

2. The width w is 75 feet and the perimeter P is 300 feet.

Use the formula $2x + 3y = 6$ to find y when

3. $x = 3$
4. $x = -2$
5. $x = 0$
6. $x = -3$

Use the formula $2x - 5y = 20$ to find x when

7. $y = 2$
8. $y = -4$
9. $y = 0$
10. $y = -6$

Use the equation $y = (x + 1)^2 - 3$ to find the value of y when

11. $x = -2$
12. $x = -1$
13. $x = 1$
14. $x = 2$

15. Use the formula $y = \dfrac{20}{x}$ to find y when

    a. $x = 10$

    b. $x = 5$

16. Use the formula $y = 2x^2$ to find y when

    a. $x = 5$

    b. $x = -6$

17. Use the formula $y = Kx$ to find K when

    a. $y = 15$ and $x = 3$

    b. $y = 72$ and $x = 4$

18. Use the formula $y = Kx^2$ to find K when

    a. $y = 32$ and $x = 4$

    b. $y = 45$ and $x = 3$

B Solve each of the following for the indicated variable.

19. $A = lw$ for l

20. $d = rt$ for r

21. $V = lwh$ for h

22. $PV = nRT$ for P

23. $P = a + b + c$ for a

24. $P = a + b + c$ for b

25. $x - 3y = -1$ for x

26. $x + 3y = 2$ for x

27. $-3x + y = 6$ for y

28. $2x + y = -17$ for y

29. $2x + 3y = 6$ for y

30. $4x + 5y = 20$ for y

31. $y - 3 = -2(x + 4)$ for y

32. $y + 5 = 2(x + 2)$ for y

33. $y - 3 = -\frac{2}{3}(x + 3)$ for y

34. $y - 1 = -\frac{1}{2}(x + 4)$ for y

35. $P = 2l + 2w$ for w

36. $P = 2l + 2w$ for l

37. $h = vt + 16t^2$ for v

38. $h = vt - 16t^2$ for v

39. $A = \pi r^2 + 2\pi rh$ for h

40. $A = 2\pi r^2 + 2\pi rh$ for h

41. $I = prt$ for t

42. $PV = nrT$ for T

43. Solve for y.

a. $\dfrac{y - 1}{x} = \dfrac{3}{5}$

b. $\dfrac{y - 2}{x} = \dfrac{1}{2}$

c. $\dfrac{y - 3}{x} = 4$

44. Solve for y.

a. $\dfrac{y + 1}{x} = -\dfrac{3}{5}$

b. $\dfrac{y + 2}{x} = -\dfrac{1}{2}$

c. $\dfrac{y + 3}{x} = -4$

45. Solve for y.

a. $\dfrac{y + 2}{x} = -\dfrac{1}{2}$

b. $\dfrac{y - 2}{x} = -3$

c. $\dfrac{y + 1}{x} = -\dfrac{1}{3}$

46. Solve for y.

a. $\dfrac{y - 1}{x} = -\dfrac{3}{4}$

b. $\dfrac{y + 5}{x} = -7$

c. $\dfrac{y + 2}{x} = -3$

Solve each formula for y.

47. $\dfrac{x}{7} - \dfrac{y}{3} = 1$

48. $\dfrac{x}{5} - \dfrac{y}{9} = 1$

49. $-\dfrac{1}{4}x + \dfrac{1}{8}y = 1$

50. $-\dfrac{1}{9}x + \dfrac{1}{3}y = 1$

C Find the complement and the supplement of each angle.

51. $30°$

52. $60°$

53. $45°$

54. $15°$

D Translate each of the following into an equation, and then solve that equation.

55. What number is 25% of 40?

56. What number is 75% of 40?

57. What number is 12% of 2,000?

58. What number is 9% of 3,000?

59. What percent of 28 is 7?

60. What percent of 28 is 21?

61. What percent of 40 is 14?

62. What percent of 20 is 14?

63. 32 is 50% of what number?

64. 16 is 50% of what number?

65. 240 is 12% of what number?

66. 360 is 12% of what number?

Applying the Concepts

More About Temperatures In the U.S. system, temperature is measured on the Fahrenheit scale. In the metric system, temperature is measured on the Celsius scale. On the Celsius scale, water boils at 100 degrees and freezes at 0 degrees. To denote a temperature of 100 degrees on the Celsius scale, we write 100°C, which is read "100 degrees Celsius."

Table 1 is intended to give you an intuitive idea of the relationship between the two temperature scales. Table 2 gives the formulas, in both symbols and words, that are used to convert between the two scales.

Table 1

Situation	Temperature	
	Fahrenheit	Celsius
Water freezes	32°F	0°C
Room temperature	68°F	20°C
Normal body temperature	98.6°F	37°C
Water boils	212°F	100°C

Table 2

To convert from	Formula in symbols	Formula in words
Fahrenheit to Celsius	$C = \frac{5}{9}(F - 32)$	Subtract 32, then multiply by $\frac{5}{9}$.
Celsius to Fahrenheit	$F = \frac{9}{5}C + 32$	Multiply by $\frac{5}{9}$, then add 32.

67. Let $F = 212$ in the formula $C = \frac{5}{9}(F - 32)$, and solve for C. Does the value of C agree with the information in Table 1?

68. Let $C = 100$ in the formula $F = \frac{9}{5}C + 32$, and solve for F. Does the value of F agree with the information in Table 1?

69. Let $F = 68$ in the formula $C = \frac{5}{9}(F - 32)$, and solve for C. Does the value of C agree with the information in Table 1?

70. Let $C = 37$ in the formula $F = \frac{9}{5}C + 32$, and solve for F. Does the value of F agree with the information in Table 1?

71. Solve the formula $F = \frac{9}{5}C + 32$ for C.

72. Solve the formula $C = \frac{5}{9}(F - 32)$ for F.

Nutrition Labels The nutrition label in Figure 2 is from a caffe latte. The label in Figure 3 is from a nonfat caffe latte. Use the information on these labels for problems 73–76. Round your answers to the nearest tenth of a percent.

CAFFE LATTE

Nutrition Facts

Serving Size 16 fl oz (453.0g)
Servings Per Container: 1

Amount Per Serving

Calories 272   Calories from fat 126

% Daily Value*

Total Fat 14.0g	22%
Saturated Fat 9.0g	45%
Cholesterol 60mg	20%
Sodium 208mg	9%
Total Carbohydrate 22.0g	7%
Fiber 0g	0%
Sugars 19.0g	
Protein 14.0g	

Vitamin A 12%	•	Vitamin C 0%
Calcium 48%	•	Iron 0%

*Percent Daily Values (DV) are based on a 2,000 calorie diet

FIGURE 2

NONFAT CAFFE LATTE

Nutrition Facts

Serving Size 16 fl oz (453.0g)
Servings Per Container: 1

Amount Per Serving

Calories 168   Calories from fat 0

% Daily Value*

Total Fat 0.3g	0%
Saturated Fat 0.0g	0%
Cholesterol 8mg	3%
Sodium 232mg	10%
Total Carbohydrate 24.0g	8%
Fiber 0g	0%
Sugars 21.0g	
Protein 16.0g	

Vitamin A 16%	•	Vitamin C 0%
Calcium 48%	•	Iron 0%

*Percent Daily Values (DV) are based on a 2,000 calorie diet

FIGURE 3

73. What percent of the calories in the caffe latte are calories from fat?

74. What percent of the calories in the nonfat caffe latte are calories from fat?

75. One nonfat caffe latte weighs 453 grams, of which 24 grams are carbohydrates. What percent are carbohydrates?

76. One caffe latte weighs 453 grams. What percent is sugar?

Circumference The circumference of a circle is given by the formula $C = 2\pi r$. Find r using the following values. (The values given for π are approximations.)

77. The circumference C is 44 meters and π is $\frac{22}{7}$.

78. The circumference C is 176 meters and π is $\frac{22}{7}$.

79. The circumference is 9.42 inches and π is 3.14.

80. The circumference is 12.56 inches and π is 3.14.

Volume The volume of a cylinder is given by the formula $V = \pi r^2 h$. Find the height h if

81. The volume V is 42 cubic feet, the radius is $\frac{7}{22}$ feet, and π is $\frac{22}{7}$.

82. The volume V is 84 cubic inches, the radius is $\frac{7}{11}$ inches, and π is $\frac{22}{7}$.

83. The volume is 6.28 cubic centimeters, the radius is 3 centimeters, and π is 3.14.

84. The volume is 12.56 cubic centimeters, the radius is 2 centimeters, and π is 3.14.

Baseball Major league baseball has various player awards at the end of the year. One of them is the Rolaids Relief Man of the Year. To compute the Relief Man standings, points are gained or taken away based on the number of wins, losses, saves, and blown saves a relief pitcher has at the end of the year. The pitcher with the most Rolaids Points is the Rolaids Relief Man of the Year. The formula $P = 3s + 2w + t - 2l - 2b$ gives the number of Rolaids points a pitcher earns, where s = saves, w = wins, t = tough saves, l = losses, and b = blown saves. Use this formula to complete the following tables.

85. National League 2010

Pitcher, Team	W	L	Saves	Tough Saves	Blown Saves	Rolaids Points
Heath Bell, San Diego	6	1	47	3	5	
Brian Wilson, San Francisco	3	3	48	5	7	
Carlos Marmol, Chicago	2	3	38	5	5	
Billy Wagner, Atlanta	7	2	37	7	0	
Francisco Cordero, Cincinati	6	5	40	8	0	

86. American League 2010

Pitcher, Team	W	L	Saves	Tough Saves	Blown Saves	Rolaids Points
Rafael Soriano, Tampa Bay	3	2	45	0	3	
Joakim Soria, Kansas City	1	2	43	0	3	
Neftali Feliz, Texas	4	3	40	1	3	
Jonathan Papelbon, Boston	5	7	37	1	8	
Kevin Gregg, Toronto	2	6	37	1	6	

Getting Ready for the Next Section

To understand all of the explanations and examples in the next section you must be able to work the problems below.

Write an equivalent expression in English. Include the words *sum* and *difference* when possible.

87. $4 + 1 = 5$

88. $7 + 3 = 10$

89. $6 - 2 = 4$

90. $8 - 1 = 7$

91. $x - 15 = -12$

92. $2x + 3 = 7$

93. $x + 3 = 4(x - 3)$

94. $2(2x - 5) = 2x - 34$

For each of the following expressions, write an equivalent equation.

95. Twice the sum of 6 and 3 is 18.

96. Four added to the product of 5 and -1 is -1.

97. The sum of twice 5 and 3 is 13.

98. Twice the difference of 8 and 2 is 12.

99. The sum of a number and five is thirteen.

100. The difference of ten and a number is negative eight.

101. Five times the sum of a number and seven is thirty.

102. Five times the difference of twice a number and six is negative twenty.

Find the Mistake

Each sentence below contains a mistake. Circle the mistake and write the correct word(s) or equation on the line provided.

1. A formula contains only one variable. _____

2. To solve for a variable in a formula, make sure the variable appears on both sides of the equation.

_____

3. To solve the formula $2ab = a + 8$ for b, multiply each side of the equation by $2a$. _____

4. The question "What percent of 108 is 9?" can be translated into the equation $\frac{x}{9} = 108$. _____

Landmark Review: Checking Your Progress

Solve each of the following equations.

1. $x + 3 = 15$

2. $x - 5 = 20$

3. $5.6 + y = 10.8$

4. $-\dfrac{1}{5} + b = \dfrac{2}{3}$

5. $4x = 16$

6. $\dfrac{1}{5}y = 2$

7. $-5x + 3 + 3x + 5 = 0$

8. $3y + 4 - 6y = -10 + 5$

9. $4(x + 3) = 24$

10. $-4 - 3(x + 2) - 3 = -25$

Solve each of the following for the indicated variable.

11. $A = lw$ for l

12. $P = a + b + c$ for c

13. $x - 4 + 3y = 14$ for x

14. $\dfrac{y - 3}{x} = -\dfrac{4}{3}$ for y

OBJECTIVES

Apply the Blueprint for Problem Solving to solve:

A Number Problems

B Age Problems

C Perimeter Problems

D Coin Problems

An artist in Vermont built a sculpture of a dinosaur called Vermontasaurus that is 25 feet tall. The artist used found junk and scrap wood to construct his controversial work of art. The controversy lies in the structure's adherence to state building codes. The state's fire safety board has demanded that Vermontasaurus visitors not walk under the structure because of its questionable structural integrity. The area has been roped off, pending approval of a building permit.

Suppose you were tasked with constructing a permanent rectangular fence around the legs of the Vermontasaurus. You know that the length of the fence is 4 feet more than three times the width, and the perimeter is 168 feet. How would you find the length and width? In this section, we will use our knowledge of linear equations to solve similar application problems.

Problem Solving

To begin this section, we list the steps used in solving application problems. We call this strategy the *Blueprint for Problem Solving.* It is an outline that will overlay the solution process we use on all application problems.

Video Examples
Section 9.5

> ### BLUEPRINT FOR PROBLEM SOLVING
>
> **Step 1:** *Read* the problem, and then mentally *list* the items that are known and the items that are unknown. Make sure to include units of measurement where appropriate.
>
> **Step 2:** *Assign a variable* to one of the unknown items. (In most cases this will amount to letting x = the item that is asked for in the problem.) Then *translate* the other *information* in the problem to expressions involving the variable. Draw a diagram or create a table, where appropriate, to help convey this data.
>
> **Step 3:** *Reread* the problem, and then *write an equation*, using the items and variables listed in steps 1 and 2, that describes the situation.
>
> **Step 4:** *Solve the equation* found in step 3.
>
> **Step 5:** *Write* your *answer* using a complete sentence.
>
> **Step 6:** *Reread* the problem, and *check* your solution with the original words in the problem.

There are a number of substeps within each of the steps in our blueprint. For instance, with steps 1 and 2 it is always a good idea to draw a diagram or picture if it helps visualize the relationship between the items in the problem. In other cases a table helps organize the information. As you gain more experience using the blueprint to solve application problems, you will find additional techniques that expand the blueprint.

To help with problems of the type shown next in Examples 1 and 2, here are some common English words and phrases and their mathematical translations.

English	Algebra
The sum of a and b	$a + b$
The difference of a and b	$a - b$
The product of a and b	$a \cdot b$
The quotient of a and b	$\frac{a}{b}$
of	$\cdot$ (multiply)
is	$=$ (equals)
A number	x
4 more than x	$x + 4$
4 times x	$4x$
4 less than x	$x - 4$

A Number Problems

EXAMPLE 1   The sum of twice a number and three is seven. Find the number.

Solution   Using the Blueprint for Problem Solving as an outline, we solve the problem as follows:

Step 1: *Read* the problem, and then mentally *list* the items that are known and the items that are unknown.

Known items: The numbers 3 and 7

Unknown items: The number in question

Step 2: *Assign a variable* to one of the unknown items. Then *translate* the other *information* in the problem to expressions involving the variable.

Let $x =$ the number asked for in the problem, then "The sum of twice a number and three" translates to $2x + 3$.

Step 3: *Reread* the problem, and then *write an equation,* using the items and variables listed in steps 1 and 2, that describes the situation.
With all word problems, the word *is* translates to $=$.

The sum of twice x and 3 is 7.

$$2x + 3 = 7$$

Step 4: *Solve the equation* found in step 3.

$$2x + 3 = 7$$
$$2x + 3 + (-3) = 7 + (-3)$$
$$2x = 4$$
$$\frac{1}{2}(2x) = \frac{1}{2}(4)$$
$$x = 2$$

Practice Problems

1. The sum of twice a number and five is eleven. Find the number.

Note   Although many problems in this section may seem contrived, they provide excellent practice for our problem solving strategy. This practice will prove useful when we encounter more difficult application problems.

Answer
1. 3

Step 5: *Write* your *answer* using a complete sentence.

> The number is 2.

Step 6: *Reread* the problem, and *check* your solution with the original words in the problem.

> The sum of twice 2 and 3 is 7; a true statement.

You may find some examples and problems in this section that you can solve without using algebra or our blueprint. It is very important that you solve these problems using the methods we are showing here. The purpose behind these problems is to give you experience using the blueprint as a guide to solving problems written in words. Your answers are much less important than the work that you show to obtain your answer. You will be able to condense the steps in the blueprint later in the course. For now, though, you need to show your work in the same detail that we are showing in the examples in this section.

EXAMPLE 2   One number is three more than twice another; their sum is eighteen. Find the numbers.

Solution

Step 1: *Read and list.*
> *Known items*: Two numbers that add to 18. One is 3 more than twice the other.
> *Unknown items*: The numbers in question.

Step 2: *Assign a variable, and translate information.*
> Let x = the first number. The other is $2x + 3$.

Step 3: *Reread and write an equation.*

> Their sum   is 18.

$$x + (2x + 3) = 18$$

Step 4: *Solve the equation.*

$$x + (2x + 3) = 18$$
$$3x + 3 = 18$$
$$3x + 3 + (-3) = 18 + (-3)$$
$$3x = 15$$
$$x = 5$$

Step 5: *Write the answer.*
> The first number is 5. The other is $2 \cdot 5 + 3 = 13$.

Step 6: *Reread and check.*
> The sum of 5 and 13 is 18, and 13 is 3 more than twice 5.

B   Age Problems

Remember as you read through the solutions to the examples in this section, step 1 is done mentally. Read the problem, and then mentally list the items that you know and the items that you don't know. The purpose of step 1 is to give you direction as you begin to work application problems. Finding the solution to an application problem is a process; it doesn't happen all at once. The first step is to read the problem with a purpose in mind. That purpose is to mentally note the items that are known and the items that are unknown.

2. If three times the difference of a number and two were decreased by six, the result would be three. Find the number.

Note   Although these applications contain more than one variable quantity, we are choosing to write our equation in terms of a single variable to make the process of finding solutions easier.

3. Pete is five years older than Mary. Five years ago Pete was twice as old as Mary. Find their ages now.

EXAMPLE 3   Bill is 6 years older than Tom. Three years ago Bill's age was four times Tom's age. Find the age of each boy now.

Solution   Applying the Blueprint for Problem Solving, we have

Step 1: *Read and list.*

Known items: Bill is 6 years older than Tom. Three years ago Bill's age was four times Tom's age.

Unknown items: Bill's age and Tom's age

Step 2: *Assign a variable, and translate information.*

Let T = Tom's age now. That makes Bill $T + 6$ years old now. A table like the one shown here can help organize the information in an age problem. Notice how we placed the T in the box that corresponds to Tom's age now.

	Three Years Ago	Now
Bill		$T + 6$
Tom		T

If Tom is T years old now, 3 years ago he was $T - 3$ years old. If Bill is $T + 6$ years old now, 3 years ago he was $T + 6 - 3 = T + 3$ years old. We use this information to fill in the remaining spaces in the table.

	Three Years Ago	Now
Bill	$T + 3$	$T + 6$
Tom	$T - 3$	T

Step 3: *Reread and write an equation.*

Reading the problem again, we see that 3 years ago Bill's age was four times Tom's age. Writing this as an equation, we have

Bill's age 3 years ago = 4 · (Tom's age 3 years ago):

$$T + 3 = 4(T - 3)$$

Step 4: *Solve the equation.*

$$T + 3 = 4(T - 3)$$
$$T + 3 = 4T - 12$$
$$T + (-x) + 3 = 4T + (-T) - 12$$
$$3 = 3T - 12$$
$$3 + 12 = 3T - 12 + 12$$
$$15 = 3T$$
$$T = 5$$

Step 5: *Write the answer.*

Tom is 5 years old. Bill is 11 years old.

Step 6: *Reread and check.*

If Tom is 5 and Bill is 11, then Bill is 6 years older than Tom. Three years ago Tom was 2 and Bill was 8. At that time, Bill's age was four times Tom's age. As you can see, the answers check with the original problem.

Answers

3. Mary is 10; Pete is 15.

C Perimeter Problems

To understand Example 4 completely, you need to recall that the perimeter of a rectangle is the sum of the lengths of the sides. The formula for the perimeter is $P = 2l + 2w$.

EXAMPLE 4   The length of a rectangle is 5 inches more than twice the width. The perimeter is 34 inches. Find the length and width.

Solution   When working problems that involve geometric figures, a sketch of the figure helps organize and visualize the problem.

Step 1: *Read and list.*
   Known items: The figure is a rectangle. The length is 5 inches more than twice the width. The perimeter is 34 inches.
   Unknown items: The length and the width

Step 2: *Assign a variable, and translate information.*
   Because the length is given in terms of the width (the length is 5 more than twice the width), we let $x = $ the width of the rectangle. The length is 5 more than twice the width, so it must be $2x + 5$. The diagram below is a visual description of the relationships we have listed so far.

$$2x + 5$$

Step 3: *Reread and write an equation.*
   The equation that describes the situation is

   Twice the length + twice the width is the perimeter

$$2(2x + 5) \quad + \quad 2x \quad = \quad 34$$

Step 4: *Solve the equation.*

$2(2x + 5) + 2x = 34$	Original equation
$4x + 10 + 2x = 34$	Distributive property
$6x + 10 = 34$	Add $4x$ and $2x$.
$6x = 24$	Add -10 to each side.
$x = 4$	Divide each side by 6.

Step 5: *Write the answer.*
   The width x is 4 inches. The length is $2x + 5 = 2(4) + 5 = 13$ inches.

Step 6: *Reread and check.*
   If the length is 13 and the width is 4, then the perimeter must be $2(13) + 2(4) = 26 + 8 = 34$, which checks with the original problem.

D Coin Problems

EXAMPLE 5   Jennifer has $2.45 in dimes and nickels. If she has 8 more dimes than nickels, how many of each coin does she have?

Solution

Step 1: *Read and list.*
   Known items: The type of coins, the total value of the coins, and that there are 8 more dimes than nickels.
   Unknown items: The number of nickels and the number of dimes

4. The length of a rectangle is three more than twice the width. The perimeter is 36 inches. Find the length and the width.

5. Amy has $1.75 in dimes and quarters. If she has 7 more dimes than quarters, how many of each coin does she have?

Answers

4. Width is 5; length is 13
5. 3 quarters, 10 dimes

Step 2: *Assign a variable, and translate information.*

If we let N = the number of nickels, then $N + 8$ = the number of dimes. Because the value of each nickel is 5 cents, the amount of money in nickels is $5N$. Similarly, because each dime is worth 10 cents, the amount of money in dimes is $10(N + 8)$. Here is a table that summarizes the information we have so far:

	Nickels	Dimes
Number	N	$N + 8$
Value (in cents)	$5N$	$10(N + 8)$

Step 3: *Reread and write an equation.*

Because the total value of all the coins is 245 cents, the equation that describes this situation is

Amount of money in nickels		Amount of money in dimes		Total amount of money
$5N$	$+$	$10(N + 8)$	$=$	245

Step 4: *Solve the equation.*

To solve the equation, we apply the distributive property first.

$5N + 10N + 80 = 245$	*Distributive property*
$15N + 80 = 245$	*Add $5N$ and $10N$.*
$15N = 165$	*Add -80 to each side.*
$N = 11$	*Divide each side by 15.*

Step 5: *Write the answer.*

The number of nickels is $N = 11$.
The number of dimes is $N + 8 = 11 + 8 = 19$.

Step 6: *Reread and check.*

To check our results

$$11 \text{ nickels are worth } 5(11) = 55 \text{ cents}$$
$$19 \text{ dimes are worth } 10(19) = 190 \text{ cents}$$

The total value is 245 cents = $2.45

When you begin working the problems in the problem set that follows, there are a few things to remember. The first is that you may have to read the problems a number of times before you begin to see how to solve them. The second thing to remember is that word problems are not always solved correctly the first time you try them. Sometimes it takes a few attempts and some wrong answers before you can set up and solve these problems correctly. Don't give up.

GETTING READY FOR CLASS

After reading through the preceding section, respond in your own words and in complete sentences.

A. Why is the first step in the Blueprint for Problem Solving done mentally?

B. What is the last thing you do when solving an application problem?

C. Why is it important to still use the blueprint and show your work even if you can solve a problem without using algebra?

D. Write an application problem whose solution depends on solving the equation $4x + 2 = 10$.

Vocabulary Review

The following is a list of steps for the Blueprint for Problem Solving. Choose the correct words to fill in the blanks below.

equation   variable   known   sentence   unknown   problem

Step 1: Read the problem and mentally list the _____ and unknown items.

Step 2: Assign a variable to one of the _____ items, and then translate the rest of the information into expressions involving the variable.

Step 3: Reread the problem and write an equation using the_____ from step 2.

Step 4: Solve the _____.

Step 5: Write your answer using a complete _____.

Step 6: Reread the _____ and check your solution.

Problems

Solve the following word problems. Follow the steps given in the Blueprint for Problem Solving.

A Number Problems

1. The sum of a number and five is thirteen. Find the number.

2. The difference of ten and a number is negative eight. Find the number.

3. The sum of twice a number and four is fourteen. Find the number.

4. The difference of four times a number and eight is sixteen. Find the number.

5. Five times the sum of a number and seven is thirty. Find the number.

6. Five times the difference of twice a number and six is negative twenty. Find the number.

7. One number is two more than another. Their sum is eight. Find both numbers.

8. One number is three less than another. Their sum is fifteen. Find the numbers.

9. One number is four less than three times another. If their sum is increased by five, the result is twenty-five. Find the numbers.

10. One number is five more than twice another. If their sum is decreased by ten, the result is twenty-two. Find the numbers.

B Age Problems

11. Shelly is 3 years older than Michele. Four years ago the sum of their ages was 67. Find the age of each person now.

	Four Years Ago	Now
Shelly	$x - 1$	$x + 3$
Michele	$x - 4$	x

12. Cary is 9 years older than Dan. In 7 years the sum of their ages will be 93. Find the age of each man now. (Begin by filling in the table.)

	Now	In Seven Years
Cary	$x + 9$	
Dan	x	$x + 7$

13. Cody is twice as old as Evan. Three years ago the sum of their ages was 27. Find the age of each boy now.

	Three Years Ago	Now
Cody		
Evan	$x - 3$	x

14. Justin is 2 years older than Ethan. In 9 years the sum of their ages will be 30. Find the age of each boy now.

	Now	In Nine Years
Justin		
Ethan	x	

15. Fred is 4 years older than Barney. Five years ago the sum of their ages was 48. How old are they now?

	Five Years Ago	Now
Fred		
Barney		x

16. Tim is 5 years older than JoAnn. Six years from now the sum of their ages will be 79. How old are they now?

	Now	Six Years From Now
Tim		
JoAnn	x	

17. Jack is twice as old as Lacy. In 3 years the sum of their ages will be 54. How old are they now?

18. John is 4 times as old as Martha. Five years ago the sum of their ages was 50. How old are they now?

19. Pat is 20 years older than his son Patrick. In 2 years Pat will be twice as old as Patrick. How old are they now?

20. Diane is 23 years older than her daughter Amy. In 6 years Diane will be twice as old as Amy. How old are they now?

C Perimeter Problems

21. The perimeter of a square is 36 inches. Find the length of one side.

22. The perimeter of a square is 44 centimeters. Find the length of one side.

23. The perimeter of a square is 60 feet. Find the length of one side.

24. The perimeter of a square is 84 meters. Find the length of one side.

25. One side of a triangle is three times the shortest side. The third side is 7 feet more than the shortest side. The perimeter is 62 feet. Find all three sides.

26. One side of a triangle is half the longest side. The third side is 10 meters less than the longest side. The perimeter is 45 meters. Find all three sides.

27. One side of a triangle is half the longest side. The third side is 12 feet less than the longest side. The perimeter is 53 feet. Find all three sides.

28. One side of a triangle is 6 meters more than twice the shortest side. The third side is 9 meters more than the shortest side. The perimeter is 75 meters. Find all three sides.

29. The length of a rectangle is 5 inches more than the width. The perimeter is 34 inches. Find the length and width.

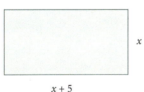

x

$x + 5$

30. The width of a rectangle is 3 feet less than the length. The perimeter is 10 feet. Find the length and width.

31. The length of a rectangle is 7 meters more than twice the width. The perimeter is 68 meters. Find the length and width.

32. The length of a rectangle is 4 inches more than three times the width. The perimeter is 72 inches. Find the length and width.

33. The length of a rectangle is 6 feet more than three times the width. The perimeter is 36 feet. Find the length and width.

34. The length of a rectangle is 3 centimeters less than twice the width. The perimeter is 54 centimeters . Find the length and width.

D Coin Problems

35. Marissa has $4.40 in quarters and dimes. If she has 5 more quarters than dimes, how many of each coin does she have?

	Dimes	Quarters
Number	x	$x + 5$
Value (in cents)	$10(x)$	$25(x + 5)$

36. Kendra has $2.75 in dimes and nickels. If she has twice as many dimes as nickels, how many of each coin does she have? (Completing the table may help you get started.)

	Nickels	Dimes
Number	x	$2x$
Value (in cents)	$5(x)$	

37. Tanner has $4.35 in nickels and quarters. If he has 15 more nickels than quarters, how many of each coin does he have?

	Nickels	Quarters
Number	$x + 15$	x
Value (in cents)		

38. Connor has $9.00 in dimes and quarters. If he has twice as many quarters as dimes, how many of each coin does he have?

	Dimes	Quarters
Number	x	$2x$
Value (in cents)		

39. Sue has $2.10 in dimes and nickels. If she has 9 more dimes than nickels, how many of each coin does she have?

	Dimes	Nickels
Number		x
Value (in cents)		

40. Mike has $1.55 in dimes and nickels. If he has 7 more nickels than dimes, how many of each coin does he have?

	Dimes	Nickels
Number	x	
Value (in cents)		

41. Katie has a collection of nickels, dimes, and quarters with a total value of $4.35. There are 3 more dimes than nickels and 5 more quarters than nickels. How many of each coin is in her collection? (*Hint:* Let N = the number of nickels.)

	Nickels	Dimes	Quarters
Number	N		
Value			

42. Mary Jo has $3.90 worth of nickels, dimes, and quarters. The number of nickels is 3 more than the number of dimes. The number of quarters is 7 more than the number of dimes. How many of each coin does she have? (*Hint:* Let D = the number of dimes.)

	Nickels	Dimes	Quarters
Number			
Value			

43. Corey has a collection of nickels, dimes, and quarters with a total value of $2.55. There are 6 more dimes than nickels and twice as many quarters as nickels. How many of each coin is in her collection?

	Nickels	Dimes	Quarters
Number	x		
Value			

44. Kelly has a collection of nickels, dimes, and quarters with a total value of $7.40. There are four more nickels than dimes and twice as many quarters as nickels. How many of each coin is in her collection?

	Nickels	Dimes	Quarters
Number			
Value			

Getting Ready for the Next Section

To understand all of the explanations and examples in the next section you must be able to work the problems below. Simplify the following expressions.

45. $x + 2x + 2x$

46. $x + 2x + 3x$

47. $x + 0.075x$

48. $x + 0.065x$

49. $0.09(x + 2{,}000)$

50. $0.06(x + 1{,}500)$

Find the Mistake

Each application problem below is followed by a corresponding equation. Write true or false if the equation accurately represents the problem. If false, write the correct equation on the line provided.

1. The sum of 7 and the quotient of m and n is 9 more than n.

Equation: $\dfrac{7}{(m + n)} = n + 9$ _____

2. Danny is 30 years older than his daughter Emma. Four years ago, Danny's age was six times Emma's age. Find Danny and Emma's age.

Equation: $x + 26 = 6(x - 4)$ _____

3. The width of a rectangle is 6 centimeters more than three times the length. The perimeter is 42 centimeters. Find the length and width.

Equation: $3x + 6 + x = 42$ _____

4. Amelie has $3.10 in nickels and dimes. If the number of dimes she has is 4 less than twice the number of nickels, how many of each coin does she have?

Equation: $2x - 4 + x = 310$ _____

OBJECTIVES

Apply the Blueprint for Problem Solving to solve:

A Consecutive Integer Problems

B Interest Problems

C Triangle Angle Problems

D Miscellaneous Problems

KEY WORDS

consecutive integers

vertices

Video Examples
Section 9.6

Note When assigning a variable or an expression to two consecutive integers, we could choose the approach shown here, or we could assign x to the larger integer and $x - 1$ to the smaller integer. Either case would yield the same answers. But for the sake of simplicity, we will continue to assign x to the first and smaller of the two consecutive integers.

©iStockphoto.com/polygrafix

In Las Vegas, the Stratosphere's hotel and casino has opened another attraction for thrill-seekers. SkyJump is a death-defying controlled free-fall from the 108th floor of the hotel to the ground. Jump Package 1 includes the jump cost plus a DVD of the jump for $114.99. A jump without the DVD costs $99.99. Suppose in one weekend, SkyJump had 15 jumpers, and receipts for that weekend's jumps totalled $1574.85. In this section, we will practice more application problems using linear equations, such as the one needed to determine how many jumpers purchased Jump Package 1 and how many paid just for the jump.

A Consecutive Integer Problems

Our first example involves consecutive integers. Here is the definition:

> **DEFINITION** consecutive integers
>
> When we ask for *consecutive integers*, we mean integers that are next to each other on the number line, like 5 and 6, or 13 and 14, or -4 and -3.

In the dictionary, consecutive is defined as following one another in uninterrupted order. If we ask for consecutive odd integers, then we mean odd integers that follow one another on the number line. For example, 3 and 5, 11 and 13, and -9 and -7 are consecutive odd integers. As you can see, to get from one odd integer to the next consecutive odd integer we add 2.

If we are asked to find two consecutive integers and we let x equal the first integer, the next one must be $x + 1$, because consecutive integers always differ by 1. Likewise, if we are asked to find two consecutive odd or even integers, and we let x equal the first integer, then the next one will be $x + 2$ because consecutive even or odd integers always differ by 2. Here is a table that summarizes this information.

In Words	Using Algebra	Example
Two consecutive integers	$x, x + 1$	The sum of two consecutive integers is 15. $x + (x + 1) = 15$ or $7 + 8 = 15$
Three consecutive integers	$x, x + 1, x + 2$	The sum of three consecutive integers is 24. $x + (x + 1) + (x + 2) = 24$ or $7 + 8 + 9 = 24$
Two consecutive odd integers	$x, x + 2$	The sum of two consecutive odd integers is 16. $x + (x + 2) = 16$ or $7 + 9 = 16$
Two consecutive even integers	$x, x + 2$	The sum of two consecutive even integers is 18. $x + (x + 2) = 18$ or $8 + 10 = 18$

EXAMPLE 1   The sum of two consecutive odd integers is 28. Find the two integers.

Solution

Step 1: *Read and list.*
   Known items: Two consecutive odd integers. Their sum is equal to 28.
   Unknown items: The numbers in question.

Step 2: *Assign a variable, and translate information.*
   If we let $x =$ the first of the two consecutive odd integers, then $x + 2$ is the next consecutive one.

Step 3: *Reread and write an equation.*
   Their sum is 28.

$$x + (x + 2) = 28$$

Step 4: *Solve the equation.*

   $2x + 2 = 28$ Simplify the left side.

   $2x = 26$ Add -2 to each side.

   $x = 13$ Multiply each side by $\frac{1}{2}$.

Step 5: *Write the answer.*
   The first of the two integers is 13. The second of the two integers will be two more than the first, which is 15.

Step 6: *Reread and check.*
   Suppose the first integer is 13. The next consecutive odd integer is 15. The sum of 15 and 13 is 28.

B Interest Problems

EXAMPLE 2   Suppose you invest a certain amount of money in an account that earns 8% in annual interest. At the same time, you invest $2,000 more than that in an account that pays 9% in annual interest. If the total interest from both accounts at the end of the year is $690, how much is invested in each account? (Hint: The formula for simple interest is $I = Prt$, where I is interest, P is principal amount invested, r is interest rate, and t is time in years.)

Solution

Step 1: *Read and list.*
   Known items: The interest rates, the total interest earned, and how much more is invested at 9%
   Unknown items: The amounts invested in each account.

Step 2: *Assign a variable, and translate information.*
   Let $P =$ the amount of money invested at 8%. From this, $P + 2,000 =$ the amount of money invested at 9%. The interest earned on P dollars invested at 8% is $0.08P$. The interest earned on $P + 2,000$ dollars invested at 9% is $0.09(P + 2,000)$.

   Here is a table that summarizes this information:

	Dollars invested at 8%	Dollars invested at 9%
Number of	P	$P + 2,000$
Interest on	$0.08P$	$0.09(P + 2,000)$

Practice Problems

1. The sum of two consecutive integers is 27. What are the two integers?

Note Now that you have worked through a number of application problems using our blueprint, you probably have noticed that step 3, in which we write an equation that describes the situation, is the key step. Anyone with experience solving application problems will tell you that there will be times when your first attempt at step 3 results in the wrong equation. Remember, mistakes are part of the process of learning to do things correctly.

2. On July 1, 1992, Alfredo opened a savings account that earned 4% in annual interest. At the same time, he put $3,000 more than he had in his savings into a mutual fund that paid 5% annual interest. One year later, the total interest from both accounts was $510. How much money did Alfredo put in his savings account?

Answers

   1. 13 and 14
   2. $4,000

Step 3: *Reread and write an equation.*

Because the total amount of interest earned from both accounts is $690, the equation that describes the situation is

Interest earned at 8%	+	Interest earned at 9%	=	Total interest earned
$0.08P$	+	$0.09(P + 2{,}000)$	=	690

Step 4: *Solve the equation.*

$$0.08P + 0.09(P + 2{,}000) = 690$$

$\qquad 0.08P + 0.09P + 180 = 690 \qquad$ *Distributive property*

$\qquad\qquad\quad\; 0.17P + 180 = 690 \qquad$ *Add $0.08P$ and $0.09P$.*

$\qquad\qquad\qquad\qquad 0.17P = 510 \qquad$ *Add -180 to each side.*

$\qquad\qquad\qquad\qquad\quad\; P = 3{,}000 \qquad$ *Divide each side by 0.17.*

Step 5: *Write the answer.*

The amount of money invested at 8% is $3,000, whereas the amount of money invested at 9% is $P + 2{,}000 = 3{,}000 + 2{,}000 = \$5{,}000$.

Step 6: *Reread and check.*

The interest at 8% is 8% of 3,000 = $0.08(3{,}000) = \$240$
The interest at 9% is 9% of 5,000 = $0.09(5{,}000) = \$450$

The total interest = $690

C Triangle Angle Problems

Labeling Triangles and the Sum of the Angles

One way to label the important parts of a triangle is to label the vertices with capital letters and the sides with small letters, as shown in Figure 1.

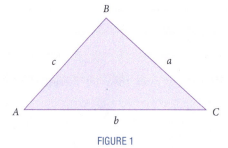

FIGURE 1

In Figure 1, notice that side *a* is opposite vertex *A*, side *b* is opposite vertex *B*, and side *c* is opposite vertex *C*. Also, because each vertex is the vertex of one of the angles of the triangle, we refer to the three interior angles as *A*, *B*, and *C*.

In any triangle, the sum of the interior angles is 180°. For the triangle shown in Figure 1, the relationship is written

$$A + B + C = 180°$$

EXAMPLE 3   The angles in a triangle are such that one angle is twice the smallest angle, whereas the third angle is three times as large as the smallest angle. Find the measure of all three angles.

Solution

Step 1: *Read and list.*

Known items: The sum of all three angles is 180°, one angle is twice the smallest angle, the largest angle is three times the smallest angle.

Unknown items: The measure of each angle

Step 2: *Assign a variable, and translate information.*

Let *A* be the smallest angle, then 2*A* will be the measure of another angle and 3*A* will be the measure of the largest angle.

Step 3: *Reread and write an equation.*

When working with geometric objects, drawing a generic diagram sometimes will help us visualize what it is that we are asked to find. In Figure 2, we draw a triangle with angles *A*, *B*, and *C*.

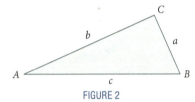

FIGURE 2

We can write the value of angle *B* in terms of *A* as *B* = 2*A*, and the value of angle *C* as *C* = 3*A*. We know that the sum of angles *A*, *B*, and *C* will be 180°, so our equation becomes

$$A + 2A + 3A = 180°$$

Step 4: *Solve the equation.*

$$A + 2A + 3A = 180°$$
$$6A = 180°$$
$$A = 30°$$

Step 5: *Write the answer.*

The smallest angle *A* measures 30°

Angle *B* measures 2*A*, or 2(30°) = 60°

Angle *C* measures 3*A*, or 3(30°) = 90°

Step 6: *Reread and check.*

The angles must add to 180°.
$$A + B + C = 180°$$
$$30° + 60° + 90° \stackrel{?}{=} 180°$$
$$180° = 180°   \textit{Our answers check.}$$

D Miscellaneous Problems

EXAMPLE 4   Tickets to a school play cost $6 for adults and $4 for children. A total of 200 tickets were sold, and $940 was collected in receipts. How many adult tickets and how many child tickets were sold?

Solution

Step 1: *Read and list.*

Known items: 200 tickets were sold, $940 was collected

Unknown items: Number of adult tickets, number of child tickets

3. The angles in a triangle are such that one angle is three times the smallest angle, whereas the largest angle is five times the smallest angle. Find the measure of all three angles.

4. Tickets to a community theater cost $10 for adults and $6 for children. A total of $680 was collected for one evening performance. If 20 more adults attended than children, how many children and how many adults attended the theater?

Answers

3. 20°, 60°, and 100°
4. 50 adults, 30 children

Step 2: *Assign a variable, and translate information.*

Let x be the number of adult tickets sold, then $200 - x$ will be the number of child tickets sold.

Step 3: *Reread and write an equation.*

$$\text{Price of adult} \times \text{\# Adult} + \text{Price of child} \times \text{\# Child} = \text{Total collected}$$
$$6 \cdot (x) \qquad + \qquad 4 \cdot (200 - x) \qquad = \qquad 940$$

Step 4: *Solve the equation.*

$$6(x) + 4(200 - x) = 940$$
$$6x + 800 - 4x = 940$$
$$2x + 800 = 940$$
$$2x = 140$$
$$x = 70$$

Step 5: *Write the answer.*

There were 70 adult tickets sold, and $200 - 70 = 130$ child tickets sold.

Step 6: *Reread and check.*

The total receipts must add to \$940.

$$6 \cdot (70) + 4 \cdot (130) \stackrel{?}{=} 940$$
$$420 + 520 \stackrel{?}{=} 940$$
$$\overline{940 = 940}$$

GETTING READY FOR CLASS

After reading through the preceding section, respond in your own words and in complete sentences.

A. Write an equation for the sum of two consecutive integers being 13.

B. Write an application problem whose solution depends on solving the equation $x + 0.075x = 500$.

C. How do you label triangles?

D. What rule is always true about the three angles in a triangle?

Vocabulary Review

Choose the correct words to fill in the blanks below.

even   sum   consecutive   integers

1. _____ integers are integers that are next to each other on the number line.

2. Two consecutive _____ integers would be given by x and $x + 2$.

3. Two consecutive _____ would be given by x and $x + 1$.

4. In any triangle, the _____ of the interior angles is 180°, and written by $A + B + C = 180°$.

A Consecutive Integer Problems

1. The sum of two consecutive integers is 11. Find the numbers.

2. The sum of two consecutive integers is 15. Find the numbers.

3. The sum of two consecutive integers is −9. Find the numbers.

4. The sum of two consecutive integers is −21. Find the numbers.

5. The sum of two consecutive odd integers is 28. Find the numbers.

6. The sum of two consecutive odd integers is 44. Find the numbers.

7. The sum of two consecutive even integers is 106. Find the numbers.

8. The sum of two consecutive even integers is 66. Find the numbers.

9. The sum of two consecutive even integers is −30. Find the numbers.

10. The sum of two consecutive odd integers is −76. Find the numbers.

11. The sum of three consecutive odd integers is 57. Find the numbers.

12. The sum of three consecutive odd integers is −51. Find the numbers.

13. The sum of three consecutive even integers is 132. Find the numbers.

14. The sum of three consecutive even integers is −108. Find the numbers.

B Interest Problems

15. Suppose you invest money in two accounts. One of the accounts pays 8% annual interest, whereas the other pays 9% annual interest. If you have $2,000 more invested at 9% than you have invested at 8%, how much do you have invested in each account if the total amount of interest you earn in a year is $860? (Begin by completing the following table.)

	Dollars invested at 8%	Dollars invested at 9%
Number of	x	
Interest on		

16. Suppose you invest a certain amount of money in an account that pays 11% interest annually, and $4,000 more than that in an account that pays 12% annually. How much money do you have in each account if the total interest for a year is $940?

	Dollars invested at 11%	Dollars invested at 12%
Number of	x	
Interest on		

17. Tyler has two savings accounts that his grandparents opened for him. The two accounts pay 10% and 12% in annual interest; there is $500 more in the account that pays 12% than there is in the other account. If the total interest for a year is $214, how much money does he have in each account?

18. Travis has a savings account that his parents opened for him. It pays 6% annual interest. His uncle also opened an account for him, but it pays 8% annual interest. If there is $800 more in the account that pays 6%, and the total interest from both accounts is $104, how much money is in each of the accounts?

19. A stockbroker has money in three accounts. The interest rates on the three accounts are 8%, 9%, and 10%. If she has twice as much money invested at 9% as she has invested at 8%, three times as much at 10% as she has at 8%, and the total interest for the year is $280, how much is invested at each rate? (*Hint:* Let x = the amount invested at 8%.)

20. An accountant has money in three accounts that pay 9%, 10%, and 11% in annual interest. He has twice as much invested at 9% as he does at 10% and three times as much invested at 11% as he does at 10%. If the total interest from the three accounts is $610 for the year, how much is invested at each rate? (*Hint:* Let x = the amount invested at 10%.)

C Triangle Angle Problems

21. Two angles in a triangle are equal and their sum is equal to the third angle in the triangle. What are the measures of each of the three interior angles?

22. One angle in a triangle measures twice the smallest angle, whereas the largest angle is six times the smallest angle. Find the measures of all three angles.

23. The smallest angle in a triangle is $\frac{1}{5}$ as large as the largest angle. The third angle is twice the smallest angle. Find the three angles.

24. One angle in a triangle is half the largest angle but three times the smallest. Find all three angles.

25. A right triangle has one 37° angle. Find the other two angles.

26. In a right triangle, one of the acute angles is twice as large as the other acute angle. Find the measure of the two acute angles.

27. One angle of a triangle measures 20° more than the smallest, while a third angle is twice the smallest. Find the measure of each angle.

28. One angle of a triangle measures 50° more than the smallest, while a third angle is three times the smallest. Find the measure of each angle.

D Miscellaneous Problems

29. Ticket Prices Miguel is selling tickets to a barbecue. Adult tickets cost $6.00 and children's tickets cost $4.00. He sells six more children's tickets than adult tickets. The total amount of money he collects is $184. How many adult tickets and how many children's tickets did he sell?

	Adult	Child
Number	x	$x + 6$
Income	$6(x)$	$4(x + 6)$

30. Working Two Jobs Maggie has a job working in an office for $10 an hour and another job driving a tractor for $12 an hour. One week she works in the office twice as long as she drives the tractor. Her total income for that week is $416. How many hours did she spend at each job?

Job	Office	Tractor
Hours Worked	$2x$	x
Wages Earned	$10(2x)$	$12x$

31. Phone Bill The cost of a long-distance phone call is $0.41 for the first minute and $0.32 for each additional minute. If the total charge for a long-distance call is $5.21, how many minutes was the call?

32. Phone Bill Danny, who is 1 year old, is playing with the telephone when he accidentally presses one of the buttons his mother has programmed to dial her friend Sue's number. Sue answers the phone and realizes Danny is on the other end. She talks to Danny, trying to get him to hang up. The cost for a call is $0.23 for the first minute and $0.14 for every minute after that. If the total charge for the call is $3.73, how long did it take Sue to convince Danny to hang up the phone?

33. Hourly Wages JoAnn works in the publicity office at the state university. She is paid $12 an hour for the first 35 hours she works each week and $18 an hour for every hour after that. If she makes $492 one week, how many hours did she work?

34. Hourly Wages Diane has a part-time job that pays her $8.50 an hour. During one week she works 26 hours and is paid $230.10. She realizes when she sees her check that she has been given a raise. How much per hour is that raise?

35. Office Numbers Professors Wong and Gil have offices in the mathematics building at Miami Dade College. Their office numbers are consecutive odd integers with a sum of 14,660. What are the office numbers of these two professors?

36. Cell Phone Numbers Diana and Tom buy two cell phones. The phone numbers assigned to each are consecutive integers with a sum of 11,109,295. If the smaller number is Diana's, what are their phone numbers?

37. Age Marissa and Kendra are 2 years apart. Their ages are two consecutive even integers. Kendra is the younger of the two. If Marissa's age is added to twice Kendra's age, the result is 26. How old is each girl?

38. Age Justin's and Ethan's ages form two consecutive odd integers. What is the difference of their ages?

39. Arrival Time Jeff and Carla Cole are driving separately from San Luis Obispo, California, to the north shore of Lake Tahoe, a distance of 425 miles. Jeff leaves San Luis Obispo at 11:00 a.m. and averages 55 miles per hour on the drive, Carla leaves later, at 1:00 p.m. but averages 65 miles per hour. Which person arrives in Lake Tahoe first?

40. Piano Lessons Tyler's parents pay him $0.50 to do the laundry and $1.25 to mow the lawn. In one month, he does the laundry 6 more times than he mows the lawn. If his parents pay him $13.50 that month, how many times did he mow the lawn?

At one time, the Texas Junior College Teachers Association annual conference was held in Austin. At that time a taxi ride in Austin was $1.25 for the first $\frac{1}{5}$ of a mile and $0.25 for each additional $\frac{1}{5}$ of a mile. Use this information for Problems 41 and 42.

41. Cost of a Taxi Ride If the distance from one of the convention hotels to the airport is 7.5 miles, how much will it cost to take a taxi from that hotel to the airport?

42. Cost of a Taxi Ride Suppose the distance from one of the hotels to one of the western dance clubs in Austin is 12.4 miles. If the fare meter in the taxi gives the charge for that trip as $16.50, is the meter working correctly?

43. Geometry The length and width of a rectangle are consecutive even integers. The perimeter is 44 meters. Find the length and width.

44. Geometry The length and width of a rectangle are consecutive odd integers. The perimeter is 128 meters. Find the length and width.

45. Geometry The angles of a triangle are three consecutive integers. Find the measure of each angle.

46. Geometry The angles of a triangle are three consecutive even integers. Find the measure of each angle.

47. Boards A 78-centimeter-long board is cut into 3 pieces. The longest piece is twice as long as the shortest piece. The other piece is 6 cm longer than the shortest piece. How long is each piece?

48. Rope A 52-meter-long rope is cut into three segments. The shortest piece is 6 meters shorter than the next piece, which is 13 meters shorter than the longest piece. How long is each piece?

Ike and Nancy give western dance lessons at the Elk's Lodge on Sunday nights. The lessons cost $3.00 for members of the lodge and $5.00 for nonmembers. Half of the money collected for the lesson is paid to Ike and Nancy. The Elk's Lodge keeps the other half. One Sunday night Ike counts 36 people in the dance lesson. Use this information to work Problems 47 through 50.

49. Dance Lessons What is the least amount of money Ike and Nancy will make?

50. Dance Lessons What is the largest amount of money Ike and Nancy will make?

51. Dance Lessons At the end of the evening, the Elk's Lodge gives Ike and Nancy a check for $80 to cover half of the receipts. Can this amount be correct?

52. Dance Lessons Besides the number of people in the dance lesson, what additional information does Ike need to know to always be sure he is being paid the correct amount?

53. Solve the problem given at the introduction to this section to find how many jumpers purchased package 1.

54. Solve the same problem as in problem 51, but this time assume Sky Jump had 30 jumpers with total receipts of $3,269.70.

Getting Ready for the Next Section

To understand all the explanations and examples in the next section you must be able to work the problems below.

Solve the following equations.

55. a. $x - 3 = 6$

b. $x + 3 = 6$

c. $-x - 3 = 6$

d. $-x + 3 = 6$

56. a. $x - 7 = 16$

b. $x + 7 = 16$

c. $-x - 7 = 16$

d. $-x + 7 = 16$

57. a. $\frac{x}{4} = -2$

 b. $-\frac{x}{4} = -2$

 c. $\frac{x}{4} = 2$

 d. $-\frac{x}{4} = 2$

58. a. $3a = 15$

 b. $3a = -15$

 c. $-3a = 15$

 d. $-3a = -15$

59. $2.5x - 3.48 = 4.9x + 2.07$

60. $2(1 - 3x) + 4 = 4x - 14$

61. $3(x - 4) = -2$

62. $-2(x - 3) = 8$

63. Solve $2x - 3y = 6$ for y.

64. Solve $3x + 2y = 7$ for y.

65. Solve $3x - 2y = 5$ for x.

66. Solve $2x - 5y = -10$ for x.

Find the Mistake

Each application problem below is followed by a corresponding equation. Write true or false if the equation accurately represents the problem. If false, write the correct equation on the line provided. (Hint: Some equations may appear in a simplified form.)

1. The sum of three consecutive integers divided by 2 is 36.

 Equation: $\frac{(3x + 3)}{2} = 36$ _____

2. You have money in two savings accounts that pay 7% and 8% in annual interest. If you have four times as much money in the account that pays 7%, and the total interest of both accounts is $396, how much money is in each account?

 Equation: $0.07x + 0.08(4x) = 396$ _____

3. One angle of a triangle measures 4 degrees more than the smallest angle. The largest angle is 16 degrees less than twice the smallest angle. Find the measure of each angle.

 Equation: $4x - 12 = 180$ _____

4. Tickets for a dinner fundraiser cost $15 per plate with dessert and $12 per plate without dessert. 150 tickets were sold for a total of $2070. How many of each ticket was sold?

 Equation: $15x + 12(150 - x) = 2070$ _____

OBJECTIVES

A Use the addition property for inequalities to solve an inequality and graph the solution set.

B Use the multiplication property for inequalities to solve inequalities.

C Use both the addition and multiplication properties to solve and graph inequalities.

D Solve and graph compound inequalities.

E Translate and solve application problems involving inequalities.

KEY WORDS

linear inequality

addition property for inequalities

multiplication property for inequalities

compound inequality

continued inequality

double inequality

Video Examples
Section 9.7

9.7 Linear Inequalities in One Variable

Image © sxc.hu/ckforjc, 2004

Suppose we know that the state of New Mexico has no more than 10 organic stores. We can write the following inequality to represent this information:

$$x \leq 10$$

If we know for certain of three stores in New Mexico, we can add to the original inequality and say $x + 3 \leq 10$. Up to now we have solved equalities. For example, when we say $x + 3 = 10$, we know x must be 7 since the statement is only true when $x = 7$. But if we say that $x + 3 \leq 10$, then there are many numbers that make this statement true. If x is 3, $3 + 3$ is less than 10. If x is 2.5, $2.5 + 3$ is less than 10. If x is any real number less than or equal to 7, then the statement $x + 3 \leq 10$ is true.

We also know that any number greater than 7 mark the statement false. If x is 8, then $x + 3$ is not less than 10.

When we solve an inequality, we look for all possible solutions—and there will usually be many of them. Any number that makes an inequality true is a *solution* to that inequality. All the numbers that make the inequality true are called the *solution set*.

We can express a solution set symbolically. If the solution set to an inequality is all the real numbers less than 7, we say the solution is $x < 7$. If the solution set to this inequality is all numbers less than or equal to 7, we say the solution is $x \leq 7$.

We can also express the solution set graphically using the number line. When a solution set is all numbers less than a certain number, we make that number with an open circle.

If we want to include the number (7, in this case) in the solution set, we mark that number with a closed circle.

Linear inequalities are solved by a method similar to the one used in solving linear equations. The only real differences between the methods are in the multiplication property for inequalities and in graphing the solution set.

A Addition Property for Inequalities

An inequality differs from an equation only with respect to the comparison symbol between the two quantities being compared. In place of the equal sign, we use < (less than), ≤ (less than or equal to), > (greater than), or ≥ (greater than or equal to). The addition property for inequalities is almost identical to the addition property for equality.

> **PROPERTY** Addition Property for Inequalities
>
> For any three algebraic expressions A, B, and C,
>
> if $A < B$
> then $A + C < B + C$
>
> *In words*: Adding the same quantity to both sides of an inequality will not change the solution set.

It makes no difference which inequality symbol we use to state the property. Adding the same amount to both sides always produces an inequality equivalent to the original inequality. Also, because subtraction can be thought of as addition of the opposite, this property holds for subtraction as well as addition.

EXAMPLE 1   Solve $x + 5 < 7$ and graph the solution.

Solution   To isolate x, we add -5 to both sides of the inequality

$$x + 5 < 7$$
$$x + 5 + (-5) < 7 + (-5) \qquad \text{Addition property for inequalities}$$
$$x < 2$$

We can go one step further here and graph the solution set. The solution set is all real numbers less than 2. To graph this set, we simply draw a straight line and label the center 0 (zero) for reference. Then we label the 2 on the right side of zero and extend an arrow beginning at 2 and pointing to the left. We use an open circle at 2 because it is not included in the solution set. Here is the graph.

EXAMPLE 2   Solve $x - 6 \le -3$ and graph the solution.

Solution   Adding 6 to each side will isolate x on the left side

$$x - 6 \le -3$$
$$x - 6 + 6 \le -3 + 6 \qquad \text{Add 6 to both sides.}$$
$$x \le 3$$

The graph of the solution set is

Notice that the dot at the 3 in Example 2 is darkened because 3 is included in the solution set. In this book we will use open circles on the graphs of solution sets with $<$ or $>$ and closed (darkened) circles on the graphs of solution sets with $\le$ or $\ge$.

B Multiplication Property for Inequalities

To see the idea behind the multiplication property for inequalities, we will consider three true inequality statements and explore what happens when we multiply both sides by a positive number and then what happens when we multiply by a negative number.

Consider the following three true statements:

$$3 < 5 \qquad -3 < 5 \qquad -5 < -3$$

Now multiply both sides by the positive number 4.

$$4(3) < 4(5) \qquad 4(-3) < 4(5) \qquad 4(-5) < 4(-3)$$
$$12 < 20 \qquad\quad -12 < 20 \qquad\quad -20 < -12$$

Practice Problems

1. Solve $x + 3 < 5$ and graph the solution.

2. Solve $x - 4 \le 1$ and graph the solution.

Answers
1. $x < 2$
2. $x \le 5$

Note   This discussion is intended to show why the multiplication property for inequalities is written the way it is. You may want to look ahead to the property itself and then come back to this discussion if you are having trouble making sense out of it.

In each case, the inequality symbol in the result points in the same direction it did in the original inequality. We say the "sense" of the inequality doesn't change when we multiply both sides by a positive quantity.

Notice what happens when we go through the same process but multiply both sides by -4 instead of 4:

$$3 < 5 \qquad\qquad -3 < 5 \qquad\qquad -5 < -3$$
$$-4(3) > -4(5) \qquad -4(-3) > -4(5) \qquad -4(-5) > -4(-3)$$
$$-12 > -20 \qquad\qquad 12 > -20 \qquad\qquad 20 > 12$$

In each case, we have to change the direction in which the inequality symbol points to keep each statement true. Multiplying both sides of an inequality by a negative quantity always reverses the sense of the inequality. Our results are summarized in the multiplication property for inequalities.

> **PROPERTY**   Multiplication Property for Inequalities
>
> For any three algebraic expressions A, B, and C,
>
> | if | $A < B$ | |
> |----|---------|---|
> | then | $AC < BC$ | when C is positive |
> | and | $AC > BC$ | when C is negative |
>
> *In words:* Multiplying both sides of an inequality by a positive number does not change the solution set. When multiplying both sides of an inequality by a negative number, it is necessary to reverse the inequality symbol to produce an equivalent inequality.

Note   Because division is defined in terms of multiplication, this property is also true for division. We can divide both sides of an inequality by any nonzero number we choose. If that number happens to be negative, we must also reverse the direction of the inequality symbol.

We can multiply both sides of an inequality by any nonzero number we choose. If that number happens to be negative, we must also reverse the sense of the inequality.

3. Solve $4a < 12$ and graph the solution.

EXAMPLE 3   Solve $3a < 15$ and graph the solution.

Solution   We begin by multiplying each side by $\frac{1}{3}$. Because $\frac{1}{3}$ is a positive number, we do not reverse the direction of the inequality symbol.

$$3a < 15$$
$$\frac{1}{3}(3a) < \frac{1}{3}(15) \qquad \text{Multiply each side by } \tfrac{1}{3}.$$
$$a < 5$$

4. Solve $-2x \le 10$ and graph the solution.

EXAMPLE 4   Solve $-3a \le 18$ and graph the solution.

Solution   We begin by multiplying both sides by $-\frac{1}{3}$. Because $-\frac{1}{3}$ is a negative number, we must reverse the direction of the inequality symbol at the same time that we multiply by $-\frac{1}{3}$.

$$-3a \le 18$$
$$-\frac{1}{3}(-3a) \ge -\frac{1}{3}(18) \qquad \text{Multiply both sides by } -\tfrac{1}{3} \text{ and reverse the direction of the inequality symbol.}$$
$$a \ge -6$$

Answers

3. $a < 3$

4. $x \ge -5$

EXAMPLE 5 Solve $-\dfrac{x}{4} > 2$ and graph the solution.

Solution To isolate x, we multiply each side by -4. Because -4 is a negative number, we also must reverse the direction of the inequality symbol.

$$-\dfrac{x}{4} > 2$$

$$-4\left(-\dfrac{x}{4}\right) < -4(2) \qquad \text{Multiply each side by } -4, \text{ and reverse the direction of the inequality symbol.}$$

$$x < -8$$

C Solving Linear Inequalities in One Variable

To solve more complicated inequalities, we use the following steps:

> **HOW TO** Solving Linear Inequalities in One Variable
>
> **Step 1a:** Use the distributive property to separate terms, if necessary.
>
> **1b:** If fractions are present, consider multiplying both sides by the LCD to eliminate the fractions. If decimals are present, consider multiplying both sides by a power of 10 to clear the inequality of decimals.
>
> **1c:** Combine similar terms on each side of the inequality.
>
> **Step 2:** Use the addition property for inequalities to get all variable terms on one side of the inequality and all constant terms on the other side.
>
> **Step 3:** Use the multiplication property for inequalities to get x by itself on one side of the inequality.
>
> **Step 4:** Graph the solution set.

EXAMPLE 6 Solve and graph: $2.5x - 3.48 < -4.9x + 2.07$.

Solution We have two methods we can use to solve this inequality. We can simply apply our properties to the inequality the way it is currently written and work with the decimal numbers, or we can eliminate the decimals to begin with and solve the resulting inequality.

Method 1 Working with the decimals.

$$2.5x - 3.48 < -4.9x + 2.07 \qquad \text{Original inequality}$$

$$2.5x + 4.9x - 3.48 < -4.9x + 4.9x + 2.07 \qquad \text{Add } 4.9x \text{ to each side.}$$

$$7.4x - 3.48 < 2.07$$

$$7.4x - 3.48 + 3.48 < 2.07 + 3.48 \qquad \text{Add } 3.48 \text{ to each side.}$$

$$7.4x < 5.55$$

$$\dfrac{7.4x}{7.4} < \dfrac{5.55}{7.4} \qquad \text{Divide each side by } 7.4.$$

$$x < 0.75$$

5. Solve $-\dfrac{x}{3} > 2$ and graph the solution.

$\longleftarrow\!\!\!\longrightarrow$

6. Solve and graph: $4.5x + 2.31 > 6.3x - 4.89$.

$\longleftarrow\!\!\!\longrightarrow$

Answers
5. $x < -6$
6. $x < 4$

Method 2 Eliminating the decimals in the beginning.

Because the greatest number of places to the right of the decimal point in any of the numbers is 2, we can multiply each side of the inequality by 100 and we will be left with an equivalent inequality that contains only whole numbers.

$$2.5x - 3.48 < -4.9x + 2.07 \qquad \text{Original inequality}$$

$$100(2.5x - 3.48) < 100(-4.9x + 2.07) \qquad \text{Multiply each side by 100.}$$

$$100(2.5x) - 100(3.48) < 100(-4.9x) + 100(2.07) \qquad \text{Distributive property}$$

$$250x - 348 < -490x + 207 \qquad \text{Multiply.}$$

$$740x - 348 < 207 \qquad \text{Add 490x to each side.}$$

$$740x < 555 \qquad \text{Add 348 to each side.}$$

$$\frac{740x}{740} < \frac{555}{740} \qquad \text{Divide each side by 740.}$$

$$x < 0.75$$

The solution by either method is $x < 0.75$. Here is the graph:

7. Solve and graph
$2(x - 5) \geq -6$.

EXAMPLE 7   Solve and graph: $3(x - 4) \geq -2$.

Solution
$$3x - 12 \geq -2 \qquad \text{Distributive property}$$

$$3x - 12 + 12 \geq -2 + 12 \qquad \text{Add 12 to both sides.}$$

$$3x \geq 10$$

$$\frac{1}{3}(3x) \geq \frac{1}{3}(10) \qquad \text{Multiply both sides by } \frac{1}{3}.$$

$$x \geq \frac{10}{3}$$

8. Solve and graph
$3(1 - 2x) + 5 < 3x - 1$.

EXAMPLE 8   Solve and graph: $2(1 - 3x) + 4 < 4x - 14$.

Solution
$$2 - 6x + 4 < 4x - 14 \qquad \text{Distributive property}$$

$$-6x + 6 < 4x - 14 \qquad \text{Simplify.}$$

$$-6x + 6 + (-6) < 4x - 14 + (-6) \qquad \text{Add } -6 \text{ to both sides.}$$

$$-6x < 4x - 20$$

$$-6x + (-4x) < 4x + (-4x) - 20 \qquad \text{Add } -4x \text{ to both sides.}$$

$$-10x < -20$$

$$\left(-\frac{1}{10}\right)(-10x) > \left(-\frac{1}{10}\right)(-20) \qquad \text{Multiply by } -\frac{1}{10}, \text{ reverse the direction of the inequality.}$$

$$x > 2$$

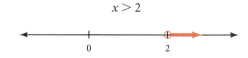

Answers

7. $x \geq 2$

8. $x > 1$

EXAMPLE 9   Solve $2x - 3y < 6$ for y.

Solution   We can solve this formula for y by first adding $-2x$ to each side and then multiplying each side by $-\frac{1}{3}$. When we multiply by $-\frac{1}{3}$ we must reverse the direction of the inequality symbol. Because this is a formula, we will not graph the solution.

$$2x - 3y < 6 \qquad \text{Original formula}$$
$$2x + (-2x) - 3y < (-2x) + 6 \qquad \text{Add } -2x \text{ to each side.}$$
$$-3y < -2x + 6$$
$$-\frac{1}{3}(-3y) > -\frac{1}{3}(-2x + 6) \qquad \text{Multiply each side by } -\frac{1}{3}.$$
$$y > \frac{2}{3}x - 2$$

9. Solve $2x - 3y < 6$ for x.

D Solving Compound Inequalities

Two inequalities considered together and connected by the words *and* or *or* are called *compound inequalities*.

EXAMPLE 10   Solve the compound inequality.

$$3t + 7 \le -4 \qquad \text{or} \qquad 3t + 7 \ge 4$$

Solution   We solve each half of the compound inequality separately, then we graph the solution set.

$$3t + 7 \le -4 \quad \text{or} \quad 3t + 7 \ge 4$$
$$3t \le -11 \quad \text{or} \quad 3t \ge -3 \qquad \text{Add } -7.$$
$$t \le -\frac{11}{3} \quad \text{or} \quad t \ge -1 \qquad \text{Multiply by } \frac{1}{3}.$$

The solution set can be graphed in the following way:

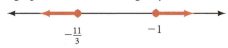

10. Solve and graph the solution for the compound inequality $3t - 6 \le -3$ or $3t - 6 \ge 3$.

Sometimes compound inequalities that use the word *and* as the connecting word can be written in a shorter form. For example, the compound inequality $-3 \le x$ and $x \le 4$ can be written $-3 \le x \le 4$. Inequalities of the form $-3 \le x \le 4$ are called *continued inequalities* or *double inequalities*. The graph of $-3 \le x \le 4$ is

EXAMPLE 11   Solve and graph $-3 \le 2x - 5 \le 3$.

Solution   We can extend our properties for addition and multiplication to cover this situation. If we add a number to the middle expression, we must add the same number to the outside expressions. We do the same for multiplication, remembering to reverse the direction of the inequality symbols if we multiply by a negative number.

$$-3 \le 2x - 5 \le 3$$
$$2 \le 2x \le 8 \qquad \text{Add 5 to all three members.}$$
$$1 \le x \le 4 \qquad \text{Multiply through by } \frac{1}{2}.$$

Here is how we graph this solution set:

11. Solve and graph $-7 \le 2x + 1 \le 7$.

Answers
9. $x < \frac{3}{2}y + 3$
10. $t \le -\frac{11}{3}$ or $t \ge -1$
11. $1 \le x \le 4$

E Applications

When working application problems that involve inequalities, the phrases "at least" and "at most" translate as follows:

In Words	In Symbols
x is at least 30	$x \geq 30$
x is at most 20	$x \leq 20$

Our next example is similar to an example done earlier in this chapter. This time it involves an inequality instead of an equation.

We can modify our Blueprint for Problem Solving to solve application problems whose solutions depend on writing and then solving inequalities.

EXAMPLE 12   The sum of two consecutive odd integers is at most 28. What are the possibilities for the first of the two integers?

Solution   When we use the phrase "their sum is at most 28," we mean that their sum is less than or equal to 28.

Step 1: *Read and list.*
  Known items: Two consecutive odd integers. Their sum is less than or equal to 28.
  Unknown items: The numbers in question.

Step 2: *Assign a variable, and translate information.*
  If we let $x =$ the first of the two consecutive odd integers, then $x + 2$ is the next consecutive one.

Step 3: *Reread and write an inequality.*
  Their sum is at most 28.
$$x + (x + 2) \leq 28$$

Step 4: *Solve the inequality.*

$$2x + 2 \leq 28 \qquad \text{Simplify the left side.}$$
$$2x \leq 26 \qquad \text{Add } -2 \text{ to each side.}$$
$$x \leq 13 \qquad \text{Multiply each side by } \tfrac{1}{2}.$$

Step 5: *Write the answer.*
  The first of the two integers must be an odd integer that is less than or equal to 13. The second of the two integers will be two more than whatever the first one is.

Step 6: *Reread and check.*
  Suppose the first integer is 13. The next consecutive odd integer is 15. The sum of 15 and 13 is 28. If the first odd integer is less than 13, the sum of it and the next consecutive odd integer will be less than 28.

EXAMPLE 13   Monica wants to invest money into an account that pays $6\frac{1}{4}$% annual interest. She wants to be able to count on at least $500 in interest at the end of the year. How much money should she invest to get $500 or more at the end of the year?

Solution   The phrase "or more" tells us this is an inequality problem. Once we find out the amount needed to get $500 interest, we can solve the problem and answer the question.

Step 1: *Read and list.*
  Known items: The interest rate if $6\frac{1}{4}$%. The minimum acceptable interest amount is $500.
  Unknown items: The investment amount.

12. The sum of two consecutive integers is at least 27. What are the possibilities for the first of the two integers?

13. Re-work Example 13 if she gets $700 or more in interest.

Step 2: *Assign a variable, and translate information.*

We will let $x =$ the amount to invest.

In words: $6\frac{1}{4}\%$ of x must be at least $500.

Step 3: *Reread and write an inequality.*

Our sentence from step 2 can be translated directly into the inequality.

$$0.0625x \geq 500$$

Step 4: *Solve the inequality.*

$$0.0625x \geq 500 \qquad \textcolor{green}{\text{Original inequality}}$$

$$x \geq 8{,}000 \qquad \textcolor{green}{\text{Divide both sides by 0.0625.}}$$

Step 5: *Write the answer.*

We now know that if Monica invests $8,000, her interest will be $500. Any more than $8,000 will give her more than $500 so the answer is that Monica must invest at least $8,000.

Step 6: *Reread and check.*

Put in any numbers greater than or equal to 8,000 in the inequality and check to make sure the result is at least 500.

$$x = 9{,}000$$

$$0.0625(9{,}000) = 562.5, \text{ which is } > \$500.$$

GETTING READY FOR CLASS

After reading through the preceding section, respond in your own words and in complete sentences.

A. State the addition property for inequalities.

B. When do you darken the circle on the graph of a solution set?

C. How is the multiplication property for inequalities different from the multiplication property of equality?

D. When do you reverse the direction of an inequality symbol?

EXERCISE SET 9.7

Vocabulary Review

Choose the correct words to fill in the blanks below.

<blockquote>
graph   compound   positive   inequalities   negative   compound
</blockquote>

1. The addition property for _____ states that $A + C < B + C$ if $A < B$.

2. The multiplication property for inequalities states that multiplying both sides of an inequality by a _____ number does not change the solution set.

3. When multiplying both sides of an inequality by a _____ number, it is necessary to reverse the inequality symbol.

4. The last step to solving a linear inequality in one variable is to _____ the solution set.

5. A _____ inequality is more than one inequality considered together using the words *and* or *or*.

Problems

A Solve the following inequalities using the addition property of inequalities. Graph each solution set.

1. $x - 5 < 7$

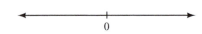

2. $x + 3 < -5$

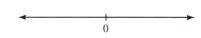

3. $a - 4 \leq 8$

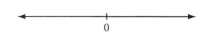

4. $a + 3 \leq 10$

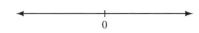

5. $x - 4.3 > 8.7$

6. $x - 2.6 > 10.4$

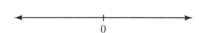

7. $y + 6 \geq 10$

8. $y + 3 \geq 12$

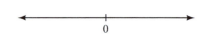

9. $2 < x - 7$

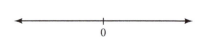

10. $3 < x + 8$

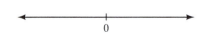

11. $-4 > x + 2$

12. $-5 \geq x - 2$

B Solve the following inequalities using the multiplication property of inequalities. If you multiply both sides by a negative number, be sure to reverse the direction of the inequality symbol. Graph the solution set.

13. $3x < 6$

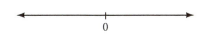

14. $2x < 14$

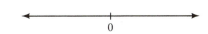

15. $5a \leq 25$

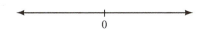

16. $4a \leq 16$

17. $\dfrac{x}{3} > 5$

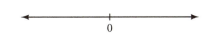

18. $\dfrac{x}{7} > 1$

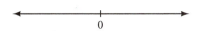

19. $-2x > 6$

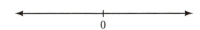

20. $-3x \geq 9$

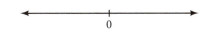

21. $-3x \geq -18$

22. $-8x \geq -24$

23. $-\dfrac{x}{5} \leq 10$

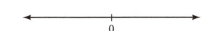

24. $-\dfrac{x}{9} \geq -1$

25. $-\dfrac{2}{3}y > 4$

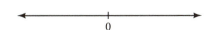

26. $-\dfrac{3}{4}y > 6$

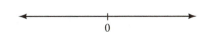

27. $-\dfrac{x}{4} > 1$

28. $-\dfrac{1}{3}x \leq -2$

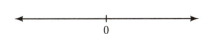

29. $-\dfrac{5}{6}x < 10$

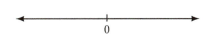

30. $-\dfrac{2}{3}x < -6$

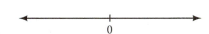

C Solve the following inequalities. Graph the solution set in each case.

31. $2x - 3 < 9$

32. $3x - 4 < 17$

33. $-\dfrac{1}{5}y - \dfrac{1}{3} \leq \dfrac{2}{3}$

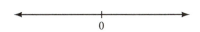

34. $-\dfrac{1}{6}y - \dfrac{1}{2} \leq \dfrac{2}{3}$

35. $-7.2x + 1.8 > -19.8$

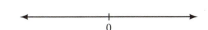

36. $-7.8x - 1.3 > 22.1$

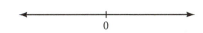

37. $\dfrac{2}{3}x - 5 \leq 7$

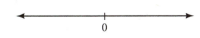

38. $\dfrac{3}{4}x - 8 \leq 1$

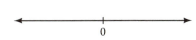

39. $-\dfrac{2}{5}a - 3 > 5$

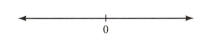

40. $-\frac{4}{5}a - 2 > 10$

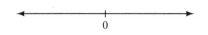

41. $5 - \frac{3}{5}y > -10$

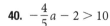

42. $4 - \frac{5}{6}y > -11$

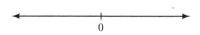

43. $0.3(a + 1) \leq 1.2$

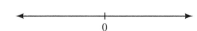

44. $0.4(a - 2) \leq 0.4$

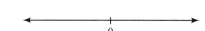

45. $2(5 - 2x) \leq -20$

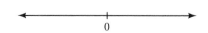

46. $7(8 - 2x) > 28$

47. $3x - 5 > 8x$

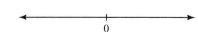

48. $8x - 4 > 6x$

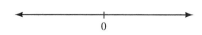

49. $\frac{1}{3}y - \frac{1}{2} \leq \frac{5}{6}y + \frac{1}{2}$

50. $\frac{7}{6}y + \frac{4}{3} \leq \frac{11}{6}y - \frac{7}{6}$

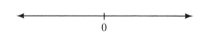

51. $-2.8x + 8.4 < -14x - 2.8$

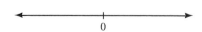

52. $-7.2x - 2.4 < -2.4x + 12$

53. $3(m - 2) - 4 \geq 7m + 14$

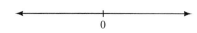

54. $2(3m - 1) + 5 \geq 8m - 7$

55. $3 - 4(x - 2) \leq -5x + 6$

56. $8 - 6(x - 3) \leq -4x + 12$

57. $4 - 3(x - 2) \geq -x + 4$

58. $5 - 2(2x - 1) \geq -2x + 4$

59. $-3x + 9 < 4 - 6(2x - 5)$

60. $-2x + 4 < 5 - 3(2x - 6)$

Solve each of the following formulas for *y*.

61. $3x + 2y < 6$

62. $-3x + 2y < 6$

63. $2x - 5y > 10$

64. $-2x - 5y > 5$

65. $-3x + 7y \leq 21$

66. $-7x + 3y \leq 21$

67. $2x - 4y \geq -4$

68. $4x - 2y \geq -8$

The next two problems are intended to give you practice reading, and paying attention to, the instructions that accompany the problems you are working.

69. Work each problem according to the instructions given.

   a. Evaluate $-5x + 3$ when $x = 0$.

   b. Solve: $-5x + 3 = -7$.

   c. Is 0 a solution to $-5x + 3 < -7$?

   d. Solve: $-5x + 3 < -7$.

70. Work each problem according to the instructions given.

   a. Evaluate $-2x - 5$ when $x = 0$.

   b. Solve: $-2x - 5 = 1$.

   c. Is 0 a solution to $-2x - 5 > 1$?

   d. Solve: $-2x - 5 > 1$.

For each graph below, write an inequality whose solution is the graph.

71.

72.

73.

74.

D Solve and graph the solution sets for the following compound inequalities.

75. $x + 5 \leq -2$ or $x + 5 \geq 2$

76. $3x + 2 < -3$ or $3x + 2 > 3$

77. $5y + 1 \leq -4$ or $5y + 1 \geq 4$

78. $7y - 5 \leq -2$ or $7y - 5 \geq 2$

79. $2x + 5 < 3x - 1$ or $x - 4 > 2x + 6$

80. $3x - 1 > 2x + 4$ or $5x - 2 < 3x + 4$

81. $3x + 1 < -8$ or $-2x + 1 \leq -3$

82. $2x - 5 \leq -1$ or $-3x - 6 < -15$

Solve the following continued inequalities. Use a line graph to write each solution set.

83. $-2 \leq m - 5 \leq 2$

84. $-3 \leq m + 1 \leq 3$

85. $-40 < 20a + 20 < 20$

86. $-60 < 50a - 40 < 60$

87. $0.5 \leq 0.3a - 0.7 \leq 1.1$

88. $0.1 \leq 0.4a + 0.1 \leq 0.3$

89. $3 < \frac{1}{2}x + 5 < 6$

90. $5 < \frac{1}{4}x + 1 < 9$

91. $4 < 6 + \frac{2}{3}x < 8$

92. $3 < 7 + \frac{4}{5}x < 15$

93. $-2 < -\frac{1}{2}x + 1 \leq 1$

94. $-3 \leq -\frac{1}{3}x - 1 < 2$

95. $-\frac{1}{2} \leq \frac{3x + 1}{2} \leq \frac{1}{2}$

96. $-\frac{5}{6} \leq \frac{2x + 5}{3} \leq \frac{5}{6}$

97. $-1.5 \leq \frac{2x - 3}{4} \leq 3.5$

98. $-1.25 \leq \frac{2x + 3}{4} \leq .75$

99. $-\frac{3}{4} \leq \frac{4x - 3}{2} \leq 1.5$

100. $-\frac{5}{6} \leq \frac{5x + 2}{3} \leq \frac{5}{6}$

Applying the Concepts

101. Consecutive Integers The sum of two consecutive integers is at least 583. What are the possibilities for the first of the two integers?

102. Consecutive Integers The sum of two consecutive integers is at most 583. What are the possibilities for the first of the two integers?

103. Number Problems The sum of twice a number and six is less than ten. Find all solutions.

104. Number Problems Twice the difference of a number and three is greater than or equal to the number increased by five. Find all solutions.

105. Number Problems The product of a number and four is greater than the number minus eight. Find the solution set.

106. Number Problems The quotient of a number and five is less than the sum of seven and two. Find the solution set.

107. Geometry Problems The length of a rectangle is 3 times the width. If the perimeter is to be at least 48 meters, what are the possible values for the width? (If the perimeter is at least 48 meters, then it is greater than or equal to 48 meters.)

108. Geometry Problems The length of a rectangle is 3 more than twice the width. If the perimeter is to be at least 51 meters, what are the possible values for the width? (If the perimeter is at least 51 meters, then it is greater than or equal to 51 meters.)

109. Geometry Problems The numerical values of the three sides of a triangle are given by three consecutive even integers. If the perimeter is greater than 24 inches, what are the possibilities for the shortest side?

110. Geometry Problems The numerical values of the three sides of a triangle are given by three consecutive odd integers. If the perimeter is greater than 27 inches, what are the possibilities for the shortest side?

111. Car Heaters If you have ever gotten in a cold car early in the morning you know that the heater does not work until the engine warms up. This is because the heater relies on the heat coming off the engine. Write an equation using an inequality sign to express when the heater will work if the heater works only after the engine is 100°F.

112. Exercise When Kate exercises, she either swims or runs. She wants to spend a minimum of 8 hours a week exercising, and she wants to swim 3 times the amount she runs. What is the minimum amount of time she must spend doing each exercise?

113. Profit and Loss Movie theaters pay a certain price for the movies that you and I see. Suppose a theater pays $1,500 for each showing of a popular movie. If they charge $7.50 for each ticket they sell, then they will lose money if ticket sales are less than $1,500. However, they will make a profit if ticket sales are greater than $1,500. What is the range of tickets they can sell and still lose money? What is the range of tickets they can sell and make a profit?

114. Stock Sales Suppose you purchase x shares of a stock at $12 per share. After 6 months you decide to sell all your shares at $20 per share. Your broker charges you $15 for the trade. If your profit is at least $3,985, how many shares did you purchase in the first place?

Maintaining Your Skills

The problems that follow review some of the more important skills you have learned in previous sections and chapters. You can consider the time you spend working these problems as time spent studying for exams.

Answer the following percent problems.

115. What number is 25% of 32?

116. What number is 15% of 75?

117. What number is 20% of 120?

118. What number is 125% of 300?

119. What percent of 36 is 9?

120. What percent of 16 is 9?

121. What percent of 50 is 5?

122. What percent of 140 is 35?

123. 16 is 20% of what number?

124. 6 is 3% of what number?

125. 8 is 2% of what number?

126. 70 is 175% of what number?

Simplify each expression.

127. $-|-5|$

128. $\left(-\dfrac{2}{3}\right)^3$

129. $-3 - 4(-2)$

130. $2^4 + 3^3 \div 9 - 4^2$

131. $5|3 - 8| - 6|2 - 5|$

132. $7 - 3(2 - 6)$

133. $5 - 2[-3(5 - 7) - 8]$

134. $\dfrac{5 + 3(7 - 2)}{2(-3) - 4}$

135. Find the difference of -3 and -9.

136. If you add -4 to the product of -3 and 5, what number results?

137. Apply the distributive property to $\dfrac{1}{2}(4x - 6)$.

138. Use the associative property to simplify $-6\left(\dfrac{1}{3}x\right)$.

Find the Mistake

Each sentence below contains a mistake. Circle the mistake and write the correct word(s) or number on the line provided.

1. If the solution set to an inequality is all the real numbers greater than 9, the solution is $x \le 9$. _____

2. To include the number 14 in the solution to the inequality $x \ge 14$, we mark the number on the graph with an open circle. _____

3. The last step in solving an inequality is to use the multiplication property for inequalities. _____

4. Upon solving the inequality $-5y < 4x + 9$ for y, the inequality symbol remains the same. _____

Show Me the Money

Starting a business involves making important financial decisions. These decisions require a lot of math. For this project, create a business model for a new business. The business can be one that you operate out of your home, or one that requires renting a commercial space from which you offer services or sell your inventory. Research the financial information you would need to start this business. What mathematical concepts would you use to interpret this information? Create linear equations that will give you calculated quantities for cost, revenue, and profit. How would linear inequalities give you knowledge about your business?

Chapter 9 Summary

Solution Set [9.1]

1. The solution set for the equation $x + 2 = 5$ is {3} because when x is 3 the equation is $3 + 2 = 5$, or $5 = 5$.

The *solution set* for an equation (or inequality) is all the numbers that, when used in place of the variable, make the equation (or inequality) a true statement.

Equivalent Equations [9.1]

2. The equation $a - 4 = 3$ and $a - 2 = 5$ are equivalent because both have solution set {7}.

Two equations are called *equivalent* if they have the same solution set.

Addition Property of Equality [9.1]

3. Solve $x - 5 = 12$.
$$x - 5\,(+\,5) = 12\,(+\,5)$$
$$x + 0 = 17$$
$$x = 17$$

When the same quantity is added to both sides of an equation, the solution set for the equation is unchanged. Adding the same amount to both sides of an equation produces an equivalent equation.

Multiplication Property of Equality [9.2]

4. Solve $3x = 18$.
$$\frac{1}{3}(3x) = \frac{1}{3}(18)$$
$$x = 6$$

If both sides of an equation are multiplied by the same nonzero number, the solution set is unchanged. Multiplying both sides of an equation by a nonzero quantity produces an equivalent equation.

Linear Equations [9.3]

A linear equation in one variable is any equation that can be put in the form $Ax + B = C$, where A, B and C are real numbers and a is not zero.

Strategy for Solving Linear Equations in One Variable [9.3]

5. Solve $2(x + 3) = 10$.
$$2x + 6 = 10$$
$$2x + 6 + (-6) = 10 + (-6)$$
$$2x = 4$$
$$\frac{1}{2}(2x) = \frac{1}{2}(4)$$
$$x = 2$$

Step 1a: Use the distributive property to separate terms, if necessary.

  1b: If fractions are present, consider multiplying both sides by the LCD to eliminate the fractions. If decimals are present, consider multiplying both sides by a power of 10 to clear the equation of decimals.

  1c: Combine similar terms on each side of the equation.

Step 2: Use the addition property of equality to get all variable terms on one side of the equation and all constant terms on the other side. A variable term is a term that contains the variable (for example, 5x). A constant term is a term that does not contain the variable (the number 3, for example).

Step 3: Use the multiplication property of equality to get x (that is, $1x$) by itself on one side of the equation.

Step 4: Check your solution in the original equation to be sure that you have not made a mistake in the solution process.

Formulas [9.4]

A formula is an equation with more than one variable. To solve a formula for one of its variables, we use the addition and multiplication properties of equality to move everything except the variable in question to one side of the equal sign so the variable in question is alone on the other side.

6. Solving $P = 2l + 2w$ for l, we have
$$P - 2w = 2l$$

$$\frac{P - 2w}{2} = l$$

Blueprint for Problem Solving [9.5, 9.6]

Step 1: *Read* the problem, and then mentally *list* the items that are known and the items that are unknown.

Step 2: *Assign a variable* to one of the unknown items. (In most cases this will amount to letting $x =$ the item that is asked for in the problem.) Then *translate* the other *information* in the problem to expressions involving the variable.

Step 3: *Reread* the problem, and then *write an equation,* using the items and variables listed in steps 1 and 2, that describes the situation.

Step 4: *Solve the equation* found in step 3.

Step 5: *Write* your *answer* using a complete sentence.

Step 6: *Reread* the problem, and *check* your solution with the original words in the problem.

Addition Property for Inequalities [9.7]

Adding the same quantity to both sides of an inequality produces an equivalent inequality, one with the same solution set.

7.   Solve $x + 5 < 7$.
$$x + 5 + (-5) < 7 + (-5)$$
$$x < 2$$

Multiplication Property for Inequalities [9.7]

Multiplying both sides of an inequality by a positive number never changes the solution set. If both sides are multiplied by a negative number, the sign of the inequality must be reversed to produce an equivalent inequality.

8. Solve $-3a \le 18$.
$$-\frac{1}{3}(-3a) \ge -\frac{1}{3}(18)$$
$$a \ge -6$$

Strategy for Solving Linear Inequalities in One Variable [9.7]

Step 1a: Use the distributive property to separate terms, if necessary.

1b: If fractions are present, consider multiplying both sides by the LCD to eliminate the fractions. If decimals are present, consider multiplying both sides by a power of 10 to clear the inequality of decimals.

1c: Combine similar terms on each side of the inequality.

Step 2: Use the addition property for inequalities to get all variable terms on one side of the inequality and all constant terms on the other side.

Step 3: Use the multiplication property for inequalities to get x by itself on one side of the inequality.

Step 4: Graph the solution set.

9. Solve $3(x - 4) \ge -2$.
$$3x - 12 \ge -2$$
$$3x - 12\,(+\,12) \ge -2\,(+\,12)$$
$$3x \ge 10$$
$$\frac{1}{3}(3x) \ge \frac{1}{3}(10)$$
$$x \ge \frac{10}{3}$$

Is 2 a solution to each of the following equations? [9.1]

1. $4x + 7 = 8$　　　　**2.** $5y - 3 = 7$

Is 3 a solution to each of the following equations? [9.1]

3. $2n + 4 = 10$　　　　**4.** $4a - 3 = 5$

Solve the following equations. [9.1]

5. $x - 5 = 10$

6. $y + 4 = 15$

7. $n - 5.4 = 10.6$

Solve the following equations. [9.2, 9.3]

8. $4x - 3 = -1$　　　　**9.** $5y - 24 = y$

10. $\frac{1}{4}x - \frac{1}{16} = \frac{1}{8} + \frac{1}{2}x$　　**11.** $-4(2x - 7) + 6 = -6$

12. $4x - 9 = 7$

13. $0.04 - 0.03(200 - x) = 5.2$

14. $3(t - 7) + 4(t + 5) = 5t + 5$

15. $5x - 3(4x - 5) = -9x + 11$

16. $3(x + 5) - 5x + 15(x + 7) = 211$

17. $2(n + 4) + 8n - 5(n + 3) = 43$

18. If $6x - 8y = 24$, find y when $x = 3$. [9.4]

19. If $6x - 8y = 24$, find x when $y = -3$. [9.4]

20. Solve $3x - 2y = 6$ for y. [9.4]

21. Solve $9 = \frac{k(T_1 - T_2)}{x}$ for T_1. [9.4]

Solve each word problem. [9.5, 9.6]

22. Age Problem Brian is 4 years older than Michelle. Five years ago the sum of their ages was 48. How old are they now?

23. Geometry The length of a rectangle is 1 inch more than twice the width. The perimeter is 44 inches. What are the length and width?

24. Coin Problem A man has $1.15 in quarters and dimes in his pocket. If he has one more dime than he has quarters, how many of each coin does he have?

25. Investing A woman has money in two accounts. One account pays 7% annual interest, whereas the other pays 15% annual interest. If she has $1,500 more invested at 15% than she does at 7% and her total interest for a year is $1,545, how much does she have in each account?

Solve each inequality, and graph the solution. [9.7]

26. $\frac{2}{3}x + 5 \le -3$　　　　**27.** $-5y > 30$

28. $0.5 - 0.4x < 2.1$　　　　**29.** $2 + 3(n + 3) \ge 17$

30. $3x - 4 < 2x$　　　　**31.** $3x - 6 \le 15$

Write an inequality whose solution is the given graph. [9.7]

32.

33.

34.

The illustration below shows the number of iPad only apps in the most popular categories. If there were 3,500 apps, what percentage of apps were in the categories below? Round to the nearest tenth of a percent.

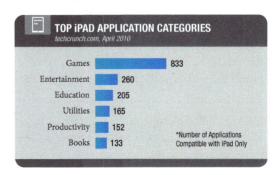

35. Games　　　　　**36.** Entertainment

Simplify.

1. $4{,}310 + 734$

2. $\frac{4}{7} + \frac{1}{4}$

3. $15.15 - 4.631$

4. $5\frac{5}{6} - 3\frac{1}{3}$

5. $6{,}314(321)$

6. $0.014(60)$

7. $143\overline{)5{,}148}$

8. $\frac{7}{28} \div \frac{14}{36}$

9. Round the number 536,204 to the nearest ten thousand.

10. Write 0.36 as a fraction in lowest terms.

11. Change $\frac{61}{8}$ to a mixed number in lowest terms.

12. Find the difference of 0.64 and $\frac{3}{8}$.

13. Write the decimal 0.37 as a percent.

14. Write 136% as a fraction or mixed number in lowest terms.

15. What percent of 74 is 25.9?

Simplify.

16. $\left(\frac{1}{2}\right)^4 + \left(\frac{1}{8}\right)^2$

17. $7x - 4 + 5x + 7$

18. $-|-9|$

19. $\dfrac{-4(-5) + 3(-7)}{14 - 11}$

20. $21 - 4(6 - 2)$

21. $\sqrt{36} - \sqrt{81}$

Solve.

22. $\frac{4}{7}y = 24$

23. $-4(3x + 5) = 3(2x + 9)$

24. $\dfrac{4.8}{4} = \dfrac{7.2}{x}$

25. Write the following ratio as a fraction in lowest terms: 0.06 to 0.54

26. Subtract -9 from 4.

27. Surface Area Find the surface area of a rectangular solid with length 9 inches, width 4 inches, and height 1 inches.

28. Age Ben is 5 years older than Ryan. In 6 years the sum of their ages will be 69. How old are they now?

29. Gas Mileage A truck travels 690 miles on 30 gallons of gas. What is the gas mileage in miles per gallon?

30. Discount A surfboard that usually sells for $560 is marked down to $420. What is the discount? What is the discount rate?

31. Geometry Find the length of the hypotenuse of a right triangle with sides of 3 and 4 meters.

32. Wildflower Seeds C.J. works in a nursery, and one of his tasks is filling packets of wildflower seeds. If each packet is to contain $\frac{1}{5}$ pound of seeds, how many packets can be filled from 12 pounds of seeds?

33. Commission A car stereo salesperson receives a commission of 6% on all units he sells. If his total sales for March are $7,800, how much money in commission will he make?

34. Volume How many 12-fluid ounce glasses of water will it take to fill a 12-gallon aquarium?

The illustration shows the most popular dogs registered by the American Kennel Club. Use the information to answer the following questions.

MOST POPULAR REGISTERED DOGS
American Kennel Club

Labrador Retriever	123,760
Yorkshire Terrier	48,346
German Shepherd Dog	43,575
Golden Retriever	42,962
Beagle	39,484

35. How many more Yorkshire Terriers are registered than Beagles? What is the percent increase between Terriers and Beagles? Round to the nearest hundredth of a percent if necessary.

36. Round the number of Labrador Retrievers and German Shepherds to the nearest thousand. Use the rounded numbers to find the ratio of Retrievers to Shepherds.

Is 4 a solution to each of the following equations? [9.1]

1. $5x + 4 = 8$

2. $3y - 5 = 7$

Is 7 a solution to each of the following equations? [9.1]

3. $4n + 3 = 31$

4. $6a - 6 = 5$

Solve the following equations. [9.1]

5. $x - 7 = 10$

6. $y + 2 = 7$

7. $n + 4.2 = 9.4$

Solve the following equations. [9.2, 9.3]

8. $3x - 2 = 7$

9. $4y + 15 = y$

10. $\dfrac{1}{4}x - \dfrac{1}{12} = \dfrac{1}{3}x + \dfrac{1}{6}$

11. $-3(3 - 2x) - 7 = 8$

12. $3x - 9 = -6$

13. $0.05 + 0.07(100 - x) = 3.2$

14. $4(t - 3) + 2(t + 4) = 2t - 16$

15. $4x - 2(3x - 1) = 2x - 8$

16. $5(x + 2) - 3x + 3(x + 4) = 32$

17. $3(a + 7) + 6a - 3(a + 2) = 45$

18. If $3x - 4y = 16$, find y when $x = 4$. [9.4]

19. If $3x - 4y = 16$, find x when $y = 2$. [9.4]

20. Solve $2x + 6y = 12$ for y. [9.4]

21. Solve $x^2 = v^2 + 2ad$ for a. [9.4]

Solve each word problem. [9.5, 9.6]

22. Age Problem Paul is twice as old as Becca. Five years ago the sum of their ages was 44. How old are they now?

23. Geometry The length of a rectangle is 5 centimeters less than 3 times the width. The perimeter is 150 centimeters. What are the length and width?

24. Coin Problem A man has a collection of dimes and nickels with a total value of $1.70. If he has 8 more dimes than he has nickels, how many of each coin does he have?

25. Investing A woman has money in two accounts. One account pays 6% annual interest, whereas the other pays 12% annual interest. If she has $500 more invested at 12% than she does at 6% and her total interest for a year is $186, how much does she have in each account?

Solve each inequality, and graph the solution. [9.7]

26. $\dfrac{1}{2}x - 2 > 3$

27. $-6y \le 24$

28. $0.3 - 0.2x < 1.1$

29. $3 - 2(n - 1) \ge 9$

30. $5x - 3 < 2x$

31. $-3 \le 2x - 7$

Write an inequality whose solution is the given graph. [9.7]

32.

33.

34.

The snapshot shows the number of hours per day the average commute is for several cities throughout the United States. Use the information to determine the percentage of each day (24 hours) are spent commuting in a car for the following cities. Round to the nearest tenth of a percent.

LIFE IN A CAR
U.S. Census Bureau

City	Hours
New York City, New York	6.7
Chicago, Illinois	5.7
Philadelphia, Pennsylvania	5.3
Los Angeles, California	4.9
Phoenix, Arizona	4.3
Dallas, Texas	4.4
Houston, Texas	4.4
San Diego, California	3.9

35. New York City

36. Los Angeles

Graphing Linear Equations and Inequalities in Two Variables

10

Google

Ligurian Sea, Italy
Image © 2010 European Space Imaging

Italy's coast of the Ligurian Sea is known for its rugged cliffs and historic settlements. If you are lucky enough to visit, you will find five fishing villages connected by a walking trail, known as Sentiero Azzurro ("Light Blue Trail"). Constructed over the centuries as a connection between these villages, the trail has become a must-see for visitors. The walk varies in difficulty and takes approximately 5 hours. Some sections are easy strolls, others are very challenging hikes. One leg of the trail begins with an easy walk and finishes with a climb of 368 stairs. It winds through olive orchards and vineyards and can be rough in places, but offers spectacular views of the bay and the unique coastline. The difference in height for the Sentiero Azzurro from bottom to top is more than 600 meters and the length is more than 10 km.

©iStockphoto.com/danbreckwoldt

From the above information, we know that the steepness of the trail varies along its length. However, we can use the knowledge of the overall change in height from the bottom of the trail to the top to potentially determine an average steepness, or slope. In mathematics, slope is the difference between the rise of a path compared to the distance over which that rise takes place. Calculation of slope is one of the topics we will explore in this chapter.

A Create a scatter diagram or line graph from a table of data.

B Graph ordered pairs on a rectangular coordinate system.

10.1 The Rectangular Coordinate System

Image © Katherine Heistand Shields, 2010

KEY WORDS

scatter diagram

line graph

x-coordinate

y-coordinate

axis

quadrant

Video Examples
Section 10.1

In New Jersey, a fast food restaurant called Cluck-U Chicken markets a wide range of hot sauces for its famous buffalo wings. Their 911 hot sauce is so spicy, the restaurant requires all tasters to sign a waiver upon order. According to Cluck-U's corporate office, the Scoville rating (a scale to measure the spicy heat of the chili pepper) of the 911 sauce is 450,000 units. Look at Table 1 below, which compares other peppers with the 911 hot sauce. Figure 1 is a bar graph, which is a visual representation of the same information. The ratings in the table and graph are approximate because each entry on the Scoville scale typically appears as a range.

Table 1

Spicy Peppers

Pepper or Pepper-Infused Item	Scoville Units
Green pepper	0
Carolina cayenne pepper	100,000
Bird's eye pepper	200,000
Habanero pepper	300,000
Cluck-U's 911 sauce	450,000

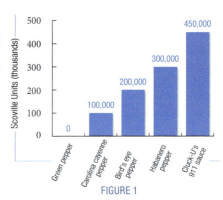

FIGURE 1

A Scatter Diagrams and Line Graphs

The information in the table in the introduction can be visualized with a *scatter diagram*, also called a *scatter plot* and *line graph* as well. Figure 2 is a scatter diagram of the information in Table 1. We use dots instead of the bars shown in Figure 1 to show the Scoville rating of each pepper item. Figure 3 is called a *line graph*. It is constructed by taking the dots in Figure 2 and connecting each one to the next with a straight line. Notice that we have labeled the axes in these two figures a little differently than we did with the bar chart by making the axes intersect at the number 0.

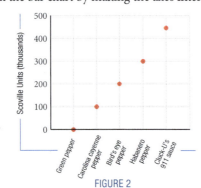

FIGURE 2

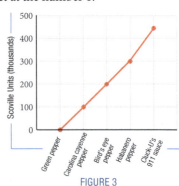

FIGURE 3

EXAMPLE 1   The tables below give the first five odd numbers and the first five counting numbers squared as paired data. In each case construct a scatter diagram.

Odd Numbers		Squares	
Position	Value	Position	Value
1	1	1	1
2	3	2	4
3	5	3	9
4	7	4	16
5	9	5	25

Solution   The two scatter diagrams are based on the data from these tables shown here. Notice how the dots in Figure 4 seem to line up in a straight line, whereas the dots in Figure 5 give the impression of a curve. We say the points in Figure 4 suggest a linear relationship between the two sets of data, whereas the points in Figure 5 suggest a nonlinear relationship.

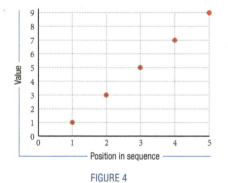

FIGURE 4

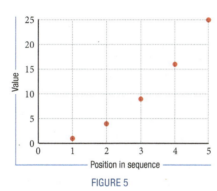

FIGURE 5

B Graphing Ordered Pairs

As you know, each dot in Figures 4 and 5 corresponds to a pair of numbers, one of which is associated with the horizontal axis and the other with the vertical axis. Paired data play a very important role in the equations we will solve in the next section. To prepare ourselves for those equations, we need to expand the concept of paired data to include negative numbers. At the same time, we want to standardize the position of the axes in the diagrams that we use to visualize paired data.

Our ability to graph paired data is due to the invention of the rectangular coordinate system also caled the *Cartesian coordinate system* after the French philosopher René Descartes (1595–1650). Descartes is the person usually credited with the invention of the rectangular coordinate system. As a philosopher, Descartes is responsible for the statement "I think, therefore I am." Until Descartes invented his coordinate system in 1637, algebra and geometry were treated as separate subjects. The rectangular coordinate system allows us to connect algebra and geometry by associating geometric shapes with algebraic equations.

Here is the formal definition for a graph's coordinates:

> **DEFINITION** *x-coordinate, y-coordinate*
>
> A pair of numbers enclosed in parentheses and separated by a comma, such as $(-2, 1)$, is called an ordered pair of numbers. The first number in the pair is called the x-coordinate of the ordered pair; the second number is called the y-coordinate. For the ordered pair $(-2, 1)$, the x-coordinate is -2 and the y-coordinate is 1.

Practice Problems

1. The table below gives the position and value of the first five counting numbers cubed. Construct a scatter diagram using these data.

Cubes	
Position	Value
1	1
2	8
3	27
4	64
5	125

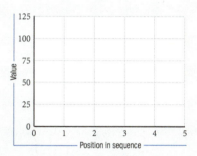

Ordered pairs of numbers are important in the study of mathematics because they give us a way to visualize solutions to equations. To see the visual component of ordered pairs, we need the diagram shown in Figure 6. It is called the *rectangular coordinate system*.

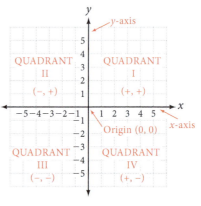

FIGURE 6

The rectangular coordinate system is built from two number lines oriented perpendicular to each other. The horizontal number line is exactly the same as our real number line and is called the *x-axis*. The vertical number line is also the same as our real number line with the positive direction up and the negative direction down. It is called the *y-axis*. The point where the two axes intersect is called the *origin*. As you can see from Figure 6, the axes divide the plane into four *quadrants*, which are numbered I through IV in a counterclockwise direction.

To graph the ordered pair (a, b), we start at the origin and move a units forward or back (forward if a is positive and back if a is negative). Then we move b units up or down (up if b is positive, down if b is negative). The point where we end up is the graph of the ordered pair (a, b). To graph the ordered pair (5, 2), we start at the origin and move 5 units to the right. Then, from that position, we move 2 units up.

> *Note*   It is very important that you graph ordered pairs quickly and accurately. Remember, the first coordinate goes with the horizontal axis and the second coordinate goes with the vertical axis.

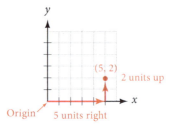

2. Graph the ordered pairs (2, 3), (2, −3), (−2, 3), and (−2, −3).

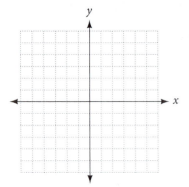

EXAMPLE 2   Graph the ordered pairs (3, 4), (3, −4), (−3, 4), and (−3, −4).

Solution

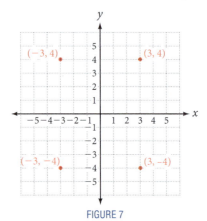

FIGURE 7

We can see in Figure 7 that when we graph ordered pairs, the x-coordinate corresponds to movement parallel to the x-axis (horizontal) and the y-coordinate corresponds to movement parallel to the y-axis (vertical).

EXAMPLE 3   Graph the ordered pairs $(-1, 3)$, $(2, 5)$, $(0, 0)$, $(0, -3)$, and $(4, 0)$.

Solution

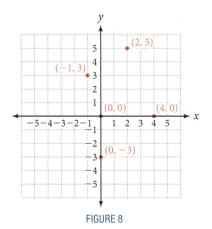

FIGURE 8

3. Graph the ordered pairs $(-2, 1)$, $(3, 5)$, $(0, 2)$, $(-5, 0)$, and $(-3, -3)$.

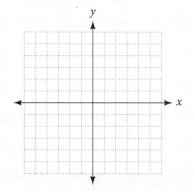

Note   If we do not label the axes of a coordinate system, we assume that each square is one unit long and one unit wide.

Review the three points in Example 3 that contain a zero as either an *x*- or a *y*-coordinate. It is important to note that ordered pairs that sit on an axis or at the origin are not associated with any quadrant.

GETTING READY FOR CLASS

After reading through the preceding section, respond in your own words and in complete sentences.

A. What is an ordered pair of numbers?

B. Explain in words how you would graph the ordered pair (3, 4).

C. How does a scatter diagram differ from a line graph?

D. Where is the origin on a rectangular coordinate system and why is it important?

Vocabulary Review

Choose the correct words to fill in the blanks below.

x-axis   line graph   ordered pair   scatter diagram/plot

y-axis   origin   quadrants

1. A _____ uses dots on a graph to identify data.

2. Connecting the dots of a scatter diagram gives a_____.

3. A pair of numbers enclosed in parentheses and separated by a comma is called an _____.

4. A rectangular coordinate system has a horizontal number line called the _____ and a vertical number line called the _____.

5. The x-axis and y-axis divide the rectangular coordinate system into four _____.

6. The point where the x-and y-axis intersect is called the _____.

Problems

A 1. **Non-Camera Phone Sales** The table here shows what are the estimated sales of non-camera phones for the years 2006–2010. Use the information from the table and chart to construct a line graph.

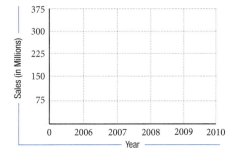

Year	Sales (in millions)
2006	300
2007	250
2008	175
2009	150
2010	125

2. **Camera Phone Sales** The table here shows the estimated sales of camera phones from 2006 to 2010. Use the information from the table to construct a line graph.

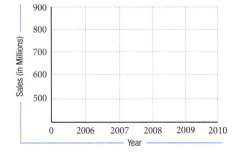

Year	Sales (in millions)
2006	500
2007	650
2008	750
2009	875
2010	900

B Graph the following ordered pairs.

3. $(3, 2)$ **4.** $(3, -2)$ **5.** $(-3, 2)$ **6.** $(-3, -2)$ **7.** $(5, 1)$ **8.** $(5, -1)$

9. $(1, 5)$ **10.** $(1, -5)$ **11.** $(-1, 5)$ **12.** $(-1, -5)$ **13.** $\left(2, \frac{1}{2}\right)$ **14.** $\left(3, \frac{3}{2}\right)$

15. $\left(-4, -\frac{5}{2}\right)$ **16.** $\left(-5, -\frac{3}{2}\right)$ **17.** $(3, 0)$ **18.** $(-2, 0)$ **19.** $(0, 5)$ **20.** $(0, 0)$

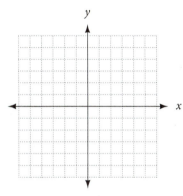

ODD-NUMBERED PROBLEMS

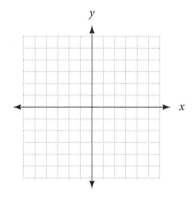

EVEN-NUMBERED PROBLEMS

Give the coordinates of each numbered point in the figure and list which quadrant each point is in.

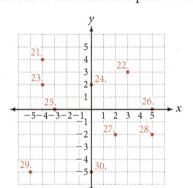

21. **22.**

23. **24.**

25. **26.**

27. **28.**

29. **30.**

Graph the points $(4, 3)$ and $(-4, -1)$, and draw a straight line that passes through both of them. Then answer the following questions.

31. Does the graph of $(2, 2)$ lie on the line?

32. Does the graph of $(-2, 0)$ lie on the line?

33. Does the graph of $(0, -2)$ lie on the line?

34. Does the graph of $(-6, 2)$ lie on the line?

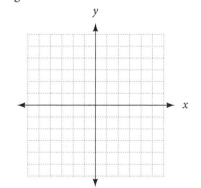

Graph the points $(-2, 4)$ and $(2, -4)$, and draw a straight line that passes through both of them. Then answer the following questions.

35. Does the graph of $(0, 0)$ lie on the line?

36. Does the graph of $(-1, 2)$ lie on the line?

37. Does the graph of $(2, -1)$ lie on the line?

38. Does the graph of $(1, -2)$ lie on the line?

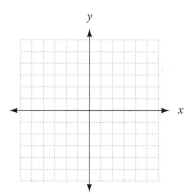

Draw a straight line that passes through the points $(3, 4)$ and $(3, -4)$. Then answer the following questions.

39. Is the graph of $(3, 0)$ on this line?

40. Is the graph of $(0, 3)$ on this line?

41. Is there any point on this line with an x-coordinate other than 3?

42. If you extended the line, would it pass through a point with a y-coordinate of 10?

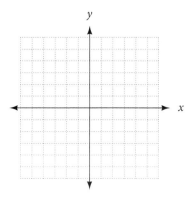

Draw a straight line that passes through the points $(3, 4)$ and $(-3, 4)$. Then answer the following questions.

43. Is the graph of $(4, 0)$ on this line?

44. Is the graph of $(0, 4)$ on this line?

45. Is there any point on this line with a y-coordinate other than 4?

46. If you extended the line, would it pass through a point with an x-coordinate of 10?

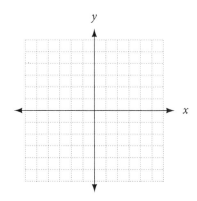

Applying the Concepts

47. Hourly Wages Jane takes a job at the local Marcy's department store. Her job pays $8.00 per hour. The graph shows how much Jane earns for working from 0 to 40 hours in a week.

a. List three ordered pairs that lie on the line graph.

b. How much will she earn for working 40 hours?

c. If her check for one week is $240, how many hours did she work?

d. She works 35 hours one week, but her paycheck before deductions are subtracted out is for $260. Is this correct? Explain.

48. Hourly Wages Judy takes a job at Gigi's boutique. Her job pays $6.00 per hour plus $50 per week in commission. The graph shows how much Judy earns for working from 0 to 40 hours in a week.

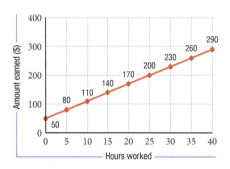

a. List three ordered pairs that lie on the line graph.

b. How much will she earn for working 40 hours?

c. If her check for one week is $230, how many hours did she work?

d. She works 35 hours one week, but her paycheck before deductions are subtracted out is for $260. Is this correct? Explain.

49. Baseball Attendance The graph gives the attendance at Major League Baseball games over a ten-year period. If *x* represents the year in question and *y* represents attendance in millions, write five ordered pairs that describe the information in the table.

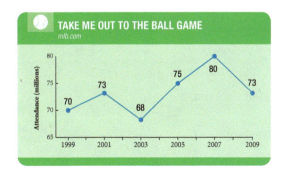

50. FIFA World Cup Goals The graph shows the number of goals scored by the US Men's National Team in FIFA World Cup competition. Write five ordered pairs that lie on the graph.

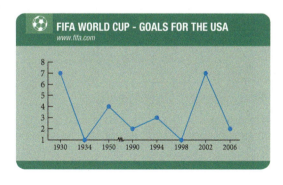

51. Right triangle ABC has legs of length 5. Point C is the ordered pair $(6, 2)$. Find the coordinates of A and B.

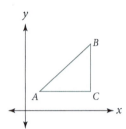

52. Right triangle ABC has legs of length 7. Point C is the ordered pair $(-8, -3)$. Find the coordinates of A and B.

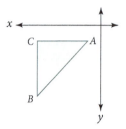

53. Rectangle $ABCD$ has a length of 5 and a width of 3. Point D is the ordered pair $(7, 2)$. Find points A, B, and C.

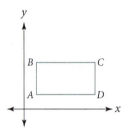

54. Rectangle $ABCD$ has a length of 5 and a width of 3. Point D is the ordered pair $(-1, 1)$. Find points A, B, and C.

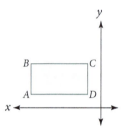

Getting Ready for the Next Section

55. Let $2x + 3y = 6$.
  a. Find x if $y = 4$.

  b. Find x if $y = -2$.

  c. Find y if $x = 3$.

  d. Find y if $x = 9$.

56. Let $2x - 5y = 20$.
  a. Find x if $y = 0$.

  b. Find x if $y = -6$.

  c. Find y if $x = 0$.

  d. Find y if $x = 5$.

57. Let $y = 2x - 1$.
  a. Find x if $y = 7$.

  b. Find x if $y = 3$.

  c. Find y if $x = 0$.

  d. Find y if $x = 5$.

58. Let $y = 3x - 2$.
  a. Find x if $y = 4$.

  b. Find x if $y = 3$.

  c. Find y if $x = 2$.

  d. Find y if $x = -3$.

59. Find y when x is 4 in the formula $3x + 2y = 6$.

60. Find y when x is 0 in the formula $3x + 2y = 6$.

61. Find y when x is 0 in $y = -\frac{1}{3}x + 2$.

62. Find y when x is 3 in $y = -\frac{1}{3}x + 2$.

63. Find y when x is 2 in $y = \frac{3}{2}x - 3$.

64. Find y when x is 4 in $y = \frac{3}{2}x - 3$.

65. Solve $5x + y = 4$ for y.

66. Solve $-3x + y = 5$ for y.

67. Solve $3x - 2y = 6$ for y.

68. Solve $2x - 3y = 6$ for y.

Find the Mistake

Each sentence below contains a mistake. Circle the mistake and write the correct word or number on the line provided.

1. A scatter diagram uses bars to show data in the form of order pairs. _____

2. In the ordered pair $(-3, 6)$, the x–coordinate is 6. _____

3. To graph the ordered pair $(7, 3)$, start at the origin, move 7 units to the left, and 3 units up. _____

4. The ordered pair $(-2, 5)$ appears in the first quadrant. _____

Navigation Skills: Prepare, Study, Achieve

The chapters in this course are organized such that each chapter builds on the previous chapters. So you already have learned the tools to master the final topics in this book and be successful on the final exam. However, studying for the final exam may still seem like an overwhelming task. To ease that anxiety, begin now to lay out a study plan. Dedicate time in your day to do the following:

- Stay calm and maintain a positive attitude.
- Scan each section of the book to review headers, graphics, definitions, rules, properties, formulas, italicized words, and margin notes. Take new notes as you scan.
- Review your notes and homework.
- Rework problems in each section and from your difficult problems list.
- Review chapter summaries, and make your own outlines for each chapter.
- Make and review flashcards of definitions, properties, or formulas.
- Explain concepts to a friend or out loud to yourself.
- Schedule time to meet with a study partner or group.

Another way to prepare is to visualize yourself being successful on the exam. Close your eyes and picture yourself arriving on exam day, receiving the test, staying calm, and working through difficult problems successfully. Now picture yourself receiving a high score on the exam. Complete this visualization multiple times before the day of the exam. If you picture yourself achieving success, you are more likely to be successful.

Video Examples
Section 10.2

Image © Katherine Heistand Shields, 2010

A new robot named EMILY (Emergency Integrated Lifesaving Lanyard) patrols dangerous ocean waters off the coast of Malibu, California. Currently, a lifeguard controls EMILY by remote. However, developers will soon equip the robot with sonar and an autonomous system able to detect swimmers in distress without the help of a lifeguard on shore. EMILY's electric-powered impeller drives the robot through even the roughest surf at a speed 6 times faster than that of a human. A swimmer in distress can use EMILY as a flotation device until further help arrives, or EMILY can tow the swimmer back to shore.

Using the information about EMILY's speed, we can write the following equation, which uses two variables:

$$y = 6x$$

where x is a human's swimming speed and y is EMILY's speed. In this section, we will begin to investigate equations in two variables, such as the one above. As you will see, equations in two variables have pairs of numbers for solutions. Because we know how to use paired data to construct tables, histograms, and other charts, we can take our work with paired data further by using equations in two variables to construct tables of paired data.

A Solving Linear Equations

Let's begin this section by reviewing the relationship between equations in one variable and their solutions.

If we solve the equation $3x - 2 = 10$, the solution is $x = 4$. If we graph this solution, we simply draw the real number line and place a dot at the point whose coordinate is 4. The relationship between linear equations in one variable, their solutions, and the graphs of those solutions look like this:

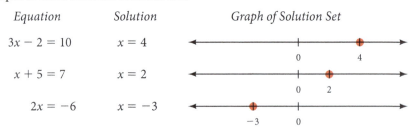

Equation	Solution	Graph of Solution Set
$3x - 2 = 10$	$x = 4$	
$x + 5 = 7$	$x = 2$	
$2x = -6$	$x = -3$	

When the equation has one variable, the solution is a single number whose graph is a point on a line.

Now, consider the equation $2x + y = 3$. The first thing we notice is that there are two variables instead of one. Therefore, a solution to the equation $2x + y = 3$ will be not a single number but a pair of numbers, one for x and one for y, that makes the equation a true statement. One pair of numbers that works is $x = 2$, $y = -1$ because when we substitute them for x and y in the equation, we get a true statement.

$$2(2) + (-1) \stackrel{?}{=} 3$$
$$4 - 1 = 3$$
$$3 = 3 \qquad \text{A true statement}$$

The pair of numbers $x = 2$, $y = -1$ is written as $(2, -1)$. As you know from Section 3.1, $(2, -1)$ is called an *ordered pair* because it is a pair of numbers written in a specific order. The first number is always associated with the variable x, and the second number is always associated with the variable y. We call the first number in the ordered pair the *x-coordinate* and the second number the *y-coordinate* of the ordered pair.

Let's look back to the equation $2x + y = 3$. The ordered pair $(2, -1)$ is not the only solution. Another solution is $(0, 3)$ because when we substitute 0 for x and 3 for y we get

$$2(0) + 3 \stackrel{?}{=} 3$$
$$0 + 3 = 3$$
$$3 = 3 \qquad \text{A true statement}$$

Still another solution is the ordered pair $(5, -7)$ because

$$2(5) + (-7) \stackrel{?}{=} 3$$
$$10 - 7 = 3$$
$$3 = 3 \qquad \text{A true statement}$$

As a matter of fact, for any number we want to use for x, there is another number we can use for y that will make the equation a true statement. There is an infinite number of ordered pairs that satisfy (are solutions to) the equation $2x + y = 3$; we have listed just a few of them.

EXAMPLE 1 Given the equation $2x + 3y = 6$, complete the following ordered pairs so they will be solutions to the equation: $(0, \ )$, $(\ , 1)$, $(3, \ )$.

Solution To complete the ordered pair $(0, \ )$, we substitute 0 for x in the equation and then solve for y.

$$2(0) + 3y = 6$$
$$3y = 6$$
$$y = 2$$

The ordered pair is $(0, 2)$.

To complete the ordered pair $(\ , 1)$, we substitute 1 for y in the equation and solve for x.

$$2x + 3(1) = 6$$
$$2x + 3 = 6$$
$$2x = 3$$
$$x = \frac{3}{2}$$

The ordered pair is $\left(\frac{3}{2}, 1\right)$.

To complete the ordered pair $(3, \ )$, we substitute 3 for x in the equation and solve for y.

$$2(3) + 3y = 6$$
$$6 + 3y = 6$$
$$3y = 0$$
$$y = 0$$

The ordered pair is $(3, 0)$.

Note If this discussion seems a little long and confusing, you may want to look over some of the examples first and then come back and read this. Remember, it isn't always easy to read material in mathematics. What is important is that you understand what you are doing when you work problems. The reading is intended to assist you in understanding what you are doing. It is important to read everything in the book, but you don't always have to read it in the order it is written.

Practice Problems

1. For the equation $2x + 5y = 10$, complete the ordered pairs $(0, \ )$, $(\ , 1)$, and $(5, \ )$.

Answers

1. $(0, 2)$, $\left(\frac{5}{2}, 1\right)$, $(5, 0)$

2. Complete the table for
$3x - 2y = 12$.

x	y
0	
	3
	0
-3	

Notice in each case that once we have used a number in place of one of the variables, the equation becomes a linear equation in one variable. We then use the method explained in Chapter 9 to solve for that variable.

EXAMPLE 2   Complete the following table for the equation $2x - 5y = 20$.

x	y
0	
	2
	0
-5	

Solution   Filling in the table is equivalent to completing the following ordered pairs: $(0, \ ), (\ , 2), (\ , 0), (-5, \ )$. So we proceed as in Example 1.

When $x = 0$, we have

$$2(0) - 5y = 20$$
$$0 - 5y = 20$$
$$-5y = 20$$
$$y = -4$$

When $y = 2$, we have

$$2x - 5(2) = 20$$
$$2x - 10 = 20$$
$$2x = 30$$
$$x = 15$$

When $y = 0$, we have

$$2x - 5(0) = 20$$
$$2x - 0 = 20$$
$$2x = 20$$
$$x = 10$$

When $x = -5$, we have

$$2(-5) - 5y = 20$$
$$-10 - 5y = 20$$
$$-5y = 30$$
$$y = -6$$

The completed table looks like this:

x	y
0	-4
15	2
10	0
-5	-6

The table above is equivalent to the ordered pairs $(0, -4)$, $(15, 2)$, $(10, 0)$, and $(-5, -6)$.

3. Complete the table for
$y = 3x - 2$.

x	y
0	
2	
	7
	3

EXAMPLE 3   Complete the following table for the equation $y = 2x - 1$.

x	y
0	
5	
	7
	3

Solution   When $x = 0$, we have

$$y = 2(0) - 1$$
$$y = 0 - 1$$
$$y = -1$$

When $x = 5$, we have

$$y = 2(5) - 1$$
$$y = 10 - 1$$
$$y = 9$$

When $y = 7$, we have $\qquad$ When $y = 3$, we have

$$7 = 2x - 1 \qquad\qquad 3 = 2x - 1$$
$$8 = 2x \qquad\qquad\quad 4 = 2x$$
$$4 = x \qquad\qquad\quad\; 2 = x$$

The completed table is

x	y
0	−1
5	9
4	7
2	3

which means the ordered pairs $(0, -1)$, $(5, 9)$, $(4, 7)$, and $(2, 3)$ are among the solutions to the equation $y = 2x - 1$.

B Determining if an Ordered Pair is a Solution

EXAMPLE 4 Which of the ordered pairs $(2, 3)$, $(1, 5)$, and $(-2, -4)$ are solutions to the equation $y = 3x + 2$?

Solution If an ordered pair is a solution to the equation, then it must satisfy the equation; that is, when the coordinates are used in place of the variables in the equation, the equation becomes a true statement.

Try $(2, 3)$ in $y = 3x + 2$:

$$3 \stackrel{?}{=} 3(2) + 2$$
$$3 = 6 + 2$$
$$3 = 8 \qquad\qquad \text{A false statement}$$

Try $(1, 5)$ in $y = 3x + 2$:

$$5 \stackrel{?}{=} 3(1) + 2$$
$$5 = 3 + 2$$
$$5 = 5 \qquad\qquad \text{A true statement}$$

Try $(-2, -4)$ in $y = 3x + 2$:

$$-4 \stackrel{?}{=} 3(-2) + 2$$
$$-4 = -6 + 2$$
$$-4 = -4 \qquad\qquad \text{A true statement}$$

The ordered pairs $(1, 5)$ and $(-2, -4)$ are solutions to the equation $y = 3x + 2$, and $(2, 3)$ is not.

C Graphing Linear Equations

At the end of the previous section we used a line graph to obtain a visual picture of *some* of the solutions to the equations $y = 1.5x + 15$. In this section we will use the rectangular coordinate system introduced in the previous section to obtain a visual picture of *all* solutions to a linear equation in two variables. The process we use to obtain a visual picture of all solutions to an equation is called *graphing*. The picture itself is called the *graph* of the equation.

4. Which of the following ordered pairs is a solution to $y = 4x + 1$? $(0, 1)$, $(3, 11)$, $(2, 9)$

Answers

4. $(0, 1)$ and $(2, 9)$

5. Graph the equation $x + y = 3$.

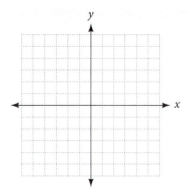

EXAMPLE 5 Graph the solution set for $x + y = 5$.

Solution We know from the previous section that an infinite number of ordered pairs are solutions to the equation $x + y = 5$. We can't possibly list them all. What we can do is list a few of them and see if there is any pattern to their graphs.

Some ordered pairs that are solutions to $x + y = 5$ are (0, 5), (2, 3), (3, 2), (5, 0). The graph of each is shown in Figure 1.

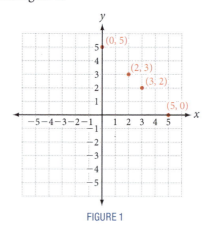

FIGURE 1

Now, by passing a straight line through these points we can graph the solution set for the equation $x + y = 5$. Linear equations in two variables always have graphs that are straight lines. The graph of the solution set for $x + y = 5$ is shown in Figure 2.

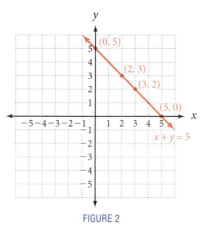

FIGURE 2

Every ordered pair that satisfies $x + y = 5$ has its graph on the line, and any point on the line has coordinates that satisfy the equation. So there is a one-to-one correspondence between points on the line and solutions to the equation. To summarize, the set of all points of a graph that satisfy an equation is the equation's solution set.

Here is the precise definition for a linear equation in two variables.

DEFINITION linear equation in two variables

Any equation that can be put in the form $Ax + By = C$, where A, B, and C are real numbers and A and B are not both 0, is called a ***linear equation in two variables***. The graph of any equation of this form is a straight line (that is why these equations are called "linear"). The form $Ax + By = C$ is called ***standard form***.

To graph a linear equation in two variables, as we did in Example 5, we simply graph its solution set; that is, we draw a line through all the points whose coordinates satisfy the equation. Here are the steps to follow:

> **HOW TO** **To Graph a Linear Equation in Two Variables**
>
> **Step 1:** Find any three ordered pairs that satisfy the equation. This can be done by using a convenient number for one variable and solving for the other variable.
>
> **Step 2:** Graph the three ordered pairs found in step 1. Actually, we need only two points to graph a straight line. The third point serves as a check. If all three points do not line up, there is a mistake in our work.
>
> **Step 3:** Draw a straight line through the three points graphed in step 2.

Note The meaning of the convenient numbers referred to in step 1 will become clear as you read the next two examples.

EXAMPLE 6   Graph the equation $y = 3x - 1$.

6. Graph the equation $y = 2x + 3$.

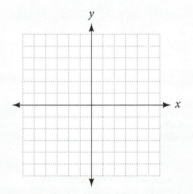

Solution   Because $y = 3x - 1$ can be put in the form $ax + by = c$, it is a linear equation in two variables. Hence, the graph of its solution set is a straight line. We can find some specific solutions by substituting numbers for x and then solving for the corresponding values of y. We are free to choose any numbers for x, so let's use 0, 2, and -1.

Let $x = 0$: $y = 3(0) - 1$
$y = 0 - 1$
$y = -1$

In table form

x	y
0	−1
2	5
−1	−4

The ordered pair $(0, -1)$ is one solution.

Let $x = 2$: $y = 3(2) - 1$
$y = 6 - 1$
$y = 5$

The ordered pair $(2, 5)$ is a second solution.

Let $x = -1$: $y = 3(-1) - 1$
$y = -3 - 1$
$y = -4$

The ordered pair $(-1, -4)$ is a third solution.

Next, we graph the ordered pairs $(0, -1)$, $(2, 5)$, $(-1, -4)$ and draw a straight line through them. The line we have drawn in Figure 3 is the graph of $y = 3x - 1$.

Note It may seem that we have simply picked the numbers 0, 2, and −1 out of the air and used them for x. In fact we have done just that. Could we have used numbers other than these? The answer is yes, we can substitute any number for x; for a linear equation in two variables, there will always be a value of y to go with it.

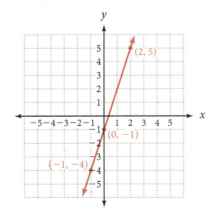

FIGURE 3

7. Graph $y = \frac{3}{2}x - 3$.

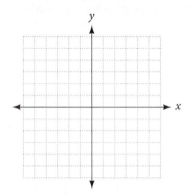

Note   In Example 7 the values of x we used, $-3, 0$, and 3, are referred to as convenient values of x because they are easier to work with than some other numbers. For instance, if we let $x = 2$ in our original equation, we would have to add $-\frac{2}{3}$ and 2 to find the corresponding value of y. Not only would the arithmetic be more difficult but also the ordered pair we obtained would have a fraction for its y-coordinate, making it more difficult to graph accurately.

Example 6 illustrates the connection between algebra and geometry. Descartes' rectangular coordinate system allows us to associate the equation $y = 3x - 1$ (an algebraic concept) with a specific straight line (a geometric concept). The study of the relationship between equations in algebra and their associated geometric figures is called *analytic geometry*.

EXAMPLE 7   Graph the equation $y = -\frac{1}{3}x + 2$.

Solution   We need to find three ordered pairs that satisfy the equation. To do so, we can let x equal any numbers we choose and find corresponding values of y. But, every value of x we substitute into the equation is going to be multiplied by $-\frac{1}{3}$. For our convenience, let's use numbers for x that are divisible by 3, like $-3, 0$, and 3. That way, when we multiply them by $-\frac{1}{3}$, the result will be an integer.

Let $x = -3$: $y = -\frac{1}{3}(-3) + 2$

$$y = 1 + 2$$
$$y = 3$$

The ordered pair $(-3, 3)$ is one solution.

Let $x = 0$: $y = -\frac{1}{3}(0) + 2$

$$y = 0 + 2$$
$$y = 2$$

The ordered pair $(0, 2)$ is a second solution.

Let $x = 3$: $y = -\frac{1}{3}(3) + 2$

$$y = -1 + 2$$
$$y = 1$$

The ordered pair $(3, 1)$ is a third solution.

In table form

x	y
-3	3
0	2
3	1

Graphing the ordered pairs $(-3, 3)$, $(0, 2)$, and $(3, 1)$ and drawing a straight line through their graphs, we have the graph of the equation $y = -\frac{1}{3}x + 2$, as shown in Figure 4.

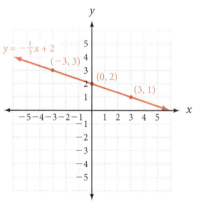

FIGURE 4

8. Graph the solution set for $2x - 4y = 8$.

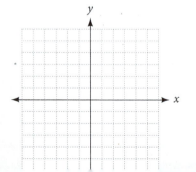

EXAMPLE 8   Graph the solution set for $3x - 2y = 6$.

Solution   It will be easier to find convenient values of x to use in the equation if we first solve the equation for y. To do so, we add $-3x$ to each side, and then we multiply each side by $-\frac{1}{2}$.

$$3x - 2y = 6 \qquad \text{Original equation}$$
$$-2y = -3x + 6 \qquad \text{Add } -3x \text{ to each side.}$$
$$-\frac{1}{2}(-2y) = -\frac{1}{2}(-3x + 6) \qquad \text{Multiply each side by } -\frac{1}{2}.$$

$$y = \frac{3}{2}x - 3 \qquad \textit{Simplify each side.}$$

Now, because each value of x will be multiplied by $\frac{3}{2}$, it will be to our advantage to choose values of x that are divisible by 2. That way, we will obtain values of y that do not contain fractions. This time, let's use 0, 2, and 4 for x.

$$\text{When } x = 0: \qquad y = \frac{3}{2}(0) - 3$$
$$y = 0 - 3$$
$$y = -3$$

The ordered pair $(0, -3)$ is one solution.

$$\text{When } x = 2: \qquad y = \frac{3}{2}(2) - 3$$
$$y = 3 - 3$$
$$y = 0$$

The ordered pair $(2, 0)$ is a second solution.

$$\text{When } x = 4: \qquad y = \frac{3}{2}(4) - 3$$
$$y = 6 - 3$$
$$y = 3$$

The ordered pair $(4, 3)$ is a third solution.

Graphing the ordered pairs $(0, -3)$, $(2, 0)$, and $(4, 3)$ and drawing a line through them, we have the graph shown in Figure 5.

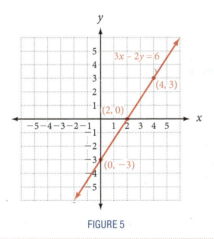

FIGURE 5

> *Note* After reading through Example 8, many students ask why we didn't use -2 for x when we were finding ordered pairs that were solutions to the original equation. The answer is, we could have. If we were to let $x = -2$, the corresponding value of y would have been -6. As you can see by looking at the graph in Figure 5, the ordered pair $(-2, -6)$ is on the graph.

D Horizontal and Vertical Lines; Lines Through The Origin

EXAMPLE 9   Graph each of the following lines.

a. $y = \frac{1}{2}x$ **b.** $x = 3$ **c.** $y = -2$

Solution

a. The line $y = \frac{1}{2}x$ passes through the origin because $(0, 0)$ satisfies the equation. To sketch the graph we need at least one more point on the line. When x is 2, we obtain the point $(2, 1)$, and when x is -4, we obtain the point $(-4, -2)$. The graph of $y = \frac{1}{2}x$ is shown in Figure 6A.

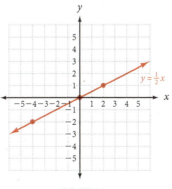

FIGURE 6A

9. Graph each of the following lines.

     a. $y = \frac{2}{3}x$

     b. $x = 3$

     c. $y = -2$

Note The equations in parts b and c are linear equations in two variables where the second variable is not explicit; in other words, $x = 3$ is the same as

$$0y + x = 3$$

and $y = -2$ is the same as

$$0x + y = -2$$

b. The line $x = 3$ is the set of all points whose x-coordinate is 3. The variable y does not appear in the equation, so the y-coordinate can be any number. Note that we can write our equation as a linear equation in two variables by writing it as $x + 0y = 3$. Because the product of 0 and y will always be 0, y can be any number. The graph of $x = 3$ is the vertical line shown in Figure 6B.

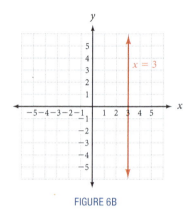

FIGURE 6B

c. The line $y = -2$ is the set of all points whose y-coordinate is -2. The variable x does not appear in the equation, so the x-coordinate can be any number. Again, we can write our equation as a linear equation in two variables by writing it as $0x + y = -2$. Because the product of 0 and x will always be 0, x can be any number. The graph of $y = -2$ is the horizontal line shown in Figure 6C.

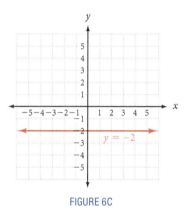

FIGURE 6C

FACTS FROM GEOMETRY Special Equations and Their Graphs

For the equations below, m, a, and b are real numbers.

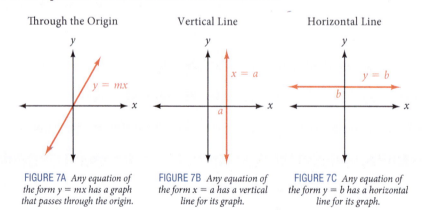

Through the Origin   Vertical Line   Horizontal Line

FIGURE 7A *Any equation of the form $y = mx$ has a graph that passes through the origin.*

FIGURE 7B *Any equation of the form $x = a$ has a vertical line for its graph.*

FIGURE 7C *Any equation of the form $y = b$ has a horizontal line for its graph.*

GETTING READY FOR CLASS

After reading through the preceding section, respond in your own words and in complete sentences.

A. How can you tell if an ordered pair is a solution to an equation?

B. How would you find a solution to $y = 3x - 5$?

C. How many solutions are there to an equation that contains two variables?

D. What type of equation has a line that passes through the origin?

Vocabulary Review

Choose the correct words to fill in the blanks below.

solution	graphing	graph	first	vertical
linear	ordered pairs	straight line	second	horizontal

1. The _____ number of an ordered pair is called the x-coordinate.

2. The _____ number of an ordered pair is called the y-coordinate.

3. An ordered pair is a _____ to an equation if it satisfies the equation.

4. The process of obtaining a visual picture of all solutions to an equation is called _____.

5. A _____ equation in two variables is any equation that can be put into the form $ax + bx = c$.

6. The first step to graph a linear equation in two variables is to find any three _____ that satisfy the equation.

7. The second step to graph a linear equation in two variables is to _____ the three ordered pairs that satisfy the equation.

8. The third step to graph a linear equation in two variables is to draw a _____ through the three ordered pairs that satisfy the equation.

9. Any equation of the form $y = b$ has a _____ line for it's graph, whereas any equation of the form $x = a$ has a _____ line for it's graph.

Problems

A For each equation, complete the given ordered pairs.

1. $2x + y = 6$ $(0, \ ), (\ , 0), (\ , -6)$

2. $3x - y = 5$ $(0, \ ), (1, \ ), (\ , 5)$

3. $3x + 4y = 12$ $(0, \ ), (\ , 0), (-4, \ )$

4. $5x - 5y = 20$ $(0, \ ), (\ , -2), (1, \ )$

5. $y = 4x - 3$ $(1, \ ), (\ , 0), (5, \ )$

6. $y = 3x - 5$ $(\ , 13), (0, \ ), (-2, \ )$

7. $y = 7x - 1$ $(2, \ ), (\ , 6), (0, \ )$

8. $y = 8x + 2$ $(3, \ ), (\ , 0), (\ , -6)$

9. $x = -5$ $(\ , 4), (\ , -3), (\ , 0)$

10. $y = 2$ $(5, \ ), (-8, \ ), \left(\dfrac{1}{2}, \ \right)$

For each of the following equations, complete the given table.

11. $y = 3x$

x	y
1	3
-3	
	12
	18

12. $y = -2x$

x	y
-4	
0	
	10
	12

13. $y = 4x$

x	y
0	
	-2
-3	
	12

14. $y = -5x$

x	y
3	
	0
-2	
	-20

15. $x + y = 5$

x	y
2	
3	
	0
	-4

16. $x - y = 8$

x	y
0	
4	
	-3
	-2

17. $2x - y = 4$

x	y
	0
	2
1	
-3	

18. $3x - y = 9$

x	y
	0
	-9
5	
-4	

19. $y = 6x - 1$

x	y
0	
	-7
-3	
	8

20. $y = 5x + 7$

x	y
0	
-2	
-4	
	-8

21. $y = -2x + 3$

x	y
0	
-2	
2	
	-7

22. $y = -3x + 1$

x	y
0	
	7
	-5
-3	

B　For the following equations, tell which of the given ordered pairs are solutions.

23. $2x - 5y = 10$　$(2, 3), (0, -2), \left(\frac{5}{2}, 1\right)$

24. $3x + 7y = 21$　$(0, 3), (7, 0), (1, 2)$

25. $y = 7x - 2$　$(1, 5), (0, -2), (-2, -16)$

26. $y = 8x - 3$　$(0, 3), (5, 16), (1, 5)$

27. $y = 6x$　$(1, 6), (-2, 12), (0, 0)$

28. $y = -4x$　$(0, 0), (2, 4), (-3, 12)$

29. $x + y = 0$　$(1, 1), (2, -2), (3, 3)$

30. $x - y = 1$　$(0, 1), (0, -1), (1, 2)$

31. $x = 3$　$(3, 0), (3, -3), (5, 3)$

32. $y = -4$　$(3, -4), (-4, 4), (0, -4)$

C D For the following equations, complete the given ordered pairs, and use the results to graph the solution set for the equation.

33. $x + y = 4$ (0,), (2,), (, 0)

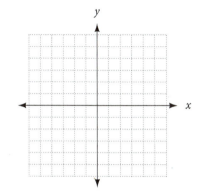

34. $x - y = 3$ (0,), (2,), (, 0)

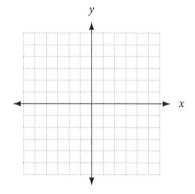

35. $x + y = 3$ (0,), (2,), (, −1)

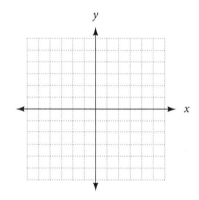

36. $x - y = 4$ (1,), (−1,), (, 0)

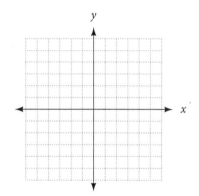

37. $y = 2x$ (0,), (−2,), (2,)

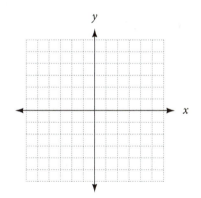

38. $y = \dfrac{1}{2}x$ (0,), (−2,), (2,)

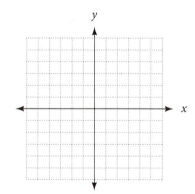

39. $y = \dfrac{1}{3}x$ (−3,), (0,), (3,)

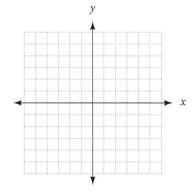

40. $y = 3x$ (−2,), (0,), (2,)

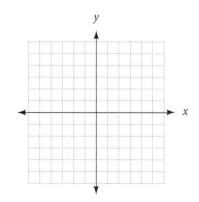

41. $y = 2x + 1$ $(0,\ ), (-1,\ ), (1,\ )$

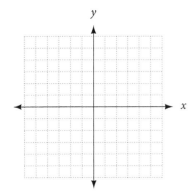

42. $y = -2x + 1$ $(0,\ ), (-1,\ ), (1,\ )$

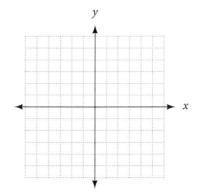

43. $y = 4$ $(0,\ ), (-1,\ ), (2,\ )$

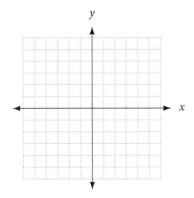

44. $x = 3$ $(\ , -2), (\ , 0), (\ , 5)$

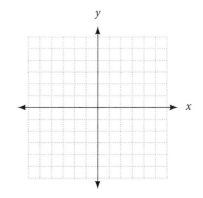

45. $y = \frac{1}{2}x + 3$ $(-2,\ ), (0,\ ), (2,\ )$

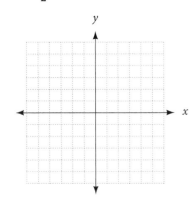

46. $y = \frac{1}{2}x - 3$ $(-2,\ ), (0,\ ), (2,\ )$

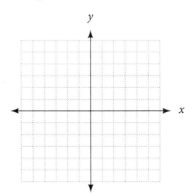

47. $y = -\frac{2}{3}x + 1$ $(-3,\ ), (0,\ ), (3,\ )$

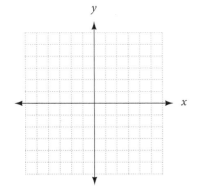

48. $y = -\frac{2}{3}x - 1$ $(-3,\ ), (0,\ ), (3,\ )$

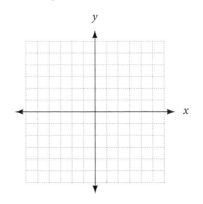

C Solve each equation for *y*. Then, complete the given ordered pairs, and use them to draw the graph.

49. $2x + y = 3$ $(-1, \ ), (0, \ ), (1, \ )$

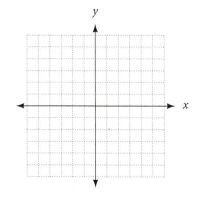

50. $3x + y = 2$ $(-1, \ ), (0, \ ), (1, \ )$

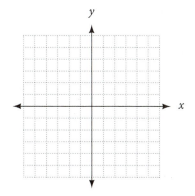

51. $3x + 2y = 6$ $(0, \ ), (2, \ ), (4, \ )$

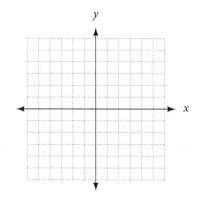

52. $2x + 3y = 6$ $(0, \ ), (3, \ ), (6, \ )$

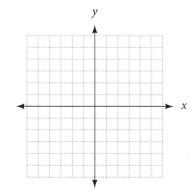

53. $-x + 2y = 6$ $(-2, \ ), (0, \ ), (2, \ )$

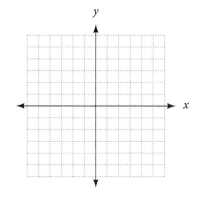

54. $-x + 3y = 6$ $(-3, \ ), (0, \ ), (3, \ )$

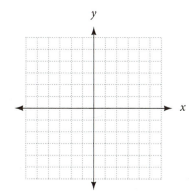

Find three solutions to each of the following equations, and then graph the solution set.

55. $y = -\dfrac{1}{2}x$

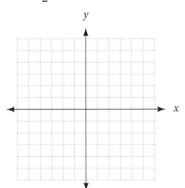

56. $y = -2x$

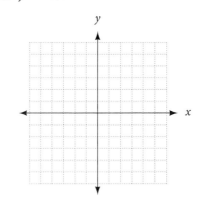

57. $y = 3x - 1$

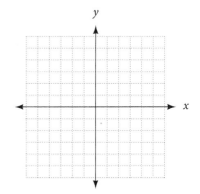

58. $y = -3x - 1$

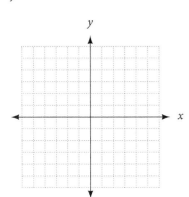

59. $-2x + y = 1$

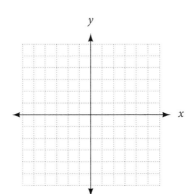

60. $-3x + y = 1$

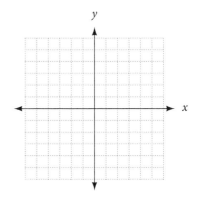

61. $3x + 4y = 8$

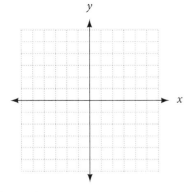

62. $3x - 4y = 8$

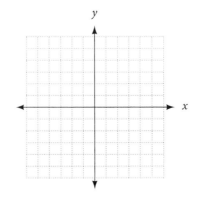

63. $x = -2$

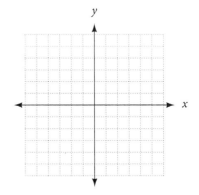

64. $y = 3$

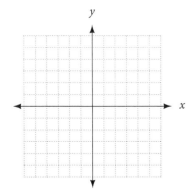

65. $y = 2$

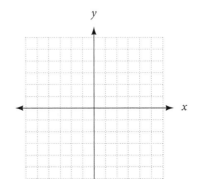

66. $x = -3$

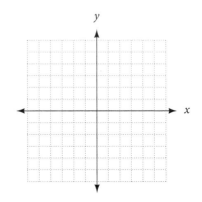

Graph each equation.

67. $y = \dfrac{3}{4}x + 1$

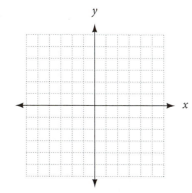

68. $y = \dfrac{2}{3}x + 1$

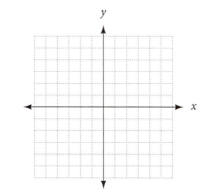

69. $y = \dfrac{1}{3}x + \dfrac{2}{3}$

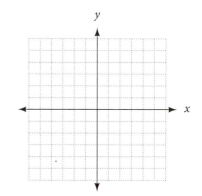

70. $y = \dfrac{1}{2}x + \dfrac{1}{2}$

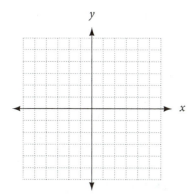

71. $y = \dfrac{2}{3}x + \dfrac{2}{3}$

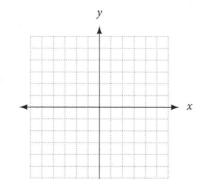

72. $y = -\dfrac{3}{4}x + \dfrac{3}{2}$

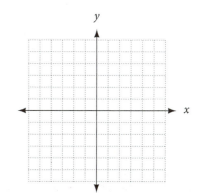

D For each equation in each table below, indicate whether the graph is horizontal (H), or vertical (V), or whether it passes through the origin (O).

73.

Equation	H, V, and/or O
$x = 3$	
$y = 3$	
$y = 3x$	
$y = 0$	

74.

Equation	H, V, and/or O
$x = \dfrac{1}{2}$	
$y = \dfrac{1}{2}$	
$y = \dfrac{1}{2}x$	
$x = 0$	

75.

Equation	H, V, and/or O
$x = -\frac{3}{5}$	
$y = -\frac{3}{5}$	
$y = -\frac{3}{5}x$	
$x = 0$	

76.

Equation	H, V, and/or O
$x = -4$	
$y = -4$	
$y = -4x$	
$y = 0$	

77. Use the graph to complete the table.

x	y
	-3
-2	
0	
	0
6	

78. Use the graph at the right to complete the table. (*Hint:* Some parts have two answers.)

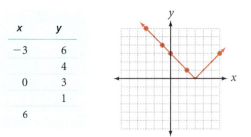

x	y
-3	6
	4
0	3
	1
6	

Applying the Concepts

79. Perimeter If the perimeter of a rectangle is 30 inches, then the relationship between the length l and the width w is given by the equation

$$2l + 2w = 30$$

What is the length when the width is 3 inches?

80. Perimeter The relationship between the perimeter P of a square and the length of its side s is given by the formula

$$P = 4s$$

If each side of a square is 5 inches, what is the perimeter? If the perimeter of a square is 28 inches, how long is a side?

81. Janai earns $12 per hour working as a math tutor. We can express the amount she earns each week, y, for working x hours with the equation $y = 12x$. Indicate with a yes or no, which of the following could be one of Janai's paychecks before deductions are taken out. If you answer no, explain your answer.

a. $60 for working five hours

b. $100 for working nine hours

c. $80 for working seven hours

d. $168 for working 14 hours

82. Erin earns $15 per hour working as a graphic designer. We can express the amount she earns each week, y, for working x hours with the equation $y = 15x$. Indicate with a yes or no which of the following could be one of Erin's paychecks before deductions are taken out. If you answer no, explain your answer.

a. $75 for working five hours

b. $125 for working nine hours

c. $90 for working six hours

d. $500 for working 35 hours

83. The equation $V = -45,000t + 600,000$, can be used to find the value, V, of a small crane at the end of t years.

   a. What is the value of the crane at the end of five years?

   b. When is the crane worth $330,000?

   c. Is it true that the crane with be worth $150,000 after nine years?

   d. How much did the crane cost?

84. The equation $V = -400t + 2,500$, can be used to find the value, V, of a notebook computer at the end of t years.

   a. What is the value of the notebook computer at the end of four years?

   b. When is the notebook computer worth $1,700?

   c. Is it true that the notebook computer with be worth $100 after five years?

   d. How much did the notebook computer cost?

Getting Ready for the Next Section

85. Let $3x + 2y = 6$.

   a. Find x when $y = 0$.

   b. Find y when $x = 0$.

88. Let $3x - y = 6$.

   a. Find x when $y = 0$.

   b. Find y when $x = 0$.

86. Let $2x - 5y = 10$.

   a. Find x when $y = 0$.

   b. Find y when $x = 0$.

89. Let $y = -\dfrac{1}{3}x + 2$.

   a. Find x when $y = 0$.

   b. Find y when $x = 0$.

87. Let $-x + 2y = 4$.

   a. Find x when $y = 0$.

   b. Find y when $x = 0$.

90. Let $y = \dfrac{3}{2}x - 3$.

   a. Find x when $y = 0$.

   b. Find y when $x = 0$.

Find the Mistake

Each sentence below contains a mistake. Circle the mistake and write the correct word(s) or number on the line provided.

   1. A solution to a linear equation in two variables will be a single number that makes the equation a true statement.

   _____

   2. A solution to the equation $3y = 6x + 9$ is $(2, 6)$. _____

   3. To graph a linear equation in two variables, you must find only one solution. _____

   4. The line $y = \dfrac{1}{2}x$ crosses the x-axis but not the y-axis. _____

**Video Examples
Section 10.3**

©iStockphoto.com/dlanier

Some talented water skiers ski on their bare feet while being towed behind a motor boat. During training, the barefoot skiers may use foot skis, which are small thin skis just a few inches longer and a little wider than the skier's foot. With foot skis, the skier can learn maneuvers faster and with fewer wipeouts than without. Normal water skiing speeds reach 30-45 miles per hour. An appropriate speed for barefoot water skiing needs to be faster and depends on the skier's body weight using the following linear equation:

$$y = \frac{x}{10} + 20$$

where *x* is the skier's weight in pounds. In this section, we will practice finding *x*- and *y*-intercepts for equations similar to the barefoot skier's equation.

A Intercepts

In this section we continue our work with graphing lines by finding the points where a line crosses the axes of our coordinate system. To do so, we use the fact that any point on the *x*-axis has a *y*-coordinate of 0 and any point on the *y*-axis has an *x*-coordinate of 0. We begin with the following definition:

> **DEFINITION** *x*-intercept, *y*-intercept
>
> The ***x*-intercept** of a straight line is the *x*-coordinate of the point where the graph intersects the *x*-axis and the *y*-coordinate is zero. The ***y*-intercept** is defined similarly. It is the *y*-coordinate of the point where the graph intersects the *y*-axis and the *x*-coordinate is zero.

B Using Intercepts to Graph Lines

If the *x*-intercept is *a*, then the point (*a*, 0) lies on the graph. (This is true because any point on the *x*-axis has a *y*-coordinate of 0.)

If the *y*-intercept is *b*, then the point (0, *b*) lies on the graph. (This is true because any point on the *y*-axis has an *x*-coordinate of 0.)

Graphically, the relationship is shown in Figure 1.

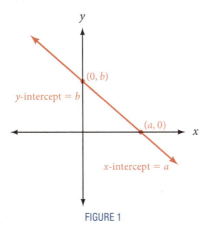

FIGURE 1

EXAMPLE 1 Find the x- and y-intercepts for $3x - 2y = 6$, and then use them to draw the graph.

Solution To find where the graph crosses the x-axis, we let $y = 0$. (The y-coordinate of any point on the x-axis is 0.)

x-intercept:

$$\begin{aligned} \text{When} \rightarrow \qquad\qquad y &= 0 \\ \text{the equation} \rightarrow \qquad 3x - 2y &= 6 \\ \text{becomes} \rightarrow \qquad 3x - 2(0) &= 6 \\ 3x - 0 &= 6 \\ x &= 2 \qquad \text{Multiply each side by } \tfrac{1}{3}. \end{aligned}$$

The graph crosses the x-axis at $(2, 0)$, which means the x-intercept is 2.

y-intercept:

$$\begin{aligned} \text{When} \rightarrow \qquad\qquad x &= 0 \\ \text{the equation} \rightarrow \qquad 3x - 2y &= 6 \\ \text{becomes} \rightarrow \qquad 3(0) - 2y &= 6 \\ 0 - 2y &= 6 \\ -2y &= 6 \\ y &= -3 \qquad \text{Multiply each side by } -\tfrac{1}{2}. \end{aligned}$$

The graph crosses the y-axis at $(0, -3)$, which means the y-intercept is -3.

Plotting the x- and y-intercepts and then drawing a line through them, we have the graph of $3x - 2y = 6$, as shown in Figure 2.

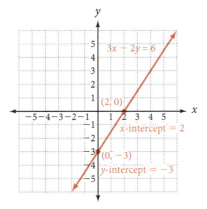

FIGURE 2

Practice Problems

1. Find the x- and y-intercepts for $2x - 5y = 10$, and use them to draw the graph.

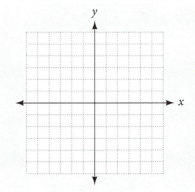

2. Graph $3x - y = 6$ by first finding the intercepts.

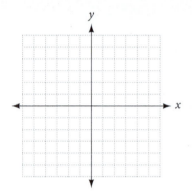

EXAMPLE 2   Graph $-x + 2y = 4$ by finding the intercepts and using them to draw the graph.

Solution   Again, we find the x-intercept by letting $y = 0$ in the equation and solving for x. Similarly, we find the y-intercept by letting $x = 0$ and solving for y.

x-intercept:

$$
\begin{aligned}
\text{When} &\rightarrow & y &= 0 \\
\text{the equation} &\rightarrow & -x + 2y &= 4 \\
\text{becomes} &\rightarrow & -x + 2(0) &= 4 \\
& & -x + 0 &= 4 \\
& & -x &= 4 \\
& & x &= -4 \qquad \text{\textit{Multiply each side by} } -1.
\end{aligned}
$$

The x-intercept is -4, indicating that the point $(-4, 0)$, is on the graph of $-x + 2y = 4$.

y-intercept:

$$
\begin{aligned}
\text{When} &\rightarrow & x &= 0 \\
\text{the equation} &\rightarrow & -x + 2y &= 4 \\
\text{becomes} &\rightarrow & -0 + 2y &= 4 \\
& & 2y &= 4 \\
& & y &= 2 \qquad \text{\textit{Multiply each side by} } \tfrac{1}{2}.
\end{aligned}
$$

The y-intercept is 2, indicating that the point $(0, 2)$ is on the graph of $-x + 2y = 4$.

Plotting the intercepts and drawing a line through them, we have the graph of $-x + 2y = 4$, as shown in Figure 3.

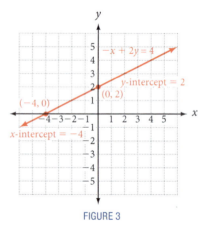

FIGURE 3

Graphing a line by finding the intercepts, as we have done in Examples 1 and 2, is an easy method of graphing if the equation has the form $ax + by = c$ and both the numbers a and b divide the number c evenly.

In our next example, we use the intercepts to graph a line in which y is given in terms of x.

EXAMPLE 3   Use the intercepts for $y = -\frac{1}{3}x + 2$ to draw its graph.

Solution   We graphed this line previously in Example 7 of the last section by substituting three different values of x into the equation and solving for y. This time we will graph the line by finding the intercepts.

x-intercept:

$$\text{When} \rightarrow \qquad y = 0$$

$$\text{the equation} \rightarrow \qquad y = -\frac{1}{3}x + 2$$

$$\text{becomes} \rightarrow \qquad 0 = -\frac{1}{3}x + 2$$

$$-2 = -\frac{1}{3}x \qquad \textcolor{green}{\text{Add } -2 \text{ to each side.}}$$

$$6 = x \qquad \textcolor{green}{\text{Multiply each side by } -3.}$$

The x-intercept is 6, which means the graph passes through the point (6, 0).

y-intercept:

$$\text{When} \rightarrow \qquad x = 0$$

$$\text{the equation} \rightarrow \qquad y = -\frac{1}{3}x + 2$$

$$\text{becomes} \rightarrow \qquad y = -\frac{1}{3}(0) + 2$$

$$y = 2$$

The y-intercept is 2, which means the graph passes through the point (0, 2). The graph of $y = -\frac{1}{3}x + 2$ is shown in Figure 4. Compare this graph, and the method used to obtain it, with Example 7 in the previous section.

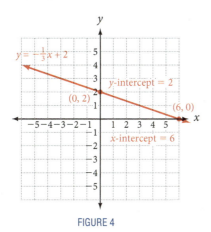

FIGURE 4

3. Use the intercepts to graph $y = \frac{3}{2}x - 3$.

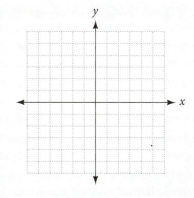

GETTING READY FOR CLASS

After reading through the preceding section, respond in your own words and in complete sentences.

A. What is the x-intercept for a graph?

B. What is the y-intercept for a graph?

C. How do you find the y-intercept for a line from the equation $3x - y = 6$?

D. How do you graph a line for $y = -\frac{1}{2}x + 3$ using its intercepts?

EXERCISE SET 10.3

Problems

A Complete each table.

1.

Equation	x-intercept	y-intercept
$3x + 4y = 12$		
$3x + 4y = 4$		
$3x + 4y = 3$		
$3x + 4y = 2$		

2.

Equation	x-intercept	y-intercept
$-2x + 3y = 6$		
$-2x + 3y = 3$		
$-2x + 3y = 2$		
$-2x + 3y = 1$		

3.

Equation	x-intercept	y-intercept
$x - 3y = 2$		
$y = \frac{1}{3}x - \frac{2}{3}$		
$x - 3y = 0$		
$y = \frac{1}{3}x$		

4.

Equation	x-intercept	y-intercept
$x - 2y = 1$		
$y = \frac{1}{2}x - \frac{1}{2}$		
$x - 2y = 0$		
$y = \frac{1}{2}x$		

B Find the x- and y-intercepts for the following equations. Then use the intercepts to graph each equation.

5. $2x + y = 4$

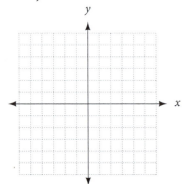

6. $2x + y = 2$

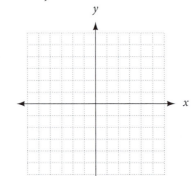

7. $-x + y = 3$

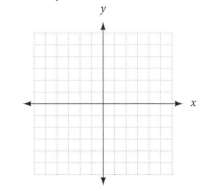

8. $-x + y = 4$

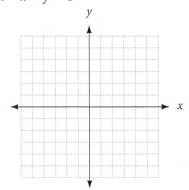

9. $-x + 2y = 2$

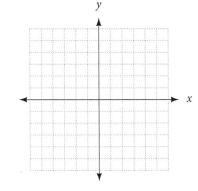

10. $-x + 2y = 4$

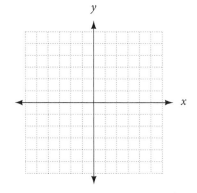

11. $5x + 2y = 10$

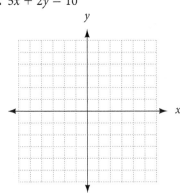

12. $2x + 5y = 10$

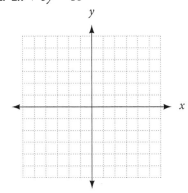

13. $4x - 2y = 8$

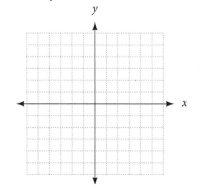

14. $2x - 4y = 8$

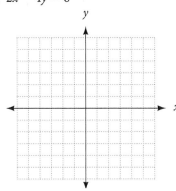

15. $-4x + 5y = 20$

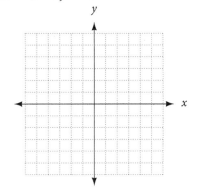

16. $-5x + 4y = 20$

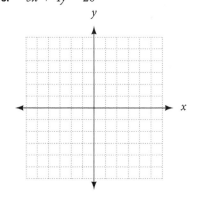

17. $y = 2x - 6$

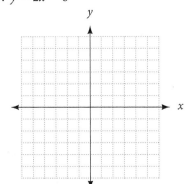

18. $y = 2x + 6$

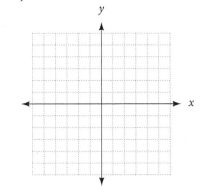

19. $y = 2x + 2$

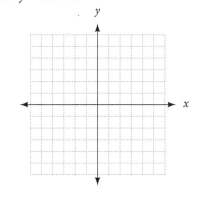

20. $y = -2x + 2$

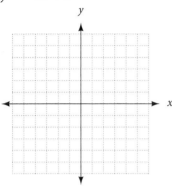

21. $y = 2x - 1$

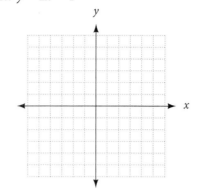

22. $y = -2x - 1$

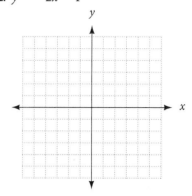

23. $y = \frac{1}{2}x + 3$

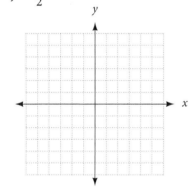

24. $y = \frac{1}{2}x - 3$

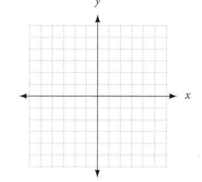

25. $y = -\frac{1}{3}x - 2$

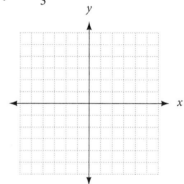

26. $y = -\frac{1}{3}x + 2$

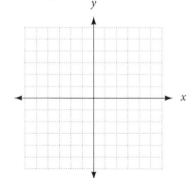

27. $y = -\frac{2}{3}x + 4$

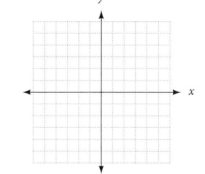

28. $y = -\frac{1}{2}x + 2$

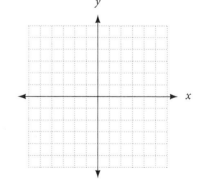

For each of the following lines the x-intercept and the y-intercept are both 0, which means the graph of each will go through the origin, $(0, 0)$. Graph each line by finding a point on each, other than the origin, and then drawing a line through that point and the origin.

29. $y = -2x$

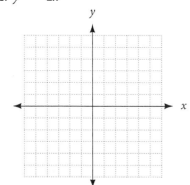

30. $y = \frac{1}{2}x$

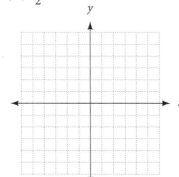

31. $y = -\frac{1}{3}x$

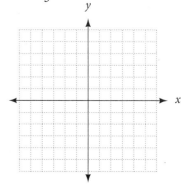

32. $y = -3x$

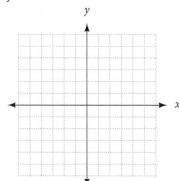

33. $y = \frac{2}{3}x$

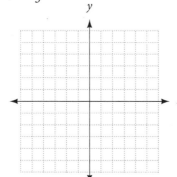

34. $y = \frac{3}{2}x$

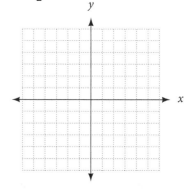

The next two problems are intended to give you practice reading, and paying attention to, the instructions that accompany the problems you are working. Working these problems is an excellent way to get ready for a test or a quiz.

35. Work each problem according to the instructions.

  a. Solve: $2x - 3 = -3$.

  b. Find the x-intercept: $2x - 3y = -3$.

  c. Find y when x is 0: $2x - 3y = -3$.

  d. Graph: $2x - 3y = -3$.

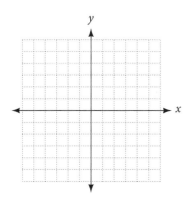

  e. Solve for y: $2x - 3y = -3$.

36. Work each problem according to the instructions.

  a. Solve: $3x - 4 = -4$.

  b. Find the y-intercept: $3x - 4y = -4$.

  c. Find x when y is 0: $3x - 4y = -4$.

  d. Graph: $3x - 4y = -4$.

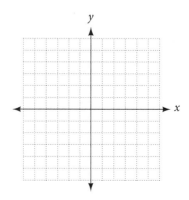

  e. Solve for y: $3x - 4y = -4$.

From the graphs below, find the *x*- and *y*-intercepts for each line.

37.

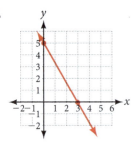

38.

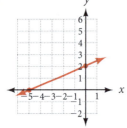

39.

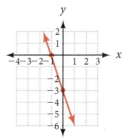

40.

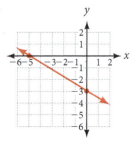

41. Graph the line that passes through the point $(-4, 4)$ and has an *x*-intercept of -2. What is the *y*-intercept of this line?

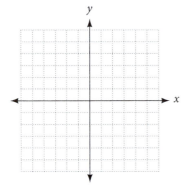

42. Graph the line that passes through the point $(-3, 4)$ and has a *y*-intercept of 1. What is the *x*-intercept of this line?

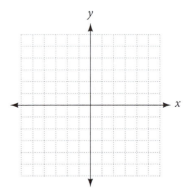

43. A line passes through the point $(1, 4)$ and has a *y*-intercept of 3. Graph the line and name its *x*-intercept.

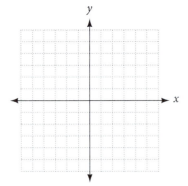

44. A line passes through the point $(3, 4)$ and has an *x*-intercept of 1. Graph the line and name its *y*-intercept.

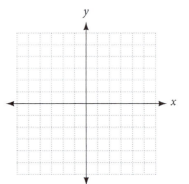

45. Graph the line that passes through the points $(-2, 5)$ and $(5, -2)$. What are the *x*- and *y*-intercepts for this line?

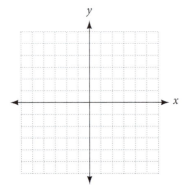

46. Graph the line that passes through the points $(5, 3)$ and $(-3, -5)$. What are the *x*- and *y*-intercepts for this line?

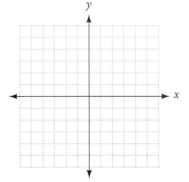

47. The vertical line $x = 3$ has only one intercept. Graph $x = 3$, and name its intercept. [Remember, ordered pairs (x, y) that are solutions to the equation $x = 3$ are ordered pairs with an x-coordinate of 3 and any y-coordinate.]

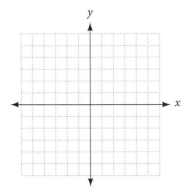

48. Graph the vertical line $x = -2$. Then name its intercept.

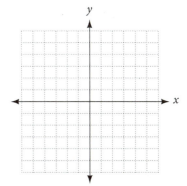

49. The horizontal line $y = 4$ has only one intercept. Graph $y = 4$, and name its intercept. [Ordered pairs (x, y) that are solutions to the equation $y = 4$ are ordered pairs with a y-coordinate of 4 and any x-coordinate.]

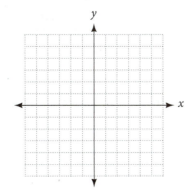

50. Graph the horizontal line $y = -3$. Then name its intercept.

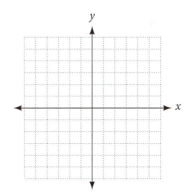

Applying the Concepts

51. Seating Capacity A theater has a capacity of 500 people. Because a different price is charged for students than for general admission, the owners would like to know what combinations of students and general tickets can be sold when the show is sold out. The relationship between student and general tickets can be written as

$$x + y = 500$$

Graph this equation where the x-axis is student tickets and the y-axis is general tickets. Find the intercepts first, and limit your graph to the first quadrant.

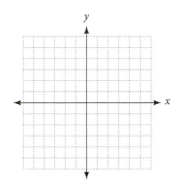

52. **Ladder Angles** A 13-foot ladder is placed against a barn. As the top of the ladder moves up and down the wall, the angles made by the ladder and the wall, α, and the ladder and the ground, β, can be written as

$$\alpha + \beta = 90.$$

Graph this equation on a coordinate system where α is the y-axis and β is the x-axis. Find the intercepts first, and limit your graph to the first quadrant only.

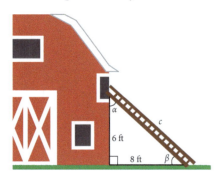 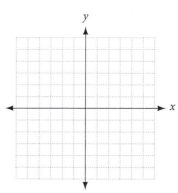

53. **Working Two Jobs** Maggie has a job working in an office for $10 an hour and another job driving a tractor pays $12 an hour. Maggie works at both jobs and earns $480 in one week.

a. Write an equation that gives the number of hours x, she worked for $10 per hour and the number of hours, y, she worked for $12 per hour.

b. Find the x- and y-intercepts for this equation.

c. Graph this equation from the intercepts, using only the first quadrant.

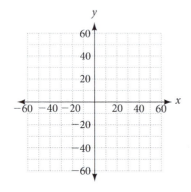

d. From the graph, find how many hours she worked at $12, if she worked 36 hours at $10 per hour?

e. From the graph, find how many hours she worked at $10, if she worked 25 hours at $12 per hour?

54. **Ticket prices** Devin is selling tickets to a BBQ. Adult tickets cost $6.00 and children's tickets cost $4.00. The total amount of money he collects is $240.

a. Write an equation that gives the number of adult tickets x, he sold for $6 and the children's tickets, y, he sold for $4

b. Find the x- and y-intercepts for this equation.

c. Graph this equation from the intercepts, using only the first quadrant.

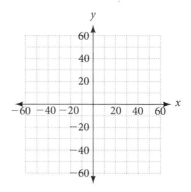

d. From the graph, find how many children's tickets were sold if 30 adult tickets sold?

e. From the graph, find how many adult tickets were sold if 30 children's tickets sold?

55. Brian drives a certain amount of time at 50 mph and another amount at 60 mph. Together he drives 450 miles.

   a. Write an equation involving x (time driving at 50 mph) and y (time driving at 60 mph).

   b. Find the x- and y- intercepts for the equation in part a.

   c. Graph this equation in the first quadrant only.

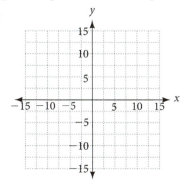

   d. From the graph, find how many hours Brian drove at 60 mph if he drove a total of 8 hours?

   e. If Brian drove at 50 mph for 4 hours, how long did he drive at 60 mph?

56. Juan tutors in the math lab for $8 per hour and privately for $12 per hour. He earns $240 in one week.

   a. Write and equation involving x (number of math lab hours) and y (number of private hours).

   b. Find the x- and y- intercepts for this equation.

   c. Graph this equation in the first quadrant only.

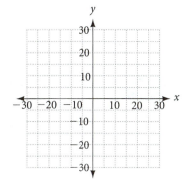

   d. From the graph, find how many hours did Juan tutor in the math lab if he tutored privately for 10 hours?

   e. If Juan is limited to tutoring 19 hours in the math lab, how many hours must he tutor privately?

Getting Ready for the Next Section

57. Evaluate

   a. $\dfrac{5-2}{3-1}$

   b. $\dfrac{2-5}{1-3}$

58. Evaluate

   a. $\dfrac{-4-1}{5-(-2)}$

   b. $\dfrac{1+4}{-2-5}$

59. Evaluate the following expressions when $x = 3$, and $y = 5$.

   a. $\dfrac{y-2}{x-1}$

   b. $\dfrac{2-y}{1-x}$

60. Evaluate the following expressions when $x = 4$, and $y = -1$.

   a. $\dfrac{-4-y}{5-x}$

   b. $\dfrac{y+4}{x-5}$

Find the Mistake

Each sentence below contains a mistake. Circle the mistake and write the correct word or number on the line provided.

1. The y–intercept is where the graph of a line crosses the x–axis at the point $(0, y)$. _____

2. The x–intercept for the linear equation $9x = \frac{1}{3}y + 5$ is $(-15, 0)$. _____

3. The y–intercept for the linear equation $2x + 7y = 3$ is $\left(0, \frac{3}{2}\right)$. _____

4. The intercepts for the linear equation $15 = 3y - 5x$ are $(5, 0)$ and $(0, -3)$. _____

KEY WORDS

slope

rise

run

**Video Examples
Section 10.4**

10.4 Graphing Linear Equations Using Slope

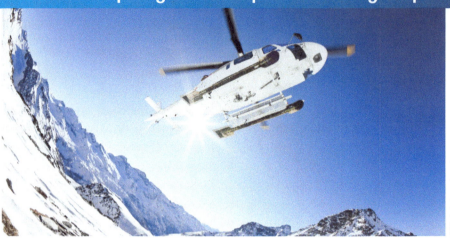

©iStockphoto.com/ rcaucino

Heli-skiing is a sport where a helicopter transports a skier to the top of the mountain he will ski down. The skier leaps out of the helicopter's cabin and onto an untracked slope, often inaccessible from the ground.

Let's suppose for every 30 feet of horizontal distance the heli-skier covers, his elevation drops 10 feet. In this section, we will learn how to use these numbers to assign a numerical value to the slope of the hill.

In defining the slope of a straight line, we are looking for a number to associate with a straight line that does two things. First of all, we want the slope of a line to measure the "steepness" of the line; that is, in comparing two lines, the slope of the steeper line should have the larger numerical value. Second, we want a line that *rises* going from left to right to have a *positive* slope. We want a line that *falls* going from left to right to have a *negative* slope. (A horizontal line that neither rises nor falls going from left to right must, therefore, have 0 slope; whereas, a vertical line has an undefined slope.) These are illustrated in Figure 1.

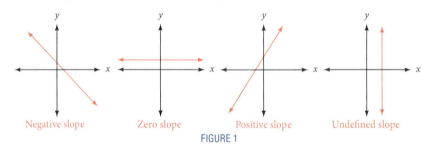

Negative slope   Zero slope   Positive slope   Undefined slope

FIGURE 1

Suppose we know the coordinates of two points on a line. Because we are trying to develop a general formula for the slope of a line, we will use general points—let's call the two points $P_1(x_1, y_1)$ and $P_2(x_2, y_2)$. They represent the coordinates of any two different points on our line. We define the *slope* of our line to be the ratio of the vertical change of *y*-coordinates to the horizontal change of *x*-coordinates as we move from point (x_1, y_1) to point (x_2, y_2) on the line. (See Figure 2 on next page.)

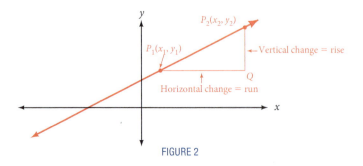

FIGURE 2

We call the vertical change the *rise* in the graph and the horizontal change the *run* in the graph. The slope, then, is

$$\text{Slope} = \frac{\text{vertical change}}{\text{horizontal change}} = \frac{\text{rise}}{\text{run}}$$

We would like to have a numerical value to associate with the rise in the graph and a numerical value to associate with the run in the graph. A quick study of Figure 2 shows that the coordinates of point Q must be (x_2, y_1), because Q is directly below point P_2 and right across from point P_1. We can draw our diagram again in the manner shown in Figure 3. It is apparent from this graph that the rise can be expressed as $(y_2 - y_1)$ and the run as $(x_2 - x_1)$. We usually denote the slope of a line by the letter m. The complete definition of slope follows along with a diagram (Figure 3) that illustrates the definition.

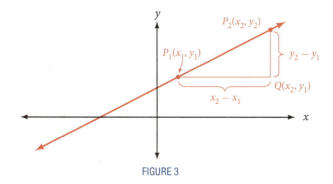

FIGURE 3

> **DEFINITION** slope
>
> If points (x_1, y_1) and (x_2, y_2) are any two different points and $x_1 \neq x_2$, then the *slope* of the line on which they lie is, denoted by m,
>
> $$\text{Slope} = m = \frac{\text{rise}}{\text{run}} = \frac{y_2 - y_1}{x_2 - x_1}$$

This definition of the *slope* of a line does just what we want it to do. If the line rises going from left to right, the slope will be positive. If the line falls from left to right, the slope will be negative. Also, the steeper the line, the larger numerical value the slope will have.

A Finding the Slope Given Two Points on a Line

EXAMPLE 1 Find the slope of the line between the points $(1, 2)$ and $(3, 5)$.

Solution We can let

$$(x_1, y_1) = (1, 2)$$

and

$$(x_2, y_2) = (3, 5)$$

then

$$m = \frac{y_2 - y_1}{x_2 - x_1} = \frac{5 - 2}{3 - 1} = \frac{3}{2}$$

The slope is $\frac{3}{2}$. For every vertical change of 3 units, there will be a corresponding horizontal change of 2 units. (See Figure 4.)

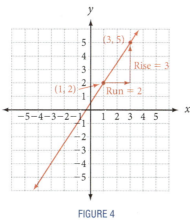

FIGURE 4

Practice Problems

1. Find the slope of the line between the points $(4, -3)$ and $(-2, 4)$.

Answer

1. $-\frac{7}{6}$

2. Find the slope of the line through $(3, 4)$ and $(1, -2)$.

EXAMPLE 2   Find the slope of the line through $(-2, 1)$ and $(5, -4)$.

Solution   It makes no difference which ordered pair we call (x_1, y_1) and which we call (x_2, y_2).

$$\text{Slope} = m = \frac{y_2 - y_1}{x_2 - x_1} = \frac{-4 - 1}{5 - (-2)} = -\frac{5}{7}$$

The slope is $-\frac{5}{7}$. Every vertical change of -5 units (down 5 units) is accompanied by a horizontal change of 7 units (to the right 7 units). (See Figure 5.)

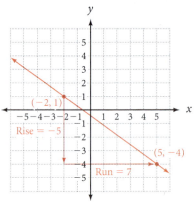

FIGURE 5

B Graphing a Line from Its Slope

3. Graph the line with slope $\frac{1}{4}$ and y-intercept -3.

EXAMPLE 3   Graph the line with slope $\frac{3}{2}$ and y-intercept 1.

Solution   Because the y-intercept is 1, we know that one point on the line is $(0, 1)$. So, we begin by plotting the point $(0, 1)$, as shown in Figure 6.

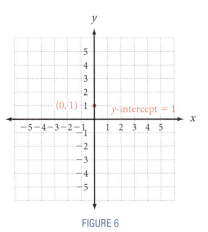

FIGURE 6

There are many lines that pass through the point shown in Figure 6, but only one of those lines has a slope of $\frac{3}{2}$. The slope, $\frac{3}{2}$, can be thought of as the rise in the graph divided by the run in the graph. Therefore, if we start at the point $(0, 1)$ and move 3 units up (that's a rise of 3) and then 2 units to the right (a run of 2), we will be at another point on the graph. Figure 7 shows that the point we reach by doing so is the point $(2, 4)$.

Answer

2. 3

$$\text{Slope} = m = \frac{\text{rise}}{\text{run}} = \frac{3}{2}$$

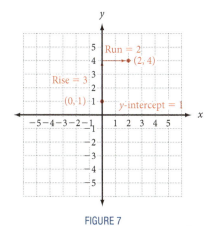

FIGURE 7

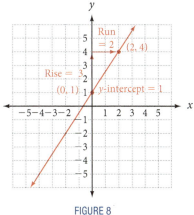

FIGURE 8

To graph the line with slope $\frac{3}{2}$ and y-intercept 1, we simply draw a line through the two points in Figure 7 to obtain the graph shown in Figure 8.

EXAMPLE 4 Find the slope of the line containing $(3, -1)$ and $(3, 4)$.

Solution Using the definition for slope, we have

$$m = \frac{y_2 - y_1}{x_2 - x_1} = \frac{4 - (-1)}{3 - 3} = \frac{5}{0}$$

The expression $\frac{5}{0}$ is undefined; that is, there is no real number to associate with it. In this case, we say the line *has an undefined slope*.

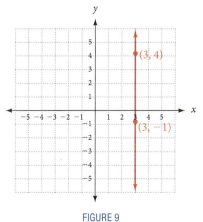

FIGURE 9

The graph of our line is shown in Figure 9. All vertical lines have an undefined slope. (And all horizontal lines, as we mentioned earlier, have a slope of zero.)

As a final note, the following summary reminds us that all horizontal lines have equations of the form $y = b$ and slopes of 0. Because they cross the y-axis at b, the y-intercept is b; there is no x-intercept. Vertical lines have an undefined slope and equations of the form $x = a$. Each will have an x-intercept at a and no y-intercept. Finally, equations of the form $y = mx$ have graphs that pass through the origin. The slope is always m and both the x-intercept and the y-intercept are 0.

4. Find the slope of the line through $(2, -3)$ and $(-1, -3)$.

Note Exceptions to this discussion are the lines $y = 0$ and $x = 0$. For $y = 0$, the x-intercepts are infinite; whereas $x = 0$ has infinite y-intercepts.

Answer

4. 0

FACTS FROM GEOMETRY Special Equations and Their Graphs, Slopes, and Intercepts

For the equations below, m, a, and b are real numbers.

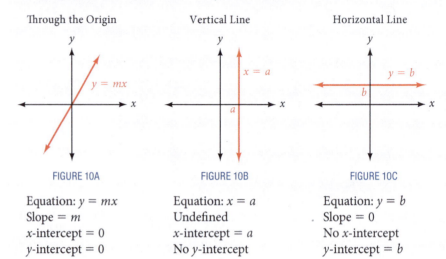

Through the Origin

FIGURE 10A

Vertical Line

FIGURE 10B

Horizontal Line

FIGURE 10C

Equation: $y = mx$
Slope $= m$
x-intercept $= 0$
y-intercept $= 0$

Equation: $x = a$
Undefined
x-intercept $= a$
No y-intercept

Equation: $y = b$
Slope $= 0$
No x-intercept
y-intercept $= b$

GETTING READY FOR CLASS

After reading through the preceding section, respond in your own words and in complete sentences.

A. Using x- and y-coordinates, how do you find the slope of a line?

B. Would you rather climb a hill with a slope of 1 or a slope of 3? Explain why.

C. Describe how you would graph a line from its slope and y-intercept.

D. Describe how to obtain the slope of a line if you know the coordinates of two points on the line.

Vocabulary Review

Choose the correct words to fill in the blanks below.

vertical   undefined   ratio   zero   horizontal   slope

1. The slope of a line is the _____ of the vertical change of y-coordinates to the horizontal change of x-coordinates as we move from point (x_1, y_1) to point (x_2, y_2) on the line.

2. The _____ change between one point on a straight line to another point is called the rise.

3. The _____ change between one point on a straight line to another point is called the run.

4. The _____ of a line is given by the equation $m = \frac{y_2 - y_1}{x_2 - x_1}$.

5. A horizontal line has a(n) _____ slope.

6. A vertical line has a(n) _____ slope.

Problems

A Find the slope of the line through the following pairs of points. Then plot each pair of points, draw a line through them, and indicate the rise and run in the graph in the same manner shown in Examples 1 and 2.

1. $(2, 1), (4, 4)$ **2.** $(3, 1), (5, 4)$ **3.** $(1, 4), (5, 2)$

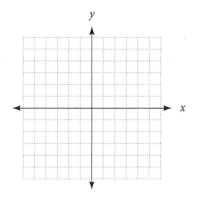 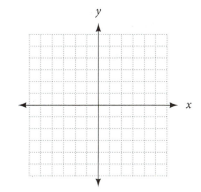 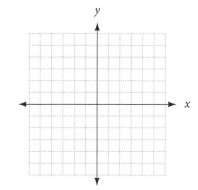

4. $(1, 3), (5, 2)$ **5.** $(1, -3), (4, 2)$ **6.** $(2, -3), (5, 2)$

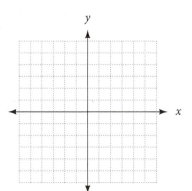 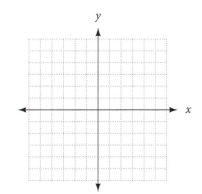 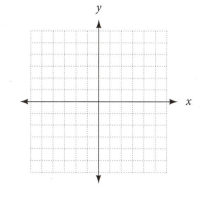

7. $(-3, -2), (1, 3)$

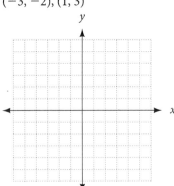

8. $(-3, -1), (1, 4)$

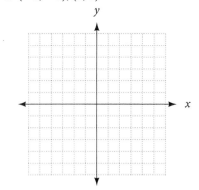

9. $(-3, 2), (3, -2)$

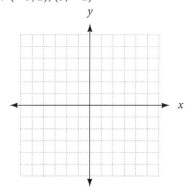

10. $(-3, 3), (3, -1)$

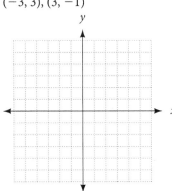

11. $(2, -5), (3, -2)$

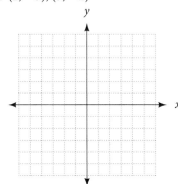

12. $(2, -4), (3, -1)$

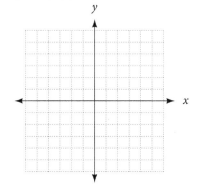

13. $(2, 4), (2, -3)$

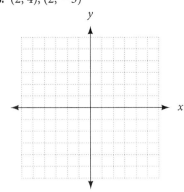

14. $(2, 3), (-1, 3)$

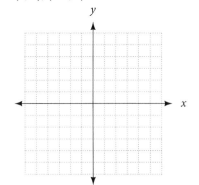

B In each of the following problems, graph the line with the given slope and y-intercept b.

15. $m = \dfrac{2}{3}, b = 1$

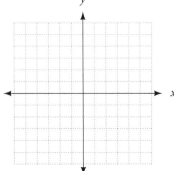

16. $m = \dfrac{3}{4}, b = -2$

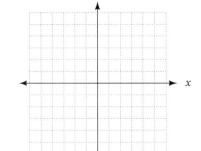

17. $m = \dfrac{3}{2}, b = -3$

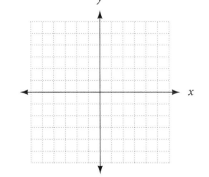

18. $m = \frac{4}{3}, b = 2$

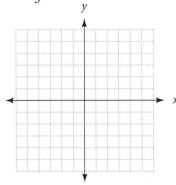

19. $m = -\frac{4}{3}, b = 5$

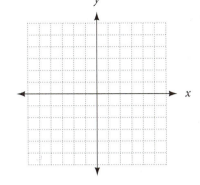

20. $m = -\frac{3}{5}, b = 4$

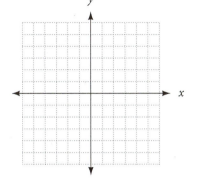

21. $m = 2, b = 1$

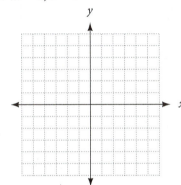

22. $m = -2, b = 4$

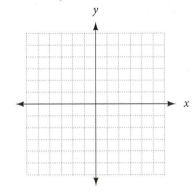

23. $m = 3, b = -1$

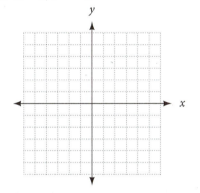

24. $m = 3, b = -2$

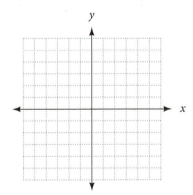

25. $m = 1, b = 0$

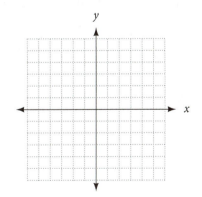

26. $m = -1, b = 0$

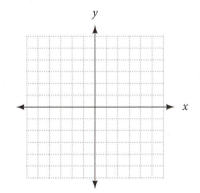

Find the slope and y-intercept for each line.

27.

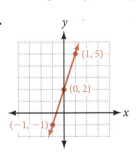

28.

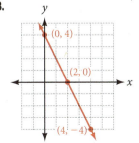

29.

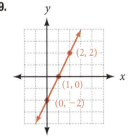

30.

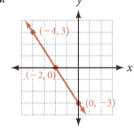

31. Graph the line that has an x-intercept of 3 and a y-intercept of -2. What is the slope of this line?

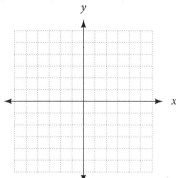

32. Graph the line that has an x-intercept of 2 and a y-intercept of -3. What is the slope of this line?

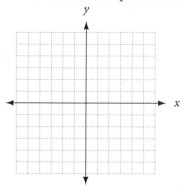

33. Graph the line with x-intercept 4 and y-intercept 2. What is the slope of this line?

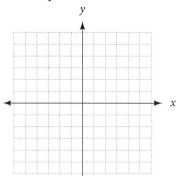

34. Graph the line with x-intercept -4 and y-intercept -2. What is the slope of this line?

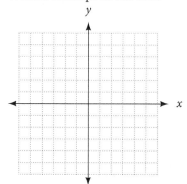

35. Graph the line $y = 2x - 3$, then name the slope and y-intercept by looking at the graph.

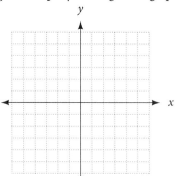

36. Graph the line $y = -2x + 3$, then name the slope and y-intercept by looking at the graph.

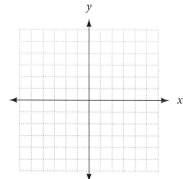

37. Graph the line $y = \frac{1}{2}x + 1$, then name the slope and y-intercept by looking at the graph.

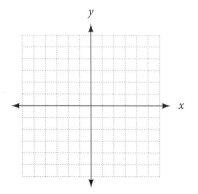

38. Graph the line $y = -\frac{1}{2}x - 2$, then name the slope and y-intercept by looking at the graph.

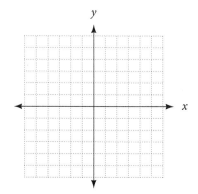

39. Find y if the line through $(4, 2)$ and $(6, y)$ has a slope of 2.

40. Find y if the line through $(1, y)$ and $(7, 3)$ has a slope of 6.

41. Find y if the line through $(3, 2)$ and $(3, y)$ has an undefined slope.

42. Find y if the line through $(2, 3)$ and $(5, y)$ has a slope of 0.

For each equation in each table, give the slope of the graph.

43.

Equation	Slope
$x = 3$	
$y = 3$	
$y = 3x$	

44.

Equation	Slope
$y = \frac{3}{2}$	
$x = \frac{3}{2}$	
$y = \frac{3}{2}x$	

45.

Equation	Slope
$y = -\frac{2}{3}$	
$x = -\frac{2}{3}$	
$y = -\frac{2}{3}x$	

46.

Equation	Slope
$x = -2$	
$y = -2$	
$y = -2x$	

Applying the Concepts

47. Garbage Production The table and completed line graph gives the annual production of garbage in the United States for some specific years. Find the slope of each of the four line segments, A, B, C, and D.

Year	Garbage (millions of tons)
1960	88
1970	121
1980	152
1990	205
2000	224
2007	254

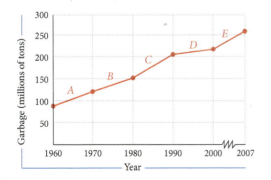

48. Grass Height The table and completed line graph gives the growth of a certain grass species over time. Find the slopes of the line segments labeled A, B, and C.

Day	Plant Height
0	0
2	1
4	3
6	6
8	13
10	23

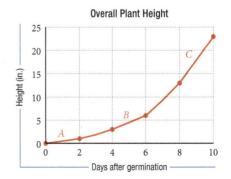

49. Non-Camera Phone Sales The table and line graph here each show the estimated non-camera phone sales each year through 2010. Find the slope of each of the three line segments, A, B, and C.

Year	Sales (in millions)
2006	300
2007	250
2008	175
2009	150
2010	125

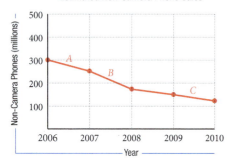

50. Camera Phone Sales The table and line graph here each show the estimated sales of camera phones from 2006 to 2010. Find the slopes of line segments A, B, and C.

Year	Sales (in millions)
2006	500
2007	650
2008	750
2009	875
2010	900

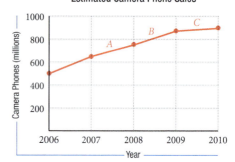

51. The federal budget deficit has become a very common political topic. The following data lists the total national outstanding debt over some selected years. Use this data to answer the questions that follow.

Year	Debt (trillions of $)
1980	0.9
1990	3.2
2000	5.7
2010	13.6

a. Using the table, sketch a line graph of A, B, and C. Add a graph.

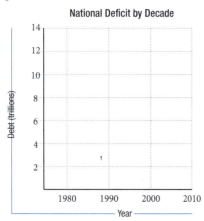

b. Calculate the slopes of the line segments.

c. Based on these slopes, is the rate of change of the debt increasing or decreasing?

Nike Women's Marathon The following chart shows the number of finishers at the Nike Women's Marathon in San Francisco, CA from 2005 to 2010.

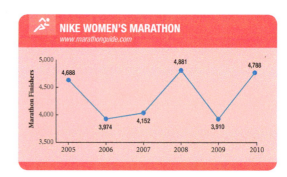

52. Find the slope of the line from 2005 to 2010. Explain in words what the slope represents.

53. Find the slope of the line from 2009 to 2010. Explain in words what the slope represents.

54. Heating a Block of Ice A block of ice with an initial temperature of −20°C is heated at a steady rate. The graph shows how the temperature changes as the ice melts to become water and the water boils to become steam and water.

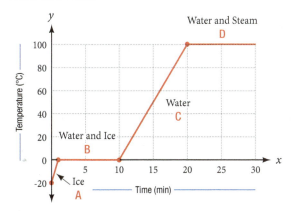

a. How long does it take all the ice to melt?

b. From the time the heat is applied to the block of ice, how long is it before the water boils?

c. Find the slope of the line segment labeled A. What units would you attach to this number?

d. Find the slope of the line segment labeled C. Be sure to attach units to your answer.

e. Is the temperature changing faster during the 1st minute or the 16th minute?

55. Slope of a Highway A sign at the top of the Cuesta Grade, outside of San Luis Obispo, reads "7% downgrade next 3 miles." The following diagram is a model of the Cuesta Grade that illustrates the information on that sign.

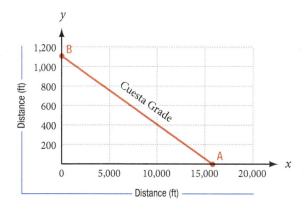

a. At point B, the graph crosses the *y*-axis at 1,106 feet. How far is it from the origin to point A?

b. What is the slope of the Cuesta Grade?

56. Facebook Users The graph below shows the number of Facebook users over a six year period. Using the graph, find the slope of the line connecting the first (2004, 1) and last (2010, 400) points on the line. Explain in words what the slope represents.

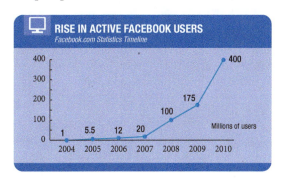

57. Mac Users The graph shows the average increase in the percentage of computer users who use Macs since 2003. Find the slope of the line connecting the first (2003, 2.2) and last (2010, 6.7) points on the line. Explain in words what the slope represents.

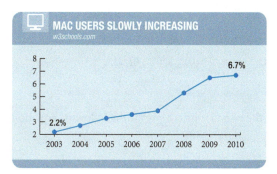

Getting Ready for the Next Section

Solve each equation for y.

58. $-2x + y = 4$ **59.** $-4x + y = -2$ **60.** $2x + y = 3$ **61.** $3x + 2y = 6$

62. $x + 2y = 4$ **63.** $x - 3y = 6$ **64.** $2x - 3y = 12$ **65.** $3x + 4y = 12$

66. $4x - 5y = 20$ **67.** $-2x - 5y = 10$ **68.** $-y - 3 = -2(x + 4)$ **69.** $-y + 5 = 2(x + 2)$

70. $-y - 3 = -\dfrac{2}{3}(x + 3)$ **71.** $-y - 1 = -\dfrac{1}{2}(x + 4)$ **72.** $-\dfrac{y - 1}{x} = \dfrac{3}{2}$ **73.** $-\dfrac{y + 1}{x} = \dfrac{3}{2}$

Find the Mistake

Each sentence below contains a mistake. Circle the mistake and write the correct word or number on the line provided.

1. A horizontal line has a positive slope. _____

2. A line that rises from the left to the right has a negative slope. _____

3. The slope of the line that passes through the points $(-1, 6)$ and $(4, -9)$ is $-\dfrac{1}{3}$. _____

4. Use the point $(3, -2)$ to graph a line with a y–intercept -6 and a slope $\dfrac{5}{2}$. _____

Landmark Review: Checking Your Progress

1. What is the name of the point with the coordinates $(0, 0)$?

2. Name the x-coordinate of any point that is on the line passing through the points $(4, 3)$ and $(4, 8)$.

3. Is the point $(5, 3)$ on the line that passes through $(2, 4)$ and $(4, 4)$?

For each equation, complete the ordered pairs.

4. $4x - y = 3$ $(0, \ ), (\ , 0), (3, \ )$

5. $2y - 3x = 15$ $(\ , 3), (1, \ ), (6, \ )$

Find the x- and y-intercepts for the following equations.

6. $x + y = 1$

7. $x - 2y = 6$

Find the slope of the line through the following pairs of points.

8. $(1, 3), (5, 4)$ **9.** $(4, 2), (3, 4)$ **10.** $(3, 4), (3, 2)$

11. Would you expect a line with a positive slope to rise or fall moving from left to right?

12. If a line has a slope of 0 and doesn't pass through the origin, what do you know about the x-intercept?

Image © sxc.hu, gundolf 2007

OBJECTIVES

A Find the equation of a line given the slope and y-intercept of the line.

B Find the slope and y-intercept of a line given the equation of the line.

C Find the equation of a line given a point on the line and the slope of the line.

D Find the equation of a line given two points on the line.

E Find the equations of parallel and perpendicular lines.

KEY WORDS

slope-intercept form

point-slope form

parallel

perpendicular

Every winter, the Angel Fire Resort in New Mexico hosts the World Shovel Racing Championships. At the top of a snow-covered slope, contestants sit straddling the handle of a stock grain scoop shovel. Then they speed down the 1,000-foot course using the handle to steer. Helmets are required because wipeouts are common. Let's say the fastest racer finished the course in 17 seconds and the runner up finished in 18 seconds. That gives us the two points (17, 1000) and (18, 1000) that could help us draw the graph for the race. In this section, we will learn ways to find the equation of a line using slope, intercepts, and given points.

To this point in the chapter, most of the problems we have worked have used the equation of a line to find different types of information about the line. For instance, given the equation of a line, we can find points on the line, the graph of the line, the intercepts, and the slope of the line. In this section, we reverse things somewhat and use information about a line, such as its slope and y-intercept, to find the line's equation.

There are three main types of problems to solve in this section.

   1. Find the equation of a line from the slope and the y-intercept.

   2. Find the equation of a line given one point on the line and the slope of the line.

   3. Find the equation of a line given two points on the line.

Examples 1 and 2 illustrate the first type of problem. Example 5 solves the second type of problem. We solve the third type of problem in Example 6.

A The Slope-Intercept Form of an Equation of a Straight Line

EXAMPLE 1   Find the equation of the line with slope $\frac{3}{2}$ and y-intercept 1.

Solution   We graphed the line with slope $\frac{3}{2}$ and y-intercept 1 in Example 3 of the previous section. Figure 1 shows that graph.

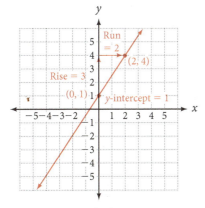

FIGURE 1   **10.5** Finding the Equation of a Line   **549**

**Video Examples
Section 10.5**

Practice Problems

   1. Find the equation of the line with slope $\frac{4}{3}$ and y-intercept 2.

Answer

1. $y = \frac{4}{3}x + 2$

What we want to do now is find the equation of the line shown in Figure 1. To do so, we take any other point (x, y) on the line and apply our slope formula to that point and the point $(0, 1)$. We set that result equal to $\frac{3}{2}$, because $\frac{3}{2}$ is the given slope of our line.

$$m = \frac{\text{rise}}{\text{run}} = \frac{y - 1}{x - 0} = \frac{3}{2} \qquad \text{Slope} = \frac{\text{vertical change}}{\text{horizontal change}}.$$

$$\frac{y - 1}{x} = \frac{3}{2} \qquad\qquad x - 0 = x.$$

$$y - 1 = \frac{3}{2}x \qquad\qquad \text{Multiply each side by } x.$$

$$y = \frac{3}{2}x + 1 \qquad\qquad \text{Add 1 to each side.}$$

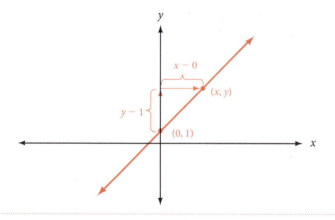

What is interesting and useful about the equation we have just found is that the number in front of x is the slope of the line and the constant term is the y-intercept. It is no coincidence that it turned out this way. Whenever an equation has the form $y = mx + b$, the graph is always a straight line with slope m and y-intercept b. To see that this is true in general, suppose we want the equation of a line with slope m and y-intercept b. Because the y-intercept is b, then the point $(0, b)$ is on the line. If (x, y) is any other point on the line, then we apply our slope formula to get

$$\frac{y - b}{x - 0} = m \qquad\qquad \text{Slope} = \frac{\text{vertical change}}{\text{horizontal change}}.$$

$$\frac{y - b}{x} = m \qquad\qquad x - 0 = x.$$

$$y - b = mx \qquad\qquad \text{Multiply each side by } x.$$

$$y = mx + b \qquad\qquad \text{Add } b \text{ to each side.}$$

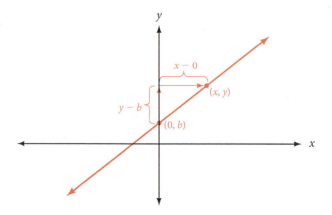

Here is a summary of what we have just found:

> **Slope-Intercept Form of the Equation of a Line**
>
> The equation of the non-vertical line with slope m and y-intercept b is always given by
>
> $$y = mx + b$$

EXAMPLE 2 Find the equation of the line with slope $-\frac{4}{3}$ and y-intercept 5. Then, graph the line.

Solution Substituting $m = -\frac{4}{3}$ and $b = 5$ into the equation $y = mx + b$, we have

$$y = -\frac{4}{3}x + 5$$

Finding the equation from the slope and y-intercept is just that easy. If the slope is m and the y-intercept is b, then the equation is always $y = mx + b$.

Because the y-intercept is 5, the graph goes through the point $(0, 5)$. To find a second point on the graph, we start at $(0, 5)$ and move 4 units down (that's a rise of -4) and 3 units to the right (a run of 3). The point we reach is $(3, 1)$. Drawing a line that passes through $(0, 5)$ and $(3, 1)$, we have the graph of our equation. (Note that we could also let the rise $= 4$ and the run $= -3$ and obtain the same graph.) The graph is shown in Figure 2.

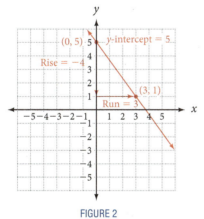

FIGURE 2

B Finding the Slope and *y*-Intercept

EXAMPLE 3 Find the slope and y-intercept for $-2x + y = -4$. Then, use them to draw the graph.

Solution To identify the slope and y-intercept from the equation, the equation must be in the form $y = mx + b$ (slope-intercept form). To write our equation in this form, we must solve the equation for y. To do so, we simply add $2x$ to each side of the equation.

$$-2x + y = -4 \qquad \text{Original equation}$$
$$y = 2x - 4 \qquad \text{Add } 2x \text{ to each side.}$$

The equation is now in slope-intercept form, so the slope must be 2 and the y-intercept must be -4. The graph, therefore, crosses the y-axis at $(0, -4)$. Because the slope is 2, we can let the rise $= 2$ and the run $= 1$ because $y = 2x - 4$ can also be written as $y = \frac{2}{1}x - 4$. We can then find a second point on the graph, which is $(1, -2)$. The graph is shown in Figure 3.

2. Find the equation of the line with slope $\frac{2}{3}$ and y-intercept 1. Then graph the line.

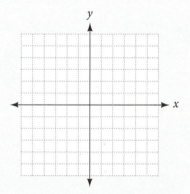

3. Find the slope and y-intercept for $3x - 2y = -6$. Then use them to draw the graph.

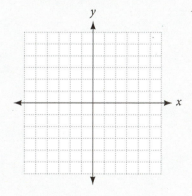

Answers

2. $y = \frac{2}{3}x + 1$

3. Slope is $\frac{3}{2}$; y-intercept is 3.

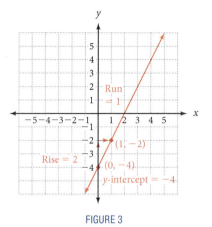

FIGURE 3

FIGURE 3

4. Find the slope and y-intercept for $4x - 2y = 12$.

EXAMPLE 4   Find the slope and y-intercept for $3x - 2y = 6$.

Solution   To find the slope and y-intercept from the equation, we must write the equation in the form $y = mx + b$. This means we must solve the equation $3x - 2y = 6$ for y.

$$3x - 2y = 6 \qquad \text{Original equation}$$

$$-2y = -3x + 6 \qquad \text{Add } -3x \text{ to each side.}$$

$$-\frac{1}{2}(-2y) = -\frac{1}{2}(-3x + 6) \qquad \text{Multiply each side by } -\frac{1}{2}.$$

$$y = \frac{3}{2}x - 3 \qquad \text{Simplify each side.}$$

Now that the equation is written in slope-intercept form, we can identify the slope as $\frac{3}{2}$ and the y-intercept as -3. We use the same process as in Example 3 to graph the line. The graph is shown in Figure 4.

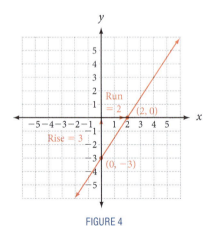

FIGURE 4

C The Point-Slope Form of an Equation of a Straight Line

A second useful form of the equation of a non-vertical straight line is the point-slope form.

Let line l contain the point (x_1, y_1) and have slope m. If (x, y) is any other point on l, then by the definition of slope we have

$$\frac{y - y_1}{x - x_1} = m$$

Multiplying both sides by $(x - x_1)$ gives us

$$(x - x_1) \cdot \frac{y - y_1}{x - x_1} = m(x - x_1)$$

$$y - y_1 = m(x - x_1)$$

Answers

4. Slope is 2; y-intercept is -6.

This last equation is known as the *point-slope form* of the equation of a straight line.

> **Point-Slope Form of the Equation of a Line**
>
> The point-slope form of the equation of the non-vertical line through (x_1, y_1) with slope m is given by
>
> $$y - y_1 = m(x - x_1)$$

This form is used to find the equation of a line, either given one point on the line and the slope, or given two points on the line.

EXAMPLE 5   Using the point-slope form, find the equation of the line with slope -2 that contains the point $(-4, 3)$. Write the answer in slope-intercept form.

Solution   Using $(x_1, y_1) = (-4, 3)$ and $m = -2$

in	$y - y_1 = m(x - x_1)$	Point-slope form
gives us	$y - 3 = -2(x + 4)$	Note: $x - (-4) = x + 4$.
	$y - 3 = -2x - 8$	Multiply out right side.
	$y = -2x - 5$	Add 3 to each side.

Figure 5 is the graph of the line that contains $(-4, 3)$ and has a slope of -2. Notice that the y-intercept on the graph matches that of the equation we found, which we wrote in slope-intercept form.

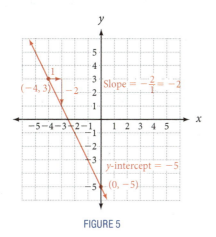

FIGURE 5

D Finding an Equation Given Two Points

EXAMPLE 6   Find the equation of the line that passes through the points $(-3, 3)$ and $(3, -1)$.

Solution   We begin by finding the slope of the line:

$$m = \frac{3 - (-1)}{-3 - 3} = \frac{4}{-6} = -\frac{2}{3}$$

Using $(x_1, y_1) = (3, -1)$ and $m = -\frac{2}{3}$ in $y - y_1 = m(x - x_1)$ yields

$$y + 1 = -\frac{2}{3}(x - 3)$$

$y + 1 = -\frac{2}{3}x + 2$	Multiply out right side.
$y = -\frac{2}{3}x + 1$	Add -1 to each side.

5. Find the equation of the line with slope 3 that contains the point $(-1, 2)$.

6. Find the equation of the line that passes through $(2, 5)$, and $(6, -3)$.

Figure 6 shows the graph of the line that passes through the points $(-3, 3)$ and $(3, -1)$. As you can see, the slope and y-intercept are $-\frac{2}{3}$ and 1, respectively.

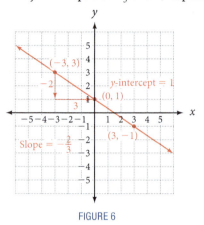

FIGURE 6

Methods of Graphing Lines

Here is a list of the graphing methods we have discovered in the last few sections.

1. Substitute convenient values of x into the equation, and find the corresponding values of y. We used this method first for equations like $y = 2x - 3$. To use this method for equations that looked like $2x - 3y = 6$, we first solved them for y.

2. Find the x- and y-intercepts. This method works best for equations of the form $3x + 2y = 6$ where the numbers in front of x and y divide the constant term evenly.

3. Find the slope and y-intercept. This method works best when the equation has the form $y = mx + b$ and b is an integer.

E Parallel and Perpendicular Lines

Now we can use what we have learned about linear equations to find the equations for parallel and perpendicular lines. Parallel lines are two lines that are nonintersecting and rise or fall at the same rate. In other words, two lines are parallel if and only if they have the same slope.

Here is a formal definition for parallel lines:

> **DEFINITION** parallel
>
> Two straight lines are *parallel* if they have the same slope. In other words, for $y = mx + b$, all equations of the form $y = mx + c$ are parallel if $b \neq c$.

7. Find the slope of a line parallel to the line that passes through $(-1, 3)$ and has a y-intercept of 1.

EXAMPLE 7   Find the slope of a line parallel to the line that passes through $(-3, 1)$ and has a y-intercept of 3.

Solution   Since the y-intercept of the line is 3, we can determine that another point on the line is $(0, 3)$. We begin by finding the slope of the line.

$$m = \frac{1 - 3}{-3 - 0} = \frac{-2}{-3} = \frac{2}{3}$$

Since we know the y-intercept is $(0, 3)$ and using the slope $m = \frac{2}{3}$, we can write an equation for our line in slope-intercept form.

$$y = \frac{2}{3}x + 3$$

Two lines are parallel if they have the same slope, therefore, any line with the slope $m = \frac{2}{3}$ will be parallel to our original line.

Although it is not as obvious, it is also true that two nonvertical lines are perpendicular if and only if the product of their slopes is −1. This is the same as saying their slopes are negative reciprocals. The following are graphs of perpendicular lines. Notice the slopes of the corresponding equations; the slope of one is the negative reciprocal of the slope of the other.

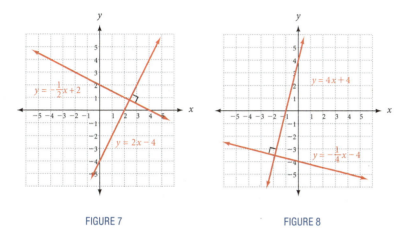

FIGURE 7   FIGURE 8

The following definition can be assumed for perpendicular lines:

> **DEFINITION** perpendicular
>
> Two non-vertical lines are *perpendicular* if the product of their slopes is −1. That is, $y = mx + b$ is perpendicular to $y = \left(-\frac{1}{m}\right)x + c$, if m is any real number except 0.

EXAMPLE 8   Find the slope of a line perpendicular to the line $y = 3x - 2$.

Solution   Since the equation is already written in slope-intercept form, we know that the slope of the first line is 3. Any line with a slope that, when multiplied by 3, results in a product of −1 is perpendicular to this line. For this to be true, the slope must be the negative reciprocal of 3, which is $-\frac{1}{3}$.

Furthermore, any line of the form $y = -\frac{1}{3}x + a$, where a is any real number, will provide a graph for our solution.

8. Find the slope of a line perpendicular to the line $y = \frac{1}{3}x - 2$.

> **GETTING READY FOR CLASS**
>
> *After reading through the preceding section, respond in your own words and in complete sentences.*
>
> **A.** What are m and b in the equation $y = mx + b$? What form is this equation?
>
> **B.** How would you find the slope and y-intercept for the line $4x - 3y = 9$?
>
> **C.** What is the point-slope form of the equation of a line?
>
> **D.** Find the equation of the line that passes through the points $(8, 1)$ and $(4, -2)$.

Answer

8. −3

Vocabulary Review

Choose the correct words to fill in the blanks below.

substitute   slope   slope-intercept   y-intercept

parallel   perpendicular   point-slope

1. The equation of a line with a slope m and y-intercept b given in _____ form is $y = mx + b$.
2. The equation of a line through (x_1, y_1) with slope m given in_____ form is $y - y_1 = m(x - x_1)$.
3. The first method used to graph lines is to _____ convenient values of x into the equation, and find the corresponding values of y.
4. The second method used to graph lines is to find the x-intercept and then _____.
5. The third method used to graph lines is to find the _____ and the y-intercept.
6. Two straight lines are _____ if they have the same slope.
7. Two straight non-vertical lines are _____ if the product of their slopes is -1.

Problems

A In each of the following problems, give the equation of the line with the given slope and y-intercept.

1. $m = \frac{2}{3}, b = 1$ **2.** $m = \frac{3}{4}, b = -2$ **3.** $m = \frac{3}{2}, b = -1$ **4.** $m = \frac{4}{3}, b = 2$

5. $m = -\frac{2}{3}, b = 3$ **6.** $m = -\frac{3}{5}, b = 4$ **7.** $m = 2, b = -4$ **8.** $m = -2, b = 4$

B Find the slope and y-intercept for each of the following equations by writing them in the form $y = mx + b$. Then, graph each equation.

9. $-2x + y = 4$ **10.** $-2x + y = 2$ **11.** $3x + y = 3$

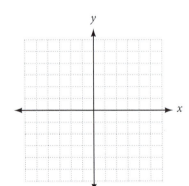 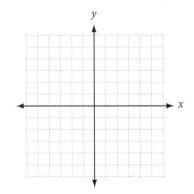 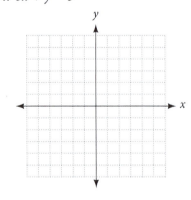

12. $3x + y = 6$

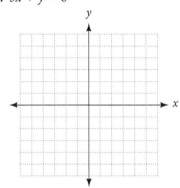

13. $3x + 2y = 6$

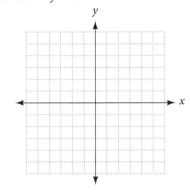

14. $2x + 3y = 6$

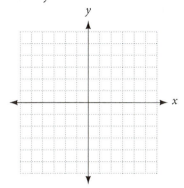

15. $4x - 5y = 20$

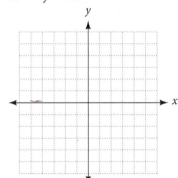

16. $2x - 5y = 10$

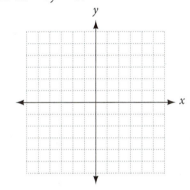

17. $-2x - 5y = 10$

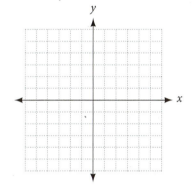

18. $-4x + 5y = 20$

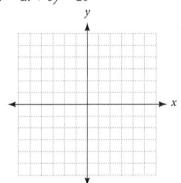

19. $-3x + 2y = -6$

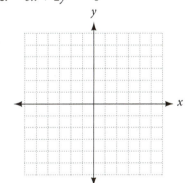

20. $2x - 3y = -6$

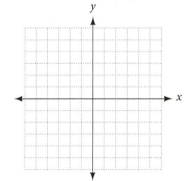

C For each of the following problems, the slope and one point on a line are given. In each case use the point-slope form to find the equation of that line. Write your answers in slope-intercept form.

21. $(-2, -5), m = 2$

22. $(-1, -5), m = 2$

23. $(-4, 1), m = -\dfrac{1}{2}$

24. $(-2, 1), m = -\dfrac{1}{2}$

25. $(2, -3)$, $m = \dfrac{3}{2}$

26. $(3, -4)$, $m = \dfrac{4}{3}$

27. $(-1, 4)$, $m = -3$

28. $(-2, 5)$, $m = -3$

29. $(3, -4)$, $m = 0$

30. $(-5, 2)$, $m = 0$

31. $(-4, -3)$, m is undefined

32. $(-5, 1)$, m is undefined

D Find the equation of the line that passes through each pair of points. Write your answers in slope-intercept form, if possible.

33. $(-2, -4), (1, -1)$

34. $(2, 4), (-3, -1)$

35. $(-1, -5), (2, 1)$

36. $(-1, 6), (1, 2)$

37. $(-3, -2), (3, 6)$

38. $(-3, 6), (3, -2)$

39. $(-3, -1), (3, -5)$

40. $(-3, -5), (3, 1)$

41. $(-4, 2), (3, 2)$

42. $(-1, 7), (1, 7)$

43. $(-5, 1), (-5, 6)$

44. $(2, -3), (2, 1)$

Find the slope and y-intercept for each line. Then write the equation of each line in slope-intercept form.

45.

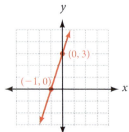

46.

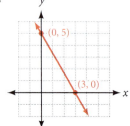

47.

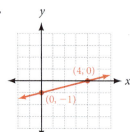

48.

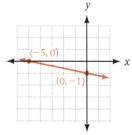

The next two problems are intended to give you practice reading, and paying attention to, the instructions that accompany the problems you are working. Working these problems is an excellent way to get ready for a test or a quiz.

49. Work each problem according to the instructions given.
  a. Solve: $-2x + 1 = 6$.

  b. Write in slope-intercept form: $-2x + y = 6$.

  c. Find the y-intercept: $-2x + y = 6$.

  d. Find the slope: $-2x + y = 6$.

  e. Graph: $-2x + y = 6$.

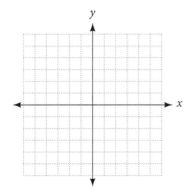

50. Work each problem according to the instructions given.
  a. Solve: $x + 3 = -6$.

  b. Write in slope-intercept form: $x + 3y = -6$.

  c. Find the y-intercept: $x + 3y = -6$.

  d. Find the slope: $x + 3y = -6$.

  e. Graph: $x + 3y = -6$.

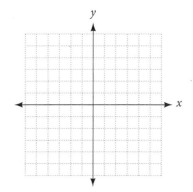

51. Find the equation of the line with x-intercept 3 and y-intercept 2.

52. Find the equation of the line with x-intercept 2 and y-intercept 3.

53. Find the equation of the line with x-intercept -2 and y-intercept -5.

54. Find the equation of the line with x-intercept -3 and y-intercept -5.

55. The equation of the vertical line that passes through the points $(3, -2)$ and $(3, 4)$ is either $x = 3$ or $y = 3$. Which one is it?

56. The equation of the horizontal line that passes through the points $(2, 3)$ and $(-1, 3)$ is either $x = 3$ or $y = 3$. Which one is it?

E **57. Parallel Lines** Provide an equation for a line parallel to the line given by $y = 4x - 2$.

58. Parallel Lines Find the slope of a line parallel to the line that crosses the point $(3, 1)$ and has a y-intercept of 2.

59. Parallel Lines Find the slope of a line parallel to the line that passes through (2, 3) and (4, 1).

60. Perpendicular Lines Find the equation of a line perpendicular to the line given by $y = \frac{1}{3}x - 2$.

61. Perpendicular Lines If a line contains the x-intercept 4 and the y-intercept –2, find the slope of any perpendicular line.

62. Perpendicular Lines One line passes through the points (3, 4) and (–3, 1). Find the slope of a second line that is perpendicular to the first.

Applying the Concepts

63. Value of a Copy Machine Cassandra buys a new color copier for her small business. It will cost $21,000 and will decrease in value each year. The graph below shows the value of the copier after the first 5 years of ownership.

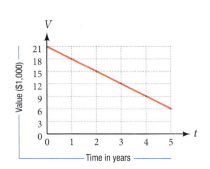

a. How much is the copier worth after 5 years?

b. After how many years is the copier worth $12,000?

c. Find the slope of this line.

d. By how many dollars per year is the copier decreasing in value?

e. Find the equation of this line where V is the value after t years.

64. Value of a Forklift Elliot buys a new forklift for his business. It will cost $140,000 and will decrease in value each year. The graph below shows the value of the forklift after the first 6 years of ownership.

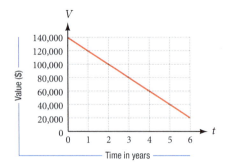

a. How much is the forklift worth after 6 years?

b. After how many years is the forklift worth $80,000?

c. Find the slope of this line.

d. By how many dollars per year is the forklift decreasing in value?

e. Find the equation of this line where V is the value after t years.

65. Salesperson's Income Kevin starts a new job in sales next month. He will earn $1,000 per month plus a certain amount for each shirt he sells. The graph below shows the amount Kevin will earn per month based on how many shirts he sells.

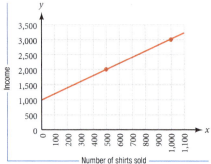

a. How much will he earn for selling 1,000 shirts?

b. How many shirts must he sell to earn $2,000 for a month?

c. Find the slope of this line.

d. How much money does Kevin earn for each shirt he sells?

e. Find the equation of this line where y is the amount he earns for selling x number of shirts.

66. Movie Prices The chart shows the price of movie tickets over a nine-year period. Let x represent the number of years past 2000 and y represent the prices shown on the graph. Use the first and last points on the graph to find the equation of the line that connects them. Then use the equation to estimate the price in 2011.

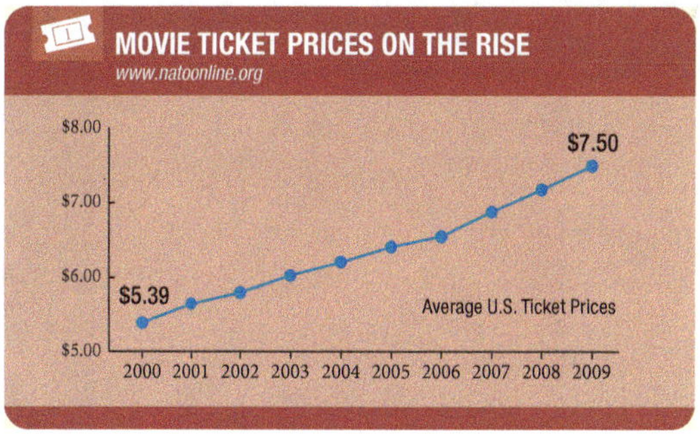

MOVIE TICKET PRICES ON THE RISE
www.natoonline.org

$8.00
$7.50
$7.00
$6.00
$5.39
Average U.S. Ticket Prices
$5.00

2000 2001 2002 2003 2004 2005 2006 2007 2008 2009

Find the Mistake

Each sentence below contains a mistake. Circle the mistake and write the correct word(s) on the line provided.

1. The equation $y = -\frac{4}{7}x + 2$ is in point-slope form. _____

2. For the equation $-8x + 2y = -4$, the slope is –2 and the y–intercept is 4. _____

3. The line represented by the equation $(y - 2) = 6(x - 4)$ has a y–intercept of 6. _____

4. The lines represented by the equations $y = \frac{7}{2}x + \frac{1}{3}$ and $y = -\frac{2}{7}x - 5$ are parallel. _____

Video Examples
Section 10.6

©iStockphoto.com/tarantas

At the U.S. National Handcar Races in Truckee, California, five-person teams face off on a railroad track. One team member gives the 1000-pound handcar a push while the other four members, standing on its flatbed, use the hand pump to propel the handcar down a 1000-foot stretch of track. The team with the fastest time wins the race. The races are held each year in honor of California's great railroad history, which aside from a few local lines began in 1869 with the completion of the First Continental Railroad. The handcar was once used for inspection and maintenance of the tracks, before an engine-powered version replaced it.

Let's assume that in 2009 the fastest handcar race time was 34 seconds, which was 7 seconds slower than the fastest time in 2010. Using this information, we can create an inequality that includes all the race times for both years. In this section, we will work with graphs of linear inequalities, such as this one.

A Graphing Linear Inequalities in Two Variables

A linear inequality in two variables is any expression that can be put in the form

$$Ax + By < C$$

where A, B, and C are real numbers (A and B not both 0). The inequality symbol can be any of the following four: $<, \leq, >, \geq$.

Some examples of linear inequalities are

$$2x + 3y < 6 \qquad y \geq 2x + 1 \qquad x - y \leq 0$$

Although not all of these inequalities have the form $Ax + By < C$, each one can be put in that form with the appropriate symbol.

In an earlier chapter, we graphed the solution set for a linear inequality in one variable on a number line. The solution set for a linear inequality in two variables is graphed as a section of the coordinate plane. The boundary for the section is found by replacing the inequality symbol with an equals sign and graphing the resulting equation using the methods summarized in the previous section. The boundary is included in the solution set (and represented with a solid line) if the inequality symbol used originally is $\leq$ or $\geq$. The boundary is not included (and is represented with a broken line) if the original symbol is $<$ or $>$.

EXAMPLE 1   Graph the linear inequality $x + y \leq 4$.

Solution   The boundary for the graph is the graph of $x + y = 4$. The boundary (a solid line) is included in the solution set because the inequality symbol is $\leq$.

The graph of the boundary is shown in Figure 1.

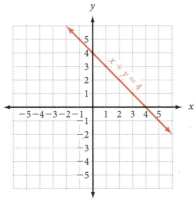

FIGURE 1

The boundary separates the coordinate plane into two sections, or regions: the region above the boundary and the region below the boundary. The solution set for $x + y \leq 4$ is one of these two regions along with the boundary. To find the correct region, we simply choose any convenient point that is *not* on the boundary. We then substitute the coordinates of the point into the original inequality $x + y \leq 4$. If the point we choose satisfies the inequality, then it is a member of the solution set, and we can assume that all points on the same side of the boundary as the chosen point are also in the solution set. If the coordinates of our point do not satisfy the original inequality, then the solution set lies on the other side of the boundary.

In this example, a convenient point not on the boundary is the origin. Substituting $(0, 0)$ into $x + y \leq 4$ gives us

$$0 + 0 \overset{?}{\leq} 4$$

$$0 \leq 4 \qquad \text{A true statement}$$

Because the origin is a solution to the inequality $x + y \leq 4$, and the origin is below the boundary, all other points below the boundary are also solutions.

The graph of $x + y \leq 4$ is shown in Figure 2.

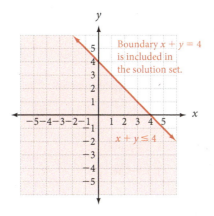

FIGURE 2

The region above the boundary is described by the inequality $x + y > 4$.

Practice Problems

1. Graph the linear inequality $x + y \geq 3$.

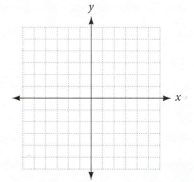

Here is a list of steps to follow when graphing the solution set for linear inequalities in two variables:

> **HOW TO**   Graph the Solution Set for Linear Inequalities in Two Variables
>
> **Step 1:** Replace the inequality symbol with an equals sign. The resulting equation represents the boundary for the solution set.
>
> **Step 2:** Graph the boundary found in step 1 using a *solid line* if the boundary is included in the solution set (that is, if the original inequality symbol was either ≤ or ≥). Use a *broken line* to graph the boundary if it is *not* included in the solution set; that is, if the original inequality was either < or >.
>
> **Step 3:** Choose any convenient point not on the boundary and substitute the coordinates into the *original* inequality. If the resulting statement is *true*, the graph lies on the *same* side of the boundary as the chosen point. If the resulting statement is *false*, the solution set lies on the *opposite* side of the boundary.

2. Graph the linear inequality for $y < \frac{1}{2}x + 3$.

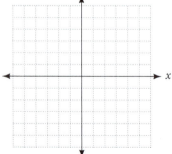

EXAMPLE 2   Graph the linear inequality $y < 2x - 3$.

Solution   The boundary is the graph of $y = 2x - 3$. The boundary is not included because the original inequality symbol is <. We therefore use a broken line to represent the boundary, as shown in Figure 3.

A convenient test point is again the origin. Using $(0, 0)$ in $y < 2x - 3$, we have

$$0 \overset{?}{<} 2(0) - 3$$

$$0 < -3 \qquad \text{A false statement}$$

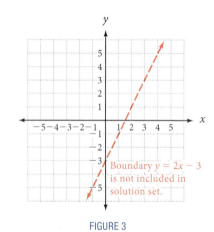

Boundary $y = 2x - 3$ is not included in solution set.

FIGURE 3

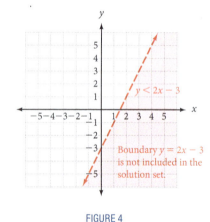

$y < 2x - 3$

Boundary $y = 2x - 3$ is not included in the solution set.

FIGURE 4

Because our test point gives us a false statement and it lies above the boundary, the solution set must lie on the other side of the boundary, as shown in Figure 4.

3. Graph the linear inequality $3x - 2y \le 6$.

EXAMPLE 3   Graph the linear inequality $2x + 3y \le 6$.

Solution   We begin by graphing the boundary $2x + 3y = 6$. The boundary is included in the solution because the inequality symbol is ≤.

If we use $(0, 0)$ as our test point, we see that it yields a true statement when its coordinates are substituted into $2x + 3y \le 6$. The graph, therefore, lies below the boundary, as shown in Figure 5.

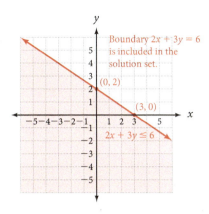

FIGURE 5

The ordered pair $(0, 0)$ is a solution to $2x + 3y \leq 6$; all points on the same side of the boundary as $(0, 0)$ also must be solutions to the inequality $2x + 3y \leq 6$.

EXAMPLE 4   Graph the linear inequality $x \leq 5$.

Solution   The boundary is $x = 5$, which is a vertical line. All points to the left have x-coordinates less than 5, and all points to the right have x-coordinates greater than 5, as shown in Figure 6.

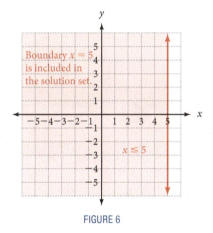

FIGURE 6

4. Graph the linear inequality $y > -2$.

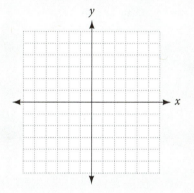

GETTING READY FOR CLASS

After reading through the preceding section, respond in your own words and in complete sentences.

A. When graphing a linear inequality in two variables, how do you find the equation of the boundary line?

B. What is the significance of a broken line in the graph of an inequality?

C. When graphing a linear inequality in two variables, how do you know which side of the boundary line to shade?

D. Why is the coordinate $(0, 0)$ a convenient test point?

EXERCISE SET 10.6

Problems

A Graph the following linear inequalities.

1. $2x - 3y < 6$

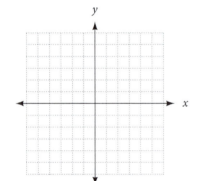

2. $3x + 2y \geq 6$

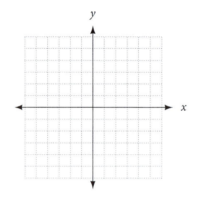

3. $x - 2y \leq 4$

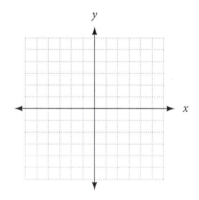

4. $2x + y > 4$

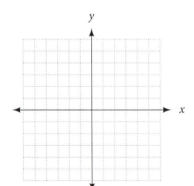

5. $x - y \leq 2$

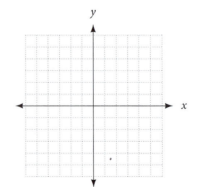

6. $x - y \leq 1$

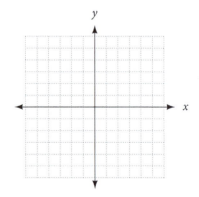

7. $3x - 4y \geq 12$

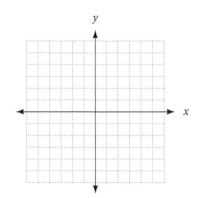

8. $4x + 3y < 12$

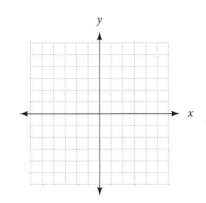

9. $5x - y \leq 5$

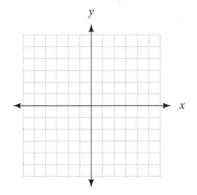

10. $4x + y > 4$

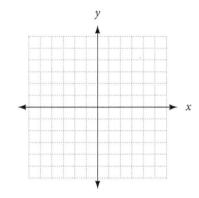

11. $2x + 6y \leq 12$

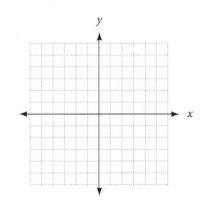

12. $x - 5y > 5$

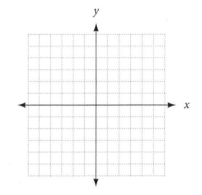

13. $x \geq 1$

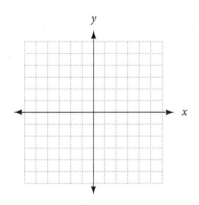

14. $x < 5$

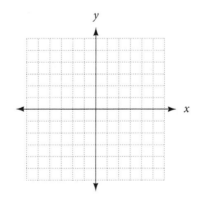

15. $x \geq -3$

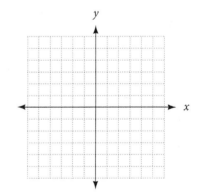

16. $y \leq -4$

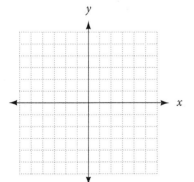

17. $y < 2$

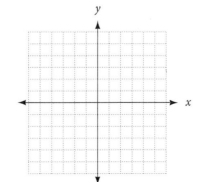

18. $3x - y > 1$

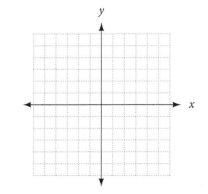

19. $2x + y > 3$

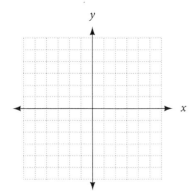

20. $5x + 2y < 2$

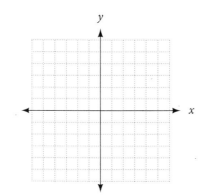

21. $y \leq 3x - 1$

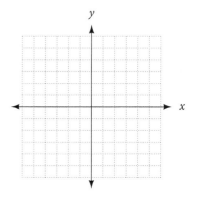

22. $y \geq 3x + 2$

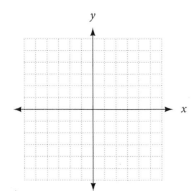

23. $y \leq -\dfrac{1}{2}x + 2$

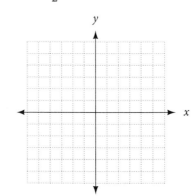

24. $y < \dfrac{1}{3}x + 3$

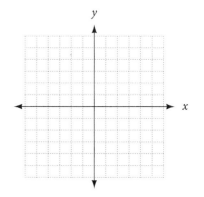

The next two problems are intended to give you practice reading, and paying attention to, the instructions that accompany the problems you are working.

25. Work each problem according to the instructions.

    a. Solve: $4 + 3y < 12$.

    b. Solve: $4 - 3y < 12$.

    c. Solve for y: $4x + 3y = 12$.

    d. Graph: $y < -\dfrac{4}{3}x + 4$.

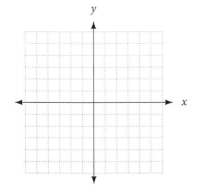

26. Work each problem according to the instructions.

    a. Solve: $3x + 2 \geq 6$.

    b. Solve: $-3x + 2 \geq 6$.

    c. Solve for y: $3x + 2y = 6$.

    d. Graph: $y \geq -\dfrac{3}{2}x + 3$.

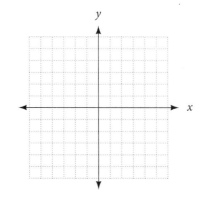

27. Find the equation of the line in part a, then use this information to find the inequalities for the graphs on parts b and c.

a.

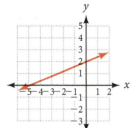

b.

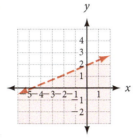

c.

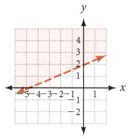

28. Find the equation of the line in part a, then use this information to find the inequalities for the graphs on parts b and c.

a.

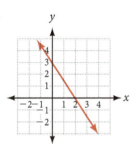

b.

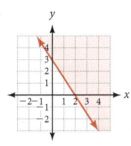

c.

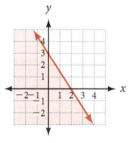

Review Problems

29. Simplify the expression $7 - 3(2x - 4) - 8$.

30. Find the value of $x^2 - 2xy + y^2$ when $x = 3$ and $y = -4$.

Solve each equation.

31. $-\dfrac{3}{2}x = 12$

32. $2x - 4 = 5x + 2$

33. $8 - 2(x + 7) = 2$

34. $3(2x - 5) - (2x - 4) = 6 - (4x + 5)$

35. Solve the formula $P = 2l + 2w$ for w.

36. Solve the formula $C = 2\pi r$ for r.

Solve each inequality, and graph the solution on a number line.

37. $-4x < 20$

38. $3 - 2x > 5$

39. $3 - 4(x - 2) \geq -5x + 6$

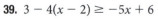

40. $4 - 6(x + 1) < -2x + 3$

41. Solve the inequality $3x - 2y \leq 12$ for y.

42. What number is 12% of 2,000?

43. Geometry The length of a rectangle is 5 inches more than 3 times the width. If the perimeter is 26 inches, find the length and width.

44. The length of a rectangle is 1 inch less than twice the width. If the perimeter is 28 inches, find the length and width.

Find the Mistake

Each sentence below contains a mistake. Circle the mistake and write the correct word(s) on the line provided.

1. The solution set for the linear inequality $y < 6x - \frac{3}{5}$ contains the points through which the boundary line passes.

_____

2. The boundary for the linear inequality $2 - 8y \geq x + 6$ is represented by a broken line.   _____

3. If a test point substituted into an original inequality results in a false statement, the solution set for the inequality lies on the same side as the chosen point.   _____

4. The solution set for the linear inequality $x < -8$, includes all points to the right of the line.   _____

CHAPTER 10 Trail Guide Project

Supplies Needed

- **A** Piece of graph paper
- **B** A pen or pencil
- **C** A ruler
- **D** Access to the internet

A Place to Call Home

Building a town from scratch is an intricate project. Research zoning codes for your hometown. Based on your research, what qualities would contribute to a successful town layout? Now imagine you have been commissioned to build and live in a new town. You will need to include the following businesses and community services:

Ambulance station	Bank
Fire station	Grocery store
Police station	Post office
Hospital	Community park
Courthouse	Your house

On a piece of graph paper, draw an x- and y-axis with the origin in the center of the paper. Each square is equivalent to one standard city block. Plot each of the above services on your graph. Log the coordinates of each service. Explain your reasoning behind the placement of each service.

Now find the distance between the locations of the following pairings, as well as the equation of the line that passes through each:

1. Police Station and Courthouse
2. Ambulance and Hospital
3. Grocery Store and Community Park
4. Bank and Post Office
5. Fire Station and Your House

EXAMPLES

1. The equation $3x + 2y = 6$ is an example of a linear equation in two variables.

Linear Equation in Two Variables [10.2]

A linear equation in two variables is any equation that can be put in the form $Ax + By = C$. The graph of every linear equation is a straight line.

Strategy for Graphing Linear Equations in Two Variables By Plotting Points [10.2]

2. The graph of $y = -\frac{2}{3}x - 1$ is shown below.

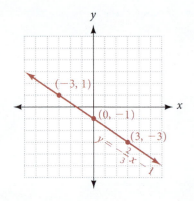

Step 1: Find any three ordered pairs that satisfy the equation. This can be done by using a convenient number for one variable and solving for the other variable.

Step 2: Graph the three ordered pairs found in step 1. Actually, we need only two points to graph a straight line. The third point serves as a check. If all three points do not line up, there is a mistake in our work.

Step 3: Draw a straight line through the three points graphed in step 2.

Intercepts [10.3]

3. To find the x-intercept for $3x + 2y = 6$, we let $y = 0$ and get

$$3x = 6$$
$$x = 2$$

In this case the x-intercept is 2, and the graph crosses the x-axis at $(2, 0)$.

The *x-intercept* of an equation is the *x-coordinate* of the point where the graph intersects the *x-axis*. The *y-intercept* is the *y-coordinate* of the point where the graph intersects the *y-axis*. We find the *y-intercept* by substituting $x = 0$ into the equation and solving for *y*. The *x-intercept* is found by letting $y = 0$ and solving for *x*.

Slope of a Line [10.4]

4. The slope of the line through $(3, -5)$ and $(-2, 1)$ is

$$m = \frac{-5 - 1}{3 - (-2)} = \frac{-6}{5} = -\frac{6}{5}$$

The *slope* of the line containing the points (x_1, y_1) and (x_2, y_2) is given by

$$\text{Slope} = m = \frac{y_2 - y_1}{x_2 - x_1} = \frac{\text{rise}}{\text{run}}$$

Slope-Intercept Form of a Straight Line [10.5]

5. The equation of the line with a slope of 2 and a y-intercept of 5 is

$$y = 2x + 5$$

The equation of the non-vertical line with a slope of m and a y-intercept of b is

$$y = mx + b$$

Point-Slope Form of a Straight Line [10.5]

If a non-vertical line has a slope of m and contains the point (x_1, y_1), the equation can be written as

$$y - y_1 = m(x - x_1)$$

6. The equation of the line through $(1, 2)$ with a slope of 3 is
$$y - 2 = 3(x - 1)$$
$$y - 2 = 3x - 3$$
$$y = 3x - 1$$

To Graph a Linear Inequality in Two Variables [10.6]

Step 1: Replace the inequality symbol with an equals sign. The resulting equation represents the boundary for the solution set.

Step 2: Graph the boundary found in step 1, using a *solid line* if the original inequality symbol was either $\leq$, or $\geq$. Use a *broken line* otherwise.

Step 3: Choose any convenient point not on the boundary and substitute the coordinates into the *original* inequality. If the resulting statement is *true*, the graph lies on the *same* side of the boundary as the chosen point. If the resulting statement is *false*, the solution set lies on the *opposite* side of the boundary.

7. Graph $x - y \geq 3$.

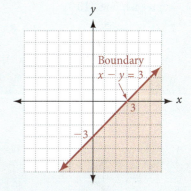

Graph the ordered pairs. [10.1]

1. $(2, -1)$

2. $\left(\frac{1}{2}, 0\right)$

3. $(-3, 5)$

4. $(1.2, -2)$

5. Fill in the following ordered pairs for the equation $2x - 5y = 10$. [10.2]

$\left(3, \quad\right) \left(-2, \quad\right) \left(\quad, \frac{2}{3}\right) \left(\frac{1}{2}, \quad\right)$

6. Which of the following ordered pairs are solutions to $y = 0.8x - 8$? [10.2]

$(10, 0) \quad (7, -2.2) \quad (8, -1.6) \quad (3, -5.8)$

Graph each line. [10.3]

7. $3x - y = 2$

8. $-y = 3$

Find the x- and y-intercepts. [10.3]

9. $2x - 12y = 12$

10. $0.3x - 2 = y$

11.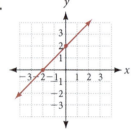

12.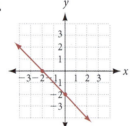

Find the slope of the line through each pair of points. [10.4]

13. $(2, 4), (-1, 3)$

14. $(0, 2), (-3, -4)$

Find the slope of each line. [10.4]

15.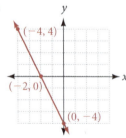

16.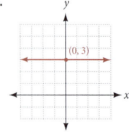

17.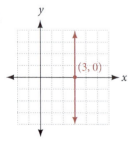

18.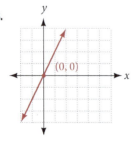

19. Find the equation of the line through $(3, -1)$ with a slope of 2. [10.5]

20. Find the equation of the line with a slope of -3 and y-intercept 6. [10.5]

21. Find the equation of the line passing through the points $(2, 2)$ and $(-3, 1)$. [10.5]

22. A straight line has an x-intercept 3 and contains the point $(-2, 6)$. Find its equation. [10.5]

23. Write the equation, in slope-intercept form, of the line whose graph is shown below. [10.5]

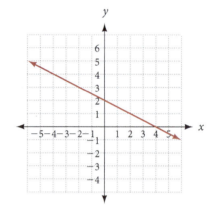

Give the slope and y-intercept of each equation. [10.5]

24. $4x - y = 3$

25. $4x - 5y = 10$

Find the equation of the line that contains the given point and has the given slope. [10.5]

26. $(3, 5), m = 3$

27. $(-2, 5), m = \frac{3}{2}$

Graph each linear inequality in two variables. [10.6]

28. $y < 4x + 2$

29. $3x - 8y \geq 12$

Simplify.

1. $9 - 5 \cdot 2$

2. $3 + 2(5 - 7)$

3. $\dfrac{2}{5}(15) - \dfrac{1}{3}(12)$

4. $2 \cdot 3 + 18 \div 9 - 2^4$

5. $(3 - 7)(-2 - 6)$

6. $2[3 + (-7)] - 4(8)$

7. $\dfrac{19 + (-41)}{8 - 8}$

8. $\dfrac{(6 - 3)^2}{6^2 - 3^2}$

9. $\dfrac{90}{126}$

10. $\dfrac{0}{-14}$

11. $4(9x)$

12. $6a + 9 - 3a - 15$

Solve each equation.

13. $7x - 3 - 6x + 9 = 4$

14. $6x + 9 = -15$

15. $-3x + 8 = 2x + 3$

16. $3(3 - t) - 2 = -5$

17. $2 = -\dfrac{1}{3}(2x - 3)$

18. $0.06 + 0.03(300 - x) = 12$

Solve each inequality, and graph the solution.

19. $-7x < 42$

20. $7 - 3x \leq -14$

Graph on a rectangular coordinate system.

21. $y = 3x - 2$

22. $2x + 3y = 6$

23. $y = 3x$

24. $y = -\dfrac{1}{4}x$

25. $3x - 4y > 6$

26. $y < 5$

27. Graph the points $(-4, -1)$ and $(1, 4)$, and draw a line that passes through both of them. Does $(2, -5)$ lie on the line? Does $(5, 8)$?

28. Graph the line through $(3, -12)$ with x-intercept 1. What is the y-intercept?

29. Find the x- and y-intercepts for $3x + 2y = 12$, and then graph the line.

30. Graph the line with slope $\dfrac{3}{5}$ and y-intercept -3.

31. Find the slope of the line through $(1, 2)$ and $(5, -2)$.

32. Find the equation of the line with slope -2 and y-intercept 5.

33. Find the slope of the line $6x - 2y = 18$.

34. Find the equation of the line with slope $\dfrac{2}{3}$ that passes through $(-3, -2)$. Write your answer in slope-intercept form.

35. Find the equation of the line through $(4, -2)$ and $(2, -5)$. Write your answer in slope-intercept form.

36. Find the equation of the line with x-intercept 5, y-intercept -3. Write your answer in slope-intercept form.

37. Complete the ordered pairs (, 1) and (0,) for $4x - 5y = 15$.

38. Which of the ordered pairs $(2, 3)$, $(-2, -5)$, and , $(6, 1)$ are solutions to $2x + 4y = 16$?

39. Write the expression, and then simplify: the difference of 24 and twice 6.

40. Twelve added to the product of 8 and -4 is what number?

Give the opposite, reciprocal, and absolute value of the given number.

41. $-\dfrac{3}{5}$

42. 4

43. Evaluate $x^2 - 3x - 9$ when $x = 3$.

44. Use the formula $6x + 5y = 11$ to find y when $x = 1$.

45. Geometry The length of a rectangle is 6 inches more than the width. The perimeter is 40 inches. Find the length and width.

46. Coin Problem Brandon has \$3.05 in dimes and quarters. If he has 6 more dimes than quarters, how many of each coin does he have?

Use the line graph to answer the following questions.

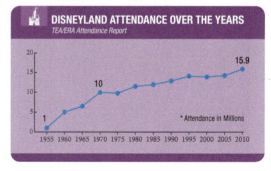

47. Find the slope of the line segment connecting the data point for 1955 and 1970, and explain its significance.

48. Find the slope of the line segment connecting the data point for 1970 and 2010, and explain its significance.

Graph the ordered pairs. [10.1]

1. $(2, -1)$ **2.** $(-4, 3)$

3. $(-3, -2)$ **4.** $(0, -4)$

5. Fill in the following ordered pairs for the equation $3x - 2y = 6$. [10.2]

$$(0, \quad) \; (\quad, 0) \; (4, \quad) \; (\quad, -6)$$

6. Which of the following ordered pairs are solutions to $y = -3x + 7$? [10.2]

$$(0, 7) \; (2, -1) \; (4, -5) \; (-5, -3)$$

Graph each line. [10.3]

7. $y = -\dfrac{1}{2}x + 4$ **8.** $x = -3$

Find the x- and y-intercepts. [10.3]

9. $8x - 4y = 16$ **10.** $y = \dfrac{3}{2}x + 6$

11.

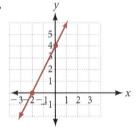

Find the slope of the line through each pair of points. [10.4]

12. $(3, 2), (-5, 6)$ **13.** $(0, 9), (7, 1)$

Find the slope of each line. [10.4]

14.

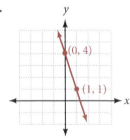

15.

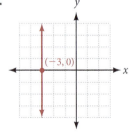

16.

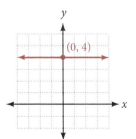

17. Find the equation of the line through $(4, 1)$ with a slope of $-\frac{1}{2}$. [10.5]

18. Find the equation of the line with a slope of 3 and y-intercept -5. [10.5]

19. Find the equation of the line passing through the points $(3, -4)$ and $(-6, 2)$. [10.5]

20. A straight line has an x-intercept -2 and passes through the point $(-4, 6)$. Find its equation. [10.5]

21. Write the equation, in slope-intercept form, of the line whose graph is shown below. [10.5]

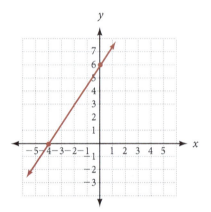

Find the equation for each line. [10.5]

22. The line through $(-6, 1)$ that has slope $m = -3$

23. The line through $(6, 8)$ and $(-3, 2)$

24. The line which contains the point $(-4, 1)$ and is perpendicular to the line $y = -\frac{2}{3}x + 4$

25. The line which contains the point $(0, -6)$ and is parallel to the line $4x - 2y = 3$

Graph each linear inequality in two variables. [10.6]

26. $y > x - 6$ **27.** $6x - 9y \le 18$

Systems of Linear Equations

f you have ever been to Las Vegas, Nevada, you have most likely seen the Stratosphere Hotel and Casino, which contains the tallest observation deck in the United States. The observation deck stands at 869 feet above the ground, offers 360-degree views, and provides visitors access to some incredible thrill rides. The Big Shot is a roller coaster circling the tower at nearly 45 miles per hour, and is the highest coaster in the United States. The second highest thrill ride in the world, Insanity, dangles riders out over The Strip on a 64-foot arm as they spin around at three times regular g-forces. The X-Scream is a seesaw that holds patrons out over the edge of the tower more than 80 stories from the ground. The newest Stratosphere attraction, the SkyJump, allows jumpers to descend over 100 stories down to street level wearing a special suit and guided by wires. The controlled descent allows visitors the experience of a sky dive without the parachute or the airplane.

Suppose a ticket to ride Insanity costs $12, a ticket for X-Scream costs $14, a ticket for Big Shot costs $13, and a ticket for SkyJump costs $110. Let's also suppose that in one day, the thrill rides were ridden 105 times bringing in a total of $2,810. How many tickets for each ride were sold? In order to answer this question, you will need to know how to work with systems of equations, which is the topic of this chapter.

OBJECTIVES

A Solve a system of linear equations in two variables by graphing.

11.1 Solving Linear Systems by Graphing

Image © sxc.hu, AmyJacobs, 2007

KEY WORDS

system of linear equations

point of intersection

consistent system

inconsistent system

dependent equations

independent equations

Each year, spectators at the QuikTrip Air and Rocket Racing Show see a variety of planes, such as rocket racers, combat aircraft, biplanes, and gliders take to the skies above an airfield in Tulsa, Oklahoma. The wide-eyed ticket holders, packed tightly into the grandstands, watch planes perform dramatic acrobatic stunts. During some acts, two or more planes fly in the air at once. It it extremely important that the planes' routines are well-planned to avoid any collisions. Suppose we were to assign a linear equation, in the form of $Ax + By = C$, to the path of each plane of a two-plane act. When we consider these two equations together, we are working with a *system of linear equations*, which is the focus of this section. Then we would be able to determine if the planes were flying directly parallel to each other or if their paths will intersect.

A Solving a System of Linear Equations by Graphing

In a system of linear equations, both equations contain two variables and have graphs that are straight lines. The following are systems of linear equations:

$$\begin{array}{ccc} x + y = 3 & y = 2x + 1 & 2x - y = 1 \\ 3x + 4y = 2 & y = 3x + 2 & 3x - 2y = 6 \end{array}$$

The solution set for a system of linear equations is all ordered pairs that are solutions to both equations. The intersection of the graphs are a point whose coordinates are solutions to a system; that is, if we graph both equations on a coordinate system, we can read the coordinates of the point of intersection as a solution to our system. Here is an example.

EXAMPLE 1 Solve the following system by graphing.

$$\begin{array}{l} x + y = 4 \\ x - y = -2 \end{array}$$

Solution Graph each equation separately. Figure 1 shows both graphs, without showing the work necessary to get them. We can see the graphs intersect at the point (1, 3). Therefore, the point (1, 3) must be a solution to our system because it is the only ordered pair that lies on both lines. Its coordinates satisfy both equations.

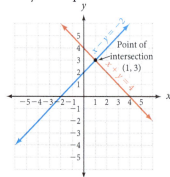

FIGURE 1

Practice Problems

1. Solve the system by graphing.

$$\begin{array}{l} x + y = 3 \\ x - y = 5 \end{array}$$

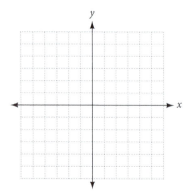

Answer

1. $(4, -1)$

We can check our results by substituting the coordinates $x = 1$, $y = 3$ into both equations to see if they work.

When →	$x = 1$	When →	$x = 1$
and →	$y = 3$	and →	$y = 3$
the equation →	$x + y = 4$	the equation →	$x - y = -2$
becomes →	$1 + 3 \overset{?}{=} 4$	becomes →	$1 - 3 \overset{?}{=} -2$
	$4 = 4$		$-2 = -2$

The point $(1, 3)$ satisfies both equations.

Here are some steps to follow when solving linear systems by graphing.

> **HOW TO** Solving a Linear System by Graphing
>
> **Step 1:** Graph the first equation by the methods described in the previous chapter
>
> **Step 2:** Graph the second equation on the same set of axes used for the first equation.
>
> **Step 3:** Read the coordinates of the point of intersection of the two graphs.
>
> **Step 4:** Check the solution in both equations.

EXAMPLE 2 Solve the following system by graphing.

$$x + 2y = 8$$
$$2x - 3y = 2$$

Solution Graphing each equation on the same coordinate system, we have the lines shown in Figure 2.

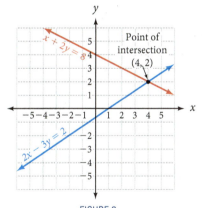

FIGURE 2

From Figure 2, we can see the solution for our system is $(4, 2)$. We check this solution as follows:

When →	$x = 4$	When →	$x = 4$
and →	$y = 2$	and →	$y = 2$
the equation →	$x + 2y = 8$	the equation →	$2x - 3y = 2$
becomes →	$4 + 2(2) \overset{?}{=} 8$	becomes →	$2(4) - 3(2) \overset{?}{=} 2$
	$4 + 4 = 8$		$8 - 6 = 2$
	$8 = 8$		$2 = 2$

The point $(4, 2)$ satisfies both equations and, therefore, must be the solution to our system.

Video Examples
Section 11.1

2. Solve the system by graphing.

$$x + 2y = 6$$
$$3x - y = 4$$

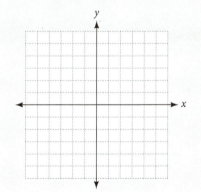

Answer
2. $(2, 2)$

3. Solve by graphing.

$$y = 3x - 1$$
$$x = 2$$

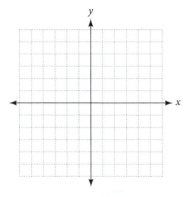

4. Solve by graphing.

$$y = x + 3$$
$$y = x - 2$$

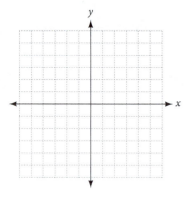

5. Graph the system.

$$6x + 2y = 12$$
$$9x + 3y = 18$$

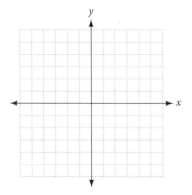

EXAMPLE 3   Solve this system by graphing.

$$y = 2x - 3$$
$$x = 3$$

Solution   Graphing both equations on the same set of axes, we have Figure 3.

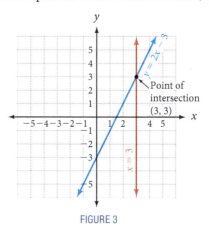

FIGURE 3

The solution to the system is the point (3, 3).

EXAMPLE 4   Solve by graphing.

$$y = x - 2$$
$$y = x + 1$$

Solution   Graphing both equations produces the lines shown in Figure 4. We can see in Figure 4 that the lines are parallel and therefore do not intersect. Our system has no ordered pair as a solution because there is no ordered pair that satisfies both equations. There is no solution to the system.

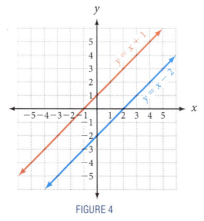

FIGURE 4

EXAMPLE 5   Graph the system.

$$2x + y = 4$$
$$4x + 2y = 8$$

Solution   Both graphs are shown in Figure 5. The two graphs coincide. The reason becomes apparent when we multiply both sides of the first equation by 2.

$$2x + y = 4$$
$$2(2x + y) = 2(4) \qquad \text{Multiply both sides by 2.}$$
$$4x + 2y = 8$$

Answers
3. (2, 5)
4. No solution.
5. The lines coincide.

The equations have the same solution set. We say the lines coincide. Any ordered pair that is a solution to one is a solution to the system. The system has an infinite number of solutions. (Any point on the line is a solution to the system.)

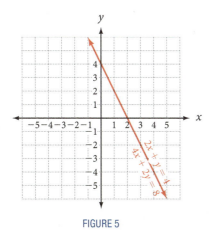

FIGURE 5

We sometimes use special vocabulary to describe the special cases shown in Examples 4 and 5. When a system of equations has no solution because the lines are parallel (as in Example 4), we say the system is *inconsistent*. When the lines coincide (as in Example 5), we say the equations are *dependent*.

The special cases illustrated in the previous two examples do not happen often. Usually, a system has a single ordered pair as a solution in which case we call the system *consistent* and the equations *independent*. Solving a system of linear equations by graphing is useful only when the ordered pair in the solution set has integers for coordinates. Two other solution methods work well in all cases. We will develop the other two methods in the next two sections.

Here is a summary of three possible types of solutions to a system of equations in two variables:

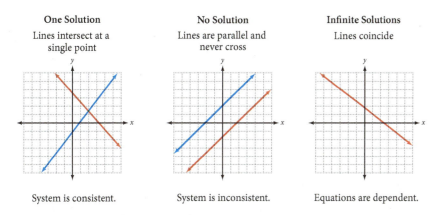

One Solution	No Solution	Infinite Solutions
Lines intersect at a single point	Lines are parallel and never cross	Lines coincide

System is consistent.   System is inconsistent.   Equations are dependent.

EXERCISE SET 11.1

Problems

A Solve the following systems of linear equations by graphing. If the system is inconsistent or dependent, state so.

1. $x + y = 3$
$x - y = 1$

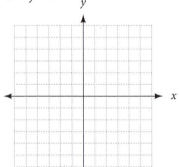

2. $x + y = 2$
$x - y = 4$

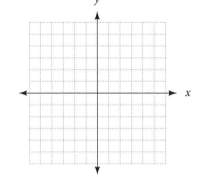

3. $x + y = 1$
$-x + y = 3$

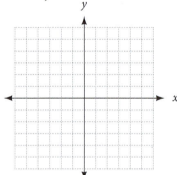

4. $x + y = 1$
$x - y = -5$

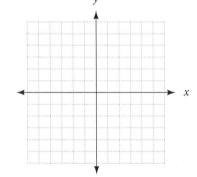

5. $x + y = 8$
$-x + y = 2$

6. $x + y = 6$
$-x + y = -2$

7. $3x - 2y = 6$
$x - y = 1$

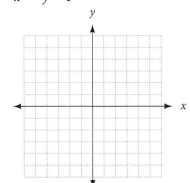

8. $5x - 2y = 10$
$x - y = -1$

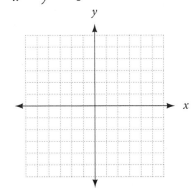

9. $6x - 2y = 12$
$3x + y = -6$

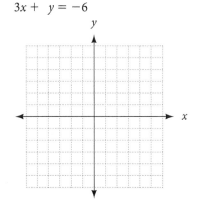

10. $4x - 2y = 8$
$2x + y = -4$

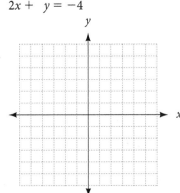

11. $4x + y = 4$
$3x - y = 3$

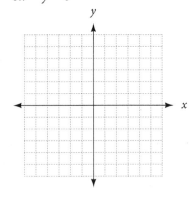

12. $5x - y = 10$
$2x + y = 4$

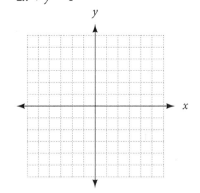

13. $x + 2y = 0$
$2x - y = 0$

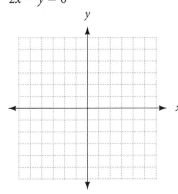

14. $3x + y = 0$
$5x - y = 0$

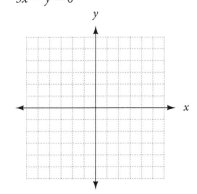

15. $3x - 5y = 15$
$-2x + y = 4$

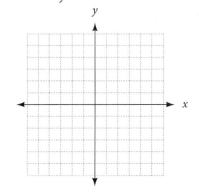

16. $2x - 4y = 8$
$2x - y = -1$

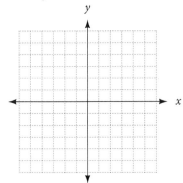

17. $y = 2x + 1$
$y = -2x - 3$

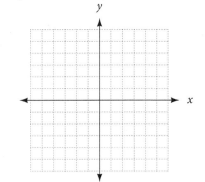

18. $y = 3x - 4$
$y = -2x + 1$

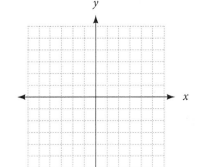

19. $x + 3y = 3$
$y = x + 5$

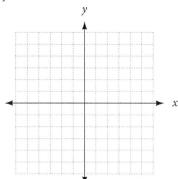

20. $2x + y = -2$
$y = x + 4$

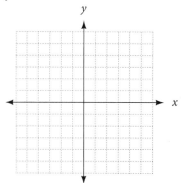

21. $x + y = 2$
$x = -3$

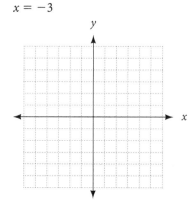

22. $x + y = 6$
$y = 2$

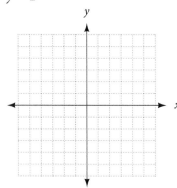

23. $x = -4$
$y = 6$

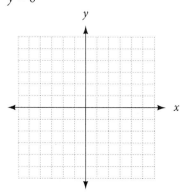

24. $x = 5$
$y = -1$

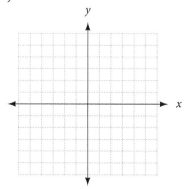

25. $x + y = 4$
$2x + 2y = -6$

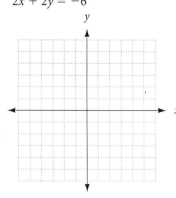

26. $x - y = 3$
$2x - 2y = 6$

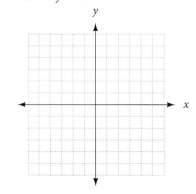

27. $4x - 2y = 8$
$2x - y = 4$

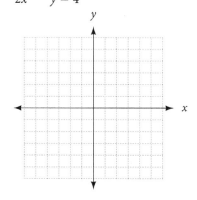

28. $3x - 6y = 6$
$x - 2y = 4$

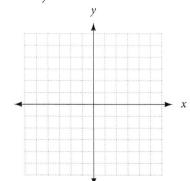

29. $y = 3x + 1$
$y = -2x + 1$

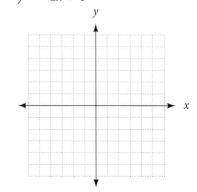

30. $y = -4x$
$y = \frac{1}{2}x$

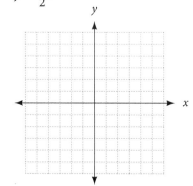

31. As you probably have guessed by now, it can be difficult to solve a system of equations by graphing if the solution to the system contains a fraction. The solution to the following system is $\left(\frac{1}{2}, 1\right)$. Solve the system by graphing.

$$y = -2x + 2$$
$$y = 4x - 1$$

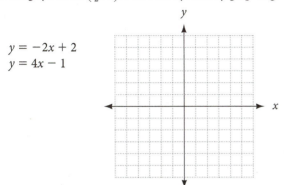

32. The solution to the following system is $\left(\frac{1}{3}, -2\right)$. Solve the system by graphing.

$$y = 3x - 3$$
$$y = -3x - 1$$

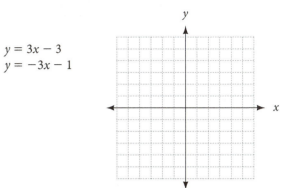

33. A second difficulty can arise in solving a system of equations by graphing if one or both of the equations is difficult to graph. The solution to the following system is $(2, 1)$. Solve the system by graphing.

$$3x - 8y = -2$$
$$x - y = 1$$

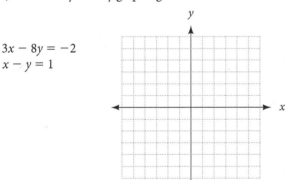

34. The solution to the following system is $(-3, 2)$. Solve the system by graphing.

$$2x + 5y = 4$$
$$x - y = -5$$

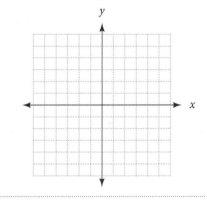

Applying the Concepts

35. Suppose you have 45 poker chips with $5 and $25 values. The total value of chips is $465. If x is the number of $5 chips and y is the number of $25 chips, then the following system of equations represents this situation:

$$x + y = 45$$
$$5x + 25y = 465$$

Solve this system by graphing each equation.

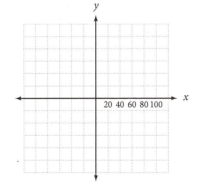

36. The stratosphere's Big Shot holds 16 passengers. Tickets can be purchased for 1 ride ($13), or for all 3 rides ($31). The total value of tickets purchased is $388. If x is the number of one-ride tickets and y is the number of three-ride tickets, then the following system of equations represents this situation:

$$x + y = 16$$
$$13x + 31y = 388$$

Solve this system by graphing each equation.

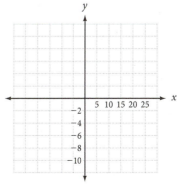

37. Job Comparison Jane is deciding between two sales positions. She can work for Marcy's and receive $8.00 per hour, or she can work for Gigi's, where she earns $6.00 per hour but also receives a $50 commission per week. The two lines in the following figure represent the money Jane will make for working at each of the jobs.

a. From the figure, how many hours would Jane have to work to earn the same amount at each of the positions?

b. If Jane expects to work less than 20 hours a week, which job should she choose?

c. If Jane expects to work more than 30 hours a week, which job should she choose?

38. Truck Rental You need to rent a moving truck for two days. Rider Moving Trucks charges $50 per day and $0.50 per mile. UMove Trucks charges $45 per day and $0.75 per mile. The following figure represents the cost of renting each of the trucks for two days.

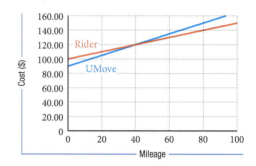

a. From the figure, after how many miles would the trucks cost the same?

b. Which company will give you a better deal if you drive less than 30 miles?

c. Which company will give you a better deal if you drive more than 60 miles?

Getting Ready for the Next Section

Solve each equation.

39. $x + (2x - 1) = 2$

40. $2x - 3(2x - 8) = 12$

41. $2(3y - 1) - 3y = 4$

42. $-2x + 4(3x + 6) = 14$

43. $4x + 2(-2x + 4) = 8$

44. $1.5x + 15 = 0.75x + 24.95$

Solve each equation for the indicated variable.

45. $x - 3y = -1$ for x

46. $-3x + y = 6$ for y

47. Let $y = 2x - 1$. If $x = 1$, find y.

48. Let $y = 2x - 8$. If $x = 5$, find y.

49. Let $x = 3y - 1$. If $y = 2$, find x.

50. Let $y = 3x + 6$. If $y = -6$, find x.

Let $y = 1.5x + 15$ for the next four problems.

51. If $x = 13$, find y.　　**52.** If $x = 14$, find y.　　**53.** If $y = 21$, find x.　　**54.** If $y = 42$, find x.

Let $y = 0.75x + 24.95$ for the next four problems.

55. If $x = 12$, find y.　　**56.** If $x = 16$, find y.　　**57.** If $y = 39.95$, find x.　　**58.** If $y = 41.45$, find x.

Find the Mistake

Each sentence below contains a mistake. Circle the mistake and write the correct word(s) or number(s) on the line provided.

1. The point of intersection for two straight line graphs is a solution to only one of the equations. ＿＿＿＿＿＿＿＿＿

＿＿＿

2. The point of intersection for the equations $y = 6$ and $y = 4x + 2$ is (2, 6). ＿＿＿＿＿＿＿＿＿＿＿＿＿＿＿＿

3. The point of intersection for the equations $x = 5$ and $x = 9$ is (5, 9). ＿＿＿＿＿＿＿＿＿＿＿＿＿＿＿

4. When lines coincide, we say that the equations are inconsistent. ＿＿＿＿＿＿＿＿＿＿＿＿＿＿＿＿＿＿＿＿

Navigation Skills: Prepare, Study, Achieve

It is important to reward yourself for working hard in this class. Recognizing your achievement will help foster success through the end of this course, and enable you to carry over the skills you've learned to future classes. What are some things you can do to reward and celebrate yourself? Also, for future classes, you could decide on these rewards at the beginning of the course. Schedule smaller short-term rewards to work towards as you complete each chapter. And then have a long-term reward that you achieve at the end of the course to celebrate your overall success. You deserve it. Congratulations!

OBJECTIVES

A Use the substitution method to solve a system of linear equations in two variables.

11.2 The Substitution Method

© iStockphoto.com/cstewart

Video Examples
Section 11.2

Art and physical endurance enthusiasts race in an annual event called the Kinetic Sculpture Grand Championship. Teams have each engineered an amphibious work of art that can travel over any terrain. The vehicle must be able to hold its pilots, who power it over sand dunes, across a harbor, up and down steep city streets, and through a swamp. The vehicles are built and decorated as if they were elaborate parade floats, many with animated parts and their pilots dressed in costume.

Let's suppose a kinetic sculpture team races its vehicle on a river. It takes 5 minutes for the vehicle to race down the river going with a current of 3 miles per hour, and 12 minutes to travel the same distance back up the river against the current. After working through this section, you will be able to set up a linear system of equations for this problem and use an algebraic method, called the substitution method, to solve the system.

A Substitution Method

EXAMPLE 1   Solve the system by substitution.

$$x + y = 2$$
$$y = 2x - 1$$

Practice Problems

1. Solve the system by substitution.

$$x + y = 3$$
$$y = x + 5$$

Note Sometimes this method of solving systems of equations is confusing the first time you see it. If you are confused, you may want to read through this first example more than once and try it on your own before attempting the previous practice problem.

Solution   If we were to solve this system by the methods used in the previous section, we would have to graph both lines and see where they intersect. There is no need to do this, however, because the second equation tells us that y is $2x - 1$. We can replace the y variable in the first equation with the expression $2x - 1$ from the second equation; that is, we *substitute* $2x - 1$ from the second equation for y in the first equation. Here is what it looks like:

$$x + (2x - 1) = 2$$

The equation we end up with contains only the variable x. The y variable has been eliminated by substitution.

Solving the resulting equation, we have

$$x + (2x - 1) = 2$$
$$3x - 1 = 2$$
$$3x = 3$$
$$x = 1$$

This is the x-coordinate of the solution to our system. To find the y-coordinate, we substitute $x = 1$ into the second equation of our system. (We could substitute $x = 1$ into the first equation also and have the same result.)

$$y = 2(1) - 1$$
$$y = 2 - 1$$
$$y = 1$$

Answer

1. $(-1, 4)$

Chapter 11 Systems of Linear Equations

The solution to our system is the ordered pair (1, 1). It satisfies both of the original equations. Figure 1 provides visual evidence that the substitution method yields the correct solution.

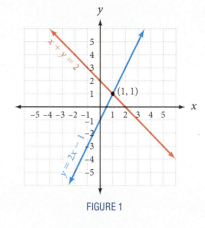

FIGURE 1

EXAMPLE 2   Solve the system by substitution.

$$2x - 3y = 12$$
$$y = 2x - 8$$

Solution   Again, the second equation says y is $2x - 8$. Because we are looking for the ordered pair that satisfies both equations, the y in the first equation must also be $2x - 8$. Substituting $2x - 8$ from the second equation for y in the first equation, we have

$$2x - 3(2x - 8) = 12$$

This equation can still be read as $2x - 3y = 12$ because $2x - 8$ is the same as y. Solving the equation, we have

$$2x - 3(2x - 8) = 12$$
$$2x - 6x + 24 = 12$$
$$-4x + 24 = 12$$
$$-4x = -12$$
$$x = 3$$

To find the y-coordinate of our solution, we substitute $x = 3$ into the second equation in the original system.

When → $x = 3$
the equation → $y = 2x - 8$
becomes → $y = 2(3) - 8$
                  $y = 6 - 8 = -2$

The solution to our system is $(3, -2)$.

2. Solve the system by substitution.
$$5x - 4y = -2$$
$$y = 2x + 2$$

EXAMPLE 3   Solve the following system by solving the first equation for x and then using the substitution method.

$$x - 3y = -1$$
$$2x - 3y = 4$$

Solution   We solve the first equation for x by adding $3y$ to both sides to get

$$x = 3y - 1$$

Using this value of x in the second equation, we have

$$2(3y - 1) - 3y = 4$$
$$6y - 2 - 3y = 4$$
$$3y - 2 = 4$$
$$3y = 6$$
$$y = 2$$

Next, we find x.

When → $y = 2$
the equation → $x = 3y - 1$
becomes → $x = 3(2) - 1$
                  $x = 6 - 1$
                  $x = 5$

The solution to our system is (5, 2)

3. Solve the first equation for x, and then substitute the result into the second equation to solve the system by substitution.
$$x - 4y = -5$$
$$3x - 2y = 5$$

Answers
2. $(-2, -2)$
3. $(3, 2)$

Here are the steps to use in solving a system of equations by the substitution method.

> **HOW TO**   Strategy for Solving a System of Linear Equations by the Substitution Method
>
> **Step 1:** Solve either one of the equations for x or y. (This step is not necessary if one of the equations is already in the correct form, as in Examples 1 and 2.)
>
> **Step 2:** Substitute the expression for the variable obtained in step 1 into the other equation and solve it.
>
> **Step 3:** Substitute the solution from step 2 into any equation in the system that contains both variables and solve it.
>
> **Step 4:** Check your results, if necessary.

4. Solve the system by substitution.

$$5x - y = 1$$
$$-2x + 3y = 10$$

EXAMPLE 4   Solve the system by substitution.

$$-2x + 4y = 14$$
$$-3x + y = 6$$

Solution   We can solve either equation for either variable. If we look at the system closely, it becomes apparent that solving the second equation for y is the easiest way to go. If we add $3x$ to both sides of the second equation, we have

$$y = 3x + 6$$

Substituting the expression $3x + 6$ back into the first equation in place of y yields the following result.

$$-2x + 4(3x + 6) = 14$$
$$-2x + 12x + 24 = 14$$
$$10x + 24 = 14$$
$$10x = -10$$
$$x = -1$$

Substituting $x = -1$ into the equation $y = 3x + 6$ leaves us with

$$y = 3(-1) + 6$$
$$y = -3 + 6$$
$$y = 3$$

The solution to our system is $(-1, 3)$.

5. Solve the system by substitution.

$$6x + 3y = 1$$
$$y = -2x - 5$$

EXAMPLE 5   Solve the system by substitution.

$$4x + 2y = 8$$
$$y = -2x + 4$$

Solution   Substituting the expression $-2x + 4$ for y from the second equation into the first equation, we have

$$4x + 2(-2x + 4) = 8$$
$$4x - 4x + 8 = 8$$
$$8 = 8 \qquad \textit{A true statement}$$

Both variables have been eliminated, and we are left with a true statement. A true statement in this situation tells us the lines coincide; that is, the equations $4x + 2y = 8$ and $y = -2x + 4$ have exactly the same graph. Any point on that graph has coordinates that satisfy both equations and is a solution to the system. This system is dependent. If the result had been a false statement, this would tell us that the system is inconsistent, that is, the lines are parallel. If the resulting statement is false, there is no solution to the system.

Answers

4. $(1, 4)$

5. No solution. The system is inconsistent.

EXAMPLE 6   The following table shows two contract rates charged by a cellular phone company. How many text messages must a person send so that the cost will be the same regardless of which plan is chosen?

	Flat Monthly Rate	Per Text Charge
Plan 1	$40	$0.15
Plan 2	$70	$0.05

Solution   If we let y = the monthly charge for x number of text messages, then the equations for each plan are

Plan 1: $y = 0.15x + 40$

Plan 2: $y = 0.05x + 70$

We can solve this system using the substitution method by replacing the variable y in Plan 2 with the expression $0.15x + 40$ from Plan 1. If we do so, we have

$0.15x + 40 = 0.05x + 70$

$0.10x + 40 = 70$

$0.10x = 30$

$x = 300$

The monthly bill is based on the number of text messages you send. We can use this value of x to gather more information by plugging it into either plan equation to find y.

Plan 1: $y = 0.15x + 40$ or   Plan 2: $y = 0.05x + 70$

$y = 0.15(300) + 40$ $y = 0.05(300) + 70$

$y = 45 + 40$ $y = 15 + 70$

$y = 85$ $y = 85$

Therefore, when you send 300 text messages in a month, the total cost for that month will be $85 regardless of which plan you used.

GETTING READY FOR CLASS

After reading through the preceding section, respond in your own words and in complete sentences.

A. What is the first step in solving a system of linear equations by substitution?

B. When would substitution be more efficient than graphing in solving two linear equations?

C. What does it mean when you solve a system of linear equations by the substitution method and you end up with the statement 8 = 8?

D. How would you begin solving the following system using the substitution method?

$$x + y = 2$$
$$y = 2x - 1$$

6. The rates for two garbage companies are given in the following table. How many bags of trash must be picked up in one month for the two companies to charge the same amount?

	Flat Rate	Per Bag Charge
Co. 1	$13.00	$1.50
Co. 2	$20.00	$1.00

Answer

6. 14 bags

Vocabulary Review

The following is a list of steps for solving a system of linear equations by the substitution method. Choose the correct words to fill in the blanks below.

substitute   variables   check   equations

Step 1: Solve either one of the _____ for x or y.

Step 2: _____ the expression for one variable into the other equation and solve it.

Step 3: Substitute the solution from Step 2 into any equation in the system that contains both _____ and solve it.

Step 4: _____ your results.

Problems

A Solve the following systems by substitution. Substitute the expression in the second equation into the first equation and solve.

1. $x + y = 11$
$y = 2x - 1$

2. $x - y = -3$
$y = 3x + 5$

3. $x + y = 20$
$y = 5x + 2$

4. $3x - y = -1$
$x = 2y - 7$

5. $-2x + y = -1$
$y = -4x + 8$

6. $4x - y = 5$
$y = -4x + 1$

7. $3x - 2y = -2$
$x = -y + 6$

8. $2x - 3y = 17$
$x = -y + 6$

9. $5x - 4y = -16$
$y = 4$

10. $6x + 2y = 18$
$x = 3$

11. $5x + 4y = 7$
$y = -3x$

12. $10x + 2y = -6$
$y = -5x$

Solve the following systems by solving one of the equations for x or y and then using the substitution method.

13. $x + 3y = 4$
$x - 2y = -1$

14. $x - y = 5$
$x + 2y = -1$

15. $2x + y = 1$
$x - 5y = 17$

16. $2x - 2y = 2$
$x - 3y = -7$

17. $3x + 5y = -3$
$x - 5y = -5$

18. $2x - 4y = -4$
$x + 2y = 8$

19. $5x + 3y = 0$
$x - 3y = -18$

20. $x - 3y = -5$
$x - 2y = 0$

21. $-3x - 9y = 7$
$x + 3y = 12$

22. $2x + 6y = -18$
$x + 3y = -9$

23. $3x + 2y = -4$
$6x + 4y = -8$

24. $y = 4x + 7$
$y = 4x - 3$

Solve the following systems using the substitution method.

25. $5x - 8y = 7$
    $y = 2x - 5$

26. $3x + 4y = 10$
    $y = 8x - 15$

27. $7x - 6y = -1$
    $x = 2y - 1$

28. $4x + 2y = 3$
    $x = 4y - 3$

29. $-3x + 2y = 6$
    $y = 3x$

30. $-2x - y = -3$
    $y = -3x$

31. $5x - 6y = -4$
    $x = y$

32. $2x - 4y = 0$
    $y = x$

33. $3x + 3y = 9$
    $y = 2x - 12$

34. $7x + 6y = -9$
    $y = -2x + 1$

35. $7x - 11y = 16$
    $y = 10$

36. $9x - 7y = -14$
    $x = 7$

37. $-4x + 4y = -8$
    $y = x - 2$

38. $-4x + 2y = -10$
    $y = 2x - 5$

39. $-6x + 3y = 4$
    $y = 2x + 1$

40. $5x - 10y = 4$
    $y = \dfrac{1}{2}x + \dfrac{2}{5}$

Solve each system by substitution. You can eliminate the decimals if you like, but you don't have to. The solution will be the same in either case.

41. $0.05x + 0.10y = 1.70$
    $y = 22 - x$

42. $0.20x + 0.50y = 3.60$
    $y = 12 - x$

Applying the Concepts

43. In the introduction to the section, it took 5 minutes to race down river with a current of 3 mph, and it took 12 minutes to travel back upriver against the current. Use the following system of equations to find the rate r of the kinetic sculpture team.

$$5(r + 3) = d$$
$$12(r - 3) = d$$

44. San Francisco Bay has a current of approximately 4 mph. It takes 20 minutes to power a boat with the current, but 35 minutes against the current. Use the following system of equations to find the rate r of the boat.

4 mph

$$20(r + 4) = d$$
$$35(r - 4) = d$$

45. Gas Mileage Daniel is trying to decide whether to buy a car or a truck. The truck he is considering will cost him $150 a month in loan payments, and it gets 20 miles per gallon in gas mileage. The car will cost $180 a month in loan payments, but it gets 35 miles per gallon in gas mileage. Daniel estimates that he will pay $2.50 per gallon for gas. This means that the monthly cost to drive the truck x miles will be $y = \frac{2.50}{20}x + 150$. The total monthly cost to drive the car x miles will be $y = \frac{2.50}{35}x + 180$. The following figure shows the graph of each equation.

a. At how many miles do the car and the truck cost the same to operate?

b. If Daniel drives more than 1,200 miles, which will be cheaper?

c. If Daniel drives fewer than 300 miles, which will be cheaper?

d. Why do the graphs appear in the first quadrant only?

46. Video Production Pat runs a small company that duplicates DVDs. The daily cost and daily revenue for a company duplicating DVDs are shown in the following figure. The daily cost for duplicating x DVDs is $y = \frac{6}{5}x + 20$; the daily revenue (the amount of money he brings in each day) for duplicating x DVDs is $y = 1.7x$. The graphs of the two lines are shown in the following figure.

a. Pat will "break even" when his cost and his revenue are equal. How many DVDs does he need to duplicate to break even?

b. Pat will incur a loss when his revenue is less than his cost. If he duplicates 30 DVDs in one day, will he incur a loss?

c. Pat will make a profit when his revenue is larger than his costs. For what values of x will Pat make a profit?

d. Why does the graph appear in the first quadrant only?

Getting Ready For the Next Section

Simplify each of the following.

47. $(x + y) + (x - y)$

48. $(x + 2y) + (-x + y)$

49. $3(2x - y) + (x + 3y)$

50. $3(2x + 3y) - 2(3x + 5y)$

51. $-4(3x + 5y) + 5(5x + 4y)$

52. $(3x + 8y) - (3x - 2y)$

53. $6\left(\dfrac{1}{2}x - \dfrac{1}{3}y\right)$

54. $12\left(\dfrac{1}{4}x + \dfrac{2}{3}y\right)$

55. $-4(3x - 2y)$

56. $-8(2x - 3y)$

57. $4(3x - 2y) + 2(5x + 4y)$

58. $-3(4x - 3y) + 4(3x - 5y)$

59. Let $x + y = 4$. If $x = 3$, find y.

60. Let $x + 2y = 4$. If $x = 3$, find y.

61. Let $x + 3y = 3$. If $x = 3$, find y.

62. Let $2x + 3y = -1$. If $y = -1$, find x.

63. Let $3x + 5y = -7$. If $x = 6$, find y.

64. Let $3x - 2y = 12$. If $y = 6$, find x.

Find the Mistake

The table below shows systems and the first step taken to solve each system by the substitution method. Circle the mistake(s) in the first step and write the correct way to begin on the line provided.

System	First Step
1. $x = 6y - 5$ $2x + 4y = 1$	Substitute the value for x from the first equation for y in the second equation. _____
2. $x - 3y = 9$ $7y - 3x = -2$	Subtract x from both sides of the first equation to isolate y. _____
3. $y = 8x + 4$ $3x = 7y$	Divide both sides of the second equation by 3 to isolate x. _____
4. $x = \dfrac{1}{2}y + \dfrac{3}{4}$ $4y - 3x = 12$	Multiply all terms in the first equation by the LCD 4. _____

11.3 The Elimination Method

OBJECTIVES

A Use the elimination method to solve a system of linear equations in two variables.

KEY WORDS

elimination

coefficients

Video Examples
Section 11.3

Practice Problems

1. Solve the following system using the elimination method.

$$x + y = 3$$
$$x - y = 5$$

Note The graphs with our first three examples are not part of their solutions. The graphs are there simply to show you that the results we obtain by the elimination method are consistent with the results we would obtain by graphing.

Answer

1. $(4, -1)$

Remember the exciting QuikTrip Air and Rocket Racing Show we discussed earlier in the chapter. Let's suppose 1720 general admission tickets were sold at the main gate for a total of $24,590. If it cost $15 for adults and $10 for children, how many of each kind of ticket were sold? In this section, we will learn another method of solving systems of linear equations in two variables, such as the one needed to answer the above question. This method is called the elimination method and makes use of the addition property.

The addition property states that if equal quantities are added to both sides of an equation, the solution set is unchanged. In the past, we have used this property to help solve equations in one variable. We will now use it to solve systems of linear equations. Here is another way to state the addition property of equality.

Let A, B, C, and D represent algebraic expressions.

$$\text{If} \rightarrow \quad A = B$$
$$\text{and} \rightarrow \quad C = D$$
$$\text{then} \rightarrow \quad A + C = B + D$$

Because C and D are equal (that is, they represent the same number), what we have done is added the same amount to both sides of the equation $A = B$. Let's see how we can use this form of the addition property of equality to solve a system of linear equations by way of the elimination, or addition, method.

A The Elimination Method

EXAMPLE 1   Solve the following system using the elimination method.

$$x + y = 4$$
$$x - y = 2$$

Solution   The system is written in the form of the addition property of equality as shown above. It looks like this:

$$A = B$$
$$C = D$$

where A is $x + y$, B is 4, C is $x - y$, and D is 2.

We use the addition property of equality to add the left sides together and the right sides together.

$$x + y = 4$$
$$\underline{x - y = 2}$$
$$2x + 0 = 6$$

We now solve the resulting equation for x.

$$2x + 0 = 6$$
$$2x = 6$$
$$x = 3$$

The value we get for x is the value of the x-coordinate of the point of intersection of the two lines $x + y = 4$ and $x - y = 2$. To find the y-coordinate, we simply substitute $x = 3$ into either of the two original equations. Using the first equation, we get

$$3 + y = 4$$
$$y = 1$$

The solution to our system is the ordered pair $(3, 1)$. It satisfies both equations.

When →	$x = 3$	When →	$x = 3$
and →	$y = 1$	and →	$y = 1$
the equation →	$x + y = 4$	the equation →	$x - y = 2$
becomes →	$3 + 1 \overset{?}{=} 4$	becomes →	$3 - 1 \overset{?}{=} 2$
	$4 = 4$		$2 = 2$

Figure 1 is visual evidence that the solution to our system is $(3, 1)$.

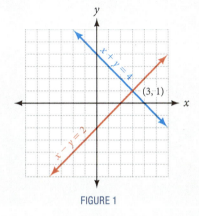

FIGURE 1

The most important part of this method of solving linear systems is eliminating one of the variables when we add the left and right sides together. In our first example, the equations were written so that the y variable was eliminated when we added the left and right sides together. If the equations are not set up this way to begin with, we have to work on one or both of them separately before we can add them together to eliminate one variable.

EXAMPLE 2   Solve the following system using the elimination method.

$$x + 2y = 4$$
$$x - y = -5$$

Solution   Notice that if we were to add the equations together as they are, the resulting equation would have terms in both x and y. Let's eliminate the variable x by multiplying both sides of the second equation by -1 before we add the equations together. (As you will see, we can choose to eliminate either the x or the y variable.) Multiplying both sides of the second equation by -1 will not change its solution, so we do not need to be concerned that we have altered the system.

$$
\begin{array}{l}
x + 2y = 4 \xrightarrow{\text{No change}} x + 2y = 4 \\
x - y = -5 \xrightarrow[\text{Multiply by } -1.]{} \underline{-x + y = 5} \\
\qquad\qquad\qquad\qquad\quad 0 + 3y = 9 \quad \text{Add left and right sides.} \\
\qquad\qquad\qquad\qquad\qquad\ 3y = 9 \\
\qquad\qquad\qquad\qquad\qquad\ \ y = 3 \quad \left\{ \begin{array}{l} \textit{y-coordinate of the} \\ \textit{point of intersection} \end{array} \right.
\end{array}
$$

Substituting $y = 3$ into either of the two original equations, we get $x = -2$. The solution to the system is $(-2, 3)$. It satisfies both equations. Figure 2 shows the solution to the system as the point where the two lines cross.

2. Solve the following system using the elimination method.

$$x + 3y = 5$$
$$x - y = 1$$

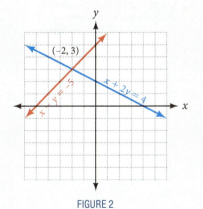

FIGURE 2

3. Solve the following system using the elimination method.

$$3x - y = 7$$
$$x + 2y = 7$$

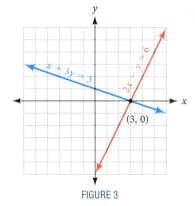

FIGURE 3

4. Solve the system using the elimination method.

$$3x + 2y = 3$$
$$2x + 5y = 13$$

Note If you are having trouble understanding this method of solution, it is probably because you can't see why we chose to multiply by 3 and -2 in the first step of Example 4. Look at the result of doing so: the $6x$ and $-6x$ will add to 0. We chose to multiply by 3 and -2 because they produce $6x$ and $-6x$, which will add to 0.

EXAMPLE 3   Solve the following system using the elimination method.

$$2x - y = 6$$
$$x + 3y = 3$$

Solution   Let's eliminate the y variable from the two equations. We can do this by multiplying the first equation by 3 and leaving the second equation unchanged.

$$2x - y = 6 \xrightarrow{\text{Multiply by 3.}} 6x - 3y = 18$$
$$x + 3y = 3 \xrightarrow{\text{No change}} x + 3y = 3$$

The important thing about our system now is that the *coefficients* (the numbers in front) of the y variables are opposites. When we add the terms on each side of the equal sign, then the terms in y will add to zero and be eliminated.

$$\begin{array}{r} 6x - 3y = 18 \\ x + 3y = 3 \\ \hline 7x = 21 \end{array} \quad \text{Add corresponding terms.}$$

This gives us $x = 3$. Using this value of x in the second equation of our original system, we have

$$3 + 3y = 3$$
$$3y = 0$$
$$y = 0$$

We could substitute $x = 3$ into any of the equations with both x and y variables and also get $y = 0$. The solution to our system is the ordered pair $(3, 0)$. Figure 3 is a picture of the system of equations showing the solution $(3, 0)$.

EXAMPLE 4   Solve the system using the elimination method.

$$2x + 3y = -1$$
$$3x + 5y = -2$$

Solution   Let's eliminate x from the two equations. If we multiply the first equation by 3 and the second by -2, the coefficients of x will be 6 and -6, respectively. The x terms in the two equations will then add to zero.

$$2x + 3y = -1 \xrightarrow{\text{Multiply by 3.}} 6x + 9y = -3$$
$$3x + 5y = -2 \xrightarrow{\text{Multiply by } -2.} -6x - 10y = 4$$

We now add the left and right sides of our new system together.

$$\begin{array}{r} 6x + 9y = -3 \\ -6x - 10y = 4 \\ \hline -y = 1 \\ y = -1 \end{array}$$

Substituting $y = -1$ into the first equation in our original system, we have

$$2x + 3(-1) = -1$$
$$2x - 3 = -1$$
$$2x = 2$$
$$x = 1$$

The solution to our system is $(1, -1)$. It is the only ordered pair that satisfies both equations.

Answers

3. $(3, 2)$

4. $(-1, 3)$

EXAMPLE 5   Solve the system using the elimination method.

$$3x + 5y = -7$$
$$5x + 4y = 10$$

Solution   Let's eliminate y by multiplying the first equation by -4 and the second equation by 5.

$$3x + 5y = -7 \xrightarrow{\text{Multiply by } -4.} -12x - 20y = 28$$
$$5x + 4y = 10 \xrightarrow{\text{Multiply by 5.}} \underline{25x + 20y = 50}$$
$$13x = 78$$
$$x = 6$$

Substitute $x = 6$ into either equation in our original system, and the result will be $y = -5$. Therefore, the solution is $(6, -5)$.

EXAMPLE 6   Solve the system using the elimination method.

$$\frac{1}{2}x - \frac{1}{3}y = 2$$
$$\frac{1}{4}x + \frac{2}{3}y = 6$$

Solution   Although we could solve this system without clearing the equations of fractions, there is probably less chance for error if we have only integer coefficients to work with. So let's begin by multiplying both sides of the top equation by 6 and both sides of the bottom equation by 12, to clear each equation of fractions.

$$\frac{1}{2}x - \frac{1}{3}y = 2 \xrightarrow{\text{Multiply by 6.}} 3x - 2y = 12$$
$$\frac{1}{4}x + \frac{2}{3}y = 6 \xrightarrow{\text{Multiply by 12.}} 3x + 8y = 72$$

Now we can eliminate x by multiplying the top equation by -1 and leaving the bottom equation unchanged.

$$3x - 2y = 12 \xrightarrow{\text{Multiply by } -1.} -3x + 2y = -12$$
$$3x + 8y = 72 \xrightarrow{\text{No change}} \underline{3x + 8y = 72}$$
$$10y = 60$$
$$y = 6$$

We can substitute $y = 6$ into any equation that contains both x and y. Let's use $3x - 2y = 12$.

$$3x - 2(6) = 12$$
$$3x - 12 = 12$$
$$3x = 24$$
$$x = 8$$

The solution to the system is $(8, 6)$.

Our next two examples will remind us what happens when we try to solve a system of equations consisting of parallel lines and a system in which the lines coincide.

5. Solve the system using the elimination method.

$$5x + 4y = -6$$
$$2x + 3y = -8$$

6. Solve the system using the elimination method.

$$\frac{1}{3}x + \frac{1}{2}y = 1$$
$$x + \frac{3}{4}y = 0$$

7. Solve the system using the elimination method.

$$x - 3y = 2$$
$$-3x + 9y = 2$$

EXAMPLE 7   Solve the system using the elimination method.

$$2x - y = 2$$
$$4x - 2y = 12$$

Solution   Let us choose to eliminate y from the system. We can do this by multiplying the first equation by -2 and leaving the second equation unchanged.

$$2x - \ y = \ 2 \xrightarrow{\text{Multiply by } -2.} -4x + 2y = -4$$

$$4x - 2y = 12 \xrightarrow[\text{No change}]{} \quad 4x - 2y = \ 12$$

If we add both sides of the resulting system, we have

$$-4x + 2y = -4$$
$$\underline{4x - 2y = \ 12}$$
$$0 + 0 = \quad 8$$

$$0 = \quad 8 \qquad \textit{A false statement.}$$

Both variables have been eliminated and we end up with the false statement $0 = 8$. We have tried to solve a system that consists of two parallel lines. There is no solution and the system is inconsistent. Figure 4 is a visual representation of the situation and is conclusive evidence that there is no solution to our system.

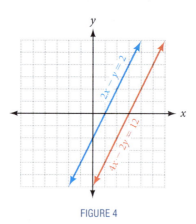

FIGURE 4

8. Solve the system using the elimination method.

$$5x - y = 1$$
$$10x - 2y = 2$$

EXAMPLE 8   Solve the system using the elimination method.

$$4x - 3y = 2$$
$$8x - 6y = 4$$

Solution   Multiplying the top equation by -2 and adding, we can eliminate the variable x.

$$4x - 3y = 2 \xrightarrow{\text{Multiply by } -2.} -8x + 6y = -4$$

$$8x - 6y = 4 \xrightarrow[\text{No change}]{} \quad \underline{8x - 6y = \quad 4}$$
$$0 = \quad 0$$

Both variables have been eliminated, and the resulting statement $0 = 0$ is true. In this case the lines coincide because the equations are equivalent. The solution set consists of all ordered pairs that satisfy either equation, and the system is dependent.

Answers

7. No solution (inconsistent)

8. Lines coincide (dependent)

The preceding two examples remind us of the the two special cases in which the graphs of the equations in the system either coincide or are parallel.

Here is a summary:

Both variables are eliminated and the resulting statement is false.	$\leftrightarrow$	The lines are parallel and there is no solution to the system.
Both variables are eliminated and the resulting statement is true.	$\leftrightarrow$	The lines coincide and there is an infinite number of solutions to the system.

The main idea in solving a system of linear equations by the elimination method is to use the multiplication property of equality on one or both of the original equations, if necessary, to make the coefficients of either variable opposites. The following box shows some steps to follow when solving a system of linear equations by the elimination method.

> **HOW TO** Solving a System of Linear Equations by the Elimination Method
>
> **Step 1:** Decide which variable to eliminate. (In some cases one variable will be easier to eliminate than the other. With some practice you will notice which one it is.)
>
> **Step 2:** Use the multiplication property of equality on each equation separately to make the coefficients of the variable that is to be eliminated opposites.
>
> **Step 3:** Add the respective left and right sides of the system together.
>
> **Step 4:** Solve for the variable remaining.
>
> **Step 5:** Substitute the value of the variable from step 4 into an equation containing both variables and solve for the other variable.
>
> **Step 6:** Check your solution in both equations, if necessary.

GETTING READY FOR CLASS

After reading through the preceding section, respond in your own words and in complete sentences.

A. How do you use the addition property of equality in the elimination method of solving a system of linear equations?

B. What happens when you use the elimination method to solve a system of linear equations consisting of two parallel lines?

C. How would you use the multiplication property of equality to solve a system of linear equations?

D. What is the first step in solving a system of linear equations that contains fractions?

Vocabulary Review

Choose the correct words to fill in the blanks below.

| substitute | solution | add | variable | solve | multiplication property of equality |

Solving a System of Linear Equations by the Elimination Method

Step 1: Decide which _____ to eliminate.

Step 2: Use the _____ on each equation separately to make the coefficients of the variable that is to be eliminated opposites.

Step 3: _____ the respective left and right sides of the system together.

Step 4: _____ for the variable remaining.

Step 5: _____ the value of the variable from step 4 into an equation containing both variable and solve for the other variable.

Step 6: Check your _____ in both equations.

Problems

A Solve the following systems of linear equations by elimination.

1. $x + y = 3$
$x - y = 1$

2. $x + y = -2$
$x - y = 6$

3. $x + y = 10$
$-x + y = 4$

4. $x - y = 1$
$-x - y = -7$

5. $x - y = 7$
$-x - y = 3$

6. $x - y = 4$
$2x + y = 8$

7. $x + y = -1$
$3x - y = -3$

8. $2x - y = -2$
$-2x - y = 2$

9. $3x + 2y = 1$
$-3x - 2y = -1$

10. $-2x - 4y = 1$
$2x + 4y = -1$

11. $3x - 2y = 4$
$-3x + 2y = 6$

12. $x - 2y = 5$
$-x + 2y = 5$

Solve each of the following systems by eliminating the y variable.

13. $3x - y = 4$
$2x + 2y = 24$

14. $2x + y = 3$
$3x + 2y = 1$

15. $5x - 3y = -2$
$10x - y = 1$

16. $4x - y = -1$
$2x + 4y = 13$

17. $11x - 4y = 11$
$5x + y = 5$

18. $3x - y = 7$
$10x - 5y = 25$

19. $2x - y = 6$
$x + 4y = 3$

20. $2x + y = 5$
$-2x + 2y = 4$

Solve each of the following systems by eliminating the x variable.

21. $3x - 5y = 7$
$-x + y = -1$

22. $4x + 2y = 32$
$x + y = -2$

23. $-x - 8y = -1$
$-2x + 4y = 13$

24. $-x + 10y = 1$
$-5x + 15y = -9$

25. $-3x - y = 7$
$6x + 7y = 11$

26. $-5x + 2y = -6$
$10x + 7y = 34$

27. $-3x + 2y = 6$
$6x - y = -9$

28. $-4x + 5y = 20$
$8x - 6y = -12$

Solve each of the following systems of linear equations by the elimination method.

29. $6x - y = -8$
$2x + y = -16$

30. $5x - 3y = -3$
$3x + 3y = -21$

31. $x + 3y = 9$
$2x - y = 4$

32. $x + 2y = 0$
$2x - y = 0$

33. $x - 6y = 3$
$4x + 3y = 21$

34. $8x + y = -1$
$4x - 5y = 16$

35. $2x + 9y = 2$
$5x + 3y = -8$

36. $5x + 2y = 11$
$7x + 8y = 7$

37. $\dfrac{1}{3}x + \dfrac{1}{4}y = \dfrac{7}{6}$

$\dfrac{3}{2}x - \dfrac{1}{3}y = \dfrac{7}{3}$

38. $\dfrac{7}{12}x - \dfrac{1}{2}y = \dfrac{1}{6}$

$\dfrac{2}{5}x - \dfrac{1}{3}y = \dfrac{11}{15}$

39. $3x + 2y = -1$

$6x + 4y = 0$

40. $8x - 2y = 2$

$4x - y = 2$

41. $11x + 6y = 17$

$5x - 4y = 1$

42. $3x - 8y = 7$

$10x - 5y = 45$

43. $\dfrac{1}{2}x + \dfrac{1}{6}y = \dfrac{1}{3}$

$-x - \dfrac{1}{3}y = -\dfrac{1}{6}$

44. $-\dfrac{1}{3}x - \dfrac{1}{2}y = -\dfrac{2}{3}$

$-\dfrac{2}{3}x - y = -\dfrac{4}{3}$

45. Multiply both sides of the second equation in the following system by 100, and then solve as usual.

$x + y = 22$

$0.05x + 0.10y = 1.70$

46. Multiply both sides of the second equation in the following system by 100, and then solve as usual.

$x + y = 15{,}000$

$0.06x + 0.07y = 980$

Applying the Concepts

47. Recall at the beginning of this section that adult tickets x for the QuikTrip Air and Rocket Racing Show cost $15 and children's tickets y cost $10 Suppose 1,720 tickets are sold at the gate for a total of $19,750.

$x + y = 1{,}720$
$15x - 10y = 19{,}750$

Solve the above system of equations to find the number of adult and children tickets sold.

48. A local community theater sells adult tickets x for $10 and children's tickets y for $5. For one evening performance of a popular musical, the theater sells out 375 seats and collects a revenue of $2,875. The system of equations for this situation is

$x + y = 375$
$10x + 5y = 2{,}875$

Solve this system to find the number of adult and children tickets sold.

Getting Ready for the Next Section

49. One number is eight more than five times another; their sum is 26. Find the numbers.

50. One number is three less than four times another; their sum is 27. Find the numbers.

51. The difference of two positive numbers is nine. The larger number is six less than twice the smaller number. Find the numbers.

52. The difference of two positive numbers is 17. The larger number is one more than twice the smaller number. Find the numbers.

53. The length of a rectangle is five inches more than three times the width. The perimeter is 58 inches. Find the length and width.

54. The length of a rectangle is three inches less than twice the width. The perimeter is 36 inches. Find the length and width.

55. John has $1.70 in nickels and dimes in his pocket. He has four more nickels than he does dimes. How many of each does he have?

56. Jamie has $2.65 in dimes and quarters in her pocket. She has two more dimes than she does quarters. How many of each does she have?

Find the Mistake

The table below shows systems and the first step taken to solve each system by the elimination method. Circle the mistake(s) in the first step and write the correct way to begin on the line provided.

System   **First Step**

1. $4x - y = -8$ Multiply both sides of the equation by -1 to eliminate the variable y.
$\quad -4x - y = 8$
$\rule{10cm}{0.4pt}$

2. $x - 7y = 11$ Multiply both sides of the first equation by 3 to eliminate x.
$\quad 3x + 2y = 15$
$\rule{10cm}{0.4pt}$

3. $5x + y = 9$ Multiply both sides of the first equation by 3 to eliminate y.
$\quad -5x + 3y = 21$
$\rule{10cm}{0.4pt}$

Circle the mistake in the sentence below and write the correct word or phrase on the line provided.

4. When solving a system by the elimination method, if both variables are eliminated and the resulting statement is true, then the lines are parallel. $\rule{6cm}{0.4pt}$

Landmark Review: Checking Your Progress

Solve the following systems.

1. $y = 2x + 5$
$\quad y = -3x - 5$

2. $2x - y = -2$
$\quad 2x - 3y = 6$

3. $2x - y = 1$
$\quad y = -2x + 5$

4. $2x - 5y = 0$
$\quad x - 5y = -9$

5. $-3x + y = 8$
$\quad x = -y + 4$

6. $-6x + 2y = -9$
$\quad y = 3x + 4$

7. $x - y = 11$
$\quad x + y = -1$

8. $-3x + y = -15$
$\quad -2x + 3y = -17$

9. $2x + 5y = 17$
$\quad x - 4y = -24$

10. $x + 4y = 5$
$\quad 3x + 8y = 14$

11.4 Applications of Systems of Equations

©iStockphoto.com / Nikada

OBJECTIVES

A Apply the Blueprint for Problem Solving to a variety of application problems involving systems of equations.

The Mad Hatter Relay Race in Boise, Idaho is the epitome of a "fun run." Two-person teams from across the state participate in 4-mile combined races, 2 miles for each team member. All participants must wear silly or theme hats while running, and are strongly encouraged to wear costumes as well. For the 2010 event, pre-registration for each team cost $28, with an additional $10 for race-day registration. Let's suppose x teams pre-registered for the race and y teams registered on race day for a total of 29 teams. Let's also suppose that the total registration fees paid for this race were $922. Can you use the information to create a linear system and solve for its variables? Read through this section and practice the application problems. You will be asked to solve this relay race problem in the problem set.

I often have heard students remark about the word problems in introductory algebra: "What does this have to do with real life?" Many of the word problems we will encounter in this section don't have much to do with "real life." We are actually just practicing. Ultimately, all problems requiring the use of algebra are word problems; that is, they are stated in words first, then translated to symbols. The problem then is solved by some system of mathematics, like algebra. If word problems frustrate you, take a deep breath and tackle them one step at a time. You can do this.

Video Examples
Section 11.4

A The Blueprint for Problem Solving

The word problems in this section have two unknown quantities. We will write two equations in two variables (each of which represents one of the unknown quantities), which of course is a system of equations. We then solve the system by one of the methods developed previously. Here are the steps to follow in solving these word problems.

> **BLUEPRINT FOR PROBLEM SOLVING** Using a System of Equations
>
> **Step 1:** *Read* the problem, and then mentally *list* the items that are known and the items that are unknown.
>
> **Step 2:** *Assign variables* to each of the unknown items; that is, let $x =$ one of the unknown items and $y =$ the other unknown item. Then *translate* the other *information* in the problem to expressions involving the two variables.
>
> **Step 3:** *Reread* the problem, and then *write a system of equations,* using the items and variables listed in steps 1 and 2, that describes the situation. Drawing a sketch or creating a table may help you organize your data.
>
> **Step 4:** *Solve the system* found in step 3.
>
> **Step 5:** *Write* your *answers* using complete sentences.
>
> **Step 6:** *Reread* the problem, and *check* your solution with the original words in the problem.

Remember, the more problems you work, the more problems you will be able to work. If you have trouble getting started on the problem set, come back to the examples and work through them yourself. The examples are similar to the problems found in the problem set.

Number Problem

EXAMPLE 1   One number is 2 more than 5 times another number. Their sum is 20. Find the two numbers.

Solution   Applying the steps in our blueprint, we have

Step 1: *Read and list.*
We know that the two numbers have a sum of 20 and that one of them is 2 more than 5 times the other. We don't know what the numbers themselves are.

Step 2: *Assign variables and translate information.*
Let x represent one of the numbers and y represent the other. "One number is 2 more than 5 times another" translates to

$$y = 5x + 2$$

"Their sum is 20" translates to

$$x + y = 20$$

Step 3: *Write a system of equations.*
The system that describes the situation must be

$$x + y = 20$$
$$y = 5x + 2$$

Step 4: *Solve the system.*
We can solve this system by substituting the expression $5x + 2$ in the second equation for y in the first equation.

$$x + (5x + 2) = 20$$
$$6x + 2 = 20$$
$$6x = 18$$
$$x = 3$$

Using $x = 3$ in either of the first two equations and then solving for y, we get $y = 17$.

Step 5: *Write answers.*
So 17 and 3 are the numbers we are looking for.

Step 6: *Read and check.*
The number 17 is 2 more than 5 times 3, and the sum of 17 and 3 is 20.

Age Problem

EXAMPLE 2   Kaci is 3 more than 10 times the age of her daughter, Amelie. The sum of their ages is 36 years. Find their ages.

Solution   Begin by mentally listing what we know and don't know.

Step 1: *Read and list*
We know that the two ages have a sum of 36, and that one of them is 3 more than 10 times the other.

Step 2: *Assign variables and translate information.*

Let x represent Amelie's age and y represent Kaci's age. "Kaci is 3 more than 10 times the age of Amelie" translates to

$$y = 10x + 3$$

"The sum of their ages is 36 years" translates to

$$x + y = 36$$

Step 3: *Write a system of equations.*
The system for this problem is

$$y = 10x + 3$$
$$x + y = 36$$

Step 4: *Solve the system.*
We can solve the system by substituting the expression $10x + 3$ in the first equation for y in the second equation.

$$x + (10x + 3) = 36$$
$$11x + 3 = 36$$
$$11x = 33$$
$$x = 3$$

Step 5: *Write answers.*
Amelie is 3 years old, and Kaci is 33 years old.

Step 6: *Read and check.*
Kaci's age of 33 is 3 more than 10 times Amelie's age of 3, and the sum of 33 and 3 is 36.

Interest Problem

EXAMPLE 3 Mr. Hicks had $15,000 to invest. He invested some at 6% and the rest at 7%. If he earns $980 in interest, how much did he invest at each rate?

Solution Remember, step 1 is done mentally.

Step 1: *Read and list.*
We do not know the specific amounts invested in the two accounts. We do know that their sum is $15,000 and that the interest rates on the two accounts are 6% and 7%.

Step 2: *Assign variables and translate information.*
Let x = the amount invested at 6% and y = the amount invested at 7%. Because Mr. Hicks invested a total of $15,000, we have

$$x + y = 15,000$$

The interest he earns comes from 6% of the amount invested at 6% and 7% of the amount invested at 7%. To find 6% of x, we multiply x by 0.06, which gives us $0.06x$. To find 7% of y, we multiply 0.07 times y and get $0.07y$.

$$\underset{\text{at 6\%}}{\text{Interest}} + \underset{\text{at 7\%}}{\text{Interest}} = \underset{\text{interest}}{\text{Total}}$$

$$0.06x + 0.07y = 980$$

Step 3: *Write a system of equations.*
The system is

$$x + y = 15,000$$
$$0.06x + 0.07y = 980$$

3. Amy has $10,000 to invest. She invests part at 6% and the rest at 7%. If she earns $630 in interest for the year, how much does she have invested at each rate?

Step 4: *Solve the system.*

We multiply the first equation by -6 and the second by 100 to eliminate x.

$$x + y = 15{,}000 \xrightarrow{\text{Multiply by } -6.} -6x - 6y = -90{,}000$$

$$0.06x + 0.07y = 980 \xrightarrow[\text{Multiply by 100.}]{} \underline{6x + 7y = 98{,}000}$$

$$y = 8{,}000$$

Substituting $y = 8{,}000$ into the first equation and solving for x, we get $x = 7{,}000$.

Step 5: *Write answers.*

He invested \$7,000 at 6% and \$8,000 at 7%.

Step 6: *Read and check.*

Checking our solutions in the original problem, we have the following: The sum of \$7,000 and \$8,000 is \$15,000, the total amount he invested. To complete our check, we find the total interest earned from the two accounts.

$$\text{The interest on \$7,000 at 6\% is } 0.06(7{,}000) = 420$$
$$\underline{\text{The interest on \$8,000 at 7\% is } 0.07(8{,}000) = 560}$$
$$\text{The total interest is \$980.}$$

Coin Problem

4. Patrick has \$1.85 in dimes and quarters. He has a total of 14 coins. How many of each kind does he have?

EXAMPLE 4 John has \$1.70 in dimes and nickels. He has a total of 22 coins. How many of each kind does he have?

Solution

Step 1: *Read and list.*

We know that John has 22 coins that are dimes and nickels. We know that a dime is worth 10 cents and a nickel is worth 5 cents. We do not know the specific number of dimes and nickels he has.

Step 2: *Assign variables and translate information.*

Let $x =$ the number of nickels and $y =$ the number of dimes. The total number of coins is 22, so

$$x + y = 22$$

The total amount of money he has is \$1.70, which comes from nickels and dimes.

$$\begin{array}{ccccc}
\text{Amount of money} & + & \text{Amount of money} & = & \text{Total amount} \\
\text{in nickels} & & \text{in dimes} & & \text{of money} \\
0.05x & + & 0.10y & = & 1.70
\end{array}$$

Step 3: *Write a system of equations.*

The system that represents the situation is

$$\begin{array}{ll}
x + y = 22 & \text{The number of coins} \\
0.05x + 0.10y = 1.70 & \text{The value of the coins}
\end{array}$$

Step 4: *Solve the system.*

We multiply the first equation by -5 and the second by 100 to eliminate the variable x.

$$x + y = 22 \xrightarrow{\text{Multiply by } -5.} -5x - 5y = -110$$

$$0.05x + 0.10y = 1.70 \xrightarrow[\text{Multiply by 100.}]{} \underline{5x + 10y = 170}$$

$$5y = 60$$

$$y = 12$$

Substituting $y = 12$ into our first equation, we get $x = 10$.

Step 5: *Write answers.*
John has 12 dimes and 10 nickels.

Step 6: *Read and check.*
Twelve dimes and 10 nickels total 22 coins.

$$12 \text{ dimes are worth } 12(0.10) = 1.20$$
$$\underline{10 \text{ nickels are worth } 10(0.05) = 0.50}$$
$$\text{The total value is } \$1.70.$$

Mixture Problem

EXAMPLE 5   How much of a 20% alcohol solution and 50% alcohol solution must be mixed to get 12 gallons of 30% alcohol solution?

Solution   To solve this problem we must first understand that a 20% alcohol solution is 20% alcohol and 80% water.

Step 1: *Read and list.*
We know there are two solutions that together must total 12 gallons. 20% of one of the solutions is alcohol and the rest is water, whereas the other solution is 50% alcohol and 50% water. We do not know how many gallons of each individual solution we need.

Step 2: *Assign variables and translate information.*
Let $x =$ the number of gallons of 20% alcohol solution needed and $y =$ the number of gallons of 50% alcohol solution needed. Because the total number of gallons we will end up with is 12, and this 12 gallons must come from the two solutions we are mixing, our first equation is

$$x + y = 12$$

To obtain our second equation, we look at the amount of alcohol in our two original solutions and our final solution. The amount of alcohol in the x gallons of 20% solution is $0.20x$, and the amount of alcohol in y gallons of 50% solution is $0.50y$. The amount of alcohol in the 12 gallons of 30% solution is $0.30(12)$. Because the amount of alcohol we start with must equal the amount of alcohol we end up with, our second equation is

$$0.20x + 0.50y = 0.30(12)$$

The information we have so far can also be summarized with a table. Sometimes by looking at a table like the one that follows, it is easier to see where the equations come from.

	20% Solution	50% Solution	Final Solution
Number of Gallons	x	y	12
Gallons of Alcohol	$0.20x$	$0.50y$	$0.30(12)$

Step 3: *Write a system of equations.*
Our system of equations is

$$x + y = 12$$
$$0.20x + 0.50y = 0.30(12)$$

5. How much 30% alcohol solution and 60% alcohol solution must be mixed to get 25 gallons of 48% solution?

Answers

5. 10 gallons of 30%, 15 gallons of 60%

Step 4: *Solve the system.*

We can solve this system by substitution. Solving the first equation for y and substituting the result into the second equation, we have

$$0.20x + 0.50(12 - x) = 0.30(12)$$

Multiplying each side by 10 gives us an equivalent equation that is a little easier to work with.

$$2x + 5(12 - x) = 3(12)$$
$$2x + 60 - 5x = 36$$
$$-3x + 60 = 36$$
$$-3x = -24$$
$$x = 8$$

If x is 8, then y must be 4 because $x + y = 12$.

Step 5: *Write answers.*

It takes 8 gallons of 20% alcohol solution and 4 gallons of 50% alcohol solution.

Step 6: *Read and check.*

Eight gallons of 20% alcohol solution plus four gallons of 50% alcohol solution will equal twelve gallons 30% alcohol solution.

$$0.20(8) + 0.50(4) = 0.30(12)$$
$$1.6 + 2 = 3.6$$
$$3.6 = 3.6 \qquad \text{True statement}$$

GETTING READY FOR CLASS

After reading through the preceding section, respond in your own words and in complete sentences.

A. What is the first step of the Blueprint for Problem Solving when using a system of equations?

B. Is the Blueprint for Problem Solving when using a system of equations any different than the Blueprint you learned for solving equations in one variable? Explain.

C. Why do you have to understand systems of linear equations to work the application problems in this section?

D. Write an application problem for which the solution depends on solving a system of equations.

Vocabulary Review

The following is a list of steps for the Blueprint for Problem Solving for systems. Choose the correct words to fill in the blanks below.

system	expressions	equations	sentences
known	problem	unknown	solution

Step 1: Read the problem and mentally list the items that are _____ and unknown.

Step 2: Assign a variable to each of the _____ items, and then translate the other information in the problem to _____ involving the variables.

Step 3: Reread the problem and write a system of _____ .

Step 4: Solve the _____ .

Step 5: Write your answers using complete _____ .

Step 6: Reread the _____ and check your _____ .

Problems

A Solve the following word problems using the Blueprint for Problem Solving. Be sure to show the equations used.

Number Problems

1. Two numbers have a sum of 25. One number is 5 more than the other. Find the numbers.

2. The difference of two numbers is 6. Their sum is 30. Find the two numbers.

3. The sum of two numbers is 15. One number is 4 times the other. Find the numbers.

4. The difference of two positive numbers is 28. One number is 3 times the other. Find the two numbers.

5. Two positive numbers have a difference of 5. The larger number is one more than twice the smaller. Find the two numbers.

6. One number is 2 more than 3 times another. Their sum is 26. Find the two numbers.

7. One number is 5 more than 4 times another. Their sum is 35. Find the two numbers.

8. The difference of two positive numbers is 8. The larger is twice the smaller decreased by 7. Find the two numbers.

Age Problems

9. Linda is 8 more than 2 times the age of her daughter, Audra. The sum of their ages is 101 years. Find their ages.

10. A mathematics professor is 3 less than 3 times the average age of his students. The difference between the professor's age and the average student is 35 years. Find the ages.

11. Kate took some time off between earning her bachelor's degree and beginning her master's program. On the first day of Kate's master's program, she was 5 less than twice the age she was when she graduated with her bachelor's. The difference between the two ages was 16 years.

12. Michael is scheduled to receive an inheritance when he is 6 more than a third the age of his grandfather. The difference between their ages will be 38 years. Find their ages at the time of inheritance.

Interest Problems

13. Mr. Wilson invested money in two accounts. His total investment was $20,000. If one account pays 6% interest and the other pays 8% interest, how much does he have in each account if he earned a total of $1,380 in interest in 1 year?

14. A total of $11,000 was invested. Part of the $11,000 was invested at 4%, and the rest was invested at 7%. If the investments earn $680 per year, how much was invested at each rate?

15. A woman invested 4 times as much at 5% as she did at 6%. The total amount of interest she earns in 1 year from both accounts is $520. How much did she invest at each rate?

16. Ms. Hagan invested twice as much money in an account that pays 7% interest as she did in an account that pays 6% interest. Her total investment pays her $1,000 a year in interest. How much did she invest at each rate?

Coin Problems

17. Ron has 14 coins with a total value of $2.30. The coins are nickels and quarters. How many of each coin does he have?

18. Diane has $0.95 in dimes and nickels. She has a total of 11 coins. How many of each kind does she have?

19. Suppose Tom has 21 coins totaling $3.45. If he has only dimes and quarters, how many of each type does he have?

20. A coin collector has 31 dimes and nickels with a total face value of $2.40. (They are actually worth a lot more.) How many of each coin does she have?

Mixture Problems

21. How many liters of 50% alcohol solution and 20% alcohol solution must be mixed to obtain 18 liters of 30% alcohol solution?

	50% Solution	20% Solution	Final Solution
Number of Liters	x	y	
Liters of Alcohol			

22. How many liters of 10% alcohol solution and 5% alcohol solution must be mixed to obtain 40 liters of 8% alcohol solution?

	10% Solution	5% Solution	Final Solution
Number of Liters	x	y	
Liters of Alcohol			

23. A mixture of 8% disinfectant solution is to be made from 10% and 7% disinfectant solutions. How much of each solution should be used if 30 gallons of 8% solution are needed?

24. How much 50% antifreeze solution and 40% antifreeze solution should be combined to give 50 gallons of 46% antifreeze solution?

Miscellaneous Problems

25. For a Saturday matinee, adult tickets cost $5.50 and kids under 12 pay only $4.00. If 70 tickets are sold for a total of $310, how many of the tickets were adult tickets and how many were sold to kids under 12?

26. The Bishop's Peak 4-H club is having its annual fundraising dinner. Adults pay $15 each and children pay $10 each. If the number of adult tickets sold is twice the number of children's tickets sold, and the total income for the dinner is $1,600, how many of each kind of ticket did the 4-H club sell?

27. A farmer has 96 feet of fence with which to make a corral. If he arranges it into a rectangle that is twice as long as it is wide, what are the dimensions?

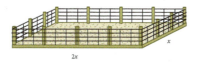

x

$2x$

28. If a 22-inch rope is to be cut into two pieces so that one piece is 3 inches longer than twice the other, how long is each piece?

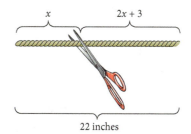

x $2x + 3$

22 inches

29. Suppose you finish a session of blackjack with $5 chips and $25 chips. You have 45 chips in all, with a total value of $465, how many of each kind of chip do you have?

30. Tyler has been saving his winning lottery tickets. He has 23 tickets that are worth a total of $175. If each ticket is worth either $5 or $10, how many of each does he have?

31. Mary Jo spends $2,550 to buy stock in two companies. She pays $11 a share to one of the companies and $20 a share to the other. If she ends up with a total of 150 shares, how many shares did she buy at $11 a share and how many did she buy at $20 a share?

32. Kelly sells 62 shares of stock she owns for a total of $433. If the stock was in two different companies, one selling at $6.50 a share and the other at $7.25 a share, how many of each did she sell?

33. Recall the relay race problem introduced in the beginning of this section. Registration costs were $28 for pre-registration and $38 for race day registration. Suppose x teams are pre-registered and y teams registered on race day, and 29 teams register for a total fee of $922. Set up a system of equations and solve it to find how many teams pre-registered and how many teams registered the day of the race.

34. Pre-event tickets for a local theater fundraiser cost $30 and $40 for at-the-door tickets. Organizers sell a total of 200 tickets and generate a total revenue of $6,650. How many pre-event and at-the-door tickets were sold?

Maintaining Your Skills

Solve each system of equations by the elimination method.

35. $2x + y = 3$
$3x - y = 7$

36. $3x - y = -6$
$4x + y = -8$

37. $4x - 5y = 1$
$x - 2y = -2$

38. $6x - 4y = 2$
$2x + y = 10$

Solve each system of equations by the substitution method.

39. $5x + 2y = 7$
$y = 3x - 2$

40. $-7x - 5y = -1$
$y = x + 5$

41. $2x - 3y = 4$
$x = 2y + 1$

42. $4x - 5y = 2$
$x = 2y - 1$

Find the Mistake

Each sentence below contains a mistake. Circle the mistake and write the correct word(s) or number(s) on the line provided.

1. The first step in the Blueprint for Problem Solving is to assign variables to each of the unknown items.

_____

2. If there are two unknown quantities, we write one equation to get our answer. _____

3. If we are solving a problem involving nickels and dimes and we know we have a total of 15 coins, we can use $0.05n + 0.10d = 15$ as one of our equations. _____

4. One number is 3 times another and their sum is 8. One of our equations will be $3x = 8$. _____

Mathematical Bees

The great mathematician Fibonacci made many mathematical discoveries. One of these, the Fibonacci sequence, can be used to explain a honey bee's family tree. How are the Fibonacci sequence and honey bee ancestry related?

Another interesting fact about honey bees is that they perform choreographed dances to communicate the location of a food source, such as a flower. The specific movements in the dance depend on the location of the sun, the hive, and the flower. The communicated information includes angles and distance measurements needed to find the food source from the hive's origin. Research this amazing process. What math concepts do the bees use to make a dance successful? Now on graph paper, create a diagram of a chosen food source in relation to the sun and the hive. Give coordinates for the three locations, and determine a system of linear equations that would represent lines drawn through the following coordinates:

1. The hive and the sun

2. The hive and the food source

3. The sun and the food source

EXAMPLES

Definitions [11.1]

1. The solution to the system
$$x + 2y = 4$$
$$x - y = 1$$
is the ordered pair (2, 1). It is the only ordered pair that satisfies both equations.

1. A *system of linear equations*, as the term is used in this book, is two linear equations that each contain the same two variables.

2. The *solution set* for a system of equations is the set of all ordered pairs that satisfy *both* equations. The solution set to a system of linear equations will contain:

Case I   One ordered pair when the graphs of the two equations intersect at only one point (this is the most common situation)

Case II   No ordered pairs when the graphs of the two equations are parallel lines

Case III   An infinite number of ordered pairs when the graphs of the two equations coincide (are the same line)

Strategy for Solving a System by Graphing [11.1]

2. Solving the system in Example 1 by graphing looks like

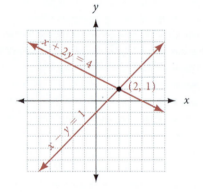

Step 1: Graph the first equation.

Step 2: Graph the second equation on the same set of axes.

Step 3: Read the coordinates of the point where the graphs cross each other (the coordinates of the point of intersection.)

Step 4: Check the solution to see that it satisfies *both* equations.

Strategy for Solving a System by the Substitution Method [11.2]

3. We can apply the substitution method to the system in Example 1 by first solving the second equation for x to get $x = y + 1$. Substituting this expression for x into the first equation, we have
$$(y + 1) + 2y = 4$$
$$3y + 1 = 4$$
$$3y = 3$$
$$y = 1$$
Using $y = 1$ in either of the original equations gives $x = 2$.

Step 1: Solve either of the equations for one of the variables (this step is not necessary if one of the equations has the correct form already.)

Step 2: Substitute the results of step 1 into the other equation, and solve.

Step 3: Substitute the results of step 2 into an equation with both x and y variables, and solve. (The equation produced in step 1 is usually a good one to use.)

Step 4: Check your solution, if necessary.

Strategy for Solving a System by the Elimination Method [11.3]

Step 1: Look the system over to decide which variable will be easier to eliminate.

Step 2: Use the multiplication property of equality on each equation separately to ensure that the coefficients of the variable to be eliminated are opposites.

Step 3: Add the left and right sides of the system produced in step 2, and solve the resulting equation.

Step 4: Substitute the solution from step 3 back into any equation with both x and y variables, and solve.

Step 5: Check your solution in both equations, if necessary.

4. We can eliminate the y variable from the system in Example 1 by multiplying both sides of the second equation by 2 and adding the result to the first equation

$$
\begin{array}{ll}
x + 2y = 4 \xrightarrow{\text{No Change}} & x + 2y = 4 \\
x - y = 1 \xrightarrow[\text{Multiply by 2}]{} & \underline{2x - 2y = 2} \\
& 3x \quad\ = 6 \\
& x \quad\ = 2
\end{array}
$$

Substituting $x = 2$ into either of the original two equations gives $y = 1$. The solution is $(2, 1)$.

Special Cases [11.1, 11.2, 11.3]

In some cases, using the elimination or substitution method eliminates both variables. The situation is interpreted as follows.

1. If the resulting statement is *false*, then the lines are parallel and there is no solution to the system. This is called an inconsistent system.

2. If the resulting statement is *true*, then the equations represent the same line (the lines coincide). In this case any ordered pair that satisfies either equation is a solution to the system. This is called a dependent system.

Blueprint for Problem Solving: Using a System of Equations [11.4]

Step 1: *Read* the problem, and then mentally *list* the items that are known and the items that are unknown.

Step 2: *Assign variables* to each of the unknown items; that is, let $x =$ one of the unknown items and $y =$ the other unknown item. Then *translate* the other *information* in the problem to expressions involving the two variables.

Step 3: *Reread* the problem, and then *write a system of equations,* using the items and variables listed in steps 1 and 2, that describes the situation.

Step 4: *Solve the system* found in step 3.

Step 5: *Write* your *answers* using complete sentences.

Step 6: *Reread* the problem, and *check* your solution with the original words in the problem.

COMMON MISTAKE

The most common mistake encountered in solving linear systems is the failure to complete the problem. Here is an example:

$$
\begin{array}{l}
x + y = 8 \\
x - y = 4 \\
\hline
2x = 12 \\
x = 6
\end{array}
$$

This is only half the solution. To find the other half, we must substitute the 6 back into one of the original equations and then solve for y.

Remember, solutions to systems of linear equations always consist of ordered pairs. We need an x-coordinate and a y-coordinate; $x = 6$ can never be a solution to a system of linear equations.

Solve the following systems by graphing. [11.1]

1. $x + y = 3$
$x - y = -5$

2. $x + y = 8$
$-x + y = 2$

3. $2x - 3y = -14$
$2x - y = -2$

4. $3x + 2y = 0$
$4x + y = 5$

5. $y = 3x + 3$
$y = 5x + 5$

6. $y = -2x - 7$
$y = x + 5$

Solve the following systems by substitution. [11.2]

7. $x + y = 8$
$y = 2x - 1$

8. $x - y = 3$
$x = 2y + 9$

9. $3x - y = -7$
$y = 2x + 6$

10. $4x - 2y = 12$
$y = 3x - 10$

11. $3x - y = -9$
$x - y = -5$

12. $4x - 3y = 5$
$x = 2y + 5$

13. $12x - 3y = -6$
$x = y + 4$

14. $4y + 8x = 6$
$-2y - 4x = -3$

Solve the following systems by the elimination method. [11.3]

15. $x - y = 3$
$2x - 5y = -3$

16. $2x + 2y = 2$
$-3x - y = 5$

17. $3x - 4y = 3$
$5x + 2y = 5$

18. $2x - 3y = -1$
$-6x + 9y = 3$

19. $-2x + y = 8$
$x - 3y = -14$

20. $2x + 5y = -9$
$-4x + 6y = 2$

21. $5x - 2y = -10$
$7x + 3y = 15$

22. $6x - 3y = 7$
$-2x + y = 5$

Solve the following word problems. Be sure to show the equations used. [11.4]

23. Number Problem The sum of two numbers is 10. If one more than twice the smaller number is the larger, find the two numbers.

24. Number Problem The difference of two positive numbers is 3. One number is $\frac{3}{4}$ of the other number. Find the two numbers.

25. Investing A total of $9,000 was invested. Part of the $9,000 was invested at 8%, and the rest was invested at 5%. If the interest for one year is $510, how much was invested at each rate?

26. Investing A total of $14,000 was invested. Part of the $14,000 was invested at 9%, and the rest was invested at 11%. If the interest for one year is $1,440, how much was invested at each rate?

27. Coin Problem Nate has $3.00 in dimes and quarters. He has a total of 21 coins. How many of each does he have?

28. Coin Problem Maria has $2.95 in dimes and quarters. She has a total of 19 coins. How many of each does she have?

29. Mixture Problem How many liters of 20% alcohol solution and 80% alcohol solution must be mixed to obtain 11 liters of a 84% alcohol solution? Round to the nearest liter.

30. Mixture Problem How many liters of 35% alcohol solution and 10% alcohol solution must be mixed to obtain 30 liters of a 20% alcohol solution? Round to the nearest liter.

Simplify.

1. $6 \cdot (-2) + 7$

2. $9 - 3(6 - 2)$

3. $5 \cdot 2^3 + 2(4 - 10)$

4. $9[-2 + 7] + 2(-8 + 2)$

5. $\dfrac{4 - 13}{3 - 3}$

6. $\dfrac{4(2 - 9) - 2(12 - 8)}{7 - 3 - 8}$

7. $\dfrac{9}{48} + \dfrac{3}{72}$

8. $\dfrac{5}{4} + \dfrac{1}{2} - \dfrac{3}{8}$

9. $3(x - 2) + 7$

10. $4 - 3(4a - 6) + 4$

Solve each equation.

11. $-4 - 3 = -3y + 5 + y$

12. $-5x = 0$

13. $4(x + 3) = -4$

14. $8 + 3(y - 6) = 11$

Solve each inequality, and graph the solution.

15. $0.2x - 0.8 \geq -3$

16. $7x - 18 \geq 9x + 2$

Graph on a rectangular coordinate system.

17. $y = \dfrac{1}{3}x - 4$

18. $y = -\dfrac{3}{5}x$

Solve each system by graphing.

19. $4x - 2y = 16$
$3x + 4y = 1$

20. $x - y = 4$
$3x - 5y = 10$

Solve each system.

21. $x + y = 6$
$3x + 3y = 18$

22. $4x + y = -4$
$4x - 2y = 2$

23. $5x + 3y = -21$
$x - y = -9$

24. $2x + 5y = -6$
$3x - y = -9$

25. $8x - 3y = -5$
$12x - 5y = -11$

26. $3x - 2y = 0$
$2x + 4y = -32$

27. $2x - y = 21$
$x + 2y = 3$

28. $5x + 3y = 11$
$6x - y = -19$

29. $5x + 2y = -1$
$y = -6x + 10$

30. $5x - 7y = -17$
$x = 9y - 11$

31. What is the quotient of -32 and 4?

32. Subtract 12 from -7.

Factor into the product of primes.

33. 270

34. 420

35. Given $2x + 5y = -1$, complete the ordered pair $(-3, \quad)$.

36. Given $y = -\dfrac{2}{3}x + 1$, complete the ordered pairs $(\quad , 3)$ and $(2, \quad)$.

37. Find the x- and y-intercepts for the line $5x - 6y = 30$.

38. Find the slope of the line $x = 4y - 8$.

Find the slope of the line through the given pair of points.

39. $(3, -4), (5, -9)$

40. $\left(8, \dfrac{1}{3}\right), \left(4, \dfrac{7}{6}\right)$

41. Find the equation of the line with slope $\dfrac{5}{4}$ and y-intercept -2.

42. Find the equation of the line with slope $\dfrac{1}{3}$ if it passes through $(3, 9)$.

43. Find the equation of the line through $(3, -4)$ and $(5, 2)$.

44. Find the x- and y-intercepts for the line that passes through the points $(-6, 2)$ and $(3, 5)$.

45. **Coin Problem**   Dave has 12 coins with a value of $2.25. The coins are quarters and dimes. How many of each does he have?

46. **Investing**   I have invested money in two accounts. One account pays 7% annual interest, and the other pays 11%. I have $300 more in the 11% account than I have in the 7% account. If the total amount of interest was $141, how much do I have in each account?

The chart shows the results of a survey of iPhones buyers. The percentages show what they reported as their previous phone. Use the information to answer the following questions.

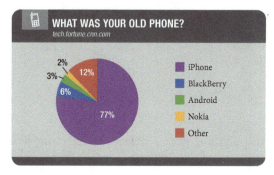

WHAT WAS YOUR OLD PHONE?
tech.fortune.cnn.com

2%
3%
12%
6%
77%

- iPhone
- BlackBerry
- Android
- Nokia
- Other

47. If 26,000 people were surveyed, how many reported replacing an Android phone with their new iPhone?

48. If 24 people say they replaced a Blackberry with their iPhone, how many people were surveyed?

1. Write the solution to the system which is graphed below. [11.1]

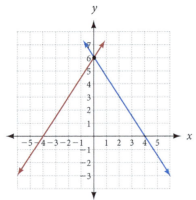

Solve each system by graphing. [11.1]

2. $4x - 2y = 8$
   $y = \dfrac{2}{3}x$

3. $3x - 2y = 13$
   $y = 4$

4. $2x - 2y = -12$
   $-3x - y = 2$

Solve each system by the substitution method. [11.2]

5. $3x - y = 12$
   $y = 2x - 8$

6. $3x + 6y = 3$
   $x = 4y - 17$

7. $2x - 3y = -18$
   $3x + y = -5$

8. $2x - 3y = 13$
   $x - 4y = -1$

Solve each system by the elimination method. [11.3]

9. $x - y = -9$
   $2x + 3y = 7$

10. $3x - y = 1$
   $5x - y = 3$

11. $2x + 3y = -3$
   $x + 6y = 12$

12. $2x + 3y = 4$
   $4x + 6y = 8$

Solve the following word problems. In each case, be sure to show the system of equations that describes the situation. [11.4]

13. Number Problem The sum of two numbers is 15. Their difference is 11. Find the numbers.

14. Number Problem The sum of two numbers is 18. One number is 2 more than 3 times the other. Find the two numbers.

15. Investing Dave has $2,000 to invest. He would like to earn $135.20 per year in interest. How much should he invest at 6% if the rest is to be invested at 7%?

16. Coin Problem Maria has 19 coins that total $1.35. If the coins are all nickels and dimes, how many of each type does she have?

17. Perimeter Problem A rancher wants to build a rectangular corral using 198 feet of fence. If the length of the corral is to be 15 feet longer than twice the width, find the dimensions of the corral.

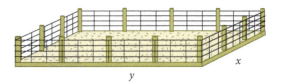

18. Moving Truck Rental One moving truck company charges $50 per day plus $0.75 per mile to rent a 28-foot truck for local use. Another company charges $90 per day plus $0.35 per mile. The total costs to use the trucks for a day and drive x miles are

Company 1: $y = 0.75x + 50$

Company 2: $y = 0.35x + 90$

The graphs of these equations are shown below.

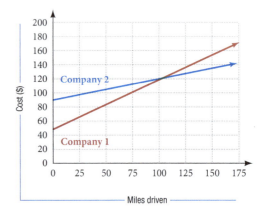

a. At how many miles do the two companies charge the same amount for the truck rental?

b. If the truck is driven less than 100 miles, which company would charge less?

c. For which values of x will Company 2 charge less than Company 1?

Answers to Odd-Numbered Problems

Chapter 1

Exercise Set 1.1
Vocabulary Review 1. origin **2.** positive **3.** less, more
4. absolute value **5.** opposites **6.** negative **7.** integers
Problems 1. 4 is less than 7. **3.** 5 is greater than -2.
5. -10 is less than -3. **7.** 0 is greater than -4 **9.** $30 > -30$
11. $-10 < 0$ **13.** $-3 > -15$ **15.** $3 < 7$ **17.** $7 > -5$
19. $-6 < 0$ **21.** $-12 < -2$ **23.** $-0.1 < -0.01$
25. $-3 < |6|$ **27.** $15 > |-4|$ **29.** $|-2| < |-7|$ **31.** 2
33. 100 **35.** 8 **37.** 231 **39.** 42 **41.** 200 **43.** 8
45. 231 **47.** -3 **49.** 2 **51.** -75 **53.** 0 **55.** 2
57. 8 **59.** -2 **61.** -8 **63.** 0 **65.** Positive
67. -100 **69.** -20 **71.** -360 **73.** $-61°$ F, $-51°$ F
75. $-5°$ F, $-15°$ F **77.** $-7°$ F **79.** 10° F and 25-mph wind

81.

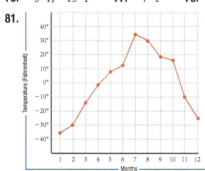

83. 25 **85.** 5
87. 6 **89.** 19

Find the Mistake 1. The expression $-4 < 1$ is read "-4 is less
than 1." **2.** The number -3 appears to the right of the number
-24 on the number line. **3.** The opposite of -18 is 18.
4. The absolute value of -36 is 36.

Exercise Set 1.2
Vocabulary Review same, absolute values, different, smaller, larger,
sign
Problems 1. 5 **3.** 1 **5.** -2 **7.** -6 **9.** 4 **11.** 4
13. -9 **15.** 15 **17.** -3 **19.** -11 **21.** -7 **23.** -3
25. -16 **27.** -8 **29.** -127 **31.** 49 **33.** 34
35.

First Number a	Second Number b	Their Sum $a + b$
5	-3	2
5	-4	1
5	-5	0
5	-6	-1
5	-7	-2

37.

First Number x	Second Number y	Their Sum $x + y$
-5	-3	-8
-5	-4	-9
-5	-5	-10
-5	-6	-11
-5	-7	-12

39. -4 **41.** 10 **43.** -445 **45.** 107 **47.** -1
49. -20 **51.** -17 **53.** -50 **55.** -7 **57.** 3
59. 50 **61.** -73 **63.** -11 **65.** 17 **67.** -21
69. -5 **71.** -4 **73.** 7 **75.** 10 **77.** a **79.** b
81. d **83.** c **85.** \$10 **87.** $\$74 + (-\$141) = -\$67$
89. $3 + (-5) = -2$ **91.** -7 and 13 **93.** -2 **95.** 4
97. 30 **99.** 2 **101.** 3

Find the Mistake 1. The problem $6 + (-10) = -4$ is interpreted
as, "Start at the origin, move 6 units in the positive direction, and then
move 10 units in the negative direction." **2.** Adding two numbers
with different signs will give an answer that has the same sign as the
number with the larger absolute value. **3.** Adding -8
and -5 gives us -13. **4.** The sum of $-2, 4, -3$, and -5 is -6.

Exercise Set 1.3
Vocabulary Review 1. subtract **2.** opposite **3.** addition
4. negative **5.** positive
Problems 1. 2 **3.** 2 **5.** -8 **7.** -5 **9.** 7 **11.** 12
13. 3 **15.** -7 **17.** -3 **19.** -13 **21.** -50
23. -100 **25.** 399 **27.** -21

29.

First Number x	Second Number y	Their Difference $x - y$
8	6	2
8	7	1
8	8	0
8	9	-1
8	10	-2

31.

First Number x	Second Number y	Their Difference $x - y$
8	-6	14
8	-7	15
8	-8	16
8	-9	17
8	-10	18

33. -7 **35.** -9 **37.** -14 **39.** -11 **41.** -400
43. 11 **45.** -4 **47.** 8 **49.** 6 **51.** b **53.** a
55. -100 **57.** b **59.** a **61.** 44°
63. $-11 - (-22) = 11°$ F **65.** $3 - (-24) = 27°$ F
67. $60 - (-26) = 86°$ F **69.** $-14 - (-26) = 12°$ F **71.** 30
73. 36 **75.** 64 **77.** 48 **79.** 41 **81.** 40

Find the Mistake 1. Subtracting 5 from 4 is the same as adding
4 and -5. **2.** To subtract -1 from 8, we must move 1 unit in
the positive direction from 8 on the number line. **3.** The problem
$11 - (-7)$ is read, "11 subtract negative 7." **4.** To find the
difference of -5 and 6, change 6 to -6, and add to -5.

Landmark Review
1. $2, -2$ **2.** 11, 11 **3.** $25, -25$ **4.** 110, 110 **5.** 7 **6.** -5
7. -3 **8.** -15 **9.** 11 **10.** -3 **11.** 2 **12.** -20
13. 4 **14.** -6 **15.** -12 **16.** 15 **17.** -9 **18.** 11

Exercise Set 1.4
Vocabulary Review 1. multiplication **2.** absolute values
3. positive **4.** negative **5.** zero
Problems 1. Base 4, Exponent 5 **3.** Base 3, Exponent 6
5. Base 8, Exponent 2 **7.** Base 9, Exponent 1
9. Base 4, Exponent 0 **11.** 36 **13.** 8 **15.** 1 **17.** 1 **19.** 81
21. 10 **23.** 12 **25.** -56 **27.** -60 **29.** 56 **31.** 81
33. -24 **35.** 24 **37.** -6 **39. a.** 16 **b.** -16
41. a. -125 **b.** -125 **43. a.** 16 **b.** -16

45.

Number x	Square x^2
-3	9
-2	4
-1	1
0	0
1	1
2	4
3	9

47.

First Number x	Second Number y	Their Product xy
6	2	12
6	1	6
6	0	0
6	-1	-6
6	-2	-12

49.

First Number a	Second Number b	Their Product ab
−5	3	−15
−5	2	−10
−5	1	−5
−5	0	0
−5	−1	5
−5	−2	10
−5	−3	15

51. −4 **53.** 50
55. 1   **57.** −35
59. −22 **61.** −30
63. −25 **65.** 9
67. −13 **69.** 19
71. 6 **73.** −6
75. −4 **77.** −17
79. A gain of $200
81. −16°

83.

Pitcher, Team	Rolaids Points
Heath Bell, San Diego	144
Brian Wilson, San Francisco	135
Carlos Marmol, Chicago	107
Billy Wagner, Atlanta	128
Francisco Cordero, Cincinnati	130

85. 70

	Product
Eagle	0
Birdie	−7
Par	0
Bogie	+3
Double Bogie	+2
Total:	−2

87. a   **89.** d
91. a   **93.** b
95. 7   **97.** 5
99. −5 **101.** 9
103. 4 **105.** 7
107. 405

Find the Mistake   **1.** Writing the problem $6(−4)$ as repeated addition gives us $(−4) + (−4) + (−4) + (−4) + (−4) + (−4)$.
2. Multiplying a negative by a positive and then by another negative will give us a positive answer.   **3.** The problem $(−5)^3$ can also be written as $(−5)(−5)(−5)$.
4. $−5(3 − 6) − 2(3 + 1) = −5(−3) − 2(4)$
$$= 15 + (−8)$$
$$= 7$$

Exercise Set 1.5
Vocabulary Review   **1.** absolute values   **2.** same   **3.** different
4. negative
Problems   **1.** −3   **3.** −5   **5.** 3   **7.** 2   **9.** −4
11. −2   **13.** 0   **15.** −5
17.

First Number a	Second Number b	The Quotient of a and b $\frac{a}{b}$
100	−5	−20
100	−10	−10
100	−25	−4
100	−50	−2

19.

First Number a	Second Number b	The Quotient of a and b $\frac{a}{b}$
−100	−5	20
−100	5	−20
100	−5	−20
100	5	20

21. 1   **23.** −6   **25.** −2   **27.** −1   **29.** −1   **31.** 2
33. −3   **35.** −7   **37.** 30   **39.** 4   **41.** −5   **43.** −20
45. −5   **47.** −5   **49.** 35   **51.** 6   **53.** −1   **55.** c
57. a   **59.** d

61.

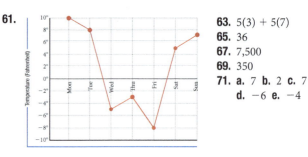

63. $5(3) + 5(7)$
65. 36
67. 7,500
69. 350
71. a. 7 **b.** 2 **c.** 7
d. −6 **e.** −4

Find the Mistake   **1.** Dividing a negative number by a positive number will give a negative number for an answer.   **2.** True.
3. $\frac{−6(−6 − 2)}{−11 − 1} = \frac{−6(−8)}{−12} = \frac{48}{−12} = −4$
4. $[(−5)(5) − 20] ÷ −3^2 = −45 ÷ (−9)$
$$= 5$$

Exercise Set 1.6
Vocabulary Review   **1.** associative   **2.** distributive
3. commutative   **4.** similar terms
Problems   **1.** $20a$ **3.** $48a$ **5.** $−18x$ **7.** $−27x$
9. $−10y$ **11.** $−60y$ **13.** $5 + x$ **15.** $13 + x$ **17.** $10 + y$
19. $8 + y$ **21.** $5x + 6$ **23.** $6y + 7$ **25.** $12a + 21$
27. $7x + 28$ **29.** $7x + 35$ **31.** $6a − 42$ **33.** $2x − 2y$
35. $20 + 4x$ **37.** $6x + 15$ **39.** $18a + 6$ **41.** $12x − 6y$
43. $35 − 20y$ **45.** $8x$ **47.** $4a$ **49.** $4x$ **51.** $5y$
53. $−10a$ **55.** $−5x$ **57.** $A = 36$ ft²; $P = 24$ ft
59. $A = 81$ in²; $P = 36$ in.   **61.** $A = 200$ in²; $P = 60$ in.
63. $A = 300$ ft²; $P = 74$ ft   **65.** 20° C   **67.** 5° C   **69.** −10° C

Find the Mistake   **1.** Simplifying $−6(3x)$ gives us $−18x$.
2. To simplify the problem $4(z − 3)$, we use the distributive property to get $4z − 12$.   **3.** To simplify $−2x + 5x + 4$, combine the similar terms $−2x$ and $5x$ to get $3x + 4$.   **4.** A rectangular hockey rink that is $12x$ meters in length and $4x$ meters in width has a perimeter of $32x$ meters.

Chapter 1 Review
1. $\frac{3}{4}, \frac{3}{4}$ **2.** −16, 16   **3.** $<$ **4.** $>$ **5.** 6   **6.** −3   **7.** 2
8. −57   **9.** 84   **10.** −14   **11.** 25   **12.** −64   **13.** −9
14. −105   **15.** $\frac{13}{2}$ **16.** −10   **17.** 9   **18.** −60   **19.** 81
20. 27   **21.** −3   **22.** 44   **23.** −44   **24.** $\frac{5}{7}$ **25.** −73
26. 18   **27.** −77   **28.** −12   **29.** 20   **30.** −27
31. $250   **32.** 50°   **33.** $441 − 1,299 = −858$ texts
34. $896 − 952 = −56$ minutes

Chapter 1 Test
1. −27, 27   **2.** $\frac{1}{4}, \frac{1}{4}$ **3.** $<$ **4.** $<$ **5.** 4   **6.** −14   **7.** −3
8. −36   **9.** 72   **10.** −21   **11.** −27   **12.** 4   **13.** 14
14. −18   **15.** $\frac{14}{3}$ **16.** −4   **17.** −43   **18.** −58   **19.** 8
20. −34   **21.** 0   **22.** 13   **23.** −3   **24.** −3   **25.** −41
26. 17   **27.** 36   **28.** −8   **29.** −15   **30.** −6   **31.** −$59
32. 17°   **33.** $2,000 − 5,000 = −3,000$ animals
34. $1,000 − 4,000 = −3,000$ animals

Chapter 2

Exercise Set 2.1
Vocabulary Review   **1.** fraction   **2.** numerator, denominator
3. proper   **4.** improper   **5.** equivalent
Problems
1. 1   **3.** −2   **5.** x **7.** $−a$ **9.** 5   **11.** 1   **13.** −12

15.

Numerator a	Denominator b	Fraction $\dfrac{a}{b}$
3	5	$\dfrac{3}{5}$
-1	7	$-\dfrac{1}{7}$
$-x$	$-y$	$\dfrac{x}{y}$
$x+1$	x	$\dfrac{x+1}{x}$

17. $\frac{3}{4}, \frac{1}{2}, \frac{9}{10}$
19. True
21. False

23. – 31. Answers on number line below

33. $\frac{1}{20} < \frac{4}{25} < \frac{3}{10} < \frac{2}{5}$ **35.** $\frac{4}{6}$ **37.** $-\frac{5}{6}$ **39.** $\frac{8}{12}$
41. $-\frac{8}{12}$ **43.** $\frac{2x}{12x}$ **45.** $\frac{6x}{12x}$ **47.** $\frac{16}{8}$ **49.** $-\frac{40}{8}$
51. Answers will vary **53.** 3 **55.** -2 **57.** 37 **59.** $\frac{3}{4}$
61. $-\frac{43}{47}$ **63.** $\frac{4}{3}$ **65.** $-\frac{13}{17}$ **67. a.** $\frac{1}{2}$ **b.** $\frac{1}{2}$ **c.** $\frac{1}{4}$ **d.** $\frac{1}{4}$

69.

How often workers send non-work-related e-mail from the office	Fraction of respondents saying yes
Never	$\dfrac{4}{25}$
1 to 5 times a day	$\dfrac{47}{100}$
5 to 10 times a day	$\dfrac{8}{25}$
More than 10 times a day	$\dfrac{1}{20}$

71. $\frac{4}{5}$ **73.** $\frac{29}{43}$ **75.** $\frac{1,121}{1,791}$ **77.** $\frac{14}{50}$ **79.** d **81.** a
83. 108 **85.** 60 **87.** 4 **89.** 5 **91.** 7 **93.** 51

Find the Mistake **1.** For the fraction $\frac{21}{7}$, the denominator is 7.
2. The fraction $\frac{90}{15}$ is considered an improper fraction.
3. The multiplication property for fractions states that if the numerator and the denominator of a fraction are multiplied by the same nonzero number, the resulting fraction is equivalent to the original fraction. **4.** If we divide the numerator and denominator of the fraction $\frac{8}{12}$ by 4, then we get the equivalent fraction $\frac{2}{3}$.

Exercise Set 2.2
Vocabulary Review **1.** prime **2.** remainder **3.** composite
4. lowest terms
Problems **1.** Prime **3.** Composite; 3, 5, and 7 are factors
5. Composite; 3 is a factor **7.** Prime **9.** $2^2 \cdot 3$ **11.** 3^4
13. $5 \cdot 43$ **15.** $3 \cdot 5$ **17.** $\frac{1}{2}$ **19.** $\frac{2}{3}$ **21.** $\frac{4}{5}$ **23.** $\frac{9}{5}$
25. $\frac{7}{11}$ **27.** $\frac{3}{5}$ **29.** $\frac{1}{7}$ **31.** $\frac{7}{9}$ **33.** $\frac{7}{5}$ **35.** $\frac{5}{7}$
37. $\frac{3}{5}$ **39.** $\frac{5}{3}$ **41.** $\frac{8}{9}$ **43.** $\frac{42}{55}$ **45.** $\frac{17}{19}$ **47.** $\frac{14}{33}$
49. a. $\frac{2}{17}$ **b.** $\frac{3}{26}$ **c.** $\frac{1}{9}$ **d.** $\frac{3}{28}$ **e.** $\frac{2}{19}$ **51. a.** $\frac{1}{45}$ **b.** $\frac{1}{30}$ **c.** $\frac{1}{18}$ **d.** $\frac{1}{15}$ **e.** $\frac{1}{10}$
53. a. $\frac{1}{3}$ **b.** $\frac{5}{6}$ **c.** $\frac{1}{5}$ **55.** $\frac{9}{16}$

57. – 59. $\frac{1}{2} = \frac{2}{4} = \frac{4}{8} = \frac{8}{16}$

$\frac{5}{4} = \frac{10}{8} = \frac{20}{16}$

61. $\frac{1}{4}$ **63.** $\frac{2}{3}$ **65.** $\frac{7}{15}$ **67.** $\frac{1}{4}$ **69.** 0 **71.** 3 **73.** 45
75. 25 **77.** $2^2 \cdot 3 \cdot 5$ **79.** $2^2 \cdot 3 \cdot 5$ **81.** 9 **83.** 25 **85.** b

Find the Mistake **1.** The number 30 is a composite number because it has 10 as a divisor. **2.** The number 70 factored into a product of primes is $2 \cdot 5 \cdot 7$. **3.** When reducing the fraction $\frac{32}{48}$ to lowest terms, we divide out the common factor 2^4 to get $\frac{2}{3}$.
4. Reducing the fraction $\frac{112}{14}$ to lowest terms gives us $\frac{8}{1} = 1$.

Landmark Review **1.** Numerator 3; denominator 5
2. Numerator 1; denominator 3 **3.** Numerator 7; denominator 15
4. Numerator 4; denominator x **5.** $\frac{4x}{8x}$ **6.** $\frac{6x}{8x}$ **7.** $\frac{x}{8x}$
8. $\frac{20x}{8x}$ **9.** $\frac{1}{2}$ **10.** $\frac{3}{5}$ **11.** $\frac{3}{5}$ **12.** $\frac{5}{8}$ **13.** $\frac{17}{18}$ **14.** $\frac{31}{42}$

Exercise Set 2.3
Vocabulary Review **1.** product **2.** fractions
3. multiplication **4.** triangle
Problems **1.** $\frac{8}{15}$ **3.** $\frac{7}{8}$ **5.** 1 **7.** $\frac{27}{4}$ **9.** 1 **11.** $\frac{1}{24}$
13. $\frac{24}{125}$ **15.** $\frac{105}{8}$

17.

First number x	Second number y	Their product xy
$\dfrac{1}{2}$	$\dfrac{2}{3}$	$\dfrac{1}{3}$
$\dfrac{2}{3}$	$\dfrac{3}{4}$	$\dfrac{1}{2}$
$\dfrac{3}{4}$	$\dfrac{4}{5}$	$\dfrac{3}{5}$
$\dfrac{5}{a}$	$-\dfrac{a}{6}$	$-\dfrac{5}{6}$

19.

First number x	Second number y	Their product xy
$\dfrac{1}{2}$	30	15
$\dfrac{1}{5}$	30	6
$\dfrac{1}{6}$	30	5
$\dfrac{1}{15}$	30	2

21. $\frac{3}{5}$ **23.** 9 **25.** 1 **27.** -8 **29.** $-\frac{1}{15}$ **31.** $\frac{4}{9}$
33. $\frac{9}{16}$ **35.** $\frac{1}{4}$ **37.** $\frac{8}{27}$ **39.** $\frac{1}{2}$ **41.** $\frac{9}{100}$ **43.** 3
45. 24 **47.** 4 **49.** 9 **51.** $\frac{3}{10}$; numerator should be 3, not 4.
53. $\frac{1}{4}$; answer should be positive **55.** 14 **57.** -14
59. 5 **61.** -6 **63.** 133 in^2 **65.** $\frac{4}{9}$ ft^2
67. 3 yd^2 **69.** 3,511 students **71.** 126,500 ft^3
73. About 2,121 children **75.** $\frac{1}{27}$ **77.** $\frac{8}{27}$ **79.** 2
81. 3 **83.** 2 **85.** 5 **87.** 3 **89.** $\frac{4}{3}$ **91.** 3 **93.** $\frac{1}{7}$

Find the Mistake **1.** To find the product of two fractions, multiply the numerators and multiply the denominators. **2.** To multiply $\frac{6}{7}$ by $\frac{12}{9}$, find the product of the numerators and divide it by the product of the denominators to get $\frac{72}{63}$. **3.** Simplifying $\left(\frac{5}{6}\right)^2 \cdot \frac{8}{9}$ gives $\frac{50}{81}$. **4.** The area of a triangle with a height of 14 inches and a base of 32 inches is 224 square inches.

Exercise Set 2.4
Vocabulary Review **1.** product **2.** reciprocal **3.** divisor
4. fraction

Problems 1. $\frac{15}{4}$ 3. $\frac{4}{3}$ 5. 9   7. 200   9. $\frac{3}{8}$ 11. -1
13. $\frac{49}{64}$ 15. $\frac{3}{4}$ 17. $\frac{15}{16}$ 19. $\frac{1}{6}$ 21. 6   23. $\frac{5}{18}$
25. $\frac{9}{2}$ 27. $\frac{2}{9}$ 29. 9   31. $\frac{4}{5}$ 33. $-\frac{15}{2}$ 35. 40   37. $\frac{7}{10}$
39. 13   41. 12   43. 186   45. 646   47. $\frac{3}{5}$ 49. 40
51. $3 \cdot 5 = 15$; $3 \div \frac{1}{5} = 3 \cdot \frac{5}{1} = 15$ 53. 14 blankets   55. 48 bags
57. 6   59. $\frac{7}{16}$ 61. 1,778 students   63. 28 cartons
Find the Mistake   1. Two numbers whose product is 1 are said to
be reciprocals.   2. Dividing the fraction $\frac{12}{7}$ by $\frac{4}{9}$, is equivalent
to $\frac{12}{7} \cdot \frac{9}{4}$.   3. To work the problem $\frac{22}{5} \div \frac{10}{3}$, multiply the first
fraction by the reciprocal of the second fraction.   4. The quotient
of $\frac{14}{11}$ and $\frac{32}{6}$ is $\frac{21}{88}$.

Chapter 2 Review
1. $\frac{5}{7}$ 2. $\frac{3}{8}$ 3. $\frac{4}{13}$ 4. $\frac{8}{15}$ 5. $\frac{2}{7}$ 6. $\frac{3}{20}$ 7. $\frac{7}{2}$
8. $\frac{1}{6}$ 9. $-\frac{4}{21}$ 10. $-\frac{3}{4}$ 11. 3   12. $\frac{6}{7}$ 13. $\frac{7}{6}$
14. $\frac{12}{5}$ 15. $-\frac{9}{4}$ 16. $-\frac{1}{12}$ 17. 25   18. 2   19. $\frac{4}{3}$
20. $\frac{10}{21}$ 21. $\frac{3}{7}$ 22. $\frac{1}{6}$ 23. $\frac{14}{3}$ 24. 2   25. $-\frac{5}{9}$
26. $-\frac{1}{3}$ 27. $\frac{3}{7}$ 28. $\frac{4}{7}$ 29. $A = 26$ ft²   30. 8 dresses
31. $\frac{15}{4}$ 32. $\frac{254}{1000}$

Chapter 2 Cumulative Review
1. $\frac{4}{5}, \frac{4}{5}$ 2. $-27, 27$ 3. $\frac{16}{81}$ 4. 20   5. 4   6. 3
7. $\frac{5}{14}$ 8. 20   9. 15   10. -3 11. $\frac{31}{6}$ 12. $\frac{3}{5}$
13. 169   14. 6   15. 7   16. 268,550   17. $\frac{3}{8}$
18. 34   19. 10   20. $-\frac{1}{75}$ 21. 4,245   22. $\frac{1}{5}$
23. $\frac{4}{21}$ 24. $\frac{3}{8}$ 25. $\frac{1}{2}$ 26. $\frac{7}{45}$ 27. 8   28. $\frac{3}{8}$
29. $\frac{7}{5}$ 30. $382   31. 10,965,000 viewers

Chapter 2 Test
1. $\frac{1}{5}$ 2. $\frac{1}{4}$ 3. $\frac{3}{14}$ 4. $\frac{3}{4}$ 5. $\frac{3}{20}$ 6. $\frac{10}{21}$ 7. $\frac{15}{8}$ 8. $\frac{1}{6}$
9. $-\frac{3}{10}$ 10. $-\frac{1}{10}$ 11. $\frac{3}{20}$ 12. $\frac{4}{5}$ 13. 4   14. $\frac{15}{32}$
15. $-\frac{4}{5}$ 16. $-\frac{1}{15}$ 17. 6   18. $\frac{4}{9}$ 19. $\frac{2}{5}$ 20. $\frac{3}{14}$
21. $\frac{10}{11}$ 22. $\frac{3}{16}$ 23. $\frac{20}{11}$ 24. $\frac{7}{18}$ 25. $-\frac{3}{8}$ 26. $-\frac{1}{2}$
27. $\frac{1}{5}$ 28. $\frac{2}{3}$ 29. $A = 15$ in²   30. 10 dresses   31. $\frac{6}{100}$
32. $\frac{4}{200}$

Chapter 3

Exercise Set 3.1
Vocabulary Review   1. least common denominator
2. equivalent   3. numerators   4. lowest terms
Problems 1. $\frac{2}{3}$ 3. $\frac{1}{4}$ 5. $\frac{1}{2}$ 7. $\frac{1}{3}$ 9. $\frac{3}{2}$ 11. $\frac{x+6}{2}$
13. $\frac{4}{5}$ 15. $\frac{10}{3}$ 17.

First Number a	Second Number b	The Sum of a and b a + b
$\frac{1}{2}$	$\frac{1}{3}$	$\frac{5}{6}$
$\frac{1}{3}$	$\frac{1}{4}$	$\frac{7}{12}$
$\frac{1}{4}$	$\frac{1}{5}$	$\frac{9}{20}$
$\frac{1}{5}$	$\frac{1}{6}$	$\frac{11}{30}$

19.

First Number a	Second Number b	The Sum of a and b a + b
$\frac{1}{12}$	$\frac{1}{2}$	$\frac{7}{12}$
$\frac{1}{12}$	$\frac{1}{3}$	$\frac{5}{12}$
$\frac{1}{12}$	$\frac{1}{4}$	$\frac{1}{3}$
$\frac{1}{12}$	$\frac{1}{6}$	$\frac{1}{4}$

21. $\frac{7}{9}$ 23. $\frac{7}{3}$ 25. $\frac{7}{4}$ 27. $\frac{7}{6}$ 29. $\frac{9}{20}$ 31. $\frac{7}{10}$
33. $\frac{19}{24}$ 35. $\frac{13}{60}$ 37. $\frac{31}{100}$ 39. $\frac{67}{144}$ 41. $\frac{29}{35}$ 43. $\frac{949}{1,260}$
45. $\frac{13}{420}$ 47. $\frac{41}{24}$ 49. $\frac{53}{60}$ 51. $\frac{5}{4}$ 53. $\frac{88}{9}$ 55. $\frac{10}{3}$
57. $\frac{3}{4}$ 59. $\frac{1}{4}$ 61. -1 63. $-\frac{11}{15}$ 65. 19   67. 3
69. 9   71. 4   73. $\frac{1}{4} < \frac{3}{8} < \frac{1}{2} < \frac{3}{4}$ 75. $\frac{160}{63}$ 77. $\frac{5}{8}$
79. $\frac{7}{3}$ 81. 3   83. $\frac{9}{2}$ pints   85. $1,325   87. $\frac{7}{18}$
89.

Grade	Number of students	Fraction of students
A	5	$\frac{1}{8}$
B	8	$\frac{1}{5}$
C	20	$\frac{1}{2}$
Below C	7	$\frac{7}{40}$
Total	40	1

91. 10 lots
93. $\frac{3}{2}$ in.
95. $\frac{9}{5}$ ft   97. 59
99. $\frac{16}{8}$ 101. $\frac{8}{8}$
103. $\frac{11}{4}$ 105. $\frac{17}{8}$
107. $\frac{9}{8}$ 109. 2 R 3
111. 8 R 16

Find the Mistake   1. The fractions $\frac{a}{c}$ and $\frac{b}{c}$ can be added to
become $\frac{a+b}{c}$ because they have a common denominator.
2. Subtracting $\frac{12}{21}$ from $\frac{18}{21}$, gives us $\frac{6}{21} = \frac{2}{7}$.   3. The least common
denominator for a set of denominators is the smallest number that
is exactly divisible by each denominator.   4. The LCD for the
fractions $\frac{4}{6}, \frac{2}{8}$ and $\frac{3}{4}$ is 24.

Exercise Set 3.2
Vocabulary Review   1. proper fraction   2. addition
3. mixed number   4. improper fraction   5. denominator
6. remainder
Problems 1. $\frac{14}{3}$ 3. $\frac{21}{4}$ 5. $\frac{13}{8}$ 7. $\frac{47}{3}$ 9. $\frac{104}{21}$ 11. $\frac{427}{33}$
13. $1\frac{1}{8}$ 15. $-4\frac{3}{4}$ 17. $4\frac{5}{6}$ 19. $3\frac{1}{4}$ 21. $4\frac{1}{27}$
23. $-28\frac{8}{15}$ 25. a. $4\frac{13}{205}$ b. $\frac{38}{65}$ 27. $6   29. $\frac{71}{12}$
31. $\frac{1,526}{5}$¢   33. $\frac{11}{4}$ 35. $\frac{37}{8}$ 37. $\frac{14}{5}$ 39. $\frac{9}{40}$ 41. $-\frac{3}{8}$
43. $\frac{32}{35}$ 45. $-\frac{4}{7}$ 47. a

Find the Mistake   1. For mixed-number notation, writing the
whole number next to the fraction implies addition.
2. A shortcut for changing a mixed number to an improper fraction
is to multiply the whole number by the denominator of the
fraction, and then add the result to the numerator of the fraction.
3. Changing $6\frac{4}{5}$ to an improper fraction gives us $\frac{34}{5}$.
4. Writing $\frac{70}{12}$ as a mixed number gives us $5\frac{5}{6}$.

Exercise Set 3.3
Vocabulary Review   1. mixed number   2. numerators
3. improper fractions   4. divisor
Problems 1. $5\frac{1}{10}$ 3. $13\frac{2}{3}$ 5. $6\frac{93}{100}$ 7. $5\frac{5}{6}$ 9. $9\frac{3}{4}$ 11. $3\frac{1}{5}$
13. $12\frac{1}{2}$ 15. $9\frac{9}{20}$ 17. $\frac{32}{45}$ 19. $1\frac{2}{3}$ 21. 4   23. $4\frac{3}{10}$

25. $\frac{1}{10}$ **27.** $-3\frac{1}{5}$ **29.** $2\frac{1}{8}$ **31.** $7\frac{1}{2}$ **33.** $\frac{11}{13}$

35. $5\frac{1}{2}$ cups   **37.** $\frac{5}{6}$ cup   **39.** $1\frac{1}{3}$ **41.** $2,441\frac{3}{5}$ cents

43. $163\frac{3}{4}$ mi   **45.** $4\frac{1}{2}$ yd   **47.** $2\frac{1}{4}$ yd^2 **49.** $\frac{3}{4} < \frac{5}{4} < 1\frac{1}{2} < 2\frac{1}{8}$

51. Can 1 contains $332\frac{1}{2}$ calories, whereas Can 2 contains 276 calories. Can 1 contains $56\frac{1}{2}$ more calories than Can 2.

53. Can 1 contains 1,325 milligrams of sodium, whereas Can 2 contains 1,095 milligrams of sodium. Can 1 contains 230 more milligrams of sodium than Can 2.

55. a. $\frac{10}{15}$ **b.** $\frac{3}{15}$ **c.** $\frac{9}{15}$ **d.** $\frac{5}{15}$ **57. a.** $\frac{5}{20}$ **b.** $\frac{12}{20}$ **c.** $\frac{18}{20}$ **d.** $\frac{2}{20}$

59. $\frac{13}{15}$ **61.** $\frac{14}{9} = 1\frac{5}{9}$ **63.** $\frac{3}{5}$ **65.** $\frac{1}{4}$ **67.** $2\frac{1}{4}$

69. $3\frac{1}{16}$ **71.** $2\frac{1}{4}$ ft^2 **73.** $3\frac{3}{8}$ ft^3

Find the Mistake   **1.** To multiply two mixed numbers, change to improper fractions, then multiply numerators and multiply denominators.   **2.** Multiplying $4\frac{3}{8}$ and $9\frac{2}{7}$ gives us the mixed number $40\frac{5}{8}$.   **3.** To divide the mixed numbers $12\frac{2}{5}$ and $3\frac{12}{5}$, change the mixed numbers to improper fractions and then multiply by the reciprocal of the divisor.   **4.** The answer to the division problem $3\frac{9}{14} \div 2$ written as a mixed number is $1\frac{23}{28}$.

Landmark Review   **1.** 1   **2.** $\frac{1}{5}$ **3.** $\frac{19}{15}$ **4.** $\frac{5}{4}$

5. $\frac{73}{105}$ **6.** $\frac{2+x}{3}$ **7.** $\frac{41}{30} = 1\frac{11}{30}$ **8.** $\frac{1}{10}$ **9.** $\frac{29}{8}$ **10.** $\frac{14}{3}$

11. $\frac{21}{2}$ **12.** $\frac{5}{4}$ **13.** $4\frac{2}{3}$ **14.** $4\frac{3}{5}$ **15.** $3\frac{1}{2}$ **16.** $2\frac{8}{17}$

17. $23\frac{3}{8}$ **18.** $9\frac{4}{9}$ **19.** $37\frac{13}{24}$ **20.** $1\frac{1}{5}$ **21.** $2\frac{5}{16}$ **22.** $1\frac{47}{64}$

Exercise Set 3.4

Vocabulary Review   **1.** addition sign   **2.** columns   **3.** LCD
4. improper, proper   **5.** borrow

Problems   **1.** $5\frac{4}{5}$ **3.** $12\frac{2}{5}$ **5.** $3\frac{4}{9}$ **7.** 12   **9.** $1\frac{3}{8}$ **11.** $14\frac{1}{6}$

13. $4\frac{1}{12}$ **15.** $2\frac{1}{12}$ **17.** $26\frac{7}{12}$ **19.** $11\frac{3}{8}$ **21.** 12   **23.** $2\frac{1}{2}$

25. $8\frac{6}{7}$ **27.** $3\frac{3}{8}$ **29.** $10\frac{4}{15}$ **31.** $2\frac{1}{15}$ **33.** 9   **35.** $18\frac{1}{10}$

37. 14   **39.** 17   **41.** $24\frac{1}{24}$ **43.** $27\frac{6}{7}$ **45.** $37\frac{3}{20}$ **47.** $33\frac{5}{24}$

49. $6\frac{1}{4}$ **51.** $9\frac{7}{10}$ **53.** $5\frac{1}{2}$ **55.** $\frac{2}{3}$ **57.** $1\frac{11}{12}$ **59.** $3\frac{11}{12}$

61. $5\frac{19}{20}$ **63.** $5\frac{1}{2}$ **65.** $\frac{13}{24}$ **67.** $10\frac{5}{12}$ **69.** $3\frac{1}{2}$ **71.** $5\frac{29}{40}$

73. $12\frac{1}{4}$ in.   **75.** $\frac{1}{8}$ mi   **77.** $31\frac{1}{8}$ in.   **79.** NFL: $P = 306\frac{2}{3}$ yd, Canadian: $P = 350$ yd, Arena: $P = 156\frac{2}{3}$ yd   **81. a.** $2\frac{1}{2}$ **b.** \$250

83. \$300   **85.** $4\frac{63}{64}$ **87.** 2   **89.** $\frac{11}{8} = 1\frac{3}{8}$ **91.** $3\frac{5}{8}$ **93.** $35\frac{3}{5}$ sec

Find the Mistake   **1.** To begin adding $3\frac{9}{14}$ and $5\frac{1}{3}$, write each mixed number with the addition sign and then apply the commutative and associative properties, such that $\left(3 + \frac{9}{14}\right) + \left(5 + \frac{1}{3}\right)$.

2. The final answer for the problem $5\frac{2}{3} + 6\frac{7}{9}$ is $12\frac{4}{9}$.

3. The first step when subtracting $8 - 3\frac{2}{7} = 4\frac{5}{7}$, is to borrow 1 from 8 in the form of $\frac{7}{7}$.   **4.** Borrow 1 from the 5 in the form of $\frac{15}{15}$ so that $4\frac{25}{15} - 2\frac{12}{15} = 2\frac{13}{15}$.

Exercise Set 3.5

Vocabulary Review   **1.** complex fraction   **2.** top, bottom
3. LCD   **4.** whole numbers   **5.** simplify

Problems   **1.** 7   **3.** 7   **5.** 2   **7.** 35   **9.** $\frac{7}{8}$ **11.** $8\frac{1}{3}$ **13.** $\frac{11}{36}$

15. $3\frac{2}{3}$ **17.** $6\frac{3}{8}$ **19.** $4\frac{5}{12}$ **21.** $\frac{8}{9}$ **23.** $\frac{1}{2}$ **25.** $1\frac{1}{10}$

27. 5   **29.** $\frac{3}{5}$ **31.** $\frac{7}{11}$ **33.** 5   **35.** $\frac{17}{28}$ **37.** $1\frac{7}{16}$ **39.** $\frac{13}{22}$

41. $\frac{5}{22}$ **43.** 14   **45.** $\frac{15}{16}$ **47.** $1\frac{5}{17}$ **49.** $\frac{3}{29}$ **51.** $1\frac{34}{67}$

53. $\frac{346}{441}$ **55.** $\frac{32}{33}$ **57.** $5\frac{2}{5}$ **59.** 8   **61.** $115\frac{2}{3}$ yd

Find the Mistake   **1.** The first step to solving the problem $\frac{1}{4} - \left(2\frac{3}{8} - 1\frac{5}{8}\right)^2$ is to subtract the fractions inside the parentheses before evaluating the exponent.   **2.** To simplify $5\frac{2}{3} + \left(10\frac{1}{3} \cdot \frac{2}{3}\right)$ you must first change $10\frac{1}{3}$ into the improper fraction $\frac{31}{3}$ before multiplying by $\frac{2}{3}$.

3. A complex fraction is a fraction in which a fraction or combination of fractions appear in the numerator and/or the denominator of the original fraction.

4. To simplify the complex fraction $\dfrac{\frac{1}{6} + \frac{2}{3}}{\frac{5}{6} + \frac{5}{12}}$, multiply the top and bottom of the fraction by 12.

Chapter 3 Review

1. $\frac{4}{3}$ **2.** $\frac{1}{2}$ **3.** $\frac{24}{5} = 4\frac{4}{5}$ **4.** $\frac{5}{8}$ **5.** $\frac{37}{30} = 1\frac{17}{30}$ **6.** $12\frac{4}{15}$

7. $2\frac{1}{10}$ **8.** $8\frac{2}{5}$ **9.** $\frac{7}{16}$ **10.** $5\frac{5}{8}$ **11.** $\frac{1}{4}$ **12.** $-1\frac{1}{2}$

13. $12\frac{1}{2}$ **14.** $\frac{1}{6}$ **15.** $4\frac{5}{24}$ **16.** $11\frac{7}{16}$ **17.** $17\frac{1}{2}$ **18.** $6\frac{7}{24}$

19. $19\frac{5}{8}$ in.   **20.** Perimeter $= 29\frac{1}{4}$ ft   **21.** $15\frac{2}{5}$ **22.** $1\frac{31}{32}$

23. $\frac{1}{2}$ **24.** $2\frac{3}{4}$ **25.** $13\frac{9}{32}$ **26.** $\frac{41}{48}$ **27.** $\frac{1}{2}$ **28.** $1\frac{21}{46}$

29. $\frac{7}{81}$ **30.** $\frac{1}{2}$ **31.** $\frac{29}{100}$ **32.** $\frac{9}{50}$

Chapter 3 Cumulative Review

1. $\frac{3}{4}$ **2.** $\frac{2}{3}$ **3.** $\frac{3}{4}$ **4.** $\frac{5}{8}$ **5.** $\frac{1}{6}$ **6.** $16\frac{1}{42}$

7. $\frac{4}{15}$ **8.** 22   **9.** 12   **10.** 4   **11.** $3\frac{7}{13}$ **12.** $\frac{5}{9}$ **13.** 196

14. 18   **15.** $\frac{11}{15}$ **16.** 630,496   **17.** $\frac{2}{3}$ **18.** 11   **19.** 12

20. $\frac{32}{75}$ **21.** 6,888   **22.** $\frac{3}{2}$ **23.** $\frac{19}{48}$ **24.** $4\frac{2}{3}$

25. $\frac{31}{24}$ **26.** $\frac{107}{60} = 1\frac{47}{60}$ **27.** $\frac{28x}{60x}$ **28.** $\frac{3}{11}$ **29.** $\frac{23}{21} = 1\frac{2}{21}$

30. 12   **31.** 12,850,361   **32.** 351,986

Chapter 3 Test

1. $\frac{1}{2}$ **2.** $\frac{3}{7}$ **3.** $\frac{45}{8} = 5\frac{5}{8}$ **4.** $1\frac{1}{2}$ **5.** $\frac{31}{24} = 1\frac{7}{24}$ **6.** $9\frac{19}{20}$

7. $1\frac{5}{9}$ **8.** $3\frac{1}{8}$ **9.** $\frac{5}{14}$ **10.** $3\frac{1}{2}$ **11.** $\frac{2}{9}$ **12.** $-1\frac{3}{8}$

13. $5\frac{1}{6}$ **14.** $\frac{5}{16}$ **15.** $1\frac{1}{6}$ **16.** $9\frac{5}{6}$ **17.** $16\frac{4}{9}$ **18.** $5\frac{25}{72}$

19. $15\frac{7}{8}$ **20.** Perimeter $= 29\frac{1}{4}$ **21.** $14\frac{1}{2}$ **22.** $13\frac{1}{8}$

23. $\frac{5}{3} = 1\frac{2}{3}$ **24.** $\frac{3}{5}$ **25.** $26\frac{9}{16}$ **26.** $\frac{17}{18}$ **27.** $\frac{3}{8}$

28. $1\frac{5}{12}$ **29.** $\frac{3}{28}$ **30.** $1\frac{1}{13}$

31. $\frac{313}{1000}$; 313 out of every 1,000 users use Windows 7 or Vista

32. $\frac{14}{125}$; 14 out of every 125 users use Macintosh or Linux

Chapter 4

Exercise Set 4.1

Vocabulary Review   **1.** fractional   **2.** decimal point   **3.** thousandths
4. tenths   **5.** hundredths   **6.** place value   **7.** left   **8.** right
Problems   **1.** Three tenths   **3.** Fifteen thousandths
5. Three and four tenths   **7.** Fifty-two and seven tenths
9. $405\frac{36}{100}$ **11.** $9\frac{9}{1,000}$ **13.** $1\frac{234}{1,000}$ **15.** $\frac{305}{100,000}$ **17.** Tens
19. Tenths   **21.** Hundred thousandths   **23.** Ones   **25.** Hundreds
27. 0.55   **29.** 6.9   **31.** 11.11   **33.** 100.02   **35.** 3,000.003

Rounded to the Nearest				
	Number	Whole	Tenth	Hundredth	Thousandth
37.	47.5479	48	47.5	47.55	47.548
39.	0.8175	1	0.8	0.82	0.818
41.	0.1562	0	0.2	0.16	0.156
43.	2,789.3241	2,789	2,789.3	2,789.32	2,789.324
45.	99.9999	100	100.0	100.00	100.000

47. Three and eleven hundredths; two and five tenths
49. 186,282.40 **51.** Fifteen hundredths
53.

Price Of 1 Gallon Of Regular Gasoline	
Date	Price (Dollars)
3/21/11	3.526
3/28/11	3.596
4/4/11	3.684
4/11/11	3.791

55. a. $<$ **b.** $>$
57. 0.002 0.005 0.02 0.025 0.05 0.052
59. 7.451 and 7.54 **61.** $\frac{1}{4}$
63. $\frac{1}{8}$ **65.** $\frac{5}{8}$ **67.** $\frac{7}{8}$
69. 9.99 **71.** 10.05 **73.** 0.05 **75.** 0.01 **77.** $6\frac{31}{100}$
79. $6\frac{23}{50}$ **81.** $18\frac{123}{1,000}$
Find the Mistake 1. To move a place value from the tens column to the hundreds column, you must multiply by ten. **2.** The decimal 0.09 can be written as the fraction $\frac{9}{100}$. **3.** The decimal 142.9643 written as a mixed number is $142\frac{9,643}{10,000}$. **4.** Rounding the decimal 0.06479 to the nearest thousandth gives us 0.065.

Exercise Set 4.2
Vocabulary Review 1. money **2.** columns **3.** decimal point
4. zeros **5.** value
Problems 1. 6.19 **3.** 1.13 **5.** −6.29 **7.** 9.042
9. −8.021 **11.** 11.7843 **13.** 24.343 **15.** 24.111
17. 258.5414 **19.** 666.66 **21.** 11.11 **23.** 3.57
25. 4.22 **27.** 120.41 **29.** 44.933 **31.** 7.673 **33.** 530.865
35. 43.55 **37.** 5.918 **39.** 27.89 **41.** 35.64 **43.** 411.438
45. 6 **47.** 1 **49.** 3.1 **51.** 5.9 **53.** 4.17 **55.** 2.8
57. −11.41 **59.** −1.9 **61.** −17.5 **63.** 3.272 **65.** 4.001
67. $116.82 **69.** $1,571.10 **71.** 4.5 in. **73.** $5.43
75. 0.03 seconds **77. a.** Less **b.** Less **c.** 3%
79. 2 in. **81.** $3.25; three $1 bills and a quarter **83.** 3.25
85. $\frac{3}{100}$ **87.** $\frac{51}{10,000}$ **89.** $1\frac{1}{2}$ **91.** 1,400 **93.** $\frac{3}{20}$
95. $\frac{147}{1,000}$ **97.** 132,980 **99.** 2,115
101. a. 8 million **b.** Between 2009 and 2010 **c.** 625 million users
Find the Mistake 1. To add 32.69 and 4.837, align the decimal point and add in columns. **2.** To add 0.004 + 5.06 + 32 by first changing each decimal to a fraction would give us the problem $\frac{4}{1,000} + 5\frac{6}{100} + 32$.
3. When subtracting 8.7 − 2.0163, we make sure to keep the digits in the correct columns by writing 8.7 as 8.7000. **4.** Subtracting 4.367 from the sum of 12.1 and 0.036 gives us 7.769.

Exercise Set 4.3
Vocabulary Review multiply, answer, digits, decimal points
Problems 1. 0.28 **3.** 0.028 **5.** 0.0027 **7.** 0.78 **9.** 0.792
11. 0.0156 **13.** 24.29821 **15.** 0.03 **17.** 187.85 **19.** 0.002
21. 27.96 **23.** 0.43 **25.** 49,940 **27.** 9,876,540 **29.** 1.89
31. 0.0025 **33.** 5.1106 **35.** 7.3485 **37.** 4.4 **39.** 2.074
41. 3.58 **43.** 187.4 **45.** 116.64 **47.** 20.75 **49.** 0.126
51. Moves it two places to the right **53.** $1,381.38 **55.** $7.10
57. $44.40 **59.** $293.04 **61.** 8,509 mm² **63.** 1.18 in²

65. 1,879 **67.** 1,516 R 4 **69.** 298 **71.** 34.8
73. 49.896 **75.** 825 **77.** No
Find the Mistake 1. To multiply 18.05 by 3.5, multiply as if the numbers were whole numbers and then place the decimal in the answer with three digits to its right. **2.** To estimate the answer for 24.9 × 7.3, round 24.9 to 25 and 7.3 to 7. **3.** To simplify (8.43 + 1.002) − (0.05)(3.2), first work the operation inside the parentheses and find the product of 0.05 and 3.2 before subtracting.
4. Lucy pays $1.52 a pound for the first three pounds of candy she buys at a candy store, and pays $3.27 for each additional pound. To find how much she will pay if she buys 5.2 pounds of candy, we must solve the problem 1.52(3) + 3.27(2.2).

Exercise Set 4.4
Vocabulary Review 1. long division **2.** above **3.** last
4. divisor, right
Problems 1. 19.7 **3.** 6.2 **5.** 5.2 **7.** 11.04 **9.** 4.8
11. 9.7 **13.** 2.63 **15.** 4.24 **17.** 2.55 **19.** 1.35
21. 6.5 **23.** 9.9 **25.** 0.05 **27.** 89 **29.** 2.2 **31.** 4.6
33. 1.35 **35.** 16.97 **37.** 0.25 **39.** 2.71 **41.** 11.69
43. 3.98 **45.** 5.98 **47.** −4.24 **49.** 7.5 mi
51.

Rank	Name	Number of Tournaments	Average per Tournament
1.	Na Yeon Choi	23	$81,360
2.	Jiyai Shin	19	$93,850
3.	Cristie Kerr	21	$76,260
4.	Yani Tseng	19	$82,820
5.	Suzann Pettersen	19	$81,960

53. $5.65/hr **55.** 22.4 mi **57.** 5 hr **59.** 7 min **61.** 2.73
63. 0.13 **65.** 0.77778 **67.** 307.20607 **69.** 0.70945 **71.** $\frac{3}{4}$
73. $\frac{2}{3}$ **75.** $\frac{3}{8}$ **77.** $\frac{19}{50}$ **79.** $\frac{60}{100}$ **81.** $\frac{500}{100}$ **83.** $\frac{12}{15}$
85. $\frac{60}{15}$ **87.** $\frac{18}{15}$ **89.** 0.75 **91.** 0.875
Find the Mistake 1. The answer to the problem 25)70.75 will have a decimal point placed with two digits to its right.
2. To work the problem 27.468 ÷ 8.4, multiply both numbers by 10 and then divide. **3.** To divide 0.6778 by 0.54, multiply both numbers by 100 to move the decimal point two places to the right.
4. Samantha earns $10.16 an hour as a cashier. She received a paycheck for $309.88. To find out how many hours she worked, you must solve the problem 309.88 ÷ 10.16.

Landmark Review
1. One and fifteen hundredths
2. Forty-five and eight hundredths **3.** Five thousandths
4. Two hundred forty-five and one hundred fifty-seven thousandths
5. 0.0067 **6.** 5.6 **7.** 23.014 **8.** 2,013.15 **9.** 28.28
10. 9.150014 **11.** 124.15831 **12.** 11.799 **13.** 3.1 **14.** 7.07
15. 13.33 **16.** 78.37 **17.** 47.35 **18.** 0.00225 **19.** 20
20. 0.4 **21.** 5.16 **22.** 11.3505 **23.** 8.8 **24.** 33.46

Exercise Set 4.5
Vocabulary Review 1. division **2.** repeats **3.** place values, reduce
Problems 1. 0.125 **3.** 0.625
5.

Fraction	$\frac{1}{5}$	$\frac{2}{5}$	$\frac{3}{5}$	$\frac{4}{5}$	$\frac{5}{5}$
Decimal	0.2	0.4	0.6	0.8	1

7. 0.5 **9.** 0.56
11. 0.5625 **13.** −0.8125
15. 0.92 **17.** 0.27
19. 0.09 **21.** 0.28
23.

Decimal	0.125	0.250	0.375	0.500	0.625	0.750	0.875
Fraction	$\frac{1}{8}$	$\frac{1}{4}$	$\frac{3}{8}$	$\frac{1}{2}$	$\frac{5}{8}$	$\frac{3}{4}$	$\frac{7}{8}$

25. $\frac{3}{20}$ **27.** $\frac{2}{25}$ **29.** $\frac{3}{8}$ **31.** $5\frac{3}{5}$ **33.** $5\frac{3}{50}$ **35.** $1\frac{11}{50}$

37. 2.4 **39.** 3.98 **41.** 3.02 **43.** 0.3 **45.** 0.072 **47.** 0.8
49. 1 **51.** 0.25 **53.** $8.42 **55.** $38.66 **57.** 9 in.
59.

Change In Stock Price		
Date	Gain ($)	As a Decimal ($)
Monday	$\frac{3}{5}$	0.60
Tuesday	$\frac{1}{2}$	0.50
Wednesday	$\frac{1}{25}$	0.04
Thursday	$\frac{1}{5}$	0.20
Friday	$\frac{1}{10}$	0.10

61. 104.625 calories
63. $10.38
65. Yes **67.** 36
69. 25 **71.** 125
73. 9 **75.** $\frac{1}{81}$
77. $\frac{25}{36}$ **79.** 0.25
81. 1.44 **83.** 25
85. 100

Find the Mistake   1. The correct way to write $\frac{6}{11}$ as a decimal is $0.\overline{54}$.
2. Writing 14.3 as a fraction gives us $14\frac{3}{10}$.   **3.** The simplified answer to the problem $\frac{12}{45(0.256 + 0.14)}$ contains only fractions or only decimals.   **4.** Simplifying the problem $\left(\frac{3}{2}\right)(0.5) + \left(\frac{1}{2}\right)^2(6.7)$ by first converting all decimals to fractions gives us $\left(\frac{3}{2}\right)\left(\frac{1}{2}\right) + \left(\frac{1}{2}\right)^2\left(\frac{67}{10}\right)$.

Exercise Set 4.6

Vocabulary Review   1. square root   **2.** radical sign
3. whole numbers   **4.** irrational   **5.** right   **6.** hypotenuse
7. Pythagorean theorem   **8.** spiral of roots
Problems   1. 8   **3.** 9   **5.** 6   **7.** 5   **9.** 15   **11.** 48   **13.** 45
15. 48   **17.** 15   **19.** 1   **21.** 78   **23.** 9   **25.** $\frac{4}{7}$
27. $\frac{3}{4}$ **29.** $\frac{2}{3}$ **31.** $\frac{1}{4}$ **33.** False   **35.** True   **37.** 1.1180
39. 11.1803   **41.** 3.46   **43.** 11.18   **45.** 0.58   **47.** 0.58
49. 12.124   **51.** 9.327   **53.** 12.124   **55.** 12.124   **57.** 10 in.
59. 13 ft   **61.** 8.06 km   **63.** 17.49 m   **65.** 30 ft   **67.** 25 ft
71.

Height h(feet)	Distance d(miles)
10	4
50	9
90	12
130	14
170	16
190	17

73. a. 8.9 million
b. 8.25 million

Find the Mistake   1. The square root of a positive number x is the number we square to get x.   **2.** The square root of 225 is 15, and can be written in symbols as $\sqrt{225} = 15$.   **3.** Simplifying the radical $\sqrt{\frac{196}{36}}$ gives us $\frac{14}{6}$ because $\left(\frac{14}{6}\right)^2 = \frac{14}{6} \cdot \frac{14}{6} = \frac{196}{36}$.
4. The Pythagorean theorem states that $c = \sqrt{a^2 + b^2}$.

Chapter 4 Review

1. Six and three hundred two thousandths   **2.** Thousandths
3. 23.5006   **4.** 72.20   **5.** 12.02   **6.** 7.904   **7.** 16.254
8. 2.482   **9.** 14.59   **10.** 7.892   **11.** −3.069   **12.** 3.091
13. $2.77   **14.** $23.81   **15.** $6.75   **16.** 0.85   **17.** $\frac{31}{50}$
18. 29.036   **19.** 7.55   **20.** 16.882   **21.** 0.06   **22.** 23.68
23. 44.876   **24.** 2   **25.** 11   **26.** 36   **27.** 25   **28.** 12.65 in.
29. 10 in.   **30.** $26,483,333.33 per month
31. $25,225,000 per month

Chapter 4 Cumulative Review

1. 5,291   **2.** 144   **3.** 17,670   **4.** 2.82   **5.** 153.568$\overline{3}$
6. $\frac{53}{42} = 1\frac{11}{42}$ **7.** 100.373   **8.** 62   **9.** $\frac{15}{4} = 3\frac{3}{4}$ **10.** 0
11. 16$\frac{1}{5}$ **12.** $\frac{31}{8}$ **13.** 25$\frac{3}{8}$

14.

Decimal	Fraction
0.125	$\frac{1}{8}$
0.250	$\frac{1}{4}$
0.375	$\frac{3}{8}$
0.500	$\frac{1}{2}$
0.625	$\frac{5}{8}$
0.750	$\frac{3}{4}$
0.875	$\frac{7}{8}$
1	$\frac{8}{8}$

15. 4   **16.** 4(2 + 3) = 20
17. 7(6 + 3) = 63   **18.** $\frac{15}{26}$
19. False   **20.** 207   **21.** 2   **22.** 8
23. 1.02   **24.** $\frac{4}{81}$ **25.** 19$\frac{49}{64}$
26. 86°   **27.** 9$\frac{1}{2}$ cups
28. $8.25

Chapter 4 Test

1. Eleven and eight hundred nineteen thousandths   **2.** Tenths
3. 73.0046   **4.** 100.91   **5.** 7.02   **6.** 11.724   **7.** 16.56
8. 5.84   **9.** 10.1   **10.** 7.89   **11.** −3.33   **12.** 0.22   **13.** $3.61
14. $18.10   **15.** $8.15   **16.** 0.68   **17.** $\frac{19}{50}$ **18.** 25.704
19. 2.38   **20.** 16.897   **21.** 0.47125   **22.** 8.92   **23.** 26.674
24. 10   **25.** 3   **26.** 15   **27.** 30   **28.** 11.18 in.   **29.** 21.93 yds.
30. $44,033,333.33 per month   **31.** $40,000,000 per month

Chapter 5

Exercise Set 5.1

Vocabulary Review   1. fraction   **2.** numerator, denominator
3. ratio   **4.** colon   **5.** complex
Problems   1. $\frac{4}{3}$ **3.** $\frac{16}{3}$ **5.** $\frac{2}{5}$ **7.** $\frac{1}{2}$ **9.** $\frac{3}{1}$ **11.** $\frac{7}{6}$ **13.** $\frac{7}{5}$
15. $\frac{5}{7}$ **17.** $\frac{8}{5}$ **19.** $\frac{1}{3}$ **21.** $\frac{1}{10}$ **23.** $\frac{3}{25}$ **25. a.** $\frac{1}{2}$ **b.** $\frac{1}{3}$ **c.** $\frac{2}{3}$
27. a. $\frac{13}{8}$ **b.** $\frac{1}{4}$ **c.** $\frac{3}{8}$ **d.** $\frac{13}{3}$ **29. a.** $\frac{73}{14}$ **b.** $\frac{26}{51}$ **c.** $\frac{7}{13}$ **d.** $\frac{14}{51}$
31. a. $\frac{3}{4}$ **b.** 12 **c.** $\frac{3}{4}$ **33.** $\frac{2,408}{2,314} \approx 1.04$ **35.** $\frac{4,722}{2,408} \approx 1.96$ **37.** 40
39. 0.2   **41.** 0.695   **43.** 3.98   **45.** 36   **47.** 0.065   **49.** 0.025
Find the Mistake   1. Writing the ratio of $\frac{2}{5}$ to $\frac{3}{8}$ is the same as writing $\frac{2}{5} \cdot \frac{8}{3}$ or $\frac{\frac{2}{5}}{\frac{3}{8}}$.   **2.** The ratio of 6 to 24 expressed in lowest terms is $\frac{1}{4}$.
3. To write the ratio of 0.04 to 0.20 as a fraction in lowest terms, you must first multiply 0.04 and 0.20 by 100 to rid the ratio of decimals.
4. A cleaning solution of bleach and water contains 100 milliliters of bleach and 150 milliliters of water. To find the ratio of water to the whole solution in lowest terms, you must write the ratio as $\frac{150}{250} = \frac{3}{5}$.

Exercise Set 5.2

Vocabulary Review   1. rate   **2.** denominator   **3.** division
4. unit pricing
Problems   1. 55 mi/hr   **3.** 84 km/hr   **5.** 0.2 gal/sec
7. 12 L/min   **9.** 19 mi/gal   **11.** 4$\frac{1}{3}$ mi/L
13. 16¢ per ounce   **15.** 4.95¢ per ounce
17. Dry Baby: 34.7¢/diaper, Happy Baby: 31.6¢/diaper, Happy Baby is better buy   **19.** 7.7 tons/year   **21.** 10.8¢ per day
23. 9.3 mi/gal   **25.** $64   **27.** $16,000   **29.** $n = 6$
31. $n = 4$ **33.** $n = 4$ **35.** $n = 65$ **37.** 100 oz is the better value.

Wisk Laundry Detergent		
	Old	New
Size	100 Ounces	80 Ounces
Container Cost	$6.99	$5.75
Price per quart	$2.24	$2.30

Find the Mistake **1.** The rate in miles per hour for a plane traveling 3000 miles in 6 hours is 500 miles per hour. **2.** If a runner can run 16 miles in 2 hours, then her ratio of miles to hours is 8 miles per hour. **3.** If a supermarket sells a package of 20 cookies for $4.27, then the unit price for each cookie is $0.21. **4.** A supermarket sells 10 packages of oatmeal for $5.33. A wholesale store sells the same oatmeal for $8.86 for 24 packages of oatmeal. Given the information, we find that the wholesale store has the lowest unit price.

Exercise Set 5.3
Vocabulary Review **1.** equation, variable **2.** division **3.** equals sign
Problems **1.** 35 **3.** 18 **5.** 14 **7.** n **9.** y **11.** b
13. $n = 2$ **15.** $x = 7$ **17.** $y = 7$ **19.** $n = 8$ **21.** $a = 8$
23. $x = 2$ **25.** $y = 1$ **27.** $a = 6$ **29.** $n = 5$ **31.** $x = 3$
33. $n = 7$ **35.** $y = 1$ **37.** $y = 9$ **39.** $n = \frac{7}{2} = 3\frac{1}{2}$
41. $x = \frac{7}{2} = 3\frac{1}{2}$ **43.** $a = \frac{12}{5} = 2\frac{2}{5}$ **45.** $y = \frac{4}{7}$ **47.** $y = \frac{10}{13}$
49. $x = \frac{5}{2} = 2\frac{1}{2}$ **51.** $n = \frac{3}{2} = 1\frac{1}{2}$ **53.** $\frac{3}{4}$ **55.** $\frac{2}{3}$ **57.** 1.2
59. 4 **61.** 6.5 **63.** 0.5

Find the Mistake **1.** To simplify $\frac{3 \cdot a \cdot 8 \cdot 11}{a \cdot 11}$, divide out a and 11 to get 24. **2.** Dividing $6 \cdot z$ by 6 gives us z. **3.** Solving the equation $6 \cdot a = 48$ for a gives us $a = 8$. **4.** Solving the equation $36 = w \cdot 12$ for w gives $w = \frac{36}{12} = 3$.

Landmark Review **1.** $\frac{5}{6}$ **2.** $\frac{10}{1}$ **3.** $\frac{1}{10}$ **4.** $\frac{1}{3}$
5. 70 mi/hr **6.** 0.125 gal/sec **7.** 16 mi/gal
8. 14.5 cents/ounce **9.** $x = 5$ **10.** $n = 4$ **11.** $y = 4$
12. $a = 3$ **13.** $b = 2$ **14.** $x = 3$ **15.** $y = \frac{1}{10}$ **16.** $x = \frac{1}{4}$

Exercise Set 5.4
Vocabulary Review **1.** proportion **2.** term **3.** first, fourth
4. second, third **5.** product **6.** fundamental property of proportions
Problems **1.** Means: 3, 5; extremes: 1, 15; products: 15
3. Means: 25, 2; extremes: 10, 5; products: 50
5. Means: $\frac{1}{2}$, 4; extremes: $\frac{1}{3}$, 6; products: 2
7. Means: 5, 1; extremes: 0.5, 10; products: 5 **9.** $x = 10$
11. $y = \frac{12}{5}$ **13.** $x = \frac{3}{2}$ **15.** $x = \frac{10}{9}$ **17.** $x = 7$ **19.** $x = 14$
21. $y = 18$ **23.** $n = 6$ **25.** $n = 40$ **27.** $x = 50$ **29.** $y = 108$
31. $x = 3$ **33.** $n = 1$ **35.** $x = 0.25$ **37.** $x = 108$ **39.** $y = 65$
41. $n = 41$ **43.** $x = 108$ **45.** 20 **47.** 300 **49.** 297.5
51. 50.4 **53.** $x = 450$ **55.** $x = 5$
Find the Mistake **1.** A statement that two ratios are equal is called a proportion. **2.** For the proportion $\frac{5}{6} = \frac{10}{x}$, the extremes are 5 and x. **3.** To solve $\frac{7}{10} = \frac{n}{0.2}$, set the product of first and fourth terms equal to the product of the second and third terms. **4.** Solving the proportion $\frac{8}{5} = \frac{n}{\frac{3}{10}}$ gives us $n = \frac{12}{25}$.

Exercise Set 5.5
Vocabulary Review **1.** word problems **2.** proportion, quantities
3. division
Problems **1.** 329 mi **3.** 360 points **5.** 15 pt **7.** 427.5 mi
9. 900 eggs **11.** 435 in. = 36.25 ft **13.** $119.70 **15.** 265 g
17. 91.3 liters **19.** 60,113 people **21.** 2 **23.** 147
25. $x = 20$ **27.** $x = 147$
Find the Mistake **1.** A basketball player scores 112 points in 8 games. The proportion to find how many points the player will score in 14 games is $\frac{112}{8} = \frac{x}{14}$. **2.** The scale of a map indicates that 2 inches corresponds to 250 miles in real life. If two cities on the map are 3.5 inches apart, they are 437.5 miles apart in real life. **3.** A jellybean company knows that for every 100 jellybeans, 4 will be misshapen.

The proportion needed to find how many jelly beans were made if 36 misshapen jelly beans are found is $\frac{4}{100} = \frac{36}{x}$. **4.** If burning 1 gallon of gasoline produces 20.2 pounds of carbon dioxide, then burning 12 gallons of gasoline produces 242.4 pounds of carbon dioxide.

Exercise Set 5.6
Vocabulary Review **1.** shape, size **2.** similar **3.** verticies
4. ratio, width
Problems **1.** $h = 9$ **3.** $y = 14$ **5.** $x = 12$ **7.** $a = 25$ **9.** $y = 32$
11.  **13.**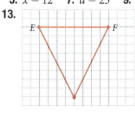

15. 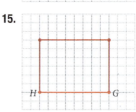 **17.**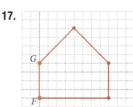

19. 250 in. **21.** 960 pixels **23.** 1,440 pixels **25.** 57 ft
27. 4 ft **29. a.** $\frac{1}{2}$ **b.** Any 18 rectangles should be shaded. **c.** $\frac{1}{3}$
Find the Mistake **1.** The two triangles below are similar. The side x is equal to 12. **2.** The two triangles below are similar. We can find x by solving the proportion $\frac{12}{x} = \frac{8}{2}$ or $\frac{4}{1}$. **3.** The width of a rectangle on graph paper is 5 squares and the length is 7 squares. If a similar rectangle has a width of 10, then the length would be 14.
4. A pocket dictionary is similar to a regular dictionary. The pocket dictionary is 4 inches wide by 6 inches long. The width of the regular dictionary is 16 inches. You must solve the proportion $\frac{6}{x} = \frac{4}{16}$ to find the remaining side length of the regular dictionary.

Chapter 5 Review
1. $\frac{9}{4}$ **2.** $\frac{4}{3}$ **3.** $\frac{4}{3}$ **4.** $\frac{12}{7}$ **5.** $\frac{7}{5}$ **6.** $\frac{15}{64}$ **7.** $\frac{40}{7}$
8. 29 mpg **9.** 20-ounce cup: 19¢/ounce; 16-ounce cup: 20¢/ounce; 20-ounce cup is the better buy **10.** $x = 56$ **11.** $x = 2$
12. 36 hits **13.** 81 miles **14.** 28.8 mg **15.** 42.88 mg
16. $h = 5$ **17.** 6,371 km **18.** 142,710 km

Chapter 5 Cumulative Review
1. 13,372 **2.** 336 **3.** 17 **4.** 48 **5.** $516\frac{3}{8}$ or 516.375
6. 256 **7.** 63 **8.** 8 **9.** 91 **10.** 27 **11.** 144
12. 29.76 **13.** 42.6 **14.** 14.66 **15.** 6.6 **16.** $\frac{1}{432}$
17. 16 **18.** $\frac{19}{24}$ **19.** 1 **20.** $12\frac{2}{7}$ **21.** $x = 25$ **22.** $x = 24$
23. $P = 35$ in.; $A = 48$ in^2 **24.** $P = 74$ cm; $A = 144$ cm^2
25. $x = 78$ cm **26.** 20 women **27.** $83\frac{1}{4}$ in. or 6 ft, $11\frac{1}{4}$ in.
28. 69 sections **29.** $\frac{4}{1}$ **30.** $\frac{1}{32}$

Chapter 5 Test
1. $\frac{8}{3}$ **2.** $\frac{5}{6}$ **3.** $\frac{7}{3}$ **4.** $\frac{7}{20}$ **5.** $\frac{7}{4}$ **6.** $\frac{64}{15}$ **7.** $\frac{25}{8}$
8. 35 mpg **9.** 16-ounce cup: 22¢/ounce; 12-ounce cup: 23¢/ounce; 16-ounce cup is the better buy **10.** $x = 42$ **11.** $x = 3$
12. 39 hits **13.** 82.5 miles **14.** 51.3 mg **15.** 15.39 mg
16. $h = 9$ **17.** $\frac{164}{197}$ **18.** $\frac{1}{1}$

Chapter 6

Exercise Set 6.1

Vocabulary Review **1.** hundred **2.** ratio **3.** left
4. right **5.** % symbol **6.** decimal

Problems **1.** $\frac{20}{100}$ **3.** $\frac{60}{100}$ **5.** $\frac{24}{100}$ **7.** $\frac{65}{100}$ **9.** 0.23 **11.** 0.92
13. 0.09 **15.** 0.034 **17.** 0.0634 **19.** 0.009 **21.** 23%
23. 92% **25.** 45% **27.** 3% **29.** 60% **31.** 80%
33. 27% **35.** 123% **37.** $\frac{3}{5}$ **39.** $\frac{3}{4}$ **41.** $\frac{1}{25}$ **43.** $\frac{53}{200}$
45. $\frac{7,187}{10,000}$ **47.** $\frac{3}{400}$ **49.** $\frac{1}{16}$ **51.** $\frac{1}{3}$ **53.** 50%
55. 75% **57.** $33\frac{1}{3}$% **59.** 80% **61.** 87.5% **63.** 14%
65. 325% **67.** 150% **69.** 48.8% **71.** 0.50; 0.75
73. a. $\frac{1}{50}, \frac{3}{100}, \frac{3}{50}, \frac{3}{25}, \frac{77}{100}$ **b.** 0.02, 0.03, 0.06, 0.12, 0.77
c. About 2 times as likely. **75.** 20%
77. Liberal Arts: 15%, Science & Math: 15%, Engineering: 27.78%,
Business: 11.11%, Architecture & Environmental Design: 11.11%,
Agriculture: 22.22%
79. 78.4% **81.** 11.8% **83.** 72.2% **85.** 8.3%; 0.2%
87. 18.5 **89.** 10.875 **91.** 0.5 **93.** 62.5 **95.** 0.5

Find the Mistake **1.** Writing 0.4% as a decimal gives us .004.
2. To write 3.21 as a percent, multiply the number by 100; that is,
move the decimal two places to the right. **3.** Writing 25% as a
fraction in lowest terms gives us $\frac{1}{4}$. **4.** To change $\frac{5}{8}$ to a
percent, we change $\frac{5}{8}$ to 0.625 and then move the decimal two
places to the right to get 62.5%.

Exercise Set 6.2

Vocabulary Review **1.** equals sign **2.** multiply **3.** variable
4. decimal **5.** fraction
Problems **1.** 8 **3.** 24 **5.** 20.52 **7.** 7.37 **9.** 50% **11.** 10%
13. 25% **15.** 75% **17.** 64 **19.** 50 **21.** 925 **23.** 400
25. 5.568 **27.** 120 **29.** 13.72 **31.** 22.5 **33.** 50%
35. 942.684 **37.** 97.8 **39.** What number is 25% of 350?
41. What percent of 24 is 16? **43.** 46 is 75% of what number?
45. 11.3% calories from fat; healthy **47.** 56.9% calories from
fat; not healthy **49.** 0.80 **51.** 0.76 **53.** 48
55. Fewer than 175 to 280 gulls of breeding age

Find the Mistake **1.** The question, "What number is 28.5% of 30?"
translates to $n = 0.285 \cdot 30$. **2.** Asking "75 is 30% of what
number?" gives us 250. **3.** To answer the question, "What
number is 45% of 90?", we can solve the proportion $\frac{x}{90} = \frac{45}{100}$.
4. Using a proportion to answer the question, "What percent of 65 is
26?" will give us $n = 40\%$.

Landmark Review
1. $\frac{15}{100}$ **2.** $\frac{27}{100}$ **3.** $\frac{14}{100}$
4. $\frac{89}{100}$ **5.** 0.17 **6.** 0.28 **7.** 0.05 **8.** 0.0637 **9.** 38%
10. 98% **11.** 9% **12.** 487% **13.** 10% **14.** 33.3%
15. 14.3% **16.** 320% **17.** 5.25 **18.** 62.35 **19.** 237.84

Exercise Set 6.3

Vocabulary Review **1.** What number is y% of x?
2. What percent of x is z? **3.** z is y% of what number?
Problems **1.** 70% **3.** 40 mL **5.** 45 mL
7. 18.2 acres for farming; 9.8 acres are not available for farming
9. 3,000 students **11.** 400 students **13.** 1,664 female students
15. 31.25% **17.** 50% **19.** 1,267 students **21.** 33 **23.** 8,685
25. 136 **27.** 0.05 **29.** 15,300 **31.** 0.15 **33.** 33.7%, to the
nearest tenth of a percent **35.** 170 hits **37.** 17 hits

Find the Mistake **1.** On a test with 110 questions, a student
answered 98 questions correctly. The percentage of questions the
student answered correctly is 89.1%. **2.** A school track team
consists of 12 boys and 10 girls. The total number of girls makes up
45.5% percent of the whole team. **3.** Suppose 39 students in a
college class of 130 students received a B on their tests. To find what
percent of students earned a B, solve the proportion $\frac{39}{130} = \frac{x}{100}$.
4. Suppose a basketball player made 120 out of 150 free throws
attempted. To find what percent of free throws the player made,
solve the proportion $\frac{120}{150} = \frac{x}{100}$.

Exercise Set 6.4

Vocabulary Review **1.** D **2.** P **3.** P **4.** D
Problems **1.** $52.50 **3.** $2.70; $47.70 **5.** $150; $156 **7.** 5%
9. $2,820 **11.** $200 **13.** 14% **15.** $11.93 **17.** 4.5%
19. $3,995 **21.** 1,100 **23.** 75 **25.** 0.16 **27.** 4 **29.** 396
31. 415.8 **33.** Sales tax = $3,180; luxury tax = $2,300 **35.** $3,180
37. You saved $1,600 on the sticker price and $150 in luxury tax.
If you lived in a state with a 6% sales tax rate, you saved an
additional 0.06($1600) = $96.

Find the Mistake **1.** Suppose the sale tax rate on a new computer
is 8%. If the computer cost $650, then the total price of purchase
would be $702. **2.** If a new shirt that costs $32 has sales tax
equal to $1.92, then the sales tax rate is 6%. **3.** A car salesman's
commission rate is 7%. To find his commission on a $15,000 sale
of a 2005 Ford truck, we must divide the product of 7 and 15,000
by 100. **4.** A saleswoman makes a commission of $6.80 on a
sale of $85 worth of clothing. To find the woman's commission rate,
solve the proportion $\frac{6.8}{85} = \frac{x}{100}$.

Exercise Set 6.5

Vocabulary Review **1.** b **2.** d **3.** a **4.** c
Problems **1.** $24,610 **3.** $3,510 **5.** $13,200 **7.** 10% **9.** 20%
11. 61% **13.** $45; $255 **15.** $381.60 **17.** $46,595.88
19. a. 51.9% **b.** 7.8% **21.** 140 **23.** 4 **25.** 152.25 **27.** 3,434.7
29. 10,150 **31.** 10,456.78 **33.** 2,140 **35.** 3,210

Find the Mistake **1.** If a new model of a car increases 12% from
and old model's price of $24,000, then the new selling price is
$26,880. **2.** A lawnmower goes on sale from $98 to $63.70. The
percent decrease of the lawnmower's price is 35%. **3.** A backpack
that normally sells for $75 is on sale. The new price of $45 shows a
percent decrease of 40%. **4.** A designer pair of sunglasses is on
sale from $125 for 20% off. If the sales tax is 6% of the sale price,
then the total bill for the glasses would be $106.

Exercise Set 6.6

Vocabulary Review **1.** principal **2.** interest rate
3. simple **4.** compound
Problems **1.** $2,160 **3.** $665 **5.** $8,560 **7.** $2,160 **9.** $5
11. $813.33 **13.** $5,618 **15.** $8,407.56, Some answers may
vary in the hundredths column depending on whether rounding is
done in the intermediate steps. **17.** $974.59
19. a. $13,468.55 **b.** $13,488.50 **c.** $12,820.37 **d.** $12,833.59
21. Percent increase in production costs: *A Beautiful Mind* to
Braveheart, 24%; *Braveheart* to *The Departed*, 25%; *The Departed* to
Lord of the Rings, 8%; *Lord of the Rings* to *Gladiator*, 6%, *Gladiator*
to *Titanic*, 94%

Find the Mistake **1.** A woman invests $1,500 into an account with
a 6% annual interest rate. She will have $1,590 in her account by
the end of one year. **2.** A business man invests $2,750 into an
account that has an 8% interest rate per year. To find out how much
money will be in the man's account after 72 days, you must multiply
the product of 2,750 and 0.08 by $\frac{72}{360}$. **3.** If a person invests

$10,000 into an account that is compounded annually at 6%, then after two years, the account will contain $11,236.   **4.** A woman deposits $4,000 into a savings account that pays 7% compounded quarterly. At the end of the year, the account contains $4,287.44.

Chapter 6 Review
1. 0.56   **2.** 0.03   **3.** 0.004   **4.** 32%   **5.** 70%   **6.** 164%
7. $\frac{17}{20}$ **8.** $1\frac{7}{25}$ **9.** $\frac{21}{250}$ **10.** 52%   **11.** 62.5%   **12.** 145%
13. 12.8   **14.** 60%   **15.** 80   **16.** 84%   **17.** $1,000
18. $290.50; 35% off   **19.** $77.31   **20.** $1,006.20   **21.** $320
22. $5,310.52   **23.** 93,100,000 iPhones   **24.** 3,800,000 iPhones

Chapter 6 Cumulative Review
1. 5,411   **2.** 1,091   **3.** 7,969   **4.** 53.4   **5.** 32   **6.** $\frac{4}{5}$
7. $\frac{27}{125}$ **8.** 9.502   **9.** 0.286   **10.** 1.19   **11.** $\frac{1}{9}$ **12.** 12
13. $\frac{29}{24} = 1\frac{5}{24}$ **14.** $\frac{23}{6} = 3\frac{5}{6}$ **15.** $\frac{29}{2} = 14\frac{1}{2}$ **16.** 6   **17.** 14
18. $3(6 + 8) = 42$ **19.** $\frac{1}{6}$ **20.** 19,008 feet   **21.** 1,134 in^2
22. 62.5%   **23.** $\frac{8}{25}$ **24.** 3.5   **25.** 38   **26.** 3.9   **27.** 35%
28. 95   **29.** $1.33   **30.** 91¢   **31.** 35°C   **32.** 9.5%
33. 375 miles   **34.** $3,187.50   **35.** $P = 24.8$ in.; $A = 38.44$ in^2
36. $8.50 per hour   **37.** 12.77 acres   **38.** 5.5 acres

Chapter 6 Test
1. 0.27   **2.** 0.06   **3.** 0.009   **4.** 64%   **5.** 30%   **6.** 149%
7. $\frac{9}{20}$ **8.** $1\frac{9}{125}$ **9.** $\frac{9}{125}$ **10.** 65%   **11.** 87.5%   **12.** 225%
13. 12   **14.** 35%   **15.** 75   **16.** .96%   **17.** $900
18. $145; 20% off   **19.** $135.32   **20.** $866.40   **21.** $140
22. $2,662.40   **23.** 22.46%   **24.** 5,863 finishers

Chapter 7

Exercise Set 7.1
Vocabulary Review   1. foot   **2.** meter   **3.** conversion factor
4. fraction bar   **5.** unit analysis   **6.** divide
Problems   1. 60 in.   **3.** 120 in.   **5.** 6 ft   **7.** 162 in.   **9.** $2\frac{1}{4}$ ft
11. 13,200 ft   **13.** $1\frac{1}{3}$ yd   **15.** 1,800 cm   **17.** 4,800 m
19. 50 cm   **21.** 0.248 km   **23.** 670 mm   **25.** 34.98 m
27. 6.34 dm   **29.** 20 yd   **31.** 80 in.   **33.** 244 cm
35. 65 mm   **37.** 2,960 chains   **39.** 120,000 μm   **41.** 7,920 ft
43. 0.12 m, 0.35 m   **45.** 0.016 kw   **47.** 80.7 ft/sec
49. 19.5 mi/hr   **51.** 1,023 mi/hr   **53.** $18,216   **55.** $157.50
57. 3,965,280 ft   **59.** 179,352 in.   **61.** 2.7 mi   **63.** 18,094,560 ft
65. 144   **67.** 8   **69.** 1,000   **71.** 3,267,000   **73.** 6
75. 0.4   **77.** 405   **79.** 450   **81.** 45   **83.** 2,200   **85.** 607.5
Find the Mistake   1. The average length for a house cat is roughly 18 inches. To show how many feet this is, multiply by the conversion factor $\frac{1 \text{ foot}}{12 \text{ inches}}$.   **2.** The length of a parking space is based on the average car length, which measures approximately 15 feet. This length converted to yards is 5 yards.   **3.** A cell phone measures 100 mm in length. To show how many meters this is, multiply 100 mm by the conversion factor $\frac{1 \text{ m}}{1,000 \text{ mm}}$.   **4.** To complete a unit analysis problem, make sure all units, except those you want to end up with, will divide out.

Exercise Set 7.2
Vocabulary Review   1. square foot   **2.** square inches
3. multiply   **4.** U.S. system   **5.** metric system   **6.** liter
Problems   1. 432 in^2 **3.** 2 ft^2 **5.** 1,306,800 ft^2 **7.** 1,280 acres
9. 3 mi^2 **11.** 108 ft^2 **13.** 1,700 mm^2 **15.** 28,000 cm^2
17. 0.0012 m^2 **19.** 500 m^2 **21.** 700 a   **23.** 3.42 ha

25. 45 ft^2 **27.** 48 fl oz   **29.** 8 qt   **31.** 20 pt   **33.** 480 fl oz
35. 8 gal   **37.** 6 qt   **39.** 9 yd^3 **41.** 5,000 mL
43. 0.127 L   **45.** 4,000,000 mL   **47.** 14,920 L
49. NFL: $A = 5,333\frac{1}{3}$ sq yd = 1.1 acres; Canadian: $A = 7,150$ sq yd = 1.48 acres; Arena: $A = 1,416\frac{2}{3}$ sq yd = 0.29 acres
51. 30 a   **53.** 5,500 bricks   **55.** 0.119 ft^2 **57.** 0.62 km^2
59. 12,200 ha   **61.** 4,200 mL   **63.** 1,500 mL   **65.** 16 cups
67. 34,560 in^3 **69.** 48 glasses   **71.** 20,288,000 acres   **73.** 3,230.93 mi^2
75. 23.35 gal   **77.** 21,492 ft^3 **79.** 285,795,000 ft^3 **81.** 192
83. 6,000   **85.** 300,000   **87.** 12,500   **89.** 12.5
Find the Mistake   1. A rectangular computer monitor is 24 inches by 30 inches. To find the area of the computer monitor in square feet, divide the product of 24 and 30 by 144.   **2.** A family bought 320 acres of land. The area of this land in square miles is 0.5 mi^2.   **3.** A 3-quart container of ice cream can hold 6 pints of ice cream.   **4.** To find how many milliliters four 2-liter soda bottles can hold, you must find the product of 8 L and $\frac{1,000 \text{ mL}}{1 \text{ L}}$.

Exercise Set 7.3
Vocabulary Review   1. U.S. System   **2.** pounds, ounces
3. metric sytem   **4.** kilograms, metric tons
Problems   1. 128 oz   **3.** 4,000 lb   **5.** 12 lb   **7.** 0.9 T
9. 32,000 oz   **11.** 56 oz   **13.** 13,000 lb   **15.** 2,000 g
17. 40 mg   **19.** 200,000 cg   **21.** 508 cg   **23.** 4.5 g
25. 47.895 cg   **27.** 1.578 g   **29.** 0.42 kg   **31.** 48 g
33. 4 g   **35.** 9.72 g   **37.** 1,540 lbs   **39.** 1,400 g
41. 250 cg   **43.** 3 L   **45.** 1.5 L   **47.** 20.32   **49.** 6.36
51. 50   **53.** 56.8   **55.** 122   **57.** 248   **59.** 38.89
61. 0.000789 kg/m^3
Find the Mistake   1. Correct   **2.** Incorrect, $\frac{2,000 \text{ lbs}}{1 \text{ T}}$
3. Incorrect, $\frac{1 \text{kg}}{1,000 \text{ g}}$ **4.** Correct

Landmark Review
1. 84 inches   **2.** 0.75 yards
3. 443,520 inches   **4.** 0.002841 miles   **5.** 3,800 centimeters
6. 1.43 decimeters   **7.** 4.3 meters   **8.** 1.15 meters
9. 1,152 square inches   **10.** 3.886 square miles   **11.** 8,960 acres
12. 45 square feet   **13.** 700 square meters   **14.** 0.00143 m^2
15. 3.5 square centimeters   **16.** 420 ares   **17.** 112 ounces
18. 0.9375 pounds   **19.** 64,000 ounces   **20.** 0.75 tons
21. 6,500 grams   **22.** 50 milligrams   **23.** 1.759 grams
24. 8.59 grams

Exercise Set 7.4
Vocabulary Review   1. centimeters   **2.** quarts
3. subtract, multiply   **4.** Celsius, Fahrenheit
Problems   1. 15.24 cm   **3.** 13.12 ft   **5.** 6.56 yd   **7.** 32,200 m
9. 5.98 yd^2 **11.** 24.7 acres   **13.** 8,195 mL   **15.** 2.12 qt
17. 75.8 L   **19.** 339.6 g   **21.** 33 lb   **23.** 365°F   **25.** 30°C
27. 3.94 in.   **29.** 7.62 m   **31.** 46.23 L   **33.** 17.67 oz
35. Answers will vary.   **37.** 91.46 m   **39.** 20.90 m^2
41. 88.55 km/hr   **43.** 2.03 m   **45.** 38.3°C   **47.** 75
49. 82   **51.** 3.25   **53.** 22   **55.** 41   **57.** 48   **59.** 195
61. 3.27   **63.** $90.00   **65.** 0.71 oz/in^3 **67.** 68.5 g
Find the Mistake   1. To convert 6 miles to km, multiply 6 miles by $\frac{1.61 \text{ km}}{1 \text{ mi}}$.   **2.** Suppose a pasta recipe requires a half gallon of water, but your measuring cup only measures milliliters. To find how many milliliters are equal to a half gallon of water, multiply $\frac{1}{2}$ gal by both conversion factors $\frac{3.79 \text{ L}}{1 \text{ gal}}$ and $\frac{1,000 \text{ mL}}{1 \text{ L}}$.   **3.** To convert 25° Celsius to Fahrenheit, use the formula $F = \frac{9}{5}C + 32$ to get 77° F.
4. A cookie recipe requires you to bake them at 325° F. This temperature in Celsius is 163°C (rounded to the nearest degree).

Exercise Set 7.5
Vocabulary Review   **1.** minutes to seconds   **2.** columns
3. borrow
Problems   **1. a.** 270 min **b.** 4.5 hr   **3. a.** 320 min **b.** 5.33 hr
5. a. 390 sec **b.** 6.5 min   **7. a.** 320 sec **b.** 5.33 min
9. a. 40 oz **b.** 2.5 lb   **11. a.** 76 oz **b.** 4.75 lb
13. a. 54 in. **b.** 4.5 ft   **15. a.** 69 in. **b.** 5.75 ft
17. a. 9 qt **b.** 2.25 gal   **19.** 11 hr   **21.** 22 ft 4 in.
23. 11 lb   **25.** 5 hr 40 min   **27.** 3 hr 47 min   **29.** 52 min
31. 8:06:58; 8:08:19   **33.** 00:00:09   **35.** $104   **37.** 10 hr
39. $150   **41.** 356.79 g   **43.** $29,293.60   **45.** 2.64 sec
Find the Mistake   **1.** Converting 6 feet 6 inches to inches gives
us 78 inches.   **2.** The correct way to write the sum of 2 hours,
55 min and 4 hours, 10 min is 7 hours and 5 minutes.
3. The correct way to subtract 27 minutes from 6 hours and 12
minutes is to borrow 60 minutes from the 6 hours to get 5 hours
and 45 minutes.   **4.** Jane is buying two cups of frozen yogurt.
One cup contains 3 ounces and the other contains 11.9 grams. If
each ounce cost $1.50, then the total purchase price will be $5.13.

Chapter 7 Review
1. 27 ft   **2.** 0.57 km   **3.** 108,900 ft^3 **4.** 5.5 ft^2
5. 16,400 mL   **6.** 8.05 km   **7.** 13.78 qt   **8.** 25° C
9. 92,000 lb   **10.** 8.5 lb   **11.** 6.88 ft^2 **12.** 730 m^2
13. 6.2 yd^3 **14.** 149° F   **15.** 0.40 gal   **16.** 22,572 in.
17. 0.02 ft^3 **18.** 88.68 L   **19.** 151.83 m   **20.** 14.15 L
21. 14.57 in.   **22.** 67.62 km   **23.** 303.15 cm^2 **24.** 56.81 acres
25. 53.77 L   **26.** 21.36 kg   **27.** 54 tiles   **28.** 72 glasses
29. a. 435 min **b.** 7.25 hr   **30.** 16 lb 2 oz
31. DiGiorno 7,136,540,789 pesos; Red Baron 3,093,637,560 pesos
32. Tombstone 5,269,534, 989, 027 dong; Totinos
2,974,315,977,207 dong

Chapter 7 Cumulative Review
1. 11,623   **2.** 1,001   **3.** 27   **4.** 162   **5.** $356\frac{7}{8} = 356.875$
6. 243   **7.** 34   **8.** 9   **9.** 56   **10.** 66   **11.** 15   **12.** 44.85
13. 21.65   **14.** 48.78   **15.** 13.5   **16.** $\frac{1}{576}$ **17.** 27   **18.** $\frac{9}{16}$
19. $2\frac{1}{4}$ **20.** $8\frac{1}{4}$ **21.** 2.1   **22.** 1.075   **23.** $14\frac{2}{3}$ **24.** $x = 4.25$
25. $y = 4.5$ **26.** $x = 21$ **27.** $P = 78$ in, $A = 228$ in^2 **28.** $3\frac{1}{20}$ cm
29. 18   **30.** 64 mi/hr   **31.** 2,064   **32.** $2 \cdot 3^2 \cdot 5 \cdot 7$ **33.** 54
34. 1.6 mi   **35.** 21.34 %   **36.** 39.59 %

Chapter 7 Test
1. 9 ft   **2.** 0.64 km   **3.** 174,240 ft^2 **4.** 6 ft^2 **5.** 12,000 mL
6. 4.83 km   **7.** 8.48 qt   **8.** 32.2° C   **9.** 5,200 lb   **10.** 7 lb
11. 4.5 ft^2 **12.** 630 m^2 **13.** 4.3 yd^3 **14.** 68° F   **15.** 0.53 gal
16. 13,032 in.   **17.** 0.22 ft^3 **18.** 61.32 L   **19.** 114.33 m
20. 10.38 L   **21.** 10.63 in.   **22.** 14.49 km   **23.** 187.05 cm^2
24. 41.99 acres   **25.** 33.96 L   **26.** 10.45 kg   **27.** 120 tiles
28. 192 glasses   **29. a.** 165 min **b.** 2.75 hr   **30.** 7 lb
31. Barq's 0.023 g; Coca-Cola Classic 0.035 g; Dr.Pepper 0.043 g;
Mountain Dew 0.054 g; Pepsi 0.037; Vault 0.07 g   **32.** Barq's
0.0008127 oz; Coca-Cola Classic 0.001237 oz; Dr.Pepper 0.001519 oz;
Mountain Dew 0.001908 oz; Pepsi 0.001307 oz; Vault 0.002509 oz

Chapter 8

Exercise Set 8.1
Vocabulary Review   **1.** polygon   **2.** perimeter
3. circumference   **4.** radius   **5.** diameter
Problems   **1.** 32 in.   **3.** 260 yd   **5.** 36 in.   **7.** $15\frac{3}{4}$ in.
9. 76 in.   **11.** 168 ft   **13.** 18.28 in.   **15.** 25.12 in.
17. 112 in.   **19.** 37.68 in.   **21.** 24,492 mi   **23.** 388 mm

25. 2.36 in.   **27.** 33.62 in.   **29.** 9 in.
31. $w = 1$ and $l = 5$; $w = 2$ and $l = 4$; $w = 3$ and $l = 3$
33. Yes, when $w = l = 5$ ft   **35.** 2 in.   **37.** 9   **39.** $4\frac{3}{8}$
41. 660.19   **43.** 30.96
Find the Mistake   **1.** To find the perimeter of a triangle with side
lengths of 4 inches, 3 inches, and 12 inches, find the sum of the
side lengths.   **2.** A standard piece of paper measures 8.5 inches
by 11 inches and has a perimeter of 39 inches.   **3.** To find the
circumference of a circle with a radius of 7.2 feet, multiply the
diameter by π to get 45.22 feet.   **4.** The circumference of a coin
with a radius of 11.25 mm is 70.65 mm.

Exercise Set 8.2
Vocabulary Review   **1.** area   **2.** square   **3.** parallelogram
4. triangle   **5.** circle
Problems   **1.** 25 cm^2 **3.** 336 m^2 **5.** 60 ft^2 **7.** 2,200 ft^2
9. 945 cm^2 **11.** 50.24 in^2 **13.** 22.28 in^2 **15. a.** 100 in^2 **b.** 314 in^2
17. 133 in^2 **19.** $\frac{4}{9}$ ft^2 **21.** 124 tiles   **23.** 1.18 in^2
25. 8,509 mm^2 **27.** 24 ft^2.   **29.** 551.27 mm^2
31. b.

Perimeters Of Squares		Areas Of Squares	
Length of each side (in cm)	Perimeter (in cm)	Length of each side (in cm)	Area (in cm²)
1	4	1	1
2	8	2	4
3	12	3	9
4	16	4	16

33. 22.6 cm   **35.** 24   **37.** 72   **39.** 4.71   **41.** 547
Find the Mistake   **1.** To find the area of a parallelogram with
a base of 6 inches and a height of 4 inches, you must multiply
the base and height measurements to get an area of 24 square
inches.   **2.** To find the area of a triangle with a base of 6 cm and a
height of 3 cm, use the formula $A = \frac{1}{2}bh$ to get an area of 9 cm^2.
3. To find the area of a circle with a diameter of 12 feet, use the
formula $A = \pi r^2$ to get an area of 113.04 ft^2.   **4.** To find the area
of the shaded region below, subtract the area of a circle from the
area of a square.

Landmark Review
1. $P = 28$ in, $A = 49$ in^2
2. $P = 108$ cm, $A = 729$ cm^2 **3.** $P = 14$ cm, $A = 10$ cm^2
4. $P = 22$ mi, $A = 24$ mi^2 **5.** $P = 25$ yd, $A = 60$ yd^2
6. $P = 24$ ft, $A = 25$ ft^2 **7.** $P = 14$ in, $A = 8$ in^2
8. $P = 24$ m, $A = 28$ m^2 **9.** $C = 31.4$ in, $A = 78.5$ in^2
10. $C = 174.84$ mm, $A = 2,461.76$ mm^2

Exercise Set 8.3
Vocabulary Review   **1.** rectagular solid   **2.** box   **3.** surface area
4. open   **5.** closed   **6.** sphere
Problems   **1.** 96 cm^2 **3.** 378 ft^2 **5.** 270 in^2 **7.** 125.6 ft^2
9. 75.36 ft^2 **11.** 50.24 mi^2 **13.** 197.82 ft^2 **15.** 1,268.95 mm^2
17. 216 m^2 **19. a.** 1.354 in. **b.** 93.5 in^2 **c.** 93.5 in^2, yes
21. 288 ft^2 **23.** 60,288,000 mi^2 **25.** $l = 13$ cm, $s = 282.6$ cm^2
27. 27   **29.** $\frac{2}{3}$ **31.** 0.294   **33.** 785
Find the Mistake   **1.** To find surface area of a rectangular solid,
we add the areas of the top, the bottom, the front, the back, and the
two sides.   **2.** The surface area of this box is 248 in^2.
3. Suppose a soup company needs new labels for their cans.
Each label will wrap around the entire side of the can. To find the
surface area of the label, use the formula $S = 2\pi rh$.
4. The surface area of a sphere with radius 4.2 in is 221.6 in^2.

Exercise Set 8.4

Vocabulary Review   1. volume   **2.** rectangular solid
3. cubic feet   **4.** cylinder   **5.** cone   **6.** sphere
Problems   1. 64 cm³   **3.** 420 ft³   **5.** 162 in³   **7.** 100.48 ft³
9. 50.24 ft³   **11.** 33.49 mi³   **13.** 226.08 ft³   **15.** 113.03 in³
17. 1,102.53 mm³   **19.** 85.27 cm³   **21.** 426.51 in³
23. 49.63 cm³   **25.** 142.72 ft³   **27.** 46,208 cm³
Find the Mistake   1. The volume of a rectangular solid with a
length of 4 mm, a height of 9 mm, and a width of 4 mm is found by
multiplying 4 · 9 · 4 to get 144 mm³.   **2.** The volume of a right
circular cylinder is found by using the formula $V = \pi r^2 h$.
3. Finding the volume of a cone with a radius of 7 cm and a height
of 10 cm would give $V = 512.87$ cm³.   **4.** To find the volume of a
tennis ball, use the formula for the volume of a sphere $V = \frac{4}{3}\pi r^3$.

Chapter 8 Review

1. 40 ft   **2.** 25.8 ft   **3.** 90 yd   **4.** 50.24 in.   **5.** 36 cm
6. 121 cm²   **7.** 66 m²   **8.** 39 ft²   **9.** 118.88 m²   **10.** 113 mm²
11. 358 m²   **12.** 282.6 ft²   **13.** 678.2 m²   **14.** 314 in²
15. 276 cm³   **16.** 412.22 ft³   **17.** 2,308.4 cm³   **18.** 732.7 in³
19. $V = 216$ ft³, $A = 228$ ft²   **20.** $52\frac{1}{2}$ ft   **21.** 540 mm³

Chapter 8 Cumulative Review

1. 5,945   **2.** 13,056   **3.** 24.6   **4.** 15.25   **5.** $\frac{1}{3}$ **6.** $\frac{25}{81}$
7. 13.813   **8.** 8.286   **9.** 2.142   **10.** $\frac{3}{8}$ **11.** $\frac{4}{3} = 1\frac{1}{3}$
12. $\frac{37}{48}$ **13.** $\frac{35}{16} = 2\frac{3}{16}$ **14.** 9   **15.** 54   **16.** 0.75
17. 0   **18.** 96   **19.** 12   **20.** 30   **21.** $\frac{21}{2} = 10\frac{1}{2}$ **22.** 24
23. $\frac{3}{2} = 1\frac{1}{2}$ **24.** $3(4) - 7 = 5$ **25.** $\frac{5}{4}$ **26.** 10.2 mi/hr
27. 1,528.7 ha   **28.** 20%   **29.** $\frac{13}{20}$ **30.** $\frac{13}{5} = 2\frac{3}{5}$
31. $60{,}000 + 5{,}000 + 200 + 9$ **32.** 228   **33.** 37.5%   **34.** 18
35. 26.3 km/hr   **36.** 146.7% more   **37.** $189.88
38. $P = 34.5$ cm, $A = 36$ cm²   **39.** $P = 52$ ft, $A = 153$ ft²

Chapter 8 Test

1. 52 in   **2.** 85 m   **3.** 86 ft   **4.** 75.36 yd   **5.** 108 mm
6. 256 in²   **7.** 112 ft²   **8.** 138 ft²   **9.** 38.72 yd²   **10.** 531 m²
11. 622 cm²   **12.** 527.52 ft²   **13.** 967.12 ft²   **14.** 50.24 cm²
15. 276.5 ft³   **16.** 378 ft³   **17.** 4,574.9 cm³   **18.** 1,130.4 in³
19. $V = 105$ yd³, $SA = 142$ yd²   **20.** $66\frac{1}{3}$ ft   **21.** 1,252 mm³

Chapter 9

Exercise Set 9.1

Vocabulary Review   1. solution set   **2.** equivalent
3. addition   **4.** subtraction
Problems   1. Yes   **3.** Yes   **5.** No   **7.** Yes   **9.** No
11. Yes   **13.** 11   **15.** 4   **17.** $-\frac{3}{4}$ **19.** -5.8 **21.** -17
23. $-\frac{1}{8}$ **25.** -4 **27.** -3.6 **29.** 1   **31.** $-\frac{7}{45}$ **33.** 3
35. $\frac{11}{8}$ **37.** 21   **39.** 7   **41.** 3.5   **43.** 22   **45.** -2 **47.** -16
49. -3 **51.** 10   **53.** -12 **55.** 4   **57.** 2   **59.** -5
61. -1 **63.** -3 **65.** 8   **67.** -8 **69.** 2   **71.** 11
73. a. 280 tickets **b.** $2,100   **75.** 67°   **77. a.** 225 **b.** $11,125
79.

t	n
32	72
50	90
44	84
96	136

81. $x + 55 + 55 = 180$; 70°   **83.** y
85. x **87.** 6   **89.** 6   **91.** -9
93. $-\frac{15}{8}$ **95.** 8   **97.** $-\frac{5}{4}$
99. $3x$ **101.** $-9x$

Find the Mistake   1. The number -2 is a solution to the equation
$3x - 9 = 8x + 1$.   **2.** Saying that the number 12 is a solution to
the equation $\frac{1}{6}y + 3 = 5$ is a true statement.   **3.** The addition
property of equality states that $A + C = B + C$ if $A = B$.
4. To solve the equation $\frac{5}{6}b + 6 = 11$ for b, begin by adding -6 to
both sides.

Exercise Set 9.2

Vocabulary Review   1. nonzero   **2.** multiplication
3. division   **4.** reciprocal
Problems   1. 2   **3.** 4   **5.** $-\frac{1}{2}$ **7.** -2 **9.** 3   **11.** 4
13. 0   **15.** 0   **17.** 6   **19.** -50 **21.** $\frac{3}{2}$ **23.** 12
25. -3 **27.** 32   **29.** -8 **31.** $\frac{1}{2}$ **33.** 4   **35.** 8
37. -12 **39.** -3 **41.** -4 **43.** 4   **45.** -15 **47.** $-\frac{1}{2}$
49. 3   **51.** 1   **53.** $\frac{1}{4}$ **55.** -3 **57.** 3   **59.** 2
61. $-\frac{3}{2}$ **63.** $-\frac{3}{2}$ **65.** 1   **67.** 1   **69.** -2 **71.** -2
73. 200 tickets   **75.** $1,390.85 per month   **77.** 31
79. 21   **81.** 20 songs   **83. a.** 20 ft. **b.** 35 ft. **c.** 75 ft.
85. 55 hours   **87.** 2   **89.** 6   **91.** 3,000   **93.** 650
95. $3x - 11$ **97.** $0.09x + 180$ **99.** $-6y + 4$ **101.** $4x - 11$
Find the Mistake   1. The multiplication property of equality states
that multiplying both sides of an equation by the same nonzero
number will not change the solution set.   **2.** To solve $-\frac{2b}{9} = 3$ for
b, we multiply both sides of the equation by the reciprocal of $-\frac{2}{9}$.
3. To begin solving the equation $\frac{5}{6}x + \frac{1}{2} = -\frac{1}{4}$, we can either add
$-\frac{1}{2}$ to each side of the equation or multiply both sides by the LCD 12.
4. To solve the equation $-\frac{8}{9}y = -3$, we begin by multiplying both
sides of the equation by 9.

Exercise Set 9.3

Vocabulary Review 1. a. distributive **b.** LCD, decimals **c.** similar
2. addition, constant   **3.** multiplication   **4.** solution
Problems   1. 3   **3.** -2 **5.** -1 **7.** 2   **9.** -4 **11.** -2
13. 0   **15.** 1   **17.** $\frac{1}{2}$ **19.** 7   **21.** 8   **23.** All real numbers
25. $\frac{3}{4}$ **27.** 75   **29.** 2   **31.** 6   **33.** 8   **35.** 0   **37.** $\frac{3}{7}$
39. No solution   **41.** 1   **43.** -1 **45.** 6   **47.** $\frac{3}{4}$
49. 3   **51.** No solution   **53.** 8   **55.** 6   **57.** -2
59. -2 **61.** 2   **63.** All real numbers   **65.** 2   **67.** 20
69. 4,000   **71.** 700   **73.** 11   **75.** 7
77. a. $\frac{5}{4} = 1.25$ **b.** $\frac{15}{2} = 7.5$ **c.** $6x + 20$ **d.** 15 **e.** $4x - 20$ **f.** $\frac{45}{2} = 22.5$
79. $x = 3$ or 3 robot events were entered   **81.** $4,200
83. 14   **85.** -3 **87.** $\frac{1}{4}$ **89.** $\frac{1}{3}$ **91.** $-\frac{3}{2}x + 3$
93. $\frac{5}{4}x - 5$ **95.** $4x - 3$ **97.** $8x + 3$
Find the Mistake   1. A linear equation in one variable is any
equation that can be put in the form of $Ax + B = C$.
2. $12 + 6y - 4 = 3y$ is a linear equation simplified to $3y + 8 = 0$.
3. To solve the linear equation $3(2x + 7) = 2(x - 1)$, begin by
applying the distributive property.   **4.** To solve a linear equation,
make sure to combine similar terms on each side.

Exercise Set 9.4

Vocabulary Review   1. variable   **2.** formula   **3.** isolate
4. complementary   **5.** supplement
Problems   1. 100 feet   **3.** 0   **5.** 2   **7.** 15   **9.** 10
11. -2 **13.** 1   **15. a.** 2 **b.** 4   **17. a.** 5 **b.** 18
19. $l = \frac{A}{w}$ **21.** $h = \frac{V}{lw}$ **23.** $a = P - b - c$
25. $x = 3y - 1$ **27.** $y = 3x + 6$ **29.** $y = -\frac{2}{3}x + 2$
31. $y = -2x - 5$ **33.** $y = -\frac{2}{3}x + 1$ **35.** $w = \frac{P - 2l}{2}$
37. $v = \frac{h - 16t^2}{t}$ **39.** $h = \frac{A - \pi r^2}{2\pi r}$ **41.** $t = \frac{I}{pr}$

43. a. $y = \frac{3}{5}x + 1$ **b.** $y = \frac{1}{2}x + 2$ **c.** $y = 4x + 3$
45. a. $y = -\frac{1}{2}x - 2$ **b.** $y = -3x + 2$ **c.** $y = -\frac{1}{3}x - 1$
47. $y = \frac{3}{7}x - 3$ **49.** $y = 2x + 8$ **51.** $60°; 150°$
53. $45°; 135°$ **55.** 10   **57.** 240   **59.** 25%   **61.** 35%
63. 64   **65.** 2,000   **67.** $100°C$; yes   **69.** $20°C$; yes
71. $C = \frac{5}{9}(F - 32)$ **73.** 46.3%   **75.** 5.3%   **77.** 7 meters
79. $\frac{3}{2}$ or 1.5 inches   **81.** 132 feet   **83.** $\frac{2}{9}$ centimeters
85.

Pitcher, Team	Rolaids Points
Heath Bell, San Diego	144
Brian Wilson, San Francisco	135
Carlos Marmol, Chicago	107
Billy Wagner, Atlanta	128
Francisco Cordero, Cincinnati	130

87. The sum of 4 and 1 is 5.   **89.** The difference of 6 and 2 is 4.
91. The difference of a number and 5 is -12.
93. The sum of a number and 3 is four times the difference of that
number and 3.   **95.** $2(6 + 3) = 18$ **97.** $2(5) + 3 = 13$
99. $x + 5 = 13$ **101.** $5(x + 7) = 30$
Find the Mistake   **1.** A formula contains more than one variable.
2. To solve for a variable in a formula, make sure the variable
is isolated on one side of the equation.   **3.** To solve the
formula $2ab = a + 8$ for b, divide each side of the equation by $2a$.
4. The question "What percent of 108 is 9?" can be translated into
the equation $x \cdot 108 = 9$.

Landmark Review   **1.** 12   **2.** 25   **3.** 5.2   **4.** $\frac{13}{15}$ **5.** 4
6. 10   **7.** 4   **8.** 3   **9.** 3   **10.** 4   **11.** $\frac{A}{w} = l$
12. $P - a - b = c$ **13.** $18 - 3y = x$ **14.** $-\frac{4}{3}x + 3 = y$

Exercise Set 9.5
Vocabulary Review   **1.** known   **2.** unknown   **3.** variable
4. equation   **5.** sentence   **6.** problem
Problems   **1.** 8   **3.** 5   **5.** -1 **7.** 3 and 5   **9.** 6 and 14
11. Shelly is 39; Michele is 36   **13.** Evan is 11; Cody is 22
15. Barney is 27; Fred is 31   **17.** Lacy is 16; Jack is 32
19. Patrick is 18; Pat is 38   **21.** $s = 9$ inches   **23.** $s = 15$ feet
25. 11 feet, 18 feet, 33 feet   **27.** 26 feet, 13 feet, 14 feet
29. $l = 11$ inches; $w = 6$ inches   **31.** $l = 25$ meters; $w = 9$ meters
33. $l = 15$ feet; $w = 3$ feet   **35.** 9 dimes; 14 quarters
37. 12 quarters; 27 nickels   **39.** 8 nickels; 17 dimes
41. 7 nickels; 10 dimes, 12 quarters   **43.** 3 nickels; 9 dimes, 6 quarters
45. $5x$ **47.** $1.075x$ **49.** $0.09x + 180$
Find the Mistake   **1.** False, $7 + \frac{m}{n} = n + 9$ **2.** True
3. False, $2(3x + 6) + 2x = 42$ **4.** False, $10(2x - 4) + 5x = 310$

Exercise Set 9.6
Vocabulary Review   **1.** consecutive   **2.** even   **3.** integers
4. sum
Problems   **1.** 5 and 6   **3.** -4 and -5 **5.** 13 and 15   **7.** 52 and 54
9. -14 and -16 **11.** 17, 19, and 21   **13.** 42, 44, and 46
15. $4,000 at 8%, $6,000 at 9%   **17.** $700 at 10%; $1,200 at 12%
19. $500 at 8%, $1,000 at 9%, $1,500 at 10%   **21.** $45°, 45°, 90°$
23. $22.5°, 45°, 112.5°$ **25.** $53°, 90°$ **27.** $80°, 60°, 40°$
29. 16 adult and 22 children's tickets   **31.** 16 minutes
33. 39 hours   **35.** They are in offices 7329 and 7331.
37. Kendra is 8 years old and Marissa is 10 years old.   **39.** Jeff
41. $10.38   **43.** $l = 12$ meters; $w = 10$ meters   **45.** $59°, 60°, 61°$
47. 18 cm, 24 cm, 36 cm   **49.** $54.00   **51.** Yes   **53.** 5

55. a. 9 **b.** 3 **c.** -9 **d.** -3 **57. a.** -8 **b.** 8 **c.** 8 **d.** -8
59. -2.3125 **61.** $\frac{10}{3}$ **63.** $y = \frac{2}{3}x - 2$ **65.** $x = \frac{2}{3}y + \frac{5}{3}$
Find the Mistake   **1.** True   **2.** False, $0.07(4x) + 0.08x = 396$
3. True   **4.** True

Exercise Set 9.7
Vocabulary Review   **1.** inequalities   **2.** positive   **3.** negative
4. graph   **5.** compound
Problems
1. $x < 12$

3. $a \leq 12$

5. $x > 13$

7. $y \geq 4$

9. $x > 9$

11. $x < -6$

13. $x < 2$

15. $a \leq 5$

17. $x > 15$

19. $x < -3$

21. $x \leq 6$

23. $x \geq -50$

25. $y < -6$

27. $x < -4$

29. $x > -12$

31. $x < 6$

33. $y \geq -5$

37. $x \leq 18$

39. $a < -20$

41. $y < 25$

43. $a \leq 3$

45. $x \geq \frac{15}{2}$

47. $x < -1$

49. $y \geq -2$

51. $x < -1$

53. $m \leq -6$

55. $x \leq -5$

57. $x \leq 3$

59. $x \leq \frac{25}{9}$

61. $y < -\frac{3}{2}x + 3$ **63.** $y < \frac{2}{5}x - 2$ **65.** $y \leq \frac{3}{7}x + 3$ **67.** $y \leq \frac{1}{2}x + 1$
69. a. 3 **b.** 2 **c.** No **d.** $x > 2$ **71.** $x < 3$ **73.** $x \leq 3$

75. $x \leq -7$ or $x \geq -3$

77. $x \leq -1$ or $x \geq \frac{3}{5}$

79. $x < -10$ or $x > 6$

81. $x < -3$ or $x \geq 2$

83. $3 \leq m \geq 7$

85. $-3 < a < 0$

87. $4 \leq a \leq 6$

89. $-4 < x < 2$

91. $-3 < x < 3$

93. $0 \leq x < 6$

95. $-\frac{2}{3} \leq x \leq 0$

97. $-\frac{3}{2} \leq x \leq \frac{17}{2}$

99. $\frac{3}{8} \leq x \leq \frac{3}{2}$

101. At least 291 **103.** $x < 2$ **105.** $x > -\frac{8}{3}$
107. $x \geq 6$; the width is at least 6 meters. **109.** $x > 6$; the shortest side is even and greater than 6 inches. **111.** $t \geq 100$
113. Lose money if they sell less than 200 tickets. Make a profit if they sell more than 200 tickets. **115.** 8 **117.** 24 **119.** 25%
121. 10% **123.** 80 **125.** 400 **127.** -5 **129.** 5 **131.** 7
133. 9 **135.** 6 **137.** $2x - 3$

Find the Mistake **1.** If the solution set to an inequality is all the real numbers greater than 9, the solution is $x > 9$. **2.** To include the number 14 in the solution to the inequality $x \geq 14$, we mark the number on the graph with a closed circle. **3.** The last step in solving an inequality is to graph the solution set.
4. Upon solving the inequality $-5y < 4x + 9$ for y, the inequality symbol is reversed.

Chapter 9 Review

1. No **2.** Yes **3.** Yes **4.** No **5.** 15 **6.** 11
7. 16 **8.** $\frac{1}{2}$ **9.** 6 **10.** $-\frac{3}{4}$ **11.** 5 **12.** 4 **13.** 372
14. 3 **15.** -2 **16.** 7 **17.** 10 **18.** $-\frac{3}{4}$ **19.** 0
20. $y = \frac{3}{2}x - 3$ **21.** $T_1 = \frac{9}{k}x + T_2$ **22.** Brian is 31, Michelle is 27
23. $l = 15$ in., $w = 7$ in. **24.** 3 quarters, 4 dimes
25. \$6,000 at 7% and \$7,500 at 15% **26.** $x \leq -12$ **27.** $y < -6$
28. $x > -4$ **29.** $n \geq 2$ **30.** $x < 4$ **31.** $x \leq 7$ **32.** $x \geq -5$
33. $x \leq 7$ **34.** $x > -3$ **35.** 23.8% **36.** 7.4%

Chapter 9 Cumulative Review

1. 5,044 **2.** $\frac{23}{28}$ **3.** 10.519 **4.** $2\frac{1}{2}$ **5.** 2,026,794
6. 0.84 **7.** 36 **8.** $\frac{9}{14}$ **9.** 540,000 **10.** $\frac{9}{25}$ **11.** $7\frac{5}{8}$
12. 0.265 **13.** 37% **14.** $1\frac{9}{25}$ **15.** 35% **16.** $\frac{5}{64}$
17. $12x + 3$ **18.** -9 **19.** $-\frac{1}{3}$ **20.** 5 **21.** -3 **22.** 42
23. $-\frac{47}{18}$ **24.** 6 **25.** $\frac{1}{9}$ **26.** 13 **27.** 98 in²
28. Ben is 31, Ryan is 26 **29.** 23 mpg **30.** \$140, 25%
31. 5 m **32.** 60 **33.** \$468 **34.** 128 **35.** 8,862, 18.33%
36. 124,000; 44,000; $\frac{31}{11}$

Chapter 9 Test

1. No **2.** Yes **3.** Yes **4.** No **5.** 17 **6.** 5
7. 5.2 **8.** 3 **9.** −5 **10.** 1 **11.** 4 **12.** 1 **13.** 55
14. −3 **15.** $\frac{5}{2}$ **16.** 2 **17.** 5 **18.** −1 **19.** 8
20. $y = -\frac{1}{3}x + 2$ **21.** $a = \frac{x^2 - v^2}{2d}$ **22.** Paul is 36, Becca is 18
23. $l = 55$ cm, $w = 20$ cm **24.** 14 dimes, 6 nickels
25. $700 at 6% and $1,200 at 12% **26.** $x > 10$ **27.** $y \geq -4$
28. $x > 4$ **29.** $n \geq -2$ **30.** $x < 1$ **31.** $x \geq 2$ **32.** $x > -3$
33. $x \geq -1$ **34.** $x < -2$ **35.** 27.9% **36.** 20.4%

Chapter 10

Exercise Set 10.1

Vocabulary Review **1.** scatter diagram/plot **2.** line graph
3. ordered pair **4.** x–axis, y–axis **5.** quadrants **6.** origin
Problems
1.

3-19.

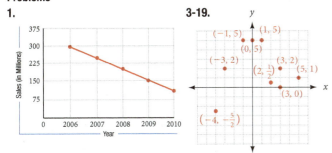

21. $(-4, 4)$; QII **23.** $(-4, 2)$; QII **25.** $(-3, 0)$; no quadrant, x-axis
27. $(2, -2)$; QIV **29.** $(-5, -5)$; QIII

31. Yes **33.** No **35.** Yes **37.** No

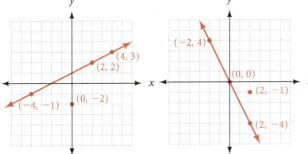

39. Yes **41.** No **43.** No **45.** No

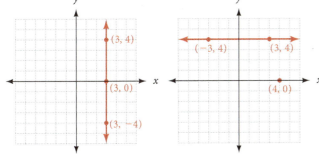

47. a. $(5, 40), (10, 80), (20, 160)$, Answers may vary **b.** $320
c. 30 hours **d.** No, if she works 35 hours, she should be paid
$280.
49. $(1999, 70), (2001, 73), (2003, 68), (2005, 75), (2007, 80) , (2009, 73)$
51. $A = (1, 2), B = (6, 7)$ **53.** $A = (2, 2), B = (2, 5), C = (7, 5)$
55. a. −3 **b.** 6 **c.** 0 **4.** −4 **57. a.** 4 **b.** 2 **c.** −1 **d.** 9

59. −3 **61.** 2 **63.** 0 **65.** $y = -5x + 4$ **67.** $y = \frac{3}{2}x - 3$
Find the Mistake **1.** A scatter diagram uses dots to show data in
the form of order pairs. **2.** In the ordered pair $(-3, 6)$, the
x-coordinate is −3. **3.** To graph the ordered pair $(7, 3)$, start
the origin, move 7 units to the right, and 3 units up. **4.** The
ordered pair $(-2, 5)$ appears in the second quadrant.

Exercise Set 10.2

Vocabulary Review **1.** first **2.** second **3.** solution
4. graphing **5.** linear **6.** ordered pairs **7.** graph
8. straight line **9.** horizontal, vertical
Problems **1.** $(0, 6), (3, 0), (6, -6)$ **3.** $(0, 3), (4, 0), (-4, 6)$
5. $(1, 1), \left(\frac{3}{4}, 0\right), (5, 17)$ **7.** $(2, 13), (1, 6), (0, -1)$
9. $(-5, 4), (-5, -3), (-5, 0)$

11.

x	y
1	3
−3	−9
4	12
6	18

13.

x	y
0	0
$-\frac{1}{2}$	−2
−3	−12
3	12

15.

x	y
2	3
3	2
5	0
9	−4

17.

x	y
2	0
3	2
1	−2
−3	−10

19.

x	y
0	−1
−1	−7
−3	−19
$\frac{3}{2}$	8

21.

x	y
0	3
−2	7
2	−1
5	−7

23. $(0, -2)$ **25.** $(1, 5), (0, -2),$ and $(-2, -16)$
27. $(1, 6)$ and $(0, 0)$ **29.** $(2, -2)$ **31.** $(3, 0)$ and $(3, -3)$

33. $(0, 4), (2, 2), (4, 0)$ **35.** $(0, 3), (2, 1), (4, -1)$

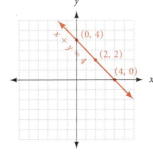 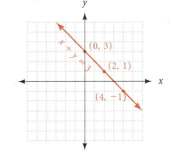

37. $(0, 0), (-2, -4), (2, 4)$ **39.** $(-3, -1), (0, 0), (3, 1)$

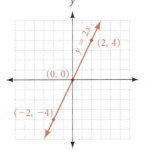 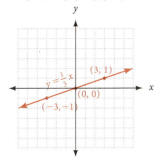

41. $(0, 1), (-1, -1), (1, 3)$

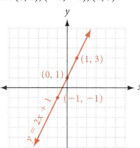

43. $(0, 4), (-1, 4), (2, 4)$

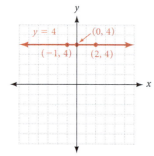

45. $(-2, 2), (0, 3), (2, 4)$

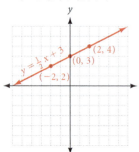

47. $(-3, 3), (0, 1), (3, -1)$

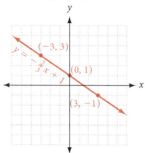

49. $(-1, 5), (0, 3), (1, 1)$

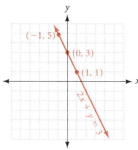

51. $(0, 3), (2, 0), (4, -3)$

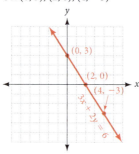

53. $(-2, 2), (0, 3), (2, 4)$

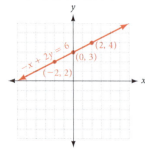

55.

57.

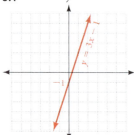

59.

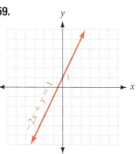

61.

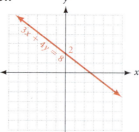

63.

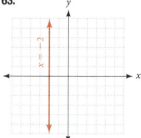

65.

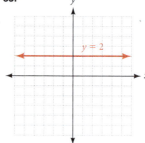

67.

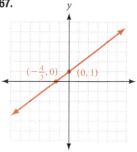

69.

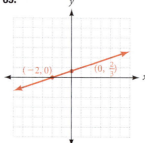

71.

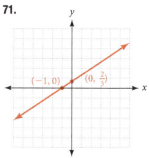

73.

Equation	H, V, and/or O
$x = 3$	V
$y = 3$	H
$y = 3x$	O
$y = 0$	O, H

75.

Equation	H, V, and/or O
$x = -\frac{3}{5}$	V
$y = -\frac{3}{5}$	H
$y = -\frac{3}{5}x$	O
$x = 0$	O, V

77.

x	y
-4	-3
-2	-2
0	-1
2	0
6	2

79. 12 inches

81. a. Yes **b.** No, she should earn $108 for working 9 hours. **c.** No, she should earn $84 for working 7 hours. **d.** Yes

83. a. $375,000 **b.** At the end of 6 years. **c.** No, the crane will be worth $195,000 after 9 years. **d.** $600,000

85. a. 2 **b.** 3 **87. a.** -4 **b.** 2

89. a. 6 **b.** 2

Find the Mistake   **1.** A solution to a linear equation in two variable will be an ordered pair that makes the equation a true statement.   **2.** A solution to the equation $3y = 6x + 9$ is $(2, 7)$.

3. To graph a linear equation in two variables, you must find two or more solutions.   **4.** The line $y = \frac{1}{2}x$ intersects both axes, or passes through the origin.

Exercise Set 10.3

Vocabulary Review **1.** x-intercept   **2.** y-intercept

3. x-axis   **4.** y-axis

Problems

1.

Equation	x-intercept	y-intercept
$3x + 4y = 12$	4	3
$3x + 4y = 4$	$\frac{4}{3}$	1
$3x + 4y = 3$	1	$\frac{3}{4}$
$3x + 4y = 2$	$\frac{2}{3}$	$\frac{1}{2}$

3.

Equation	x-intercept	y-intercept
$x - 3y = 2$	2	$-\frac{2}{3}$
$y = \frac{1}{3}x - \frac{2}{3}$	2	$-\frac{2}{3}$
$x - 3y = 0$	0	0
$y = \frac{1}{3}x$	0	0

5.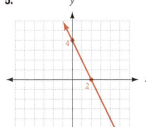

7.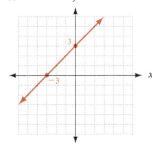

9.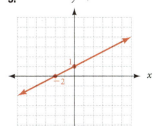

11.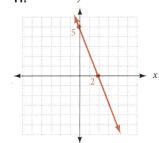

13.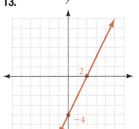

15.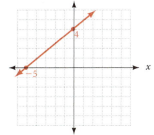

17.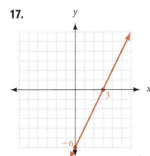

19.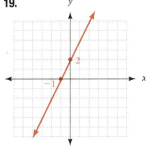

21.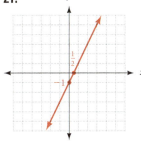

23.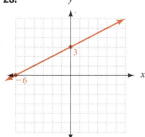

25.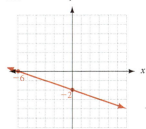

27.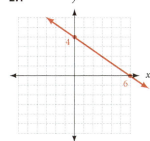

29.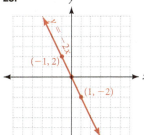

31.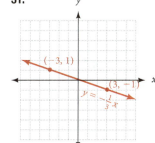

33.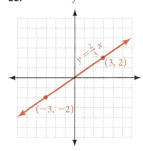

35. **a.** 0   **b.** $-\frac{3}{2}$ **c.** 1
d.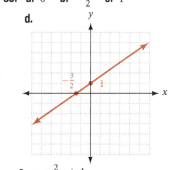

e. $y = \frac{2}{3}x + 1$

37. x-intercept $= 3$; y-intercept $= 5$
39. x-intercept $= -1$; y-intercept $= -3$

41.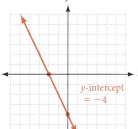

43.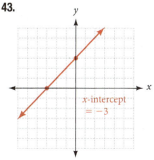

45.

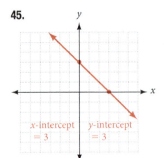

47.

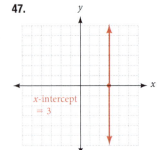

5.

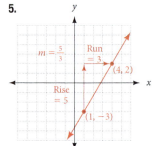

7.

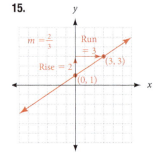

49.

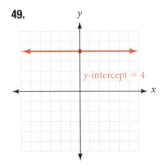

51.

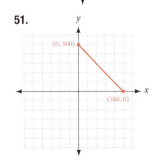

9.

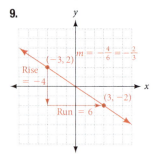

11.

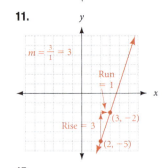

53. a. $10x + 12y = 480$
  b. x-intercept = 48;
    y-intercept = 40
  c.

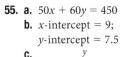

  d. 10 hours **e.** 18 hours

55. a. $50x + 60y = 450$
  b. x-intercept = 9;
    y-intercept = 7.5
  c.

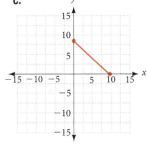

  d. 5 hours **e.** 4.17 hours

13.

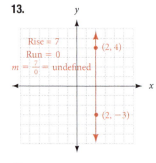

15.

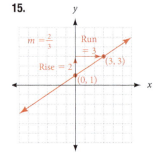

57. a. $\frac{3}{2}$ **b.** $\frac{3}{2}$ **59. a.** $\frac{3}{2}$ **b.** $\frac{3}{2}$

Find the Mistake **1.** The y-intercept is where the graph of a line crosses the y-axis at the point $(0, y)$. **2.** The x-intercept for the linear equation $9x = \frac{1}{3}y + 5$ is $\left(\frac{5}{9}, 0\right)$. **3.** The y-intercept for the linear equation $2x + 7y = 3$ is $\left(0, \frac{3}{7}\right)$. **4.** The intercepts for the linear equation $15 = 3y - 5x$ are $(-3, 0)$ and $(0, 5)$.

17.

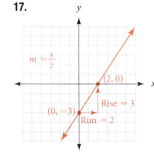

19.

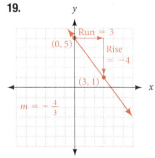

Exercise Set 10.4

Vocabulary Review **1.** ratio **2.** vertical **3.** horizontal
4. slope **5.** zero **6.** undefined
Problems **1.**

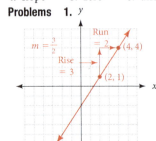

3.

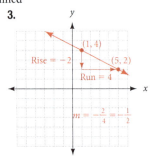

21.

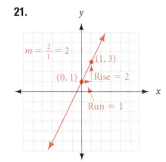

23.

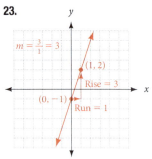

25.

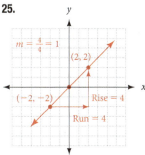

27. Slope = 3; y-intercept = 2
29. Slope = 2; y-intercept = -2
31.

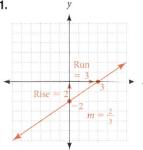

33.

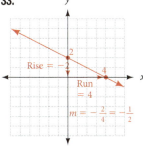

35.

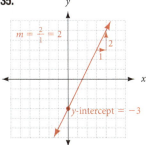

37.

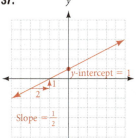

39. 6   **41.** All real numbers

43.

Equation	Slope
$x = 3$	no slope
$y = 3$	0
$y = 3x$	3

45.

Equation	Slope
$y = -\frac{2}{3}$	0
$x = -\frac{2}{3}$	no slope
$y = -\frac{2}{3}x$	$-\frac{2}{3}$

47. Slopes: A, 3.3; B, 3.1, C, 5.3, D, 1.9
49. Slopes: A, -50; B, -75, C, -25

51. a.

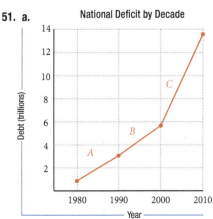

b. 0.23, 0.25, 0.79
c. The rate of change is increasing

53. 878 finishers. The number of finishers increased by 878 in one year.
55. a. 15,800 feet   **b.** $-\frac{7}{100}$
57. 0.64% per year. Between 2003 and 2010 the percentage of computer users who use Macs has increased an average of 0.64% per year.
59. $y = 4x - 2$ **61.** $y = -\frac{3}{2}x + 3$ **63.** $y = \frac{1}{3}x - 2$

65. $y = -\frac{3}{4}x + 3$ **67.** $y = -\frac{2}{5}x - 2$ **69.** $y = -2x + 1$
71. $y = \frac{1}{2}x + 1$ **73.** $y = -\frac{3}{2}x - 1$

Find the Mistake   1. A horizontal line has zero slope.
2. A line that rises from the left to the right has a positive slope.
3 The slope of the line that passes through the points $(-1, 6)$ and $(4, -9)$ is -3.   **4.** Use the point $(2, -1)$ to graph a line with a y-intercept -6 and a slope $\frac{5}{2}$.

Landmark Review   1. The origin   **2.** 4   **3.** No
4. $(0, -3), \left(\frac{3}{4}, 0\right), (3, 9)$ **5.** $(-3, 3), \left(1, \frac{21}{2}\right), \left(6, \frac{33}{2}\right)$
6. x-intercept = 1; y-intercept = 1
7. x-intercept = 6; y-intercept = -3 **8.** $\frac{1}{4}$ **9.** -2
10. Undefined slope   **11.** Rise   **12.** There isn't one

Exercise Set 10.5
Vocabulary Review   1. slope-intercept **2.** point-slope **3.** substitute
4. y-intercept   **5.** slope   **6.** parallel   **7.** perpendicular
Problems   1. $y = \frac{2}{3}x + 1$ **3.** $y = \frac{3}{2}x - 1$ **5.** $y = -\frac{2}{3}x + 3$
7. $y = 2x - 4$

9. $m = 2; b = 4$

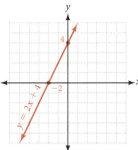

11. $m = -3; b = 3$

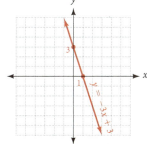

13. $m = -\frac{3}{2}; b = 3$

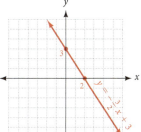

15. $m = \frac{4}{5}; b = -4$

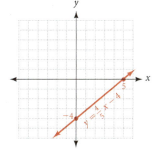

17. $m = -\frac{2}{5}; b = -2$

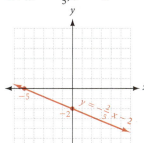

19. $m = \frac{3}{2}; b = -3$

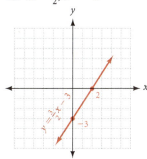

21. $y = 2x - 1$ **23.** $y = -\frac{1}{2}x - 1$ **25.** $y = \frac{3}{2}x - 6$
27. $y = -3x + 1$ **29.** $y = -4$ **31.** $x = -4$ **33.** $y = x - 2$
35. $y = 2x - 3$ **37.** $y = \frac{4}{3}x + 2$ **39.** $y = -\frac{2}{3}x - 3$
41. $y = 2$ **43.** $x = -5$ **45.** $m = 3, b = 3, y = 3x + 3$

47. $m = \frac{1}{4}, b = -1, y = \frac{1}{4}x - 1$

49. a. $-\frac{5}{2}$ **b.** $y = 2x + 6$

   c. 6   **d.** 2

   e.

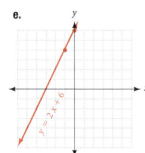

51. $y = -\frac{2}{3}x + 2$

53. $y = -\frac{5}{2}x - 5$ **55.** $x = 3$

57. Any given line by $y = 4x \pm a$

59. $m = -1$ **61.** $m = -2$

63. a. $6,000

   b. 3 years

   c. Slope $= -3,000$

   d. $3,000

   e. $V = -3,000t + 21,000$

65. a. $3,000 **b.** 500 shirts **c.** Slope $= 2$ **d.** $2 **e.** $y = 2x + 1,000$

Find the Mistake   **1.** The equation $y = -\frac{4}{7}x + 2$ is in slope–intercept form.   **2.** For the equation $-8x + 2y = -4$, the slope is 4 and the y–intercept is -2.   **3.** The line represented by the equation $(y - 2) = 6(x - 4)$ has a slope of 6.   **4.** The lines represented by the equation $y = \frac{7}{2}x + \frac{1}{3}$ and $y = -\frac{2}{7}x - 5$ are perpendicular.

Exercise Set 10.6

Vocabulary Review   **1.** linear inequality   **2.** solution set
3. boundary   **4.** solid   **5.** broken
Problems

1.

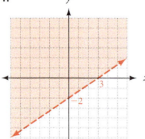

3.

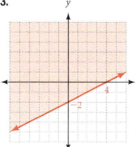

5.

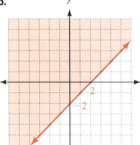

7.

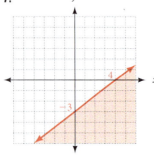

9.

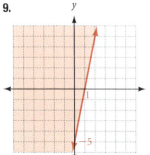

11.

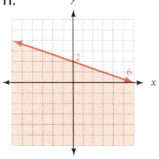

13.

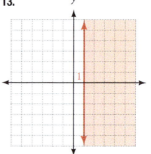

15.

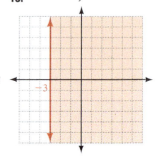

17.

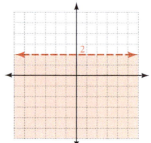

19.

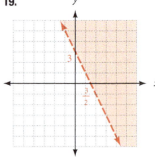

21.

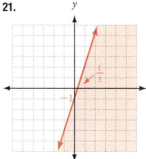

23.

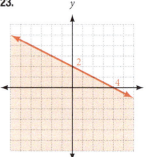

25. a. $y < \frac{8}{3}$ **b.** $y > -\frac{8}{3}$

   c. $y = -\frac{4}{3}x + 4$

   d.

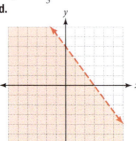
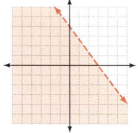

27. a. $y = \frac{2}{5}x + 2$

   b. $y < \frac{2}{5}x + 2$

   c. $y > \frac{2}{5}x + 2$

29. $-6x + 11$ **31.** -8 **33.** -4 **35.** $w = \frac{P - 2l}{2}$

37.

39.

41. $y \geq \frac{3}{2}x - 6$ **43.** width $= 2$ inches; length $= 11$ inches

Find the Mistake   **1.** The solution set for the linear inequality $y < 6x - \frac{3}{5}$ does not contain the points through which the boundary line passes.   **2.** The boundary for the linear inequality $2 - 8y \geq x + 6$ is represented by a solid line.   **3.** If a test point substituted into an original inequality results in a false statement, the solution set for the inequality lies on the opposite.   **4.** The solution set for the linear inequality $x < -8$, includes all points to the left of the line.

Chapter 10 Review

1-4.

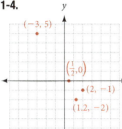

5. $\left(3, -\frac{4}{5}\right), \left(-2, -\frac{14}{5}\right)$
$\left(\frac{27}{3}, \frac{2}{3}\right), \left(\frac{1}{2}, -\frac{9}{5}\right)$

6. $(10, 0), (8, -1.6)$

7.

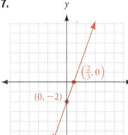

8.

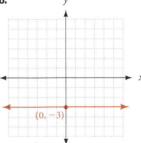

9. x-intercept $= 6$; y-intercept $= -1$

10. x-intercept $= \frac{20}{3}$; y-intercept $= -2$

11. x-intercept $= -2$; y-intercept $= 2$

12. x-intercept $= -2$; y-intercept $= -2$

13. $-\frac{1}{3}$ **14.** 2 **15.** -2 **16.** 0 **17.** Undefined slope

18. 2 **19.** $y = 2x - 7$ **20.** $y = -3x + 6$ **21.** $y = \frac{1}{5}x + \frac{8}{5}$

22. $y = -\frac{6}{5}x + \frac{18}{5}$ **23.** $y = -\frac{1}{2}x + 2$ **24.** $m = 4, b = -3$

25. $m = \frac{4}{5}, b = -2$ **26.** $y = 3x - 4$ **27.** $y = \frac{3}{2}x + 8$

28.

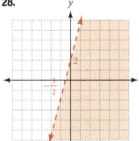

29.

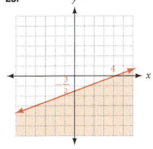

Chapter 10 Cumulative Review

1. -1 **2.** -1 **3.** 2 **4.** -8 **5.** 32 **6.** -40

7. Undefined **8.** $\frac{1}{3}$ **9.** $\frac{5}{7}$ **10.** 0 **11.** $36x$

12. $3a - 6$ **13.** -2 **14.** -4 **15.** 1 **16.** 4 **17.** $-\frac{3}{2}$

18. -98 **19.** $x > -6$ **20.** $x \geq 7$

21.

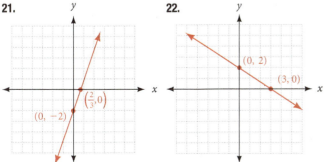

22.

23.

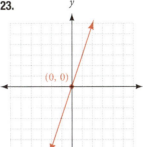

24.

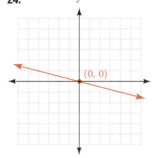

25.

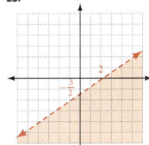

26.

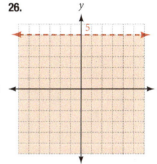

27. No, yes **28.** y-intercept $= 6$

29. x-intercept $= 4$; y-intercept $= 6$

30.

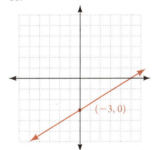

31. $m = -1$ **32.** $y = -2x + 5$

33. $m = 3$ **34.** $y = \frac{2}{3}x$

35. $y = \frac{3}{2}x - 8$

36. $y = \frac{3}{5}x - 3$

37. $(5, 1), (0, -3)$

38. $(2,3), (6, 1)$

39. $24 - 2(6) = 12$

40. -20 **41.** $\frac{3}{5}, -\frac{5}{3}, \frac{3}{5}$

42. $-4, \frac{1}{4}, 4$ **43.** -9 **44.** 1

45. length $= 13$ in.; width $= 7$ in. **46.** 7 quarters, 13 dimes

47. 0.6, The attendance from 1955 to 1970 was increasing at an average rate of 0.6 million people every year **48.** 0.1475, The attendance from 1970 to 2010 was increasing at an average rate of 0.1475 million people every year

Chapter 10 Test

1-4.

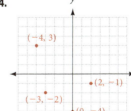
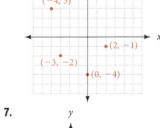

5. $(0, -3), (2, 0), (4, 3), (-2, -6)$

6. $(0, 7), (4, -5)$

7.

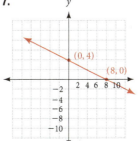

8.

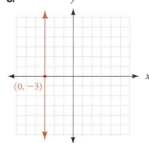

9. x-intercept $= 2$; y-intercept $= -4$

10. x-intercept $= -4$; y-intercept $= 6$

11. x-intercept $= -2$; y-intercept $= 4$

12. $-\frac{1}{2}$ **13.** $-\frac{8}{7}$ **14.** -3 **15.** Undefined slope   **16.** 0

17. $y = -\frac{1}{2}x + 3$ **18.** $y = 3x - 5$ **19.** $y = -\frac{2}{3}x - 2$

20. $y = -3x - 6$ **21.** $y = \frac{3}{2}x + 6$ **22.** $y = -3x - 17$

23. $y = \frac{3}{2}x + 4$ **24.** $y = \frac{3}{2}x + 7$ **25.** $y = 2x - 6$

26.

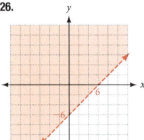

27.

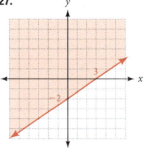

Chapter 11

Exercise Set 11.1

Vocabulary Review   **1.** system   **2.** ordered pairs   **3.** axes

4. intersection

Problems

1.

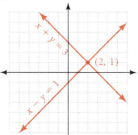

3.

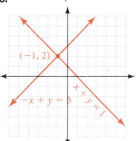

5.

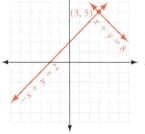

7.

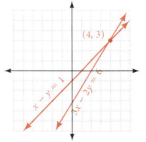

9.

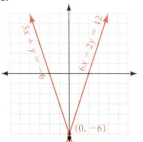

11.

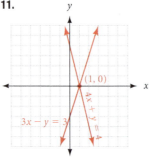

13.

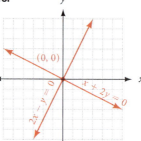

15.

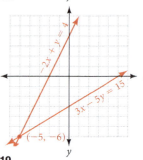

17.

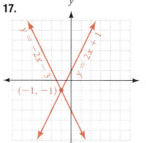

19.

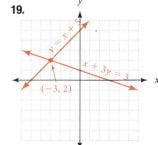

21.

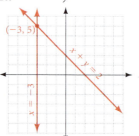

23.

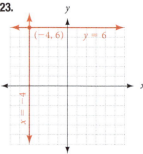

25.

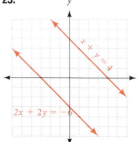

No solution (inconsistent)

27.

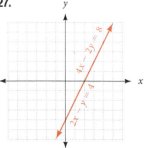

Lines coincide (dependent)

29.

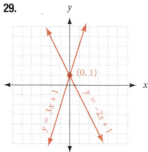

31.

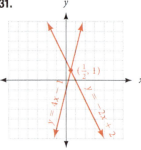

33.

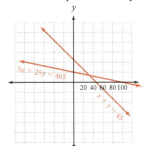

35. 33 $5 chips, 12 $25 chips

37. a. 25 hours **b.** Gigi's **c.** Marcy's   **39.** 1   **41.** 2
43. All real numbers   **45.** $x = 3y - 1$ **47.** 1   **49.** 5
51. 34.5   **53.** 4   **55.** 33.95   **57.** 20
Find the Mistake   1. The point of intersection for two straight line graphs is a solution to both equations.   **2.** The point of intersection for the equations $y = 6$ and $y = 4x + 2$ is (1, 6).
3. The point of intersection for the equations $x = 5$ and $x = 9$ is not (5, 9) because there is no point of intersection; the lines are parallel.
4. When lines coincide, we say that the equations are dependent.

Exercise Set 11.2
Vocabulary Review   1. equations   **2.** substitute   **3.** variables
4. check
Problems   1. (4, 7)   **3.** (3, 17)   **5.** $\left(\frac{3}{2}, 2\right)$ **7.** (2, 4)   **9.** (0, 4)
11. (−1, 3)   **13.** (1, 1)   **15.** (2, −3)   **17.** $\left(-2, \frac{3}{5}\right)$ **19.** (−3, 5)
21. No solution (inconsistent)   **23.** Lines coincide (dependent)
25. (3, 1)   **27.** $\left(\frac{1}{2}, \frac{3}{4}\right)$ **29.** (2, 6)   **31.** (4, 4)   **33.** (5, −2)
35. (18, 10)   **37.** Lines coincide (dependent)
39. No solution (inconsistent)   **41.** (10, 12)   **43.** 7.29 miles per hour
45. a. 560 miles   **b.** Car   **c.** Truck   **d.** We are only working with positive numbers
47. $2x$ **49.** $7x$ **51.** $13x$ **53.** $3x - 2y$ **55.** $-12x + 8y$
57. $22x$ **59.** 1   **61.** 0   **63.** −5
Find the Mistake   1. Substitute the value for x from the first equation for x in the second equation.   **2.** Add $3y$ to both sides of the first equation to isolate x.   **3.** Substitute the value for y from the first equation for y in the second equation.   **4.** Substitute the value for x from the first equation for x in the second equation.

Exercise Set 11.3
Vocabulary Review 1. variable   **2.** multiplication property of equality
3. add   **4.** solve   **5.** substitute   **6.** solution
Problems   1. (2, 1)   **3.** (3, 7)   **5.** (2, −5)   **7.** (−1, 0)
9. Lines coincide (dependent)   **11.** No solution (inconsistent)
13. (4, 8)   **15.** $\left(\frac{1}{5}, 1\right)$ **17.** (1, 0)   **19.** (3, 0)   **21.** (−1, −2)
23. $\left(-5, \frac{3}{4}\right)$ **25.** (−4, 5)   **27.** $\left(-\frac{4}{3}, 1\right)$ **29.** (−3, −10)
31. (3, 2)   **33.** $\left(5, \frac{1}{3}\right)$ **35.** $\left(-2, \frac{2}{3}\right)$ **37.** (2, 2)
39. No solution (inconsistent)   **41.** (1, 1)
43. No solution (inconsistent)   **45.** (10, 12)
47. 1,478 adult tickets, 242 children tickets   **49.** 3 and 23
51. 15 and 24   **53.** Length = 23 in.; width = 6 in.
55. 14 nickels and 10 dimes
Find the Mistake   1. Add the original equations to eliminate x and give a false statement, making the system inconsistent.
2. Multiply both sides of the first equation by −3 to eliminate x.
3. Add the original equations to eliminate x and solve for y.
4. When solving a system by the elimination method, if both variables are eliminated and the resulting statement is true, then the lines coincide.

Landmark Review   1. (−2, 1)   **2.** (−3, −4)   **3.** $\left(\frac{3}{2}, 2\right)$
4. $\left(9, \frac{18}{5}\right)$ **5.** (−1, 5)   **6.** No solution (inconsistent)
7. (5, −6)   **8.** (4, −3)   **9.** (−4, 5)   **10.** $\left(4, \frac{1}{4}\right)$

Exercise Set 11.4
Vocabulary Review   1. known   **2.** unknown, expressions
3. equations   **4.** system   **5.** sentences   **6.** problem, solution
Problems   1. 10 and 15   **3.** 3 and 12   **5.** 4 and 9   **7.** 6 and 29
9. Linda is 70 years old, and Audra is 31 years old.
11. Kate was 21 years old when she received her bachelor's, and 37 years old when she began her master's program.
13. $9,000 at 8%, $11,000 at 6%   **15.** $2,000 at 6%, $8,000 at 5%
17. 6 nickels, 8 quarters   **19.** 12 dimes, 9 quarters
21. 6 liters of 50% solution, 12 liters of 20% solutions
23. 10 gallons of 10% solution, 20 gallons of 7% solution
25. 20 adults, 50 kids   **27.** 16 feet wide, 32 feet long
29. 33 $5 chips, 12 $25 chips   **31.** 50 at $11, 100 at $20
33. 18 pre-registration, 11 race day registration
35. (2, −1)   **37.** (4, 3)   **39.** (1, 1)   **41.** (5, 2)
Find the Mistake   1. The second step in the Blueprint for Problem Solving is to assign variables to each of the unknown items.
2. If there are two unknown quantities, we write two equations to get our answer.   **3.** If we are solving a problem involving nickels and dimes and we know we have a total of 15 coins, we can use $n + d = 15$ as one of our equations.   **4.** One of our equations will be $y = 3x$ or $x + y = 8$.

Chapter 11 Review
Problems   1. (−1, 4)   **2.** (3, 5)   **3.** (2, 6)   **4.** (2, −3)
5. (−1, 0)   **6.** (−4, 1)   **7.** (3, 5)   **8.** (−3, −6)   **9.** (−1, 4)
10. (4, 2)   **11.** (−2, 3)   **12.** (−1, −3)   **13.** (−2, −6)
14. Lines coincide   **15.** (6, 3)   **16.** (−3, 4)   **17.** (1, 0)
18. Lines coincide   **19.** (−2, 4)   **20.** (−2, −1)   **21.** (0, 5)
22. Lines coincide   **23.** 7, 3   **24.** 12, 9
25. $2,000 at 8%, $7,000 at 5%   **26.** $5,000 at 9%, $9,000 at 11%
27. 15 dimes, 6 quarters   **28.** 12 dimes, 7 quarters
29. 4 liters of 20% solution, 7 liters of 80% solution
30. 12 liters of 35% solution, 18 liters of 10% solution

Chapter 11 Cumulative Review
Problems   1. −5   **2.** −3   **3.** 28   **4.** 33   **5.** Undefined
6. 9   **7.** $\frac{11}{48}$ **8.** $\frac{11}{8}$ **9.** $3x + 1$ **10.** $-12a + 26$ **11.** 6
12. 0   **13.** −4   **14.** 7   **15.** $x \geq -11$ **16.** $x \leq -10$
17.   **18.**

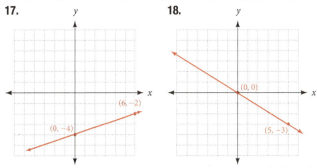

19. (3, −2)   **20.** (5, 1)   **21.** Lines coincide   **22.** $\left(-\frac{1}{2}, -2\right)$
23. (−6, 3)   **24.** (−3, 0)   **25.** (2, 7)   **26.** (−4, −6)
27. (9, −3)   **28.** (−2, 7)   **29.** (3, −8)   **30.** (−2, 1)
31. −8   **32.** −19   **33.** $2 \cdot 3^3 \cdot 5$ **34.** $2^2 \cdot 3 \cdot 5 \cdot 7$
35. (−3, 1)   **36.** (−3, 3) and $\left(2, -\frac{1}{3}\right)$

37. x-intercept $= 6$; y-intercept $= -5$ **38.** $m = \frac{1}{4}$ **39.** $m = -\frac{5}{2}$
40. $m = -\frac{5}{24}$ **41.** $y = \frac{5}{4}x - 2$ **42.** $y = \frac{1}{3}x + 8$ **43.** $y = 3x - 13$
44. x-intercept $= -12$; y-intercept $= 4$ **45.** 5 dimes, 7 quarters
46. \$600 at 7%, \$900 at 11% **47.** 780 people **48.** 400 people

Chapter 11 Test
Problems **1.** $(0, 6)$ **2.** $(3, 2)$ **3.** $(7, 4)$ **4.** $(-2, 4)$
5. $(4, 0)$ **6.** $(-5, 3)$ **7.** $(-3, 4)$ **8.** $(11, 3)$ **9.** $(-4, 5)$
10. $(1, 2)$ **11.** $(-6, 3)$ **12.** Lines coincide **13.** 2, 13
14. 4, 14 **15.** \$480 at 6% **16.** 8 dimes, 11 nickels
17. 28 ft by 71 ft **18. a.** 100 miles **b.** Company 1 **c.** $x > 100$

Index